Digital Signal Processing: Theory and Applications

Digital Signal Processing: Theory and Applications

Edited by **George Pilato**

CLANRYE INTERNATIONAL

New Jersey

Published by Clanrye International,
55 Van Reypen Street,
Jersey City, NJ 07306, USA
www.clanryeinternational.com

Digital Signal Processing: Theory and Applications
Edited by George Pilato

International Standard Book Number: 978-1-63240-146-5 (Hardback)

Printed in the United States of America.

Contents

Preface

This book was inspired by the evolution of our times; to answer the curiosity of inquisitive minds. Many developments have occurred across the globe in the recent past which has transformed the progress in the field.

This book covers the wide realm of digital signal processing. It is a compilation of chapters drafted by various specialists. It focuses on the recent developments in the field of signal processing and presents the current research going on in this field. This book targets researchers in the area of signal processing and related areas of interest. It is comprehensible to all those who have a scientific background and intend to get familiar with the contemporary developments in this field. The chapters within this book discuss the methodological development and latest practices of signal processing.

This book was developed from a mere concept to drafts to chapters and finally compiled together as a complete text to benefit the readers across all nations. To ensure the quality of the content we instilled two significant steps in our procedure. The first was to appoint an editorial team that would verify the data and statistics provided in the book and also select the most appropriate and valuable contributions from the plentiful contributions we received from authors worldwide. The next step was to appoint an expert of the topic as the Editor-in-Chief, who would head the project and finally make the necessary amendments and modifications to make the text reader-friendly. I was then commissioned to examine all the material to present the topics in the most comprehensible and productive format.

I would like to take this opportunity to thank all the contributing authors who were supportive enough to contribute their time and knowledge to this project. I also wish to convey my regards to my family who have been extremely supportive during the entire project.

Editor

Part 1

DSP in Communications

1

Complex Digital Signal Processing in Telecommunications

Zlatka Nikolova, Georgi Iliev,
Miglen Ovtcharov and Vladimir Poulkov
Technical University of Sofia
Bulgaria

1. Introduction

1.1 Complex DSP versus real DSP

Digital Signal Processing (DSP) is a vital tool for scientists and engineers, as it is of fundamental importance in many areas of engineering practice and scientific research.

The "alphabet" of DSP is mathematics and although most practical DSP problems can be solved by using real number mathematics, there are many others which can only be satisfactorily resolved or adequately described by means of complex numbers.

If real number mathematics is the language of *real* DSP, then complex number mathematics is the language of *complex* DSP. In the same way that real numbers are a part of complex numbers in mathematics, *real* DSP can be regarded as a part of *complex* DSP (Smith, 1999).

Complex mathematics manipulates complex numbers – the representation of two variables as a single number - and it may appear that *complex* DSP has no obvious connection with our everyday experience, especially since many DSP problems are explained mainly by means of real number mathematics. Nonetheless, some DSP techniques are based on complex mathematics, such as Fast Fourier Transform (FFT), z-transform, representation of periodical signals and linear systems, etc. However, the imaginary part of complex transformations is usually ignored or regarded as zero due to the inability to provide a readily comprehensible physical explanation.

One well-known practical approach to the representation of an engineering problem by means of complex numbers can be referred to as the *assembling approach*: the real and imaginary parts of a complex number are real variables and individually can represent two real physical parameters. Complex math techniques are used to process this complex entity once it is assembled. The real and imaginary parts of the resulting complex variable preserve the same real physical parameters. This approach is not universally-applicable and can only be used with problems and applications which conform to the requirements of complex math techniques. Making a complex number entirely mathematically equivalent to a substantial physical problem is the real essence of *complex* DSP. Like complex Fourier transforms, complex DSP transforms show the fundamental nature of *complex* DSP and such complex techniques often increase the power of basic DSP methods. The development and application of *complex* DSP are only just beginning to increase and for this reason some researchers have named it *theoretical* DSP.

It is evident that *complex* DSP is more complicated than *real* DSP. Complex DSP transforms are highly theoretical and mathematical; to use them efficiently and professionally requires a large amount of mathematics study and practical experience.
Complex math makes the mathematical expressions used in DSP more compact and solves the problems which real math cannot deal with. Complex DSP techniques can complement our understanding of how physical systems perform but to achieve this, we are faced with the necessity of dealing with extensive sophisticated mathematics. For DSP professionals there comes a point at which they have no real choice since the study of complex number mathematics is the foundation of DSP.

1.2 Complex representation of signals and systems

All naturally-occurring signals are real; however in some signal processing applications it is convenient to represent a signal as a complex-valued function of an independent variable. For purely mathematical reasons, the concept of complex number representation is closely connected with many of the basics of electrical engineering theory, such as voltage, current, impedance, frequency response, transfer-function, Fourier and z-transforms, etc.
Complex DSP has many areas of application, one of the most important being modern telecommunications, which very often uses narrowband analytical signals; these are complex in nature (Martin, 2003). In this field, the complex representation of signals is very useful as it provides a simple interpretation and realization of complicated processing tasks, such as modulation, sampling or quantization.
It should be remembered that a complex number could be expressed in *rectangular*, *polar* and *exponential* forms:

$$a + jb = A(\cos\theta + j\sin\theta) = Ae^{j\theta} . \tag{1}$$

The third notation of the complex number in the equation (1) is referred to as *complex exponential* and is obtained after Euler's relation is applied. The exponential form of complex numbers is at the core of *complex* DSP and enables magnitude A and phase θ components to be easily derived.
Complex numbers offer a compact representation of the most often-used waveforms in signal processing – *sine* and *cosine* waves (Proakis & Manolakis, 2006). The complex number representation of sinusoids is an elegant technique in signal and circuit analysis and synthesis, applicable when the rules of complex math techniques coincide with those of sine and cosine functions. Sinusoids are represented by complex numbers; these are then processed mathematically and the resulting complex numbers correspond to sinusoids, which match the way sine and cosine waves would perform if they were manipulated individually. The complex representation technique is possible only for sine and cosine waves of the same frequency, manipulated mathematically by linear systems.
The use of Euler's identity results in the class of complex exponential signals:

$$x(n) = A\alpha^n = |A|e^{j\phi}e^{(\sigma_0 + j\omega_0)} = x_R(n) + jx_I(n) . \tag{2}$$

$\alpha = e^{(\sigma_0 + j\omega_0)}$ and $A = |A|e^{j\phi}$ are complex numbers thus obtaining:

$$x_R(n) = |A|e^{\sigma_0 n}\cos(\omega_0 n + \phi); \qquad x_I(n) = |A|e^{\sigma_0 n}\sin(\omega_0 n + \phi). \tag{3}$$

Clearly, $x_R(n)$ and $x_I(n)$ are real discrete-time sinusoidal signals whose amplitude $|A|e^{\sigma_0 n}$ is constant (σ_0=0), increasing (σ_0>0) or decreasing (σ_0<0) exponents (Fig. 1).

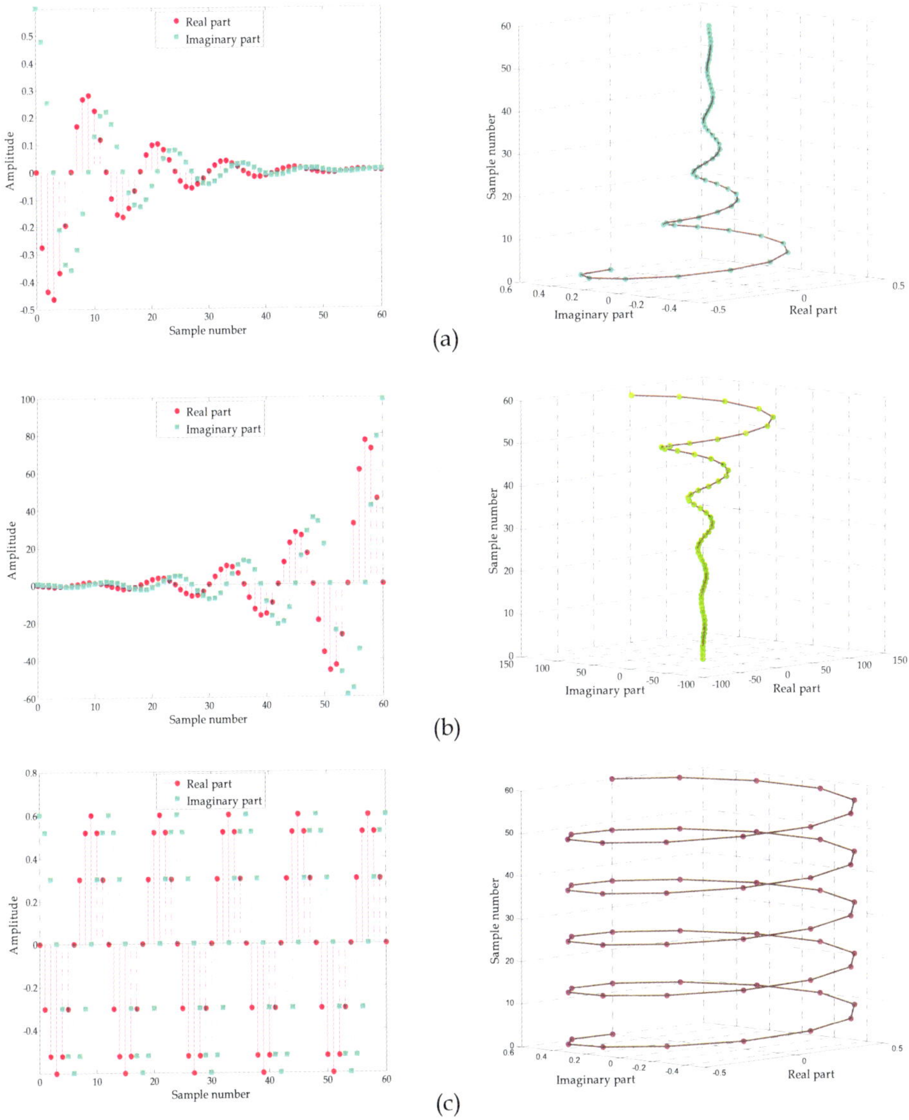

Fig. 1. Complex exponential signal $x(n)$ and its real and imaginary components $x_R(n)$ and $x_I(n)$ for (a) σ_0=-0.085; (b) σ_0=0.085 and (c) σ_0=0

The spectrum of a real discrete-time signal lies between $-\omega_s/2$ and $\omega_s/2$ (ω_s is the sampling frequency in radians per sample), while the spectrum of a complex signal is twice as narrow and is located within the positive frequency range only.

Narrowband signals are of great use in telecommunications. The determination of a signal's attributes, such as frequency, envelope, amplitude and phase are of great importance for signal processing e.g. modulation, multiplexing, signal detection, frequency transformation, etc. These attributes are easier to quantify for narrowband signals than for wideband signals (Fig. 2). This makes narrowband signals much simpler to represent as complex signals.

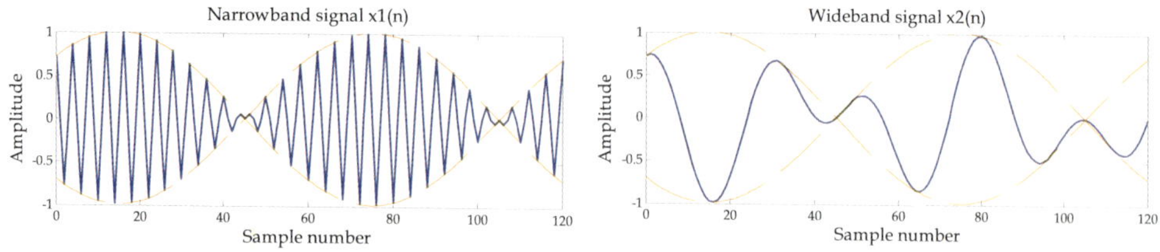

(a) (b)

Fig. 2. Narrowband signal (a) $x_1(n) = \sin(\pi/60n + \pi/4)\cos(\pi/2n)$;
wideband signal (b) $x_2(n) = \sin(\pi/60n + \pi/4)\cos(\pi/16n)$

Over the years different techniques of describing narrowband complex signals have been developed. These techniques differ from each other in the way the imaginary component is derived; the real component of the complex representation is the real signal itself.

Some authors (Fink, 1984) suggest that the imaginary part of a complex narrowband signal can be obtained from the first $x'_R(n)$ and second $x''_R(n)$ derivatives of the real signal:

$$x_I(n) = -x'_R(n)\sqrt{\frac{-x_R(n)}{x''_R(n)}}. \tag{4}$$

One disadvantage of the representation in equation (4) is that insignificant changes in the real signal $x_R(n)$ can alter the imaginary part $x_I(n)$ significantly; furthermore the second derivative can change its sign, thus removing the sense of the square root.

Another approach to deriving the imaginary component of a complex signal representation, applicable to harmonic signals, is as follows (Gallagher, 1968):

$$x_I(n) = \frac{-x_R(n)}{\omega_0}, \tag{5}$$

where ω_0 is the frequency of the real harmonic signal.

Analytical representation is another well-known approach used to obtain the imaginary part of a complex signal, named the *analytic* signal. An analytic complex signal is represented by its *inphase* (the real component) and *quadrature* (the imaginary component). The approach includes a low-frequency envelope modulation using a complex carrier signal – a complex exponent $e^{j\omega_0 n}$ named *cissoid* (Crystal & Ehrman, 1968) or *complexoid* (Martin, 2003):

$$x_R(n)\otimes e^{j\omega_0 n} \Rightarrow \quad x(n) = x_R(n)e^{j\omega_0 n} = x_R(n)[\cos\omega_0 n + j\sin\omega_0 n] = x_R(n) + jx_I(n). \tag{6}$$

In the frequency domain an analytic complex signal is:

$$X_C\left(e^{j\omega n}\right) = X_R\left(e^{j\omega n}\right) + jX_I\left(e^{j\omega n}\right). \tag{7}$$

The real signal and its Hilbert transform are respectively the real and imaginary parts of the analytic signal; these have the same amplitude and $\pi/2$ phase-shift (Fig. 3).

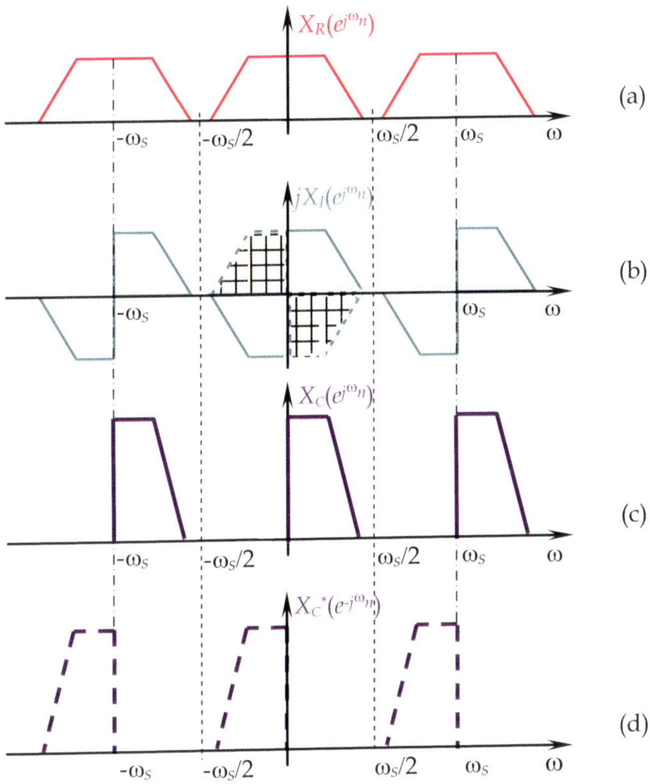

Fig. 3. Complex signal derivation using the Hilbert transformation

According to the Hilbert transformation, the components of the $X_R\left(e^{j\omega n}\right)$ spectrum are shifted by $\pi/2$ for positive frequencies and by $-\pi/2$ for negative frequencies, thus the pattern areas in Fig. 3b are obtained. The real signal $X_R\left(e^{j\omega n}\right)$ and the imaginary one $X_I\left(e^{j\omega n}\right)$ multiplied by j (square root of -1), are identical for positive frequencies and $-\pi/2$ phase shifted for negative frequencies – the solid blue line (Fig. 3b). The complex signal $X_C\left(e^{j\omega n}\right)$ occupies half of the real signal frequency band; its amplitude is the sum of the $X_R\left(e^{j\omega n}\right)$ and $jX_I\left(e^{j\omega n}\right)$ amplitudes (Fig. 3c). The spectrum of the complex conjugate analytic signal $X_C^*\left(e^{-j\omega n}\right)$ is depicted in Fig. 3d.

In the frequency domain the analytic complex signal, its complex conjugate signal, real and imaginary components are related as follows:

$$
\begin{aligned}
X_R\left(e^{j\omega n}\right) &= \frac{1}{2}\left\{X\left(e^{j\omega n}\right) + X^*\left(e^{-j\omega n}\right)\right\} \\
jX_I\left(e^{j\omega n}\right) &= \frac{1}{2}\left\{X\left(e^{j\omega n}\right) - X^*\left(e^{-j\omega n}\right)\right\} \\
X\left(e^{j\omega n}\right) &= \begin{cases} 2X_R\left(e^{j\omega n}\right) = 2jX_I\left(e^{j\omega n}\right), & 0 < \omega < \omega_S/2 \\ 0, & -\omega_S/2 \le \omega < 0 \end{cases}
\end{aligned}
\tag{8}
$$

Discrete-time complex signals are easily processed by digital complex circuits, whose transfer functions contain complex coefficients (Márquez, 2011).
An output complex signal $Y_C(z)$ is the response of a complex system with transfer function $H_C(z)$, when complex signal $X_C(z)$ is applied as an input. Being complex functions, $X_C(z)$, $Y_C(z)$ and $H_C(z)$, can be represented by their real and imaginary parts:

$$
\underbrace{Y_C(z)}_{\Downarrow} = \underbrace{H_C(z)}_{\Downarrow} \quad \underbrace{X_C(z)}_{\Downarrow}
$$
$$
\left[Y_R(z) + jY_I(z)\right] = \left[H_R(z) + jH_I(z)\right]\left[X_R(z) + jX_I(z)\right] \tag{9}
$$

After mathematical operations are applied, the complex output signal and its real and imaginary parts become:

$$
\begin{aligned}
Y_C(z) &= \left[H_R(z) + jH_I(z)\right]\left[X_R(z) + jX_I(z)\right] = \\
&= \underbrace{\left[H_R(z)X_R(z) - H_I(z)X_I(z)\right]}_{\Downarrow\; Y_R(z)} + j\underbrace{\left[H_I(z)X_R(z) + H_R(z)X_I(z)\right]}_{\Downarrow\; Y_I(z)}
\end{aligned}
\tag{10}
$$

According to equation (10), the block-diagram of a complex system will be as shown in Fig. 4.

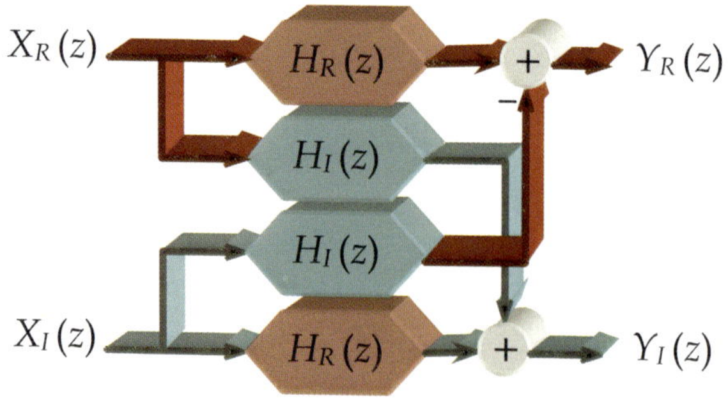

Fig. 4. Block-diagram of a complex system

1.3 Complex digital processing techniques - complex Fourier transforms

Digital systems and signals can be represented in three domains – time domain, z-domain and frequency domain. To cross from one domain to another, the Fourier and z-transforms are employed (Fig. 5). Both transforms are fundamental building-blocks of signal processing theory and exist in two formats - *forward* and *inverse* (Smith, 1999).

Fig. 5. Relationships between frequency, time, and z- domains

The Fourier transforms group contains four families, which differ from one another in the type of time-domain signal which they process - *periodic* or *aperiodic* and *discrete* or *continuous*. Discrete Fourier Transform (DFT) deals with *discrete periodic* signals, Discrete Time Fourier Transform (DTFT) with *discrete aperiodic* signals, and Fourier Series and Fourier Transform with *periodic* and *aperiodic continuous* signals respectively. In addition to having forward and inverse versions, each of these four Fourier families exists in two forms - *real* and *complex*, depending on whether real or complex number math is used. All four Fourier transform families decompose signals into sine and cosine waves; when these are expressed by complex number equations, using Euler's identity, the *complex* versions of the Fourier transforms are introduced.

DFT is the most often-used Fourier transform in DSP. The DFT family is a basic mathematical tool in various processing techniques performed in the frequency domain, for instance frequency analysis of digital systems and spectral representation of discrete signals. In this chapter, the focus is on *complex* DFT. This is more sophisticated and wide-ranging than real DFT, but is based on the more complicated complex number math. However, numerous digital signal processing techniques, such as convolution, modulation, compression, aliasing, etc. can be better described and appreciated via this extended math. (Sklar, 2001)

Complex DFT equations are shown in Table 1. The *forward complex* DFT equation is also called *analysis* equation. This calculates the frequency domain values of the discrete periodic signal, whereas the *inverse* (*synthesis*) equation computes the values in the time domain.

Complex Discrete Fourier Transform	
Forward (analysis) equation	**Inverse (synthesis) equation**
$X(k)=\frac{1}{N}\sum_{n=0}^{N-1}x(n)\left[\cos\frac{2\pi kn}{N}-j\sin\frac{2\pi kn}{N}\right]$	$x(n)=\sum_{k=0}^{N-1}\operatorname{Re}X(k)\left[\cos\frac{2\pi kn}{N}+j\sin\frac{2\pi kn}{N}\right] - \sum_{k=0}^{N-1}\operatorname{Im}X(k)\left[\sin\frac{2\pi kn}{N}-j\cos\frac{2\pi kn}{N}\right]$

Table 1. Complex DFT transforms in rectangular form

The time domain signal $x(n)$ is a complex discrete periodic signal; only an N-point unique discrete sequence from this signal, situated in a single time-interval ($0\div N$, $-N/2\div N/2$, etc.) is

considered. The forward equation multiplies the periodic time domain number series from $x(0)$ to $x(N\text{-}1)$ by a sinusoid and sums the results over the complete time-period.
The frequency domain signal $X(k)$ is an N-point complex periodic signal in a single frequency interval, such as $[0\div0.5\omega_s]$, $[-0.5\omega_s\div0]$, $[-0.25\omega_s\div0.25\omega_s]$, etc. (the sampling frequency ω_s is often used in its normalized value). The inverse equation employs all the N points in the frequency domain to calculate a particular discrete value of the time domain signal. It is clear that *complex* DFT works with finite-length data.
Both the time domain $x(n)$ and the frequency domain $X(k)$ signals are complex numbers, i.e. *complex* DFT also recognizes negative time and negative frequencies. Complex mathematics accommodates these concepts, although imaginary time and frequency have only a theoretical existence so far. *Complex* DFT is a symmetrical and mathematically comprehensive processing technology because it doesn't discriminate between negative and positive frequencies.
Fig. 6 shows how the forward *complex* DFT algorithm works in the case of a complex time-domain signal. $x_R(n)$ is a real time domain signal whose frequency spectrum has an even real part and an odd imaginary part; conversely, the frequency spectrum of the imaginary part of the time domain signal $x_I(n)$ has an odd real part and an even imaginary part. However, as can be seen in Fig. 6, the actual frequency spectrum is the sum of the two individually-calculated spectra. In reality, these two time domain signals are processed simultaneously, which is the whole point of the Fast Fourier Transform (FFT) algorithm.

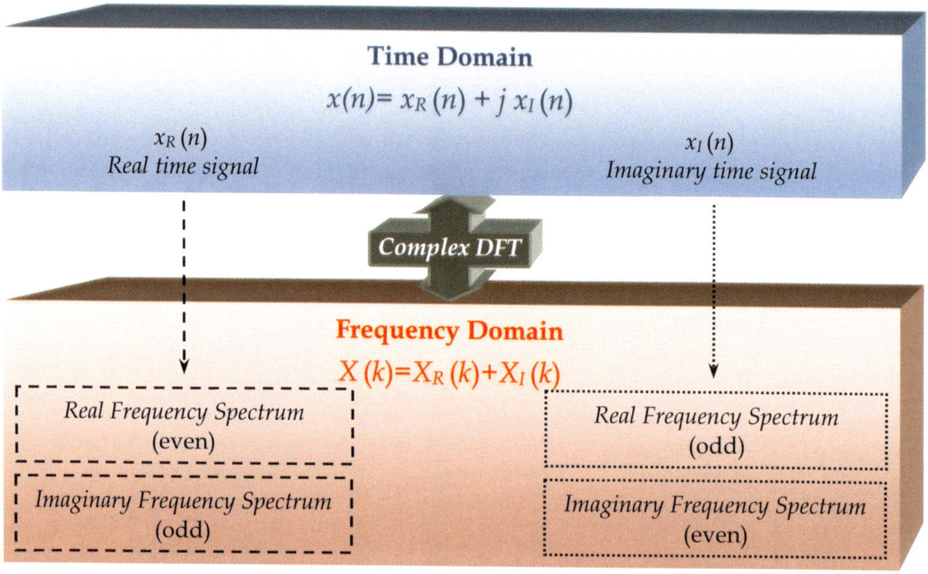

Fig. 6. Forward *complex* DFT algorithm

The imaginary part of the time-domain complex signal can be omitted and the time domain then becomes totally real, as is assumed in the numerical example shown in Fig. 7. A real sinusoidal signal with amplitude M, represented in a complex form, contains a positive ω_0 and a negative frequency $-\omega_0$. The complex spectrum $X(k)$ describes the signal in the

frequency domain. The frequency range of its real, Re $X(k)$, and imaginary part, Im $X(k)$, comprises both positive and negative frequencies simultaneously. Since the considered time domain signal is real, Re $X(k)$ is even (the spectral values A and B have the same sign), while the imaginary part Im $X(k)$ is odd (C is negative, D is positive).
The amplitude of each of the four spectral peaks is $M/2$, which is half the amplitude of the time domain signal. The single frequency interval under consideration $[-\frac{1}{4}\omega_s \div \frac{1}{4}\omega_s]$ ([-0.5÷0.5] when normalized frequency is used) is symmetric with respect to a frequency of zero. The real frequency spectrum Re $X(k)$ is used to reconstruct a cosine time domain signal, whilst the imaginary spectrum Im $X(k)$ results in a negative sine wave, both with amplitude M in accordance with the complex analysis equation (Table 1). In a way analogous to the example shown in Fig. 7, a complex frequency spectrum can also be derived.

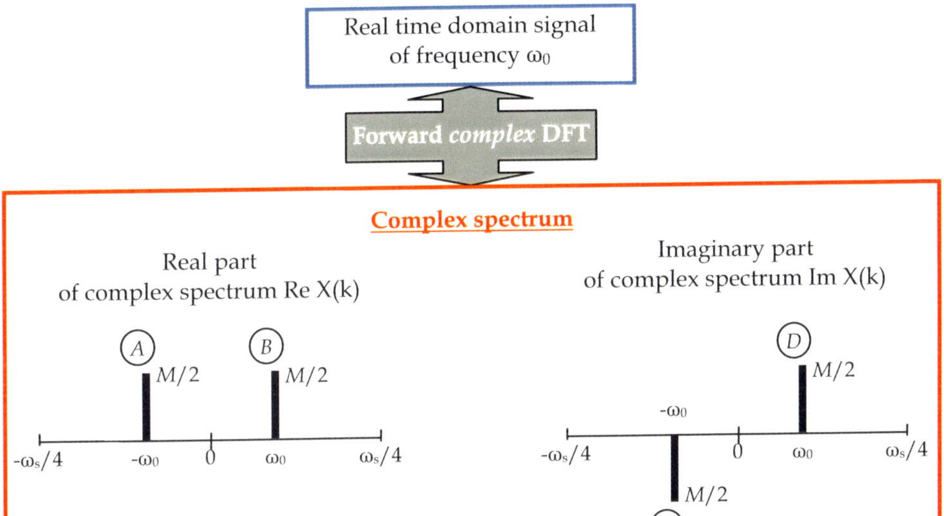

$$B \Rightarrow \left\{\frac{M}{2}\cos(\omega_0 n) + \frac{M}{2}\sin(\omega_0 n)\right\}$$
$$\oplus$$
$$A \Rightarrow \left\{\frac{M}{2}\cos(-\omega_0 n) + \frac{M}{2}\sin(-\omega_0 n)\right\}$$
$$X_R(k) \Rightarrow M\cos(\omega_0 n)$$

$$D \Rightarrow \left\{-\frac{M}{2}\sin(\omega_0 n) - \frac{M}{2}\cos(\omega_0 n)\right\}$$
$$\oplus$$
$$C \Rightarrow \left\{\frac{M}{2}\sin(-\omega_0 n) + \frac{M}{2}\cos(-\omega_0 n)\right\}$$
$$X_I(k) \Rightarrow -M\sin(\omega_0 n)$$

Fig. 7. Inverse *complex* DFT - reconstruction of a real time domain signal

Why is *complex* DFT used since it involves intricate complex number math?
Complex DFT has persuasive advantages over *real* DFT and is considered to be the more comprehensive version. *Real* DFT is mathematically simpler and offers practical solutions to real world problems; by extension, negative frequencies are disregarded. Negative frequencies are always encountered in conjunction with complex numbers.

A real DFT spectrum can be represented in a complex form. Forward *real* DFT results in cosine and sine wave terms, which then form respectively the real and imaginary parts of a complex number sequence. This substitution has the advantage of using powerful complex number math, but this is not true *complex* DFT. Despite the spectrum being in a complex form, the DFT remains *real* and j is not an integral part of the *complex representation* of *real* DFT.

Another mathematical inconvenience of *real* DFT is the absence of symmetry between analysis and synthesis equations, which is due to the exclusion of negative frequencies. In order to achieve a perfect reconstruction of the time domain signal, the first and last samples of the *real* DFT frequency spectrum, relating to zero frequency and Nyquist's frequency respectively, must have a scaling factor of $1/N$ applied to them rather than the $2/N$ used for the rest of the samples.

In contrast, *complex* DFT doesn't require a scaling factor of 2 as each value in the time domain corresponds to two spectral values located in a positive and a negative frequency; each one contributing half the time domain waveform amplitude, as shown in Fig. 7. The factor of $1/N$ is applied equally to all samples in the frequency domain. Taking the negative frequencies into account, *complex* DFT achieves a mathematically-favoured symmetry between *forward* and *inverse* equations, i.e. between time and frequency domains.

Complex DFT overcomes the theoretical imperfections of *real* DFT in a manner helpful to other basic DSP transforms, such as forward and inverse z-transforms. A bright future is confidently predicted for *complex* DSP in general and the *complex* versions of Fourier transforms in particular.

2. Complex DSP – some applications in telecommunications

DSP is making a significant contribution to progress in many diverse areas of human endeavour - science, industry, communications, health care, security and safety, commercial business, space technologies etc.

Based on powerful scientific mathematical principles, *complex* DSP has overlapping boundaries with the theory of, and is needed for many applications in, telecommunications. This chapter presents a short exploration of precisely this common area.

Modern telecommunications very often uses narrowband signals, such as NBI (Narrowband Interference), RFI (Radio Frequency Interference), etc. These signals are complex by nature and hence it is natural for *complex* DSP techniques to be used to process them (Ovtcharov et al, 2009), (Nikolova et al, 2010).

Telecommunication systems very commonly require processing to occur in real time, adaptive complex filtering being amongst the most frequently-used *complex* DSP techniques. When multiple communication channels are to be manipulated simultaneously, parallel processing systems are indicated (Nikolova et al, 2006), (Iliev et al, 2009).

An efficient Adaptive Complex Filter Bank (ACFB) scheme is presented here, together with a short exploration of its application for the mitigation of narrowband interference signals in MIMO (Multiple-Input Multiple-Output) communication systems.

2.1 Adaptive complex filtering

As pointed out previously, adaptive complex filtering is a basic and very commonly-applied DSP technique. An adaptive complex system consists of two basic building blocks:

the variable complex filter and the adaptive algorithm. Fig. 8 shows such a system based on a variable complex filter section designated LS1 (Low Sensitivity). The variable complex LS1 filter changes the central frequency and bandwidth independently (Iliev et al, 2002), (Iliev et al, 2006). The central frequency can be tuned by trimming the coefficient θ, whereas the single coefficient β adjusts the bandwidth. The LS1 variable complex filter has two very important advantages: firstly, an extremely low passband sensitivity, which offers resistance to quantization effects and secondly, independent control of both central frequency and bandwidth over a wide frequency range.
The adaptive complex system (Fig.8) has a complex input $x(n)=x_R(n)+jx_I(n)$ and provides both band-pass (BP) and band-stop (BS) complex filtering. The real and imaginary parts of the BP filter are respectively $y_R(n)$ and $y_I(n)$, whilst those of the BS filter are $e_R(n)$ and $e_I(n)$. The cost-function is the power of the BP/BS filter's output signal.
The filter coefficient θ, responsible for the central frequency, is updated by applying an adaptive algorithm, for example LMS (Least Mean Square):

$$\theta(n+1) = \theta(n) + \mu \mathrm{Re}[e(n) y'^{*}(n)]. \tag{11}$$

The step size μ controls the speed of convergence, (∗) denotes complex-conjugate, $y'(n)$ is the derivative of complex BP filter output $y(n)$ with respect to the coefficient, which is subject to adaptation.

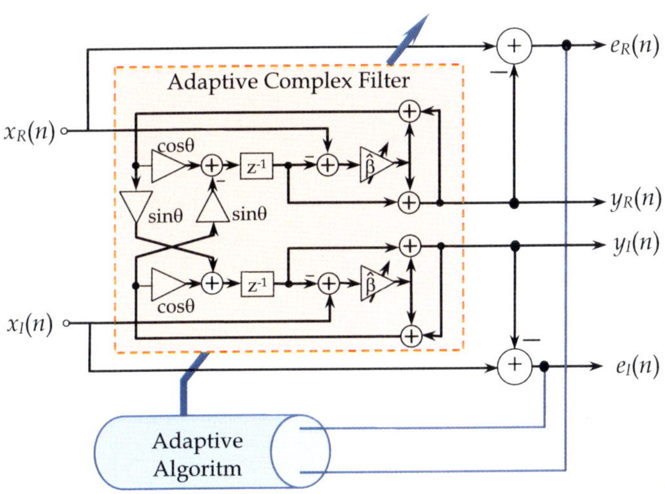

Fig. 8. Block-diagram of an LS1-based adaptive complex system

In order to ensure the stability of the adaptive algorithm, the range of the step size μ should be set according to (Douglas, 1999):

$$0 < \mu < \frac{P}{N\sigma^2}. \tag{11}$$

where N is the filter order, σ^2 is the power of the signal $y'(n)$ and P is a constant which depends on the statistical characteristics of the input signal. In most practical situations, P is approximately equal to 0.1.

The very low sensitivity of the variable complex LS1 filter section ensures the general efficiency of the adaptation and a high tuning accuracy, even with severely quantized multiplier coefficients.

This approach can easily be extended to the adaptive complex filter bank synthesis in parallel complex signal processing.

In (Nikolova et al, 2002) a narrowband ACFB is designed for the detection of multiple complex sinusoids. The filter bank, composed of three variable complex filter sections, is aimed at detecting multiple complex signals (Fig. 9).

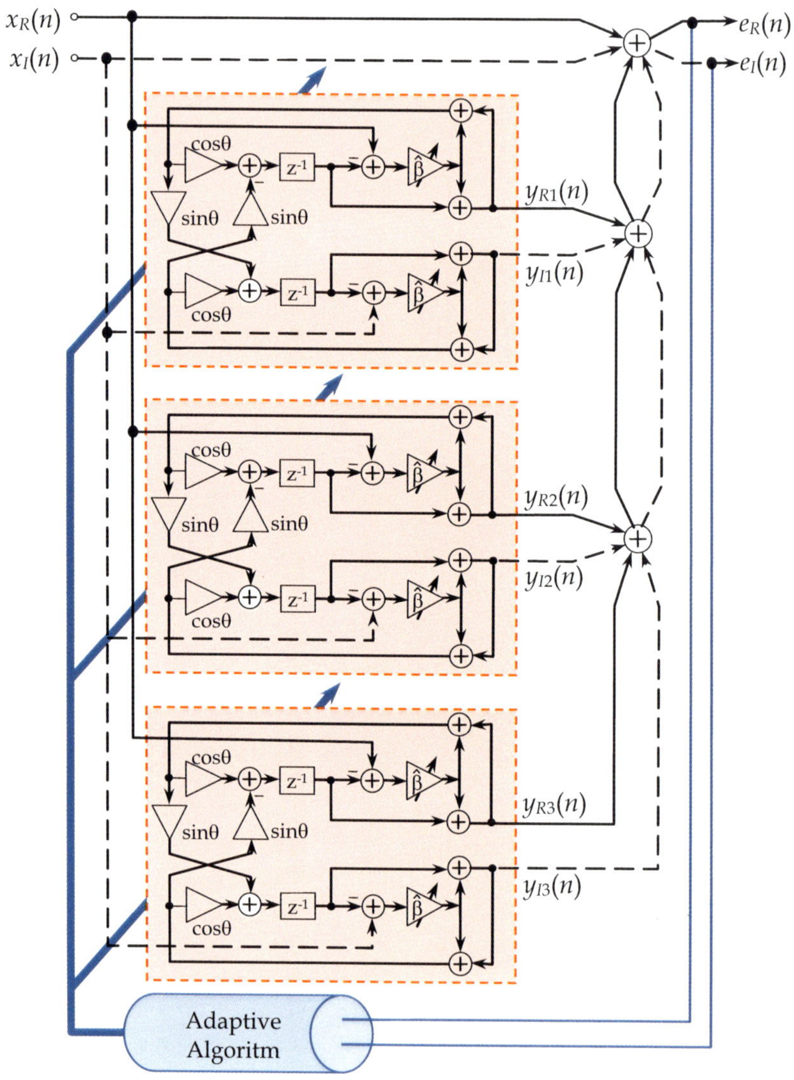

Fig. 9. Block-diagram of an adaptive complex filter bank system

The experiments are carried out with an input signal composed of three complex sine-signals of different frequencies, mixed with white noise. Fig. 10 displays learning curves for the coefficients θ_1, θ_2 and θ_3. The ACFB shows the high efficacy of the parallel filtering process. The main advantages of both the adaptive filter structure and the ACFB lie in their low computational complexity and rapid convergence of adaptation.

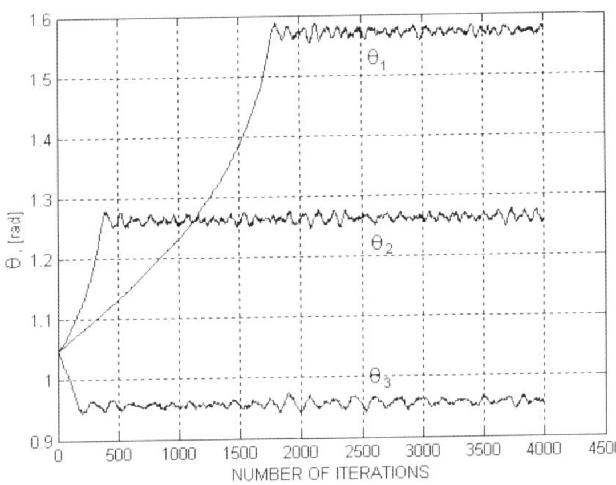

Fig. 10. Learning curves of an ACFB consisting of three complex LS1-sections

2.2 Narrowband interference suppression for MIMO systems using adaptive complex filtering

The sub-sections which follow examine the problem of narrowband interference in two particular MIMO telecommunication systems. Different NBI suppression methods are observed and experimentally compared to the *complex* DSP technique using adaptive complex filtering in the frequency domain.

2.2.1 NBI Suppression in UWB MIMO systems

Ultrawideband (UWB) systems show excellent potential benefits when used in the design of high-speed digital wireless home networks. Depending on how the available bandwidth of the system is used, UWB can be divided into two groups: *single-band* and *multi-band* (MB). Conventional UWB technology is based on *single-band* systems and employs carrier-free communications. It is implemented by directly modulating information into a sequence of impulse-like waveforms; support for multiple users is by means of time-hopping or direct sequence spreading approaches.

The UWB frequency band of *multi-band* UWB systems is divided into several sub-bands. By interleaving the symbols across sub-bands, multi-band UWB can maintain the power of the transmission as though a wide bandwidth were being utilized. The advantage of the multi-band approach is that it allows information to be processed over a much smaller bandwidth, thereby reducing overall design complexity as well as improving spectral flexibility and worldwide adherence to the relevant standards. The constantly-increasing demand for higher data transmission rates can be satisfied by exploiting both multipath- and spatial-diversity, using MIMO together with the appropriate modulation and coding techniques

(Iliev et al, 2009). The multipath energy can be captured efficiently when the OFDM (Orthogonal Frequency-Division Multiplexing) technique is used to modulate the information in each sub-band. Unlike more traditional OFDM systems, the MB-OFDM symbols are interleaved over different sub-bands across both time and frequency. Multiple access of multi-band UWB is enabled by the use of suitably-designed frequency-hopping sequences over the set of sub-bands.

In contrast to conventional MIMO OFDM systems, the performance of MIMO MB-OFDM UWB systems does not depend on the temporal correlation of the propagation channel. However, due to their relatively low transmission power, such systems are very sensitive to NBI. Because of the spectral leakage effect caused by DFT demodulation at the OFDM receiver, many subcarriers near the interference frequency suffer from serious Signal-to-Interference Ratio (SIR) degradation, which can adversely affect or even block communications (Giorgetti et al, 2005).

In comparison with the wideband information signal, the interference occupies a much narrower frequency band but has a higher-power spectral density (Park et al, 2004). On the other hand, the wideband signal usually has autocorrelation properties quite similar to those of AWGN (Adaptive Wide Gaussian Noise), so filtering in the frequency domain is possible. The *complex* DSP technique for suppressing NBI by the use of adaptive complex narrowband filtering, which is an optimal solution offering a good balance between computational complexity and interference suppression efficiency, is put forward in (Iliev et al, 2010). The method is compared experimentally with two other often-used algorithms Frequency Excision (FE) (Juang et al, 2004) and Frequency Identification and Cancellation (FIC) (Baccareli et al, 2002) for the identification and suppression of complex NBI in different types of IEEE UWB channels.

A number of simulations relative to complex baseband presentation are performed, estimating the Bit Error Ratio (BER) as a function of the SIR for the CM3 IEEE UWB channel (Molish & Foerster, 2003) and some experimental results are shown in Fig. 10.

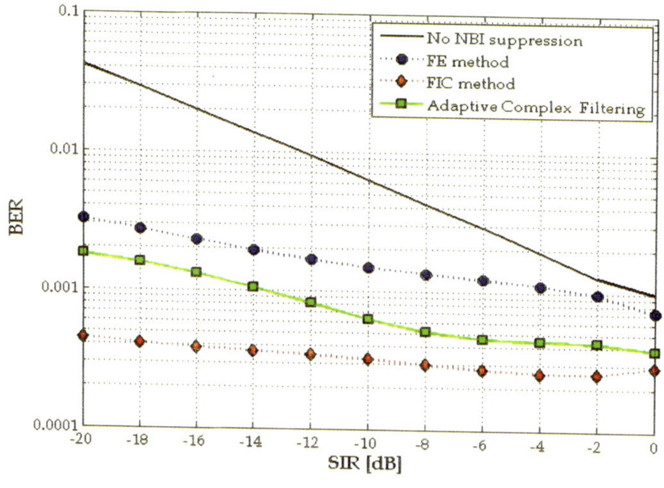

(a)

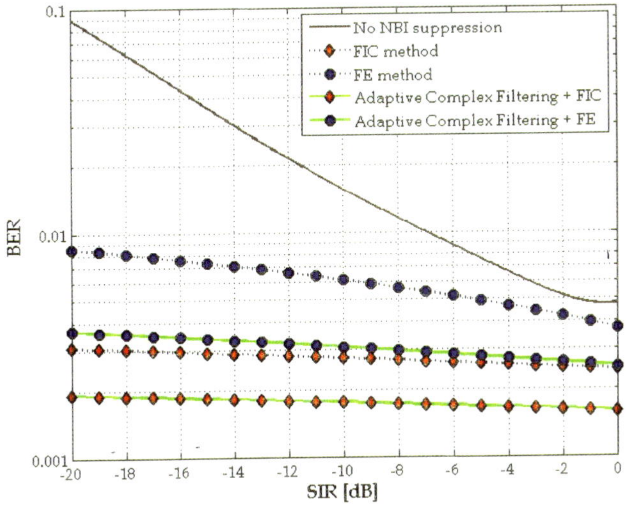

(b)

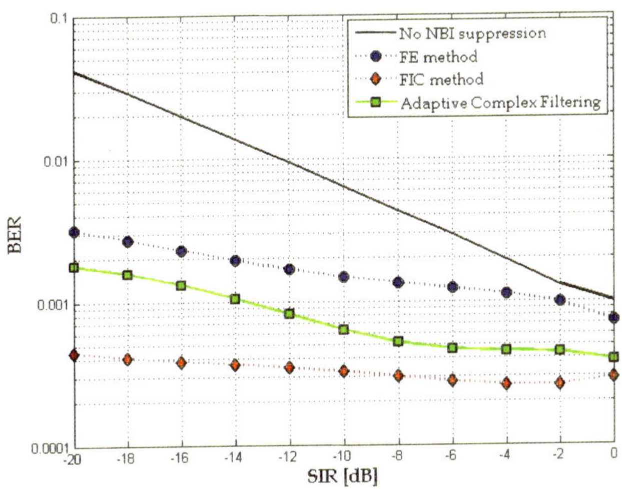

(c)

Fig. 10. BER as a function of SIR for the CM3 channel (a) complex NBI; (b) multi-tone NBI; (c) QPSK modulated NBI

The channel is subject to strong fading and, for the purposes of the experiments, background AWGN is additionally applied, so that the Signal-to-AWGN ratio at the input of the OFDM receiver is 20 dB. The SIR is varied from -20 dB to 0 dB. It can be seen (Fig. 10a) that for high NBI, i.e. where the SIR is less than 0 dB, all methods lead to a significant improvement in performance. The adaptive complex filtering scheme gives better performance than the FE method. This could be explained by the NBI spectral leakage effect caused by DFT demodulation at the OFDM receiver, when many sub-carriers near the interference frequency suffer degradation. Thus, filtering out the NBI before demodulation is better than frequency excision. The FIC algorithm achieves the best result because there is no spectrum leakage, as happens with frequency excision, and there is no amplitude and phase distortion as seen in the case of adaptive complex filtering.

It should be noted that the adaptive filtering scheme and frequency cancellation scheme lead to a degradation in the overall performance when SIR >0. This is due either to the amplitude and phase distortion of the adaptive notch filter or to a wrong estimation of NBI parameters during the identification. The degradation can be reduced by the implementation of a higher-order notch filter or by using more sophisticated identification algorithms. The degradation effect can be avoided by simply switching off the filtering when SIR > 0. Such a scheme is easily realizable, as the amplitude of the NBI can be monitored at the BP output of the filter (Fig. 8).

In Fig. 10b, the results of applying a combination of methods are presented. A multi-tone NBI (an interfering signal composed of five sine-waves) is added to the OFDM signal. One of the NBI tones is 10 dB stronger than the others. The NBI filter is adapted to track the strongest NBI tone, thus preventing the loss of resolution and *Automatic Gain Control* (AGC) saturation. It can be seen that the combination of FE and Adaptive Complex Filtering improves the performance, and the combination of FIC with Adaptive Complex Filtering is even better.

Fig. 10c shows BER as a function of SIR for the CM3 channel when QPSK modulation is used, the NBI being modelled as a complex sine wave. It is evident that the relative performance of the different NBI suppression methods is similar to the one in Fig. 10a but the BER is higher due to the fact that NBI is QPSK modulated.

The experimental results show that the FIC method achieves the highest performance. On the other hand, the extremely high computational complexity limits its application in terms of hardware resources. In this respect, Adaptive Complex Filtering turns out to be the optimal NBI suppression scheme, as it offers very good performance and reasonable complexity. The FE method shows relatively good results and its main advantage is its computational efficiency. Therefore the *complex* DSP filtering technique offers a good compromise between outstanding NBI suppression efficiency and computational complexity.

2.2.2 RFI mitigation in GDSL MIMO systems

The Gigabit Digital Subscriber Line (GDSL) system is a cost-effective solution for existing telecomunication networks as it makes use of the existing copper wires in the last distribution area segment. Crosstalk, which is usually a problem in existing DSL systems, actually becomes an enhancement in GDSL, as it allows the transmission rate to be extended to its true limits (Lee et al, 2007). A symmetric data transmission rate in excess of 1 Gbps using a set of 2 to 4 copper twisted pairs over a 300 m cable length is achievable using vectored MIMO technology, and considerably faster speeds can be achieved over shorter distances.

In order to maximize the amount of information handled by a MIMO cable channel via the cable crosstalk phenomenon, most GDSL systems employ different types of precoding algorithms, such as Orthogonal Space–Time Precoding (OSTP), Orthogonal Space–Frequency Precoding (OSFP), Optimal Linear Precoding (OLP), etc. (Perez-Cruz et al, 2008).

GDSL systems use the leading modulation technology, Discrete Multi-Tone (DMT), also known as OFDM, and are very sensitive to RFI. The presence of strong RFI causes nonlinear distortion in AGC and Analogue-to-Digital Converter (ADC) functional blocks, as well as spectral leakage in the DFT process. Many DMT tones, if they are located close to the interference frequency, will suffer serious SNR degradation. Therefore, RFI suppression is of primary importance for all types of DSL communications, including GDSL.

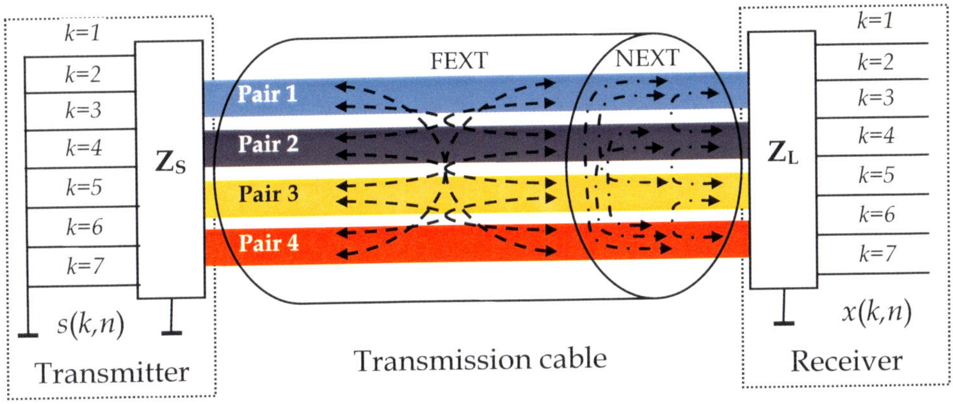

Fig. 11. MIMO GDSL Common Mode system model

The present section considers a MIMO GDSL Common Mode system, with a typical MIMO DMT receiver, using vectored MIMO DSL technology (Fig. 11) (Poulkov et al, 2009).

To achieve the outstanding data-rate of 1 Gbps, the GDSL system requires both source and load to be excited in Common Mode (Starr et al, 2003). The model of a MIMO GDSL channel depicted in Fig. 11 includes 8 wires that create k=7 channels all with the 0 wire as reference. Z_S and Z_L denote the source and load impedance matrices respectively; $s(k,n)$ is the n-th sample of k-th transmitted output, whilst $x(k,n)$ is the n-th sample of k-th received input. Wide-scale frequency variations together with standard statistics determined from measured actual Far End Crosstalk (FEXT) and Near End Crosstalk (NEXT) power transfer functions are also considered and OLP, 64-QAM demodulation and Error Correction Decoding are implemented (ITU-T Recommendation G.993.2, 2006), (ITU-T Recommendation G.996.1, 2006). As well as OLP, three major types of general RFI mitigation approaches are proposed.

The *first* one concerns various FE methods, whereby the affected frequency bins of the DMT symbol are excised or their use avoided. The frequency excision is applied to the MIMO GDSL signal with a complex RFI at each input of the receiver. The signal is converted into the frequency domain by applying an FFT at each input, oversampled by 8, and the noise peaks in the spectra are limited to the pre-determined threshold. After that, the signal is converted back to the time domain and applied to the input of the corresponding DMT demodulator. The higher the order of the FFT, the more precise the frequency excision achieved.

The *second* approach is related to the so-called Cancellation Methods, aimed at the elimination or mitigation of the effect of the RFI on the received DMT signal. In most cases, when the SIR is less than 0 dB, the degradation in a MIMO DSL receiver is beyond the reach of the FE method. Thus, mitigation techniques employing Cancellation Methods, one of which is the RFI FIC method, are recommended as a promising alternative (Juang et al, 2004). The FIC method is implemented as a two-stage algorithm with the filtering process applied independently at each receiver input. First, the complex RFI frequency is estimated by finding the maximum in the oversampled signal spectrum per each receiver's input. After that, using the Maximum Likelihood (ML) approach, the RFI amplitude and phase are estimated per input. The second stage realizes the Non-Linear Least Square (NLS) Optimization Algorithm, where the RFI complex amplitude, phase and frequency are precisely determined.

The *third* RFI mitigation approach is based on the *complex* DSP parallel adaptive complex filtering technique. A notch ACFB is connected at the receiver's inputs in order to identify and eliminate the RFI signal. The adaptation algorithm tunes the filter at each receiver input in such a way that its central frequency and bandwidth match the RFI signal spectrum (Lee et al, 2007).

Using the above-described general simulation model of a MIMO GDSL system (Fig. 11), different experiments are performed deriving the BER as a function of the SIR. The RFI is a complex single tone, the frequency of which is centrally located between two adjacent DMT tones. Depending on the number of twisted pairs used 2, 3 or 4-pair MIMO GDSL systems are considered (Fig. 12) (Poulkov et al, 2009).

The GDSL channels examined are subjected to FEXT, NEXT and a background AWGN with a flat Power Spectral Density (PSD) of - 140 dBm/Hz.

The best RFI mitigation is obtained when the *complex* DSP filtering method is applied to the highest value of channel diversity, i.e. 4-pair GDSL MIMO. The FIC method gives the highest performance but at the cost of additional computational complexity, which could limit its hardware application. The FE method has the highest computational efficiency but delivers the lowest improvement in results when SIR is low: however for high SIR its performance is good.

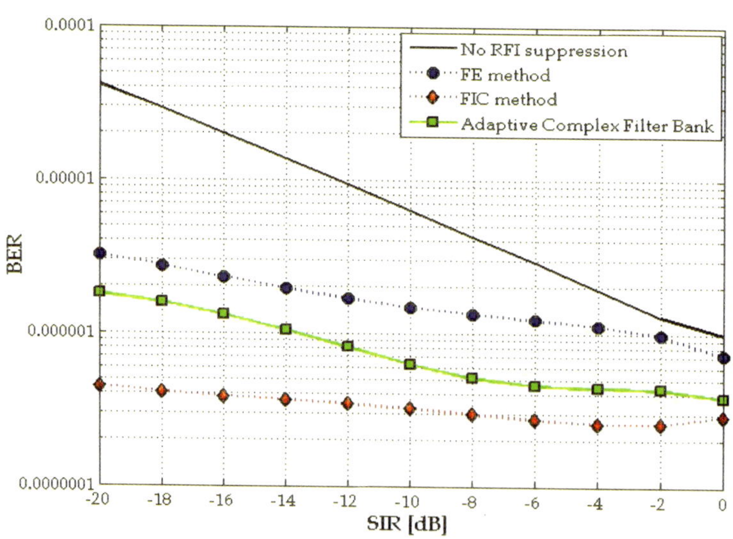

(a)

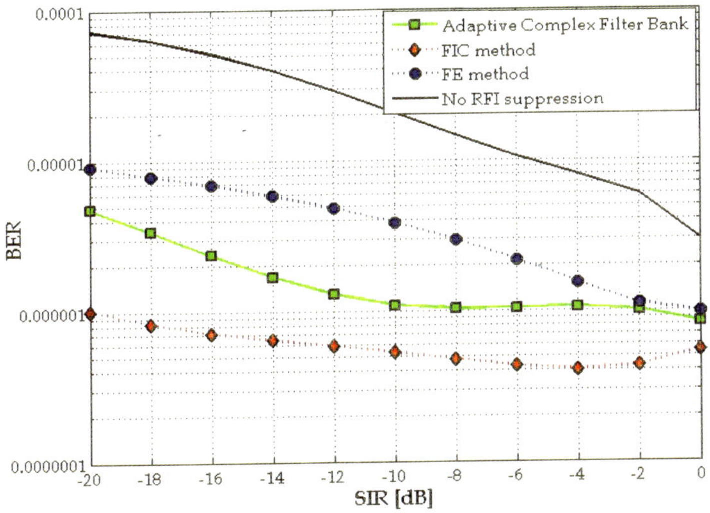

(b)

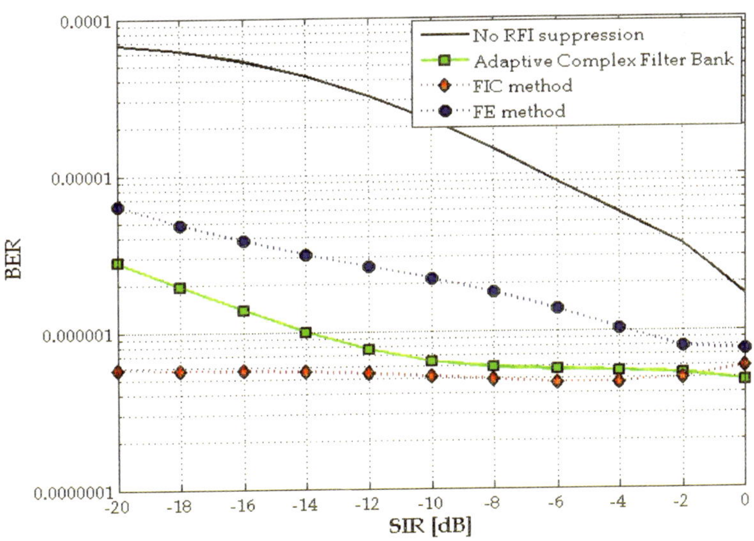

(c)

Fig. 12. BER as a function of SIR for (a) 2-pair; (b) 3-pair; (c) 4-pair GDSL MIMO channels

In this respect, *complex* DSP ACFB filtering turns out to be an optimal narrowband interference-suppression technique, offering a good balance between performance and computational complexity.

3. Conclusions

The use of complex number mathematics greatly enhances the power of DSP, offering techniques which cannot be implemented with real number mathematics alone. In comparison with *real* DSP, *complex* DSP is more abstract and theoretical, but also more powerful and comprehensive. Complex transformations and techniques, such as complex modulation, filtering, mixing, z-transform, speech analysis and synthesis, adaptive complex processing, complex Fourier transforms etc., are the essence of theoretical DSP. *Complex* Fourier transforms appear to be difficult when practical problems are to be solved but they overcome the limitations of *real* Fourier transforms in a mathematically elegant way.

Complex DSP techniques are required for many wireless high-speed telecommunication standards. In telecommunications, the complex representation of signals is very common, hence complex processing techniques are often necessary.

Adaptive complex filtering is examined in this chapter, since it is one of the most frequently-used real-time processing techniques. Adaptive complex selective structures are investigated, in order to demonstrate the high efficiency of adaptive complex digital signal processing.

The *complex* DSP filtering method, based on the developed ACFB, is applied to suppress narrowband interference signals in MIMO telecommunication systems and is then compared to other suppression methods. The study shows that different narrowband interference mitigation methods perform differently, depending on the parameters of the telecommunication system investigated, but the *complex* DSP adaptive filtering technique offers considerable benefits, including comparatively low computational complexity.

Advances in diverse areas of human endeavour, of which modern telecommunications is only one, will continue to inspire the progress of *complex* DSP.

It is indeed fair to say that complex digital signal processing techniques still contribute more to the expansion of theoretical knowledge rather than to the solution of existing practical problems - but watch this space!

4. Acknowledgment

This work was supported by the Bulgarian National Science Fund – Grant No. ДО-02-135/2008 "Research on Cross Layer Optimization of Telecommunication Resource Allocation".

5. References

Baccareli, E.; Baggi, M. & Tagilione, L. (2002). A novel approach to in-band interference mitigation in ultra wide band radio systems. *IEEE Conf. on Ultra Wide Band Systems and Technologies*, pp. 297-301, 7 Aug. 2002.

Crystal, T. & Ehrman, L. (1968). The design and applications of digital filters with complex coefficients, *IEEE Trans. on Audio and Electroacoustics*, vol. 16, Issue: 3, pp. 315-320, Sept. 1968.

Douglas, S. (1999). Adaptive filtering, in *Digital signal processing handbook*, D. Williams & V. Madisetti, Eds., Boca Raton: CRC Press LLC, pp. 451-619, 1999.

Fink L.M. (1984). *Signals, hindrances, errors*, Radio and communication, 1984.

Gallagher, R. G. (1968). *Information Theory and Reliable Communication*, New York, John Wiley and Sons, 1968.

Giorgetti, A.; Chiani, M. & Win, M. Z. (2005). The effect of narrowband interference on wideband wireless communication systems. *IEEE Trans. on Commun.*, vol. 53, No. 12, pp. 2139-2149, 2005.

Iliev, G.; Nikolova, Z.; Stoyanov, G. & Egiazarian, K. (2004). Efficient design of adaptive complex narrowband IIR filters, *Proc. of XII European Signal Proc. Conf. (EUSIPCO'04)*, pp. 1597 - 1600, Vienna, Austria, 6-10 Sept. 2004.

Iliev, G.; Nikolova, Z.; Poulkov, V. & Stoyanov, G. (2006). Noise cancellation in OFDM systems using adaptive complex narrowband IIR filtering, *IEEE Intern. Conf. on Communications (ICC-2006)*, Istanbul, Turkey, pp. 2859 – 2863, 11-15 June 2006.

Iliev, G.; Ovtcharov, M.; Poulkov, V. & Nikolova, Z. (2009). Narrowband interference suppression for MIMO OFDM systems using adaptive filter banks, *The 5th International Wireless Communications and Mobile Computing Conference (IWCMC 2009) MIMO Systems Symp.*, pp. 874 – 877, Leipzig, Germany, 21-24 June 2009.

Iliev, G.; Nikolova, Z.; Ovtcharov, M. & Poulkov, V. (2010). Narrowband interference suppression for MIMO MB-OFDM UWB communication systems, *International Journal on Advances in Telecommunications (IARIA Journals)*, ISSN 1942-2601, vol. 3, No. 1&2, pp. 1 - 8, 2010.

ITU-T Recommendation G.993.2, (2006), *Very High Speed Digital Subscriber Line* 2 (VDSL 2), Feb. 2006.

ITU-T Recommendation G.996.1, (2006), *Test Procedures for Digital Subscriber Line* (VDS) Transceivers, Feb. 2006.

Juang, J.-C.; Chang, C.-L. & Tsai, Y.-L. (2004). An interference mitigation approach against pseudolites. *The 2004 International Symposium on GNSS/GPS*, Sidney, Australia, pp. 623-634, 6-8 Dec. 2004

Lee, B.; Cioffi, J.; Jagannathan, S. & Mohseni, M. (2007). Gigabit DSL, *IEEE Trans on Communications*, print accepted, 2007.

Márquez, F. P. G.(editor) (2011). *Digital Filters*, ISBN: 978-953-307-190-9, InTech, April 2011; Chapter 9, pp. 209-239, *Complex Coefficient IIR Digital Filters*, Zlatka Nikolova, Georgi Stoyanov, Georgi Iliev and Vladimir Poulkov.

Martin, K. (2003). Complex signal processing is not – complex, *Proc. of the 29th European Conf. on Solid-State Circuits (ESSCIRC'03)*, pp. 3-14, Estoril, Portugal, 16-18 Sept. 2003.

Molish, A. F.; Foerster, J. R. (2003). Channel models for ultra wideband personal area networks. *IEEE Wireless Communications*, pp. 524-531, Dec. 2003.

Nikolova, Z.; Iliev, G.; Stoyanov, G. & Egiazarian, K. (2002). Design of adaptive complex IIR notch filter bank for detection of multiple complex sinusoids, *Proc. 2nd International Workshop on Spectral Methods and Multirate Signal Processing (SMMSP'2002)*, pp. 155 - 158, Toulouse, France, 7-8 September 2002.

Nikolova, Z.; Poulkov, V.; Iliev, G. & Stoyanov, G. (2006). Narrowband interference cancellation in multiband OFDM systems, *3rd Cost 289 Workshop "Enabling Technologies for B3G Systems"*, pp. 45-49, Aveiro, Portugal, 12-13 July 2006.

Nikolova, Z.; Poulkov, V.; Iliev, G. & Egiazarian, K. (2010). New adaptive complex IIR filters and their application in OFDM systems, *Journal of Signal, Image and Video Proc., Springer*, vol. 4, No. 2, pp. 197-207, June, 2010, ISSN: 1863-1703.

Ovtcharov, M.; Poulkov, V.; Iliev, G. & Nikolova, Z. (2009), Radio frequency interference suppression in DMT VDSL systems, "*E+E*", ISSN:0861-4717, pp. 42 - 49, 9-10/2009.

Park, S.; Shor, G. & Kim, Y. S. (2004). Interference resilient transmission scheme for multi-band OFDM system in UWB channels. *IEEE Int. Circuits and Systems Symp.*, vol. 5, Vancouver, BC, Canada, pp. 373-376, May 2004.

Perez-Cruz, F.; Rodrigues, R. D. & Verd'u, S. (2008). *Optimal precoding for multiple-input multiple-output Gaussian channels with arbitrary inputs*, preprint, 2008.

Poulkov, V.; Ovtcharov, M.; Iliev, G. & Nikolova, Z. (2009). Radio frequency interference mitigation in GDSL MIMO systems by the use of an adaptive complex narrowband filter bank, *Intern. Conf. on Telecomm. in Modern Satellite, Cable and Broadcasting Services - TELSIKS-2009*, pp. 77 – 80, Nish, Serbia, 7-9 Oct. 2009.

Proakis, J. G. & Manolakis, D. K. (2006). *Digital signal processing*, Prentice Hall; 4th edition, ISBN-10: 0131873741.

Sklar, B. (2001). *Digital communications: fundamentals and applications*, 2nd edition, Prentice Hall, 2001.

Smith, S. W. (1999). *Digital signal processing*, California Technical Publishing, ISBN 0-9660176-6-8, 1999.

Starr T.; Sorbara, M.; Cioffi, J. & Silverman, P. (2003). *DSL Advances* (Chapter 11), Prentice-Hall: Upper Saddle River, NJ, 2003.

2

Multiple-Membership Communities Detection and Its Applications for Mobile Networks

Nikolai Nefedov
Nokia Research Center
ISI Lab, Swiss Federal Institute of Technology Zurich (ETHZ)
Switzerland

1. Introduction

The recent progress in wireless technology and growing spread of smart phones equipped with various sensors make it possible to record real-world rich-content data and compliment it with on-line processing. Depending on the application, mobile data processing could help people to enrich their social interactions and improve environmental and personal health awareness. At the same time, mobile sensing data could help service providers to understand better human behavior and its dynamics, identify complex patterns of users' mobility, and to develop various service-centric and user-centric mobile applications and services on-demand. One of the first steps in analysis of rich-content mobile datasets is to find an underlying structure of users' interactions and its dynamics by clustering data according to some similarity measures.

Classification and clustering (finding groups of similar elements in data) are well-known problems which arise in many fields of sciences, e.g., (Albert & Barabási, 2002; Flake et al, 2002; Wasserman & Faust, 1994). In cases when objects are characterized by vectors of attributes, a number of efficient algorithms to find groups of similar objects based on a metric between the attribute vectors are developed. On the other hand, if data are given in the relational format (causality or dependency relations), e.g., as a network consisting of N nodes and E edges representing some relation between the nodes, then the problem of finding similar elements corresponds to detection of communities, i.e., groups of nodes which are interconnected more densely among themselves than with the rest of the network.

The growing interest to the problem of community detection was triggered by the introduction of a new clustering measure called modularity (Girvan & Newman, 2002; 2004). The modularity maximization is known as the NP-problem and currently a number of different sub-optimal algorithms are proposed, e.g., see (Fortunato, 2011) and references within. However, most of these methods address network partitions into disjoint communities.

On the other hand, in practice communities are often overlapping. It is especially visible in social networks, where only limited information is available and people are affiliated to different groups, depending on professional activities, family status, hobbies, and etc. Furthermore, social interactions are reflected in multiple dimensions, such as users activities, local proximities, geo-locations and etc. These multi-dimensional traces may be presented as multi-layer graphs. It raises the problem of overlapping communities detection at different hierarchical levels at single and multi-layer graphs.

In this chapter we present a framework for multi-membership communities detection in dynamical multi-layer graphs and its applications for missing (or hidden) link predictions/recommendations based on the network topology. In particular, we use modularity maximization with a fast greedy search (Newman, 2004) extended with a random walk approach (Lambiotte et al, 2009) to detect multi-resolution communities beyond and below the resolution provided by *max*-modularity. We generalize a random walk approach to a coupled dynamic systems (Arenas et al, 2006) and then extend it with dynamical links update to make predictions beyond the given topology. In particular, we introduce attractive and repulsive coupling that allow us to detect and predict cooperative and competitive behavior in evolving social networks.

To deal with overlapping communities we introduce a soft community detection and outline its possible applications in single and multi-layer graphs. In particular, we propose friend-recommendations in social networks, where new link recommendations are made as intra- and inter-clique communities completion and recommendations are prioritized according to topologically-based similarity measures (Liben-Nowel & Kleinberg, 2003) modified to include multiple-communities membership. We also show that the proposed prediction rules based on soft community detection are in line with the network evolution predicted by coupled dynamical systems. To test the proposed framework we use a benchmark network (Zachary, 1977) and then apply developed methods for analysis of multi-layers graphs built from real-world mobile datasets (Kiukkonen et al, 2010). The presented results show that by combining information from multi-layer graphs we can improve reliability measures of community detection and missing links predictions.

The chapter is organized as follows: in Section 2 we outline the dynamical formulation of community detection that forms the basis for the rest of the paper. Topology detection using coupled dynamical systems and its extensions to model a network evolution are described in Section 3. Soft community detection for networks with overlapping communities and its applications are addressed in Section 4, followed by combining multi-layer graphs in Section 5. Evaluation of the proposed methods in the benchmark network are presented in Section 6. Analysis of some real-world datesets collected during Nokia data collection campaign is presented in Section 7, followed by conclusions in Section 8.

2. Community detection

2.1 Modularity maximization

Let's consider the clustering problem for an undirected graph $G = (V, E)$ with $|V| = N$ nodes and E edges. Recently Newman et al (Girvan & Newman, 2002; 2004) introduced a new measure for graph clustering,, named a modularity, which is defined as a number connections within a group compared to the expected number of such connections in an equivalent null model (e.g., in an equivalent random graph). In particular, the modularity Q of a partition $\mathcal{P}$ may be written as

$$Q = \frac{1}{2m} \sum_{i,j} \left(A_{ij} - P_{ij} \right) \delta(c_i, c_j) \,, \tag{1}$$

where c_i is the i-th community., A_{ij} are elements of graph adjacency matrix; d_i is the i-th node degree, $d_i = \sum_j A_{ij}$; m is a total number of links $m = \sum_i d_i/2$; P_{ij} is a probability that nodes i and j in a null model are connected; if a random graph is taken as the null model, then

$P_{ij} = d_i d_j / 2m$.
By construction $|Q| < 1$ and $Q = 0$ means that the network under study is equivalent to the used null model (an equivalent random graph). Case $Q > 0$ indicates a presence of a community structure, i.e., more links remain within communities than would be expected in an equivalent random graph. Hence, a network partition which maximizes modularity may be used to locate communities. This maximization is NP-hard and many suboptimal algorithms are suggested, e.g., see (Fortunato, 2011) and references within.
In the following we use the basic greedy search algorithm (Newman, 2004) extended with a random walk approach, since it gives a reasonable trade-off between accuracy of community detection and scalability.

Greedy Search Algorithm

Input: a weighted graph described by $N \times N$ adjacency matrix $\mathbf{A}$.

Initialize each node i as a community c_i with modularity $Q(i) = -\left(\frac{d_i}{2m}\right)^2$.

Repeat until there is an increase in modularity:
for all nodes i **do**:
- remove i from its community c_i;
- insert i sequentially in neighboring communities c_j for all j with $A_{ij} > 0$;
- calculate modularity $Q(c_j^{(i \to c_j)})$;
- select a move (if any) of i-th node to community c_j^* with max modularity $Q(c_j^{(i \to c_j^*)}) = \max_{j \in N(i)} Q(c_j^{(i \to c_j)})$;

Stop when (a local) maximum is reached.

2.2 Communities detection with random walk

It is well-known that a network topology affects a system dynamics, it allows us to use the system dynamics to identify the underlying topology (Arenas et al, 2006; 2008; Lambiotte et al, 2009). First, we review the Laplacian dynamics formalism recently developed in (Evans & Lambiotte, 2009; Lambiotte et al, 2009).
Let's consider N independent identical Poisson processes defined on every node of a graph $G(V, E), |V| = N$, where random walkers are jumping at a constant rate from each of the nodes. We define p_n as the density of random walkers on node i at step n, then its dynamics is given by

$$p_{i,n+1} = \sum_j \frac{A_{ij}}{d_j} p_{j,n} \,. \tag{2}$$

The corresponding continuous-time process, described by (3),

$$\frac{dp_i}{dt} = \sum_j \frac{A_{ij}}{d_j} p_j - p_i = \sum_j \left(\frac{A_{ij}}{d_j} - \delta_{ij} \right) p_i \tag{3}$$

is driven by the random walk operator $\frac{A_{ij}}{d_j} - \delta_{ij}$, which in case of a discrete time process is presented by the random walk matrix $\mathbf{L}_{rw} = \mathbf{D}^{-1}\mathbf{L} = \mathbf{I} - \mathbf{D}^{-1}\mathbf{A}$, where $\mathbf{L} = \mathbf{D} - \mathbf{A}$ is a Laplacian matrix, $\mathbf{A}$ is a non-negative weighted adjacency matrix, $\mathbf{D} = \mathrm{diag}\{\mathrm{d_i}\}$, $\mathrm{i} = 1, \ldots, \mathrm{N}$. For an undirected connected network the stationary solution of (2) is given by $p_i^* = d_i/2m$.
Let's now assume that for an undirected network there exist a partition $\mathcal{P}$ with communities $c_k \in \mathcal{P},\ k = 1, \ldots, N_c$. The probability that initially, at t_0, a random walker belongs to a community c_k is $\Pr(c_k, t_0) = \sum_{j \in c_k} d_j/2m$. Probability that a random walker, which was initially in c_k, will stay in the same community at the next step $t_0 + 1$ is given by

$$\Pr(c_k, t_0, t_0 + 1) = \sum_{j \in c_k} \sum_{i \in c_k} \left(\frac{A_{ij}}{d_j} \right) \left(\frac{d_j}{2m} \right). \tag{4}$$

The assumption that dynamics is ergodic means that the memory of the initial conditions are lost at infinity, hence $\Pr(c_k, t_0, \infty)$ is equal to the probability that two independent walkers are in c_k,

$$\Pr(c_k, t_0, \infty) = \left(\sum_{i \in c_k} \frac{d_i}{2m} \right) \left(\sum_{j \in c_k} \frac{d_j}{2m} \right). \tag{5}$$

Combining (4) and (5) we may write

$$\sum_{c_k \in P} \left(\Pr(c_k, t_0, t_0 + 1) - \Pr(c_k, t_0, \infty) \right) = \frac{1}{2m} \sum_{i,j} \left(A_{ij} - \frac{d_i d_j}{2m} \right) \delta(c_i, c_j) = Q. \tag{6}$$

In general case, using (3), one may define a stability of the partition $\mathcal{P}$ as (Evans & Lambiotte, 2009; Lambiotte et al, 2009)

$$R_{\mathcal{P}}(t) = \sum_{c_k \in \mathcal{P}} \Pr(c_k, t_0, t_0 + t) - \Pr(c_k, t_0, \infty) \tag{7}$$

$$= \sum_{c_k \in \mathcal{P}} \sum_{i,j \in c_k} \left(\left(e^{t(\hat{A} - I)} \right)_{ij} \frac{d_j}{2m} - \frac{d_i d_j}{4m^2} \right), \text{ where } \hat{A}_{ij} = \frac{A_{ij}}{d_j}. \tag{8}$$

Then, as the special cases of (8) at $t = 1$, we get the expression for modularity (6).
Note that $R_{\mathcal{P}}(t)$ is non-increasing function of time: at $t = 0$ we get

$$R_{\mathcal{P}}(0) = 1 - \sum_{c_k \in \mathcal{P}} \sum_{i,j \in c_k} \frac{d_i d_j}{4m^2} \tag{9}$$

and $\max_{\mathcal{P}} R(0)$ is reached when each node is assigned to its own community. Note that (9) corresponds to collision entropy or Rényi entropy of order 2.
On the other hand, in the limit $t \to \infty$, the maximum of $R_{\mathcal{P}}(t)$ is achieved with Fiedler spectral decomposition into 2 communities. In other words, time here may be seen as a resolution parameter: with time t increasing, the $\max_{\mathcal{P}} R(t)$ results in a sequence of hierarchical

partitions $\{\mathcal{P}_t\}$ with the decreasing numbers of communities.
Furthermore, as shown in (Evans & Lambiotte, 2009), we may define a time-varying modularity $Q(t)$ by linear terms in time expansion for $R(t)$ at $t \approx 0$,

$$R(t) \approx (1-t) \cdot R(0) + t \cdot Q = Q(t)\,, \tag{10}$$

which after substitution (6) and (9) gives

$$Q(t) = (1-t) + \sum_{c_k \in \mathcal{P}} \sum_{i,j \in c_k} \left(\frac{A_{ij}}{2m} t - \frac{d_i d_j}{4m^2} \right) . \tag{11}$$

In the following we apply time-dependent modularity maximization (11) using the greedy search to find hierarchical structures in networks beyond modularity maximization Q_{max} in (1). This approach is useful in cases where maximization of (1) results in a very fragmental structure with a large number of communities. Also it allows us to evaluate the stability of communities at different resolution levels. However, since the adjacency matrix $\mathbf{A}$ is not time dependent, the time-varying modularity (11) can not be used to make predictions beyond the given topology.

3. Topology detection using coupled dynamical systems

3.1 Laplacian formulation of network dynamics

Let's consider an undirected weighted graph $G = \{V, E\}$ with N nodes and E edges, where each node represents a local dynamical system and edges correspond to local coupling. Dynamics of N locally coupled dynamical systems on the graph G is described by

$$\dot{x}_i(t) = q_i(x_i(t)) + k_c \sum_{j=1}^{N} A_{ij} \psi \left(x_j(t) - x_i(t) \right) , \tag{12}$$

where $q_i(x_i)$ describes a local dynamics of state x_i; A_{ij} is a coupling strength between nodes i and j; $\psi(\cdot)$ is a coupling function; k_c is a global coupling gain.
In case of weakly phase-coupled oscillators the dynamics of local states is described by Kuramoto model (Acebron et al, 2005; Kuramoto, 1975)

$$\dot{\theta}_i(t) = \omega_i + k_c \sum_{j=1}^{N} A_{ij} \sin \left[\theta_j(t) - \theta_i(t) \right] . \tag{13}$$

Linear approximation of coupling function $\sin(\theta) \simeq \theta$ in (13) results in the consensus model (Olfati-Saber et al, 2007)

$$\dot{\theta}_i(t) = k_c \sum_{j=1}^{N} A_{ij} \left[\theta_j(t) - \theta_i(t) \right] , \tag{14}$$

which for a connectivity graph G may be written as

$$\dot{\Theta}(t) = -\, k_c \mathbf{L}\, \Theta(t)\,, \tag{15}$$

where $\mathbf{L} = \mathbf{A} - \mathbf{D}$ is the Laplacian matrix of G. The solution of (15) in the form of normal modes $\omega_i(t)$ may be written as

$$\omega_i(t) = k_c \sum_{j=1}^{N} V_{ij}\theta_j(t) = k_c\, \omega_i(t_0)\mathrm{e}^{-\lambda_i t}\,, \tag{16}$$

where $\lambda_1, \ldots, \lambda_N$ are eigenvalues and $\mathbf{V}$ is the matrix of eigenvectors of $\mathbf{L}$. Note that (16) describes a convergence speed to a consensus for each nodes. Let's order these equations according to the descending order of their eigenvalues. Then it is easy to see that nodes are approaching the consensus in a hierarchical way, revealing in the same time a hierarchy of communities in the given network G.
Note that (15) has the same form as (3), with the difference that the random walk process (3) is based on $\mathbf{L}_{rw} = \mathbf{D}^{-1}\,\mathbf{L}$. It allows us to consider random-walk-based communities detection in the previous section as a special case of coupled oscillators synchronization.
Similarly to (15), we may derive the Laplacian presentation for locally coupled oscillators (13). In particular, the connectivity of a graph may be described by the graph incidence $(N \times E)$ matrix $\mathbf{B}$: $\{\mathbf{B}\}_{ij} = 1$ (or -1) if nodes j and i are connected, otherwise $\{\mathbf{B}\}_{ij} = 0$. In case of weighted graphs we use the weighted Laplacian defined as

$$\mathbf{L}_{\mathrm{A}} \triangleq \mathbf{B}\mathbf{D}_{\mathrm{A}}\mathbf{B}^T\,. \tag{17}$$

Based on (17) we can rewrite (13) as

$$\dot{\Theta}(t) = \mathbf{\Omega} - k_c\mathbf{B}\mathbf{D}_{\mathrm{A}} \sin\left(\mathbf{B}^T\Theta(t)\right), \tag{18}$$

where vectors and matrices are defined as follows:
$\Theta(t) \triangleq [\theta_1(t), \cdots, \theta_N(t)]^T$; $\Omega(t) \triangleq [\omega_1(t), \cdots, \omega_N(t)]^T$; $\mathbf{D}_{\mathrm{A}} \triangleq \mathrm{diag}\,\{a_1, \ldots, a_E\}$, $a_1, \ldots, a_E$ are weights A_{ij} indexed from 1 to E.
In the following we use (18) to describe different coupling scenarios.

3.2 Dynamical structures with different coupling scenarios

Let's consider local correlations between instant phases of oscillators,

$$r_{ij}(t) = \langle cos\left[\theta_j(t) - \theta_i(t)\right]\rangle, \tag{19}$$

where the average is taken over initial random phases $\theta_i(t = 0)$.
Following (Arenas et al, 2006; 2008) we may define a dynamical connectivity matrix $\mathbf{C}_t(\eta)$, where two nodes i and j are connected at time t if their local phase correlation is above a given threshold η,

$$C_t(\eta)_{ij} = 1 \quad \text{if} \quad r_{ij}(t) > \eta$$

$$C_t(\eta)_{ij} = 0 \quad \text{if} \quad r_{ij}(t) < \eta\,. \tag{20}$$

We select communities resolution level (time t) using a random walk as in Section 2. Next, by changing the threshold η, we obtain a set of connectivity matrices $\mathbf{C}_t(\eta)$ which reveal dynamical topological structures for different correlation levels. Since the local correlations $r_{ij}(t)$ are continuous and monotonic functions in time, we may also fix η and express dynamical connectivity matrix (20) in the form $\mathbf{C}_\eta(t)$ to present the evolution of connectivity in time for a fixed correlation threshold η. Using this approach we consider below several scenarios of networks evolution with dynamically changing coupling.

B.1. Attractive coupling with dynamical updates

As the first step, let's introduce dynamics into static attractive coupling (13). Using the dynamical connectivity matrix (20) we may write

$$\dot{\theta}_i(t) = \omega_i + k_c \sum_{j=1}^{N} F_{ij}^{(\eta)}(t) \sin\left[\theta_j(t) - \theta_i(t)\right], \tag{21}$$

where matrix $\mathbf{F}^{(\eta)}(t)$ describes dynamical attractive coupling, $F_{ij}^{(\eta)}(t) = A_{ij}C_\eta(t)_{ij} \geq 0$. Then, similar to (18), the attractive coupling with a dynamical update may be described as

$$\dot{\Theta}(t) = \mathbf{\Omega} - k_c\mathbf{B}(t)\mathbf{D}_\mathrm{F}(t)\sin\left(\mathbf{B}(t)^T\Theta(t)\right), \tag{22}$$

where initial conditions are defined by A_{ij}; $\mathbf{D}_\mathrm{F}(t)$ is formed from $\mathbf{D}_\mathrm{A}$ with elements $\{a_k\}$ scaled according to $\mathbf{C}_\eta(t)$.

B.2. Combination of attractive and repulsive coupling with dynamical links update

Many biological and social systems show a presence of a competition between conflicting processes. In case of coupled oscillators it may be modeled as the attractive coupling (driving oscillators into the global synchronization) combined with the repulsive coupling (forcing system into a chaotic/random behavior). To allow positive and negative interactions we use instant correlation matrix $\mathbf{R}(t) = \mathbf{R}^+(t) + \mathbf{R}^-(t)$, and separate attractive and repulsive parts

$$\dot{\theta}_i(t) = \omega_i + k_c \sum_{j=1}^{N} r_{ij}^+(t)\, A_{ij} \sin\left[\theta_j(t) - \theta_i(t)\right] - k_c \sum_{j=1}^{N} |r_{ij}^-(t)|\, A_{ij} \sin\left[\theta_j(t) - \theta_i(t)\right], \tag{23}$$

where superscripts denote positive and negative correlations [1].

Note that the total number of links in the network does not change, at a given time instant each link performs either attractive or repulsive function.

To obtain the Laplacian presentation we define a dynamical connectivity matrix $\mathbf{F}(t)$ as element-by-element matrix product

$$\mathbf{F}(t) = \mathbf{R}(t) \circ \mathbf{A} = \mathbf{F}^+(t) + \mathbf{F}^-(t), \tag{24}$$

and present dynamic Laplacian as the following

$$\mathbf{L}_\mathrm{F}(t) = \mathbf{B}(t)(\mathbf{D}_{\mathrm{F}^+}(t) + \mathbf{D}_{\mathrm{F}^-}(t))\mathbf{B}^T(t). \tag{25}$$

It allows us to write

$$\dot{\theta}_i(t) = \omega_n + k_c \sum_{m=1}^{N} F_{ij}^+(t) \sin\left[\theta_j(t) - \theta_i(t)\right] - k_c \sum_{m=1}^{N} F_{ij}^-(t) \sin\left[\theta_j(t) - \theta_i(t)\right], \tag{26}$$

[1] For presentation clarity we omit here the correlation threshold η.

or in matrix form

$$\dot{\Theta}(t) = \mathbf{\Omega} - k_c \mathbf{B}(t)\mathbf{D}_{\mathrm{F}^+}(t)\sin\left(\mathbf{B}^T(t)\Theta(t)\right) + k_c \mathbf{B}(t)\mathbf{D}_{\mathrm{F}^-}(t)\sin\left(\mathbf{B}^T(t)\Theta(t)\right). \quad (27)$$

B.3. Combination of attractive and initially neutral coupling with dynamical links update

Negative correlations (resulting in repulsive coupling) are typically assigned between nodes which are not initially connected. However, in many cases this scenario is not realistic. For example, in social networks, the absence of communications between people does not necessary indicate conflicting (negative) relations, but often has a neutral meaning. To take this observation into account we modified second term in (23) such that it sets neutral initial conditions to unconnected nodes in adjacency matrix **A**. In particular, system dynamics with links update (24) and initially neutral coupling is described by

$$\dot{\theta}_i(t) = \omega_i + k_c \sum_{j=1}^{N} F_{ij}^{+}(t)\sin\left[\theta_j(t) - \theta_i(t)\right] + k_c \sum_{j=1}^{N} F_{ij}^{-}(t)\cos\left[\theta_j(t) - \theta_i(t)\right], \quad (28)$$

or in the matrix form

$$\dot{\Theta}(t) = \mathbf{\Omega} - k_c \mathbf{B}(t)\mathbf{D}_{\mathrm{F}^+}(t)\sin\left(\mathbf{B}^T(t)\Theta(t)\right) - k_c \mathbf{B}(t)\mathbf{D}_{\mathrm{F}^-}(t)\cos\left(\mathbf{B}^T(t)\Theta(t)\right). \quad (29)$$

Then a dynamical interplay between the given network topology and local interactions drives the connectivity evolution. We evaluated the scenarios above using different clustering measures (Manning et al, 2008) and found that scenario *B.3* shows the best performance.

In the following we use coupled system dynamics approach to predict networks' evolution and to make missing links predictions and recommendations. Furthermore, the suggested approach allows us also to predict repulsive relations in the network based on the network topology and links dynamics.

4. Overlapping communities

4.1 Multi-membership

In social networks people belong to several overlapping communities depending on their families, occupations, hobbies, etc. As the result, users (presented by nodes in a graph) may have different levels of membership in different communities. This fact motivated us to consider multi-community membership as edge-weights to different communities and partition edges instead of clustering nodes.

As an example, we can measure a membership $g_j(k)$ of node k in j-th community as a number of links (or its weight for a weighted graph) between the k-th node and other nodes within the same community, $g_j(k) = \sum_{i \in c_j} w_{ki}$ Then, for each node k we assign a vector $\mathbf{g}(k) = [g_1(k), g_2(k), \ldots, g_{N_c}(k)]$, $k \in \{1, \ldots, N\}$ which presents the node membership (or participation) in all detected communities $\{c_1, \ldots, c_{N_c}\}$. In the following we refer $\mathbf{g}(k)$ as a soft community decision for the k-th node.

To illustrate the approach, overlapping communities derived from benchmark karate club social network (Zachary, 1977) and membership distributions for selected nodes are depicted at Fig.1 and Fig.2, respectively. Modularity maximization here reveals 4 communities shown by different colors. However, the multi-communities membership results in overlapping communities illustrated by overlapping ovals (Fig.1). For example, according to modality maximization, the node 1 belongs to community c_2, but it also has links to all other communities indicated by blue bars at Fig.2.

Participation of different nodes in selected communities is presented at Fig.3 and Fig.4. These graphs show that even if a node is assigned by some community detection algorithm to a certain community, it still may have significant membership in other communities. This multi-communities membership is one of the reasons why different algorithms disagree on communities partitions. In practice, e.g., in targeted advertisements, due to the "hard" decision in community detection, some users may be missed even if they are strongly related to the targeted group. For example, user '29' is assigned to c_3 (Fig.1), but it also has equally strong memberships in c_2 and c_4 (Fig.3). Using soft community detection user '29' can also be qualified for advertisements targeted to c_2 or c_4.

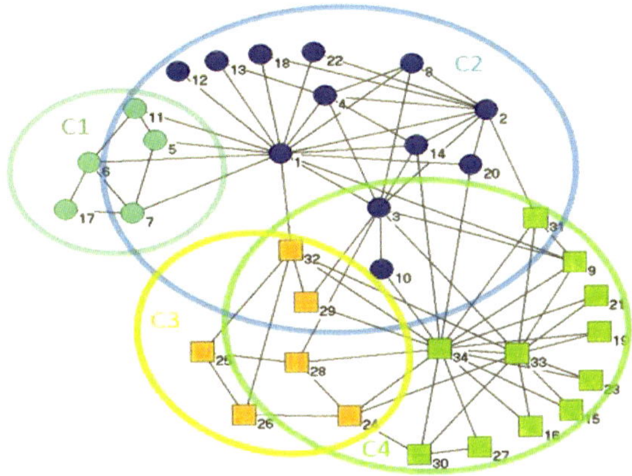

Fig. 1. Overlapping communities in karate club.

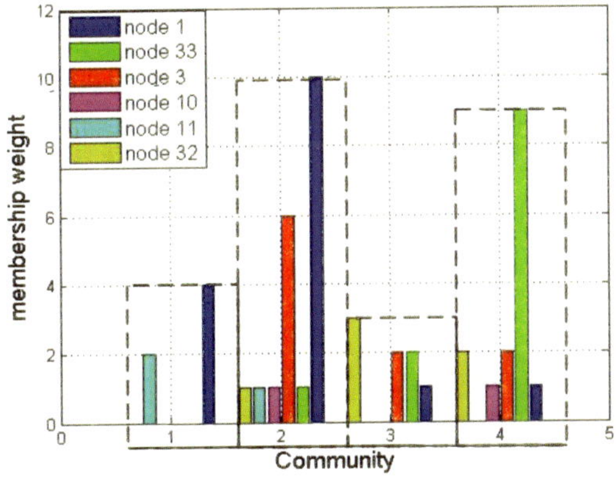

Fig. 2. Membership weight distribution for selected users in karate club social network.

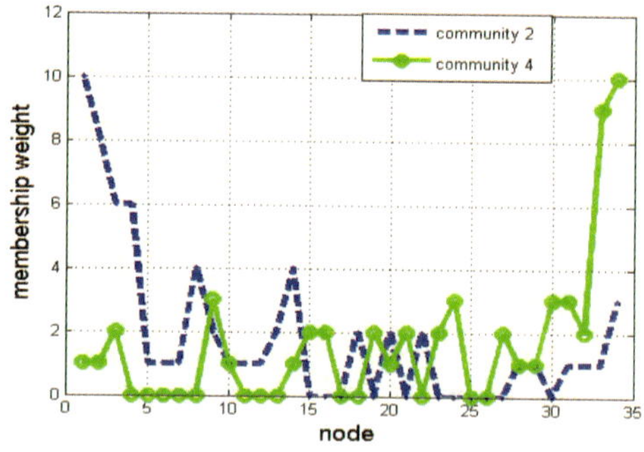

Fig. 3. Karate club: participation of users in communities c_2, c_4.

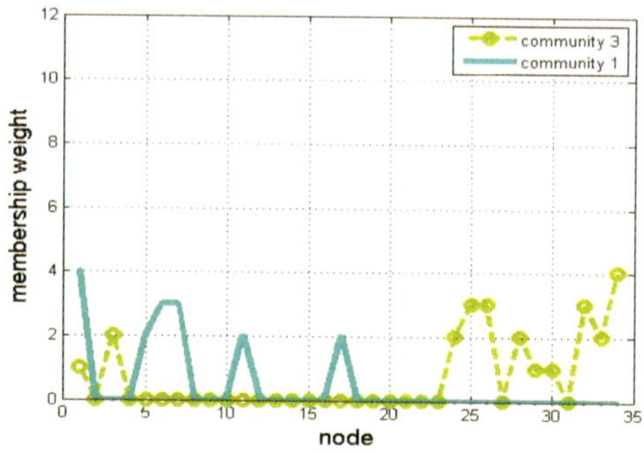

Fig. 4. Karate club: participation of users in communities c_1, c_3.

4.2 Application of soft community detection for recommendation systems

In online social networks a recommendation of new social links may be seen as an attractive service. Recently Facebook and LinkedIn introduced a service "People You May Know", which recommends new connections using the friend-of-friend (FoF) approach. However, in large networks the FoF approach may create a long and often not relevant list of recommendations, which is difficult (and also computationally expensive, in particular in mobile solutions) to navigate. Furthermore, in mobile social networks (e.g., Nokia portal Ovi Store) these kinds of recommendations are even more complicated because users' affiliations to different groups (and even its number) are not known. Hence, before making recommendations, communities are to be detected first.

Recommendations as communities completion

Based on soft communities detection we suggest to make the FoF recommendations as follows:

(i) detect communities, e.g., by using one of the methods described above;

(ii) calculate membership $g_j(k)$ in all relevant communities for each node k;

(iii) make new recommendations as communities completion following the rules below;

(iv) use multiple-membership to prioritize recommendations.

To make new link recommendations in (iii) we suggest the following rules:

- each new link creates at least one new clique (the FoF concept);
- complete cliques within the same community (intra-cliques) using the FoF concept;
- complete cliques towards to the fully-connected own community if there is no FoF links;
- complete inter-cliques (where nodes belong to different communities);
- prioritize intra-clique and inter-clique links completion according to some measure based on multi-membership.

To assign priorities we introduce several similarity measures outlined below. We will show in next sections that these rules are well in line with link predictions made by coupled dynamical systems described in Section 3.

Modified topology-based predictors

Let'a define sets of neighbors of node k, which are inside and outside of community c_i as $\Gamma_i(k) = \{\Gamma(k) \in c_i\}$ and $\Gamma_{\setminus i}(k) = \{\Gamma(k) \notin c_i\}$, respectively. This allows us to introduce a set of similarity measures by modifying topology-based base-line predictors listed in (Liben-Nowel & Kleinberg, 2003) to take into account the multiple-membership in overlapping communities.

As an example, for the intra-clique completion we may associate a quality of missing link prediction (or recommendation) between nodes k and n within c_i community by modifying the base-line predictor scores as follows:

- Preferential attachment: $S_{\mathrm{PA}}^{(i,i)}(k,n) = |\Gamma_i(k)| \cdot |\Gamma_i(n)|$;
- Jaccards score: $S_{\mathrm{JC}}^{(i,i)}(k,n) = |\Gamma_i(k) \cap \Gamma_i(n)| / |\Gamma_i(k) \cup \Gamma_i(n)|$;
- Adamic/Adar score: $S_{\mathrm{AA}}^{(i,i)}(k,n) = \sum_{z \in \Gamma_i(k) \cap \Gamma_i(n)} (\log|\Gamma(z)|)^{-1}$;
- Katz score (intra-community):

$$S_{\mathrm{KC}}^{(i,i)}(k,n) = \sum_{l=1}^{\infty} \beta^l |\mathsf{path}(k,n)^{(l)}| = \left\{(I - \beta A^{(i)})^{-1} - I\right\}_{(k,n)},$$

where $|\mathsf{path}_i(k,n)^{(l)}|$ is number of all paths of length-l from k to n within c_i; I is the identity matrix, $A^{(i)}$ is the (weighted) adjacency matrix of community c_i, β is a dumping parameter, $0 < \beta < 1$, such that $\sum_{ij} \beta A_{ij} < 1$.

Additionally to the base-line predictors above, we also use a community connectivity measure, $S_{\mathrm{CC}}^{(i,i)}(k,n) \sim \sigma_2(L_i)$, which characterizes a convergence speed of opinions to consensus within a community c_i when a link between nodes k and n is added inside the community; here $\sigma_2(L)$ is the 2nd smallest eigenvalue of the graph Laplacian $\mathbf{L}_i$ of community c_i (or the normalized Laplacian for weighted graphs, based on (17)).

The measures above consider communities as disjoint sets and may be used as the 1st order approximation for link predictions in overlapping communities. To take into account both intra- and inter-community links we use multi-community membership for nodes, $g_i(k)$. In general, for nodes $k \in c_i$ and $n \in c_j$, the inter-community relations may be asymmetric, $g_j(k) \neq g_i(n)$. In the case of undirected graphs we may use averaging and modify the base-line predictors $S(k,n)$ as

$$S^{(i,j)}(k,n) = \frac{g_j(k) + g_i(n)}{2m} S(k,n)\,. \tag{30}$$

For example, modified Jaccards and Katz scores which take into account multi-communities membership are defined as

$$S_{\mathrm{JC}}^{(i,j)}(k,n) = \frac{g_j(k)+g_i(n)}{2m} \frac{|\Gamma(k) \cap \Gamma(n)|}{|\Gamma(k) \cup \Gamma(n)|}, \tag{31}$$

$$S_{\mathrm{KC}}^{(i,j)}(k,n) = \frac{g_j(k)+g_i(n)}{2m} \left\{ (I - \beta A^{(C_{n,k})})^{-1} - I \right\}_{(k,n)}, \tag{32}$$

where $k \in c_i, n \in c_j$; $A^{(C_{n,k})}$ is an adjacency matrix formed by all communities relevant to nodes n and k.
Recommendations also may be made in the probabilistic way, e.g., to be picked up from distributions formed by modified prediction scores.

5. Multi-layer graphs

In analysis of multi-layer graphs we assume that different network layers capture different modalities of the same underlying phenomena. For example, in case of mobile networks the social relations are partly reflected in different interaction layers, such as phone and SMS communications recorded in call-logs, people "closeness" extracted from the bluetooth (BT) and WLAN proximity, common GPS locations and traveling patterns and etc. It may be expected that a proper merging of data encoded in multi-graph layers can improve the classification accuracy.
One approach to analyze multi-layer graphs is first to merge graphs according to some rules and then extract communities from the combined graph. The layers may be combined directly or using some functions defined on the graphs. For example, multiple graphs may be aggregated in spectral domain using a joint block-matrix factorization or a regularization framework (Dong et al, 2011). Another method is to extract spectral structural properties from each layer separately and then to find a common presentation shared by all layers (Tang et all, 2009).
In this paper we consider methods of combining graphs based on modularity maximization

$$\max Q = \max_{c_i,c_j} \frac{1}{2m} \sum_{i,j} \left(A_{ij} - \frac{d_i d_j}{2m} \right) \delta(c_j, c_j). \tag{33}$$

Let's define a modularity matrix $\mathbf{M}$ with elements $M_{ij} = A_{ij} - \frac{d_i d_j}{2m}$. Then the modularity in (33) may be presented as

$$Q = \frac{1}{2m} Tr \left(\mathbf{G}^T (\mathbf{A} - \frac{\mathbf{d}\mathbf{d}^T}{2m}) \mathbf{G} \right) = \frac{1}{2m} Tr(\mathbf{G}^T \mathbf{M} \mathbf{G}), \tag{34}$$

where columns of $N \times N_c$ matrix $\mathbf{G}$ describes community memberships for nodes, $g_j(i) = g_{ij} \in \{0,1\}$, $g_{ij} = 1$ if the i-th node belongs to the community c_j; N_c is a number of communities; $\mathbf{d}$ is a vector formed by degrees of nodes, $\mathbf{d} = (d_1, \cdots, d_N)^T$.
Let's consider a multi-layer graph $\mathcal{G} = \{G_1, G_2, \ldots, G_L\}$ with adjacency matrices $\mathcal{A} = \{\mathbf{A}_1, \mathbf{A}_2, \ldots, \mathbf{A}_L\}$, where L is a number of layers. Before combining. the graphs are to be normalized. In case of modularity maximization (33) it is natural to normalize each layer

according its total weight m.
The simplest method to combine multi-layer graphs is to make the average of all layers:

$$\bar{\mathbf{A}} = \frac{1}{L}\sum_{l}^{L}\mathbf{A}_l\,; \qquad \bar{\mathbf{d}} = \frac{1}{L}\sum_{l}^{L}\mathbf{d}_l\,; \qquad \bar{m} = \frac{1}{L}\sum_{l}^{L}m_l\,; \qquad \max Q = \max_{\mathbf{G}}\frac{1}{2\bar{m}}Tr(\mathbf{G}^T\bar{\mathbf{M}}\mathbf{G}) \tag{35}$$

Then the community membership matrix $\mathbf{G}$ may be found by one of community detection methods described before. By taking into account degree distributions of nodes at each graph layer, the total modularity across all layers may maximized as (Tang et all, 2009)

$$\max Q = \frac{1}{L}\sum_{l}^{L}Q_l = \max_{\mathbf{G}}\frac{1}{2L}\sum_{l}^{L}Tr\left(\mathbf{G}^T(\mathbf{A}_l - \frac{\mathbf{d}_l\mathbf{d}_l^T}{2m_l})\mathbf{G}\right) = \max_{\mathbf{G}}\frac{1}{L}\sum_{l}^{L}Tr(\mathbf{G}^T\frac{\mathbf{M}_l}{2m_l}\mathbf{G}), \tag{36}$$

Similar approach, but applied to graph Laplacian spectra and extended with a regularization, is used in (Dong et al, 2011).
Typically networks describing social relations are often undersampled, noisy and contain different amount of information at each layer. As the result, a noisy or an observable part(s) in one of the layers after averaging in (35) and (36) may deteriorate the total accuracy. A possible solution for this problem is to apply weighted superposition of layers. In particular, the more informative the layer l is, the larger weight w_l it should be given. For example, we may weight the layer l according to its modularity Q_l, hence

$$\bar{\mathbf{A}}_w = \frac{1}{L}\sum_{l}^{L}w_l\mathbf{A}_l = \frac{1}{L}\sum_{l}^{L}Q_l\mathbf{A}_l\,; \tag{37}$$

Another method to improve the robustness of nodes classification in multi-layer graphs is to extract structural properties $\mathbf{G}_l$ at each layer separately and then merge partitions (Strehl & Ghosh, 2002). The more advanced approach of processing of multi-dimensional data may be based on presenting multi-layer graphs as tensors and apply tensor decomposition algorithms (Kolda & Bader, 2009) to extract stable communities and make de-noising by lower-dimension tensor approximation. These methods are rather involved and will be considered elsewhere.

6. Simulation results for benchmark networks

To test algorithms described in the previous sections we use the karate club social network (Zachary, 1977). As mentioned before, to get different hierarchical levels beyond and below the resolution provided by *max*-modularity, we use the random walk approach. A number of detected communities in the karate club at different resolution levels is presented at Fig.5. As one can see, the *max*-modularity algorithm does not necessary result in the best partition stability. The most stable partition in case of the karate club corresponds to 2 communities (shown by squares and circles at Fig.1), which is in line with results reported by (Zachary, 1977).
Comparison of coupling scenarios *B.2* and *B.3* is presented at Fig.6 and Fig.7. Pair-wise correlations between oscillators at $t = 1$ for coupling scenarios *B.2* and *B.3* are depicted at Fig.6. Scenario *B.3* reveals clearly communities structure, while in case of *B.2* the negative coupling overwhelms the attractive coupling and forces the system into a chaotic behavior.

Dynamical connectivity matrices reordered by communities for the attractive-neural coupling $B.3$ at $t = 1$ (on the left) and $t = 10$ (on the right) are depicted at Fig.7. In case $B.3$ one can see (also cf. Fig.8) that number of connections with the attractive coupling is growing in time, while the strength of the repulsive connections is decreasing, which finally results in the global synchronization. For the scenario $B.2$ there is a dynamical balance between attractive and repulsive coupling with small fluctuations around the mean (Fig.8). Note that even the averaged strength of the repulsive connections is less than the attractive coupling, the system dynamics shows a quasi-chaotic behavior.

Fig.9 shows the adjacency matrix for Zachary karate club (red circles), detected communities by pink squares, predicted links are shown by blue dots. As expected, the dynamical methods for links prediction tend to make more connections within the established communities first, followed by merging communities and creating highly overlapped partitions at the higher hierarchical levels (the upper part at Fig.9). In case of Katz predictor (32), by increasing the dumping parameter β we take into account the larger number of paths connecting nodes in the graph, which in turn results into the larger number of suggested links above a fixed threshold. Following the concept of dynamical connectivity matrix (20), the process of growing number of links may be seen as the hierarchical community formation predicted by (32) at different values of β. This process is illustrated at Fig.9, the bottom part. Note that in case of Katz predictor, the connected graph is also approaching the fully connected graph, but the network evolution may take a different trajectory compared to the coupled dynamical systems. In particular, at small values of t and β, the network evolution is similar for both cases (cf. Fig.9(b) and Fig.9(e)), but with the time the evolution trajectories may follow different paths (cf. Fig.9(c) and Fig.9(f)), which in turn results in different predictions.

Note that in all cases of the network evolution, we may prioritize the recommended links based on the soft communities detection (Katz predictor) or the threshold η (coupled dynamical systems). We address this issue below in Section 7.

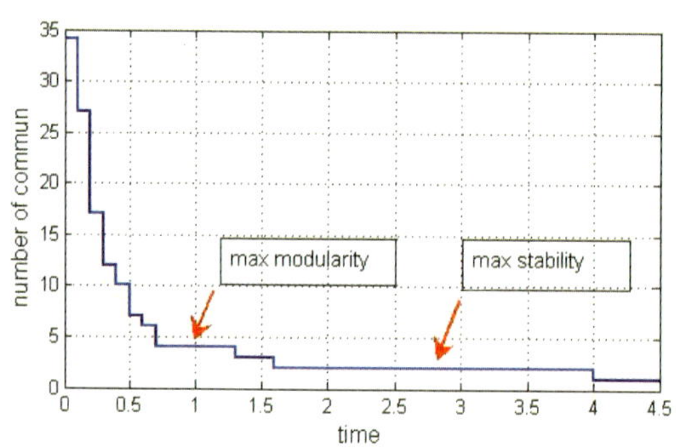

Fig. 5. Karate club: number od communities at different resolution levels.

7. Applications for real wold mobile data

7.1 Community detection in Nokia mobile datasets

To analyze mobile users behavior and study underlying social structure, Nokia Research Center/Lausanne organized mobile data collection campaign at EPFL university campus (Kiukkonen et al, 2010). Rich-content datasets (including data from mobile sensors, call-logs,

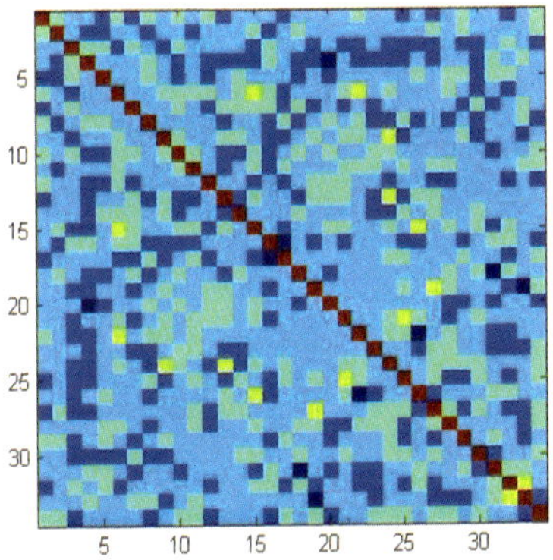

(a)

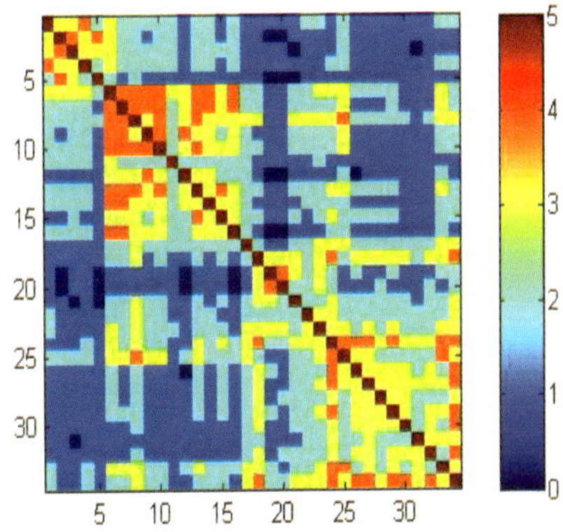

(b)

Fig. 6. Karate club: averaged pair-wise correlations (scaled by 5) between oscillators at $t = 1$ re-ordered according to communities. Coupling scenarios: (a) attractive-repulsive *B.2*; (b) attractive-neutral *B.3*.

bluetooth (BT) and WLAN proximity, GPS coordinates, information on mobile and applications usage and etc) are collected from about 200 participants for the period from June 2009 till October 2010. Besides the collected data, several surveys before and after the campaign have been conducted to profile participants and to form a basis for the ground truth. In this section we consider social affinity graphs constructed from call-logs, GPS locations and users proximity.

Fig.10 shows a weighted aggregated graph of voice-calls and SMS connections derived from corresponding datasets. This graph depicts connections among 136 users, which indicates that about 73% of participants are socially connected within the data collection campaign. To find communities in this network we first run the modularity maximization algorithm, which identifies 14 communities after the 3d iteration (Fig.10). To get the higher hierarchical levels one could represent each community by a single node and continue clustering with the new aggregated network. However, this procedure would result in a loss of underlaying structure. In particular, the hierarchical community detection with the nested communities structure poses additional constrains on the maximization process and may lead to incorrect classification at the higher layers. For example, after the 3d iteration the node "v146", shown by red arrow at Fig.10, belongs (correctly) to a community shown by white circles (3 intra-community edges and single edges to other 6 communities). After agglomeration, the node "v146" will be assigned to the community shown by white circles on the left side of the graph. However, it is easy to verify that when communities on the right are merged, the node "v146" is to be re-assigned to the community on the right side of the network. Dynamical formulation of modularity extended with the random walk allows different (not necessarily nested) allocations of nodes at different granularity (resolution) levels and helps to resolve this problem.

Fig.11 presents a number of communities at different hierarchical levels detected by the random walk for the network shown at Fig.10. As one can see, the *max*-modularity partition with 14 communities is clearly unstable and hardly could be used for reliable predictions, the

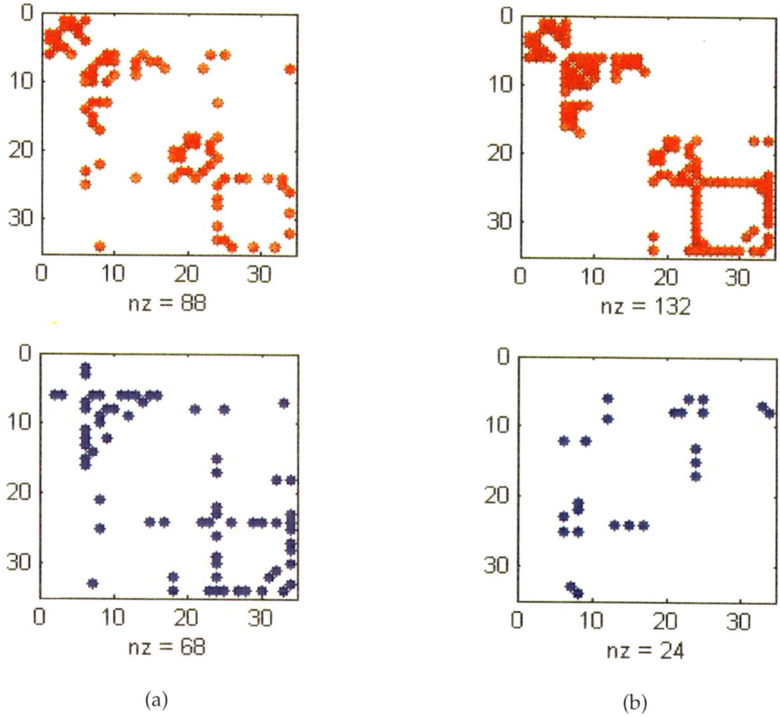

Fig. 7. Karate club: examples of dynamical connectivity matrices for attractive (shown on the top in red color) and repulsive (shown at the bottom in blue color) coupling at $t = 1$ (a) and $t = 10$ (b). Nodes are ordered according to communities. Coupling scenarios: attractive-neutral *B.3*.

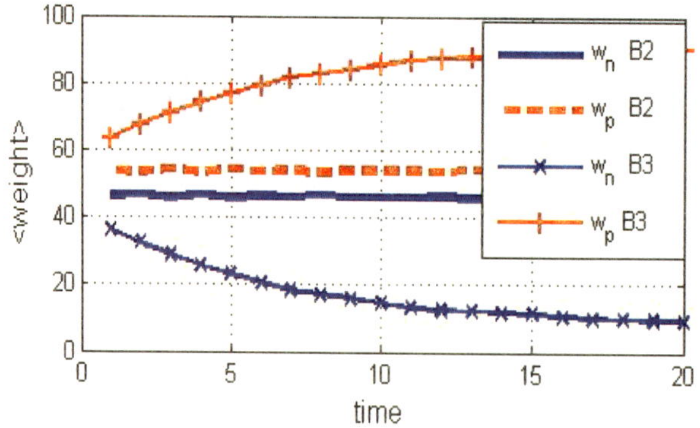

Fig. 8. Karate club: evolution of averaged attractive w_p and repulsive w_n weights for different coupling scenarios *B.2* and *B.3*; the average is made over 100 realizations.

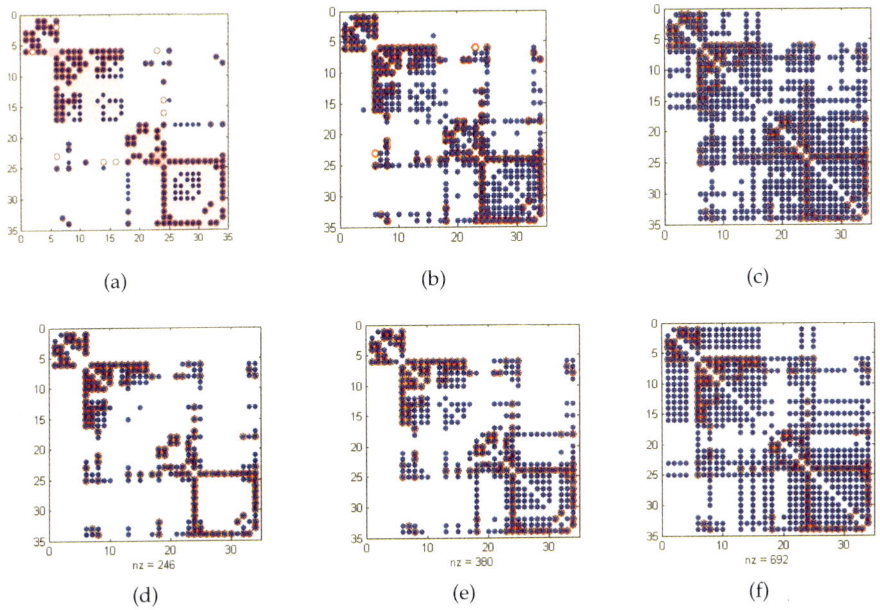

Fig. 9. Karate club: adjacency matrix is shown by red circles, detected communities by pink squares, predicted links are shown by blue dots. The upper part (a)-(c): predictions made by dynamical systems at different time scales. The bottom part (d)-(f): recommendations made by the modified Katz predictor at different values of β.

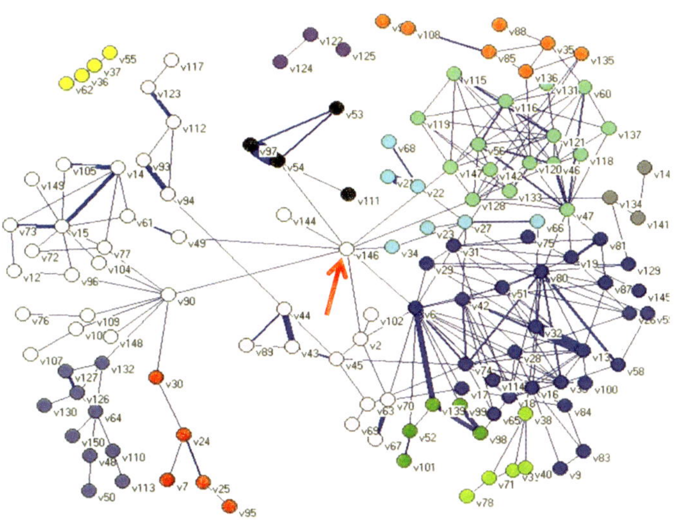

Fig. 10. Community detection based on SMS and call-logs: communities are coded by colors.

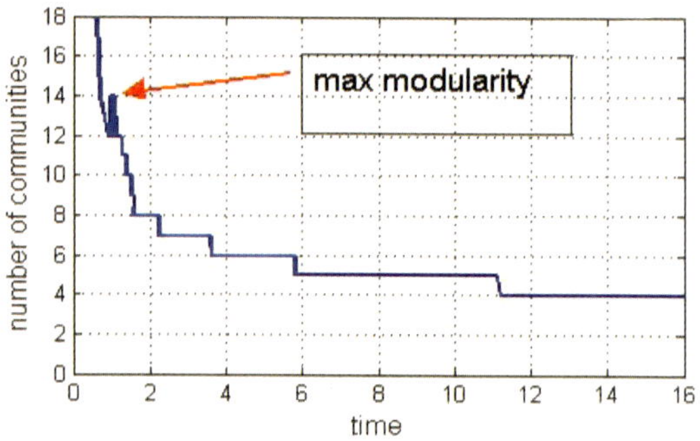

Fig. 11. Stability of communities at different resolution levels.

stable partitions appear at the higher hierarchical levels starting from 8 communities. In the following we rely on this fact to build the ground truth references for evaluation of clustering.

7.2 Applications for multi-layer graphs

Besides phone and SMS call-logs, the social affinity of participants may also be derived from other information layers, such as a local proximity of users (BT and WLAN layers) and their location information (GPS). In this case the soft communities detection may be extended to include multiple graph layers. In particular, we found that users' profiles may significantly vary across the layers. For example, a user may have dense BT connections with a multiple communities participation, while his phone call activities may be rather limited. Combining information from several graph layers can be used to improve the reliability of classification. Below we show some preliminary results, more detailed analysis of multi-layer graphs built from mobile datasets may be found in (Dong et al, 2011).

To make verification of detected communities we select a subset of 136 users with known email affiliations as the ground truth. In our case these users are allocated into 8 groups. To get the same number of communities in social affinity multi-layer graphs, we use the random walk (11) to obtain the more course resolution than provided by the modularity maximization. Fig 12 depicts communities (color coded) derived from the phone-calls graph. Single nodes here indicate users which did not make phone calls to other participants of the data collection campaign. Communities derived from the BT-proximity graph and mapped on the phone-call graph are shown at Fig.13. As expected, multi-layers graphs help us to classify users based on the additional information found in other layers. For example, users which can not be classified based on phone calls (Fig.12) are assigned to communities based on the BT proximity (Fig.13). Fig.14 shows communities detected in the combined graph formed by the BT and phone-call networks and then mapped on the phone-call network.

Next, we consider communities detected at single and combined layers with different strategies (35)-(37) described in Section 5 and compare them to the ground truth. To evaluate accuracy of community detection we use the normalized mutual information (NMI) score, purity test and Rand index (RI) (Manning et al, 2008). We found that the best graph combining is provided by weighted superposition (37) according to the *max*-modularity of layers Q. Results of the comparison are summarized in Table 1. As expected, different graph layers have a different relevance to the email affiliations and do not have fully overlapped community structures. In particular, the local proximity seems to be more relevant to professional relations

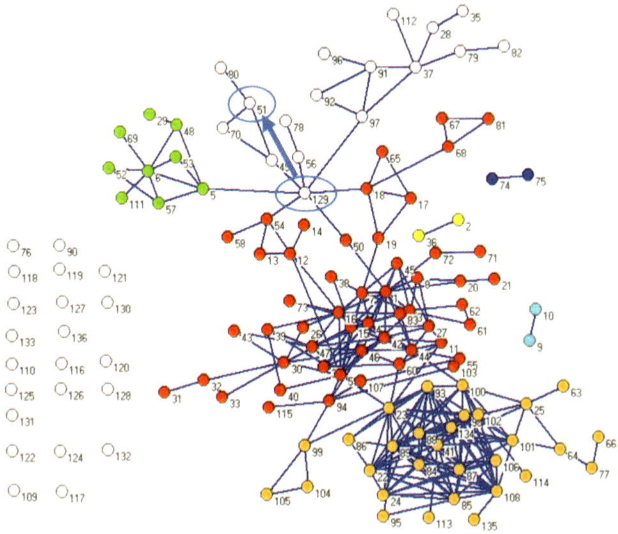

Fig. 12. Community detection using random walk in the phone-calls network.

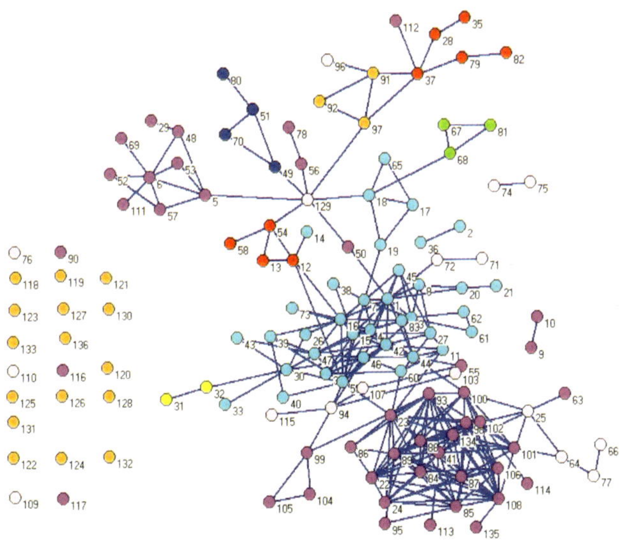

Fig. 13. Communities detected in the BT proximity network and mapped on the phone-calls network.

indicated by email affiliations, while phone calls seem to reflect more friendship and family relations. However, the detected structures are still rather close to each other (cf. columns in Table 1) reflecting underlaying social affinity. As one can see, by properly combining information from different graph layers we can improve the reliability of communities detection.

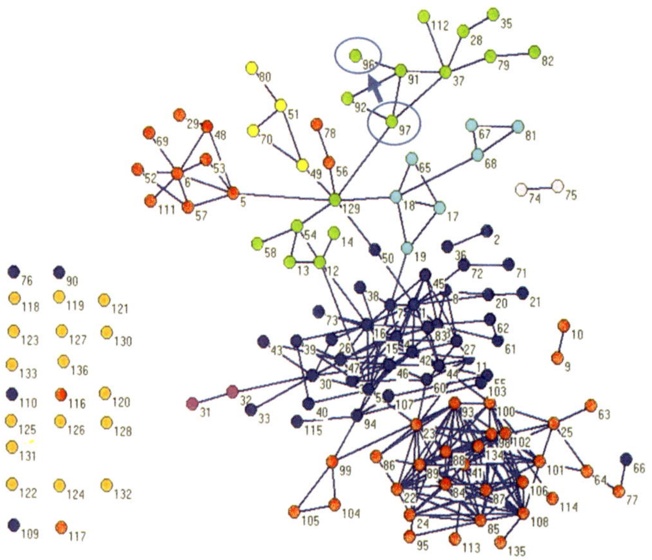

Fig. 14. Communities detected in the combined BT & phone-calls network and mapped on the phone-calls network.

	NMI	Purity	RI	Q
Phone calls	0.262	0.434	0.698	0.638
BT proximity	0.307	0.456	0.720	0.384
GPS	0.313	0.471	0.704	0.101
Phone + BT	**0.342**	0.427	**0.783**	

Table 1. Evaluation of community detection in multi-layer graphs.

7.3 Application for recommendation systems

As discussed in Section 4, one of applications of the soft communities detection and coupled systems dynamics may be seen in recommendation systems. To illustrate the approach we selected the user "129" (marked by oval) in the phone-calls network at Fig.12 and calculated proposed prediction scores for different similarity measures.

First, we consider intra-community predictions made by coupled dynamical systems. Fig.15(a) depicts pair-wise correlations (scaled by 5) between oscillators at $t = 10$ for the sub-network at Fig.12 forming the intra-community of the user "129". By changing

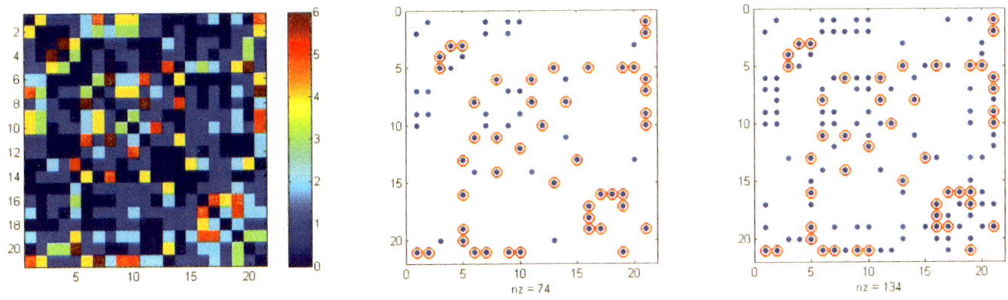

(a) (b) (c)

Fig. 15. Community of the user "129" (shown by pink color at Fig.12): averaged (scaled by 5) pair-wise correlations between oscillators at $t = 10$ (a). Intra-community adjacency matrix (red circles) and links predicted by dynamics (blue dots) at different resolution levels: $t = 15$ (b) and $t = 25$ (c).

the threshold η for the dynamical connectivity matrix $C_t(\eta)$ (which is linked to time resolution t) we obtain different connectivity matrices $C_\eta(t)$ presenting the network evolution. Connectivity matrices (blue points) corresponding to $\eta = 3$ ($t = 15$) and $\eta = 2.3$ ($t = 25$) are shown at Fig.15(b) and Fig.15(c), respectively. The community adjacency matrix is marked on the same figures by red circles. As one can see, dynamical systems first reliably detect the underlaying topology and then form new links as the result of local interactions and dynamical links update. It can be easily verified that practically all new links (e.g., 12 out of 13 at Fig.15(b)) create new cliques, hence we can interpret these new links as the Friend-of-Friend recommendations.
Calculated scores $S_{\text{DC}}^{(i,i)}(k,n)$ for dynamical systems together with the Friend-of-Friend intra-community recommendations for two predictors based on the soft community detection (Katz predictor and convergence speed to consensus, $S_{\text{CC}}^{(i,i)}(k,n)$) are summarized in Table 2. Here we list all new links together with their normalized prediction scores for the user "129" which create at least one new clique within its community (shown by pink color at Fig.12).

source	destination	$S_{\text{KC}}^{(i,i)}(s,d),\%$	$S_{\text{CC}}^{(i,i)}(s,d),\%$	$S_{\text{DC}}^{(i,i)}(s,d),\%$
129	51	10.5	**22.6**	18.6
129	78	11.1	16.3	**20.8**
129	91	**47.1**	15.4	11.6
129	70	11.3	15.3	18.9
129	92	9.6	15.3	18.8
129	37	10.5	15.1	11.4

Table 2. Scores for the FoF intra-community recommendations for user 129 according to different similarity measures for the phone-calls network at Fig.12.

Recall that both $S_{\text{CC}}^{(i,i)}(k,n)$ and $S_{\text{DC}}^{(i,i)}(k,n)$ are based on the network synchronization with closely related Laplacians. As the result, the distribution of prediction scores $S_{\text{CC}}^{(i,i)}(k,n)$ and $S_{\text{DC}}^{(i,i)}(k,n)$ are rather close to each other, compared to the the distribution of the routing-based Katz score $S_{\text{KC}}^{(i,i)}(k,n)$. Convergence of opinions to a consensus within communities in many cases is the important target in social science. As an example, the best intra-community recommendation in the phone-calls network according $S_{CC}^{(ii)}(k,n)$ is shown by the blue arrow at Fig.12. Scaled pair-wise correlations between oscillators for the whole phone-call network

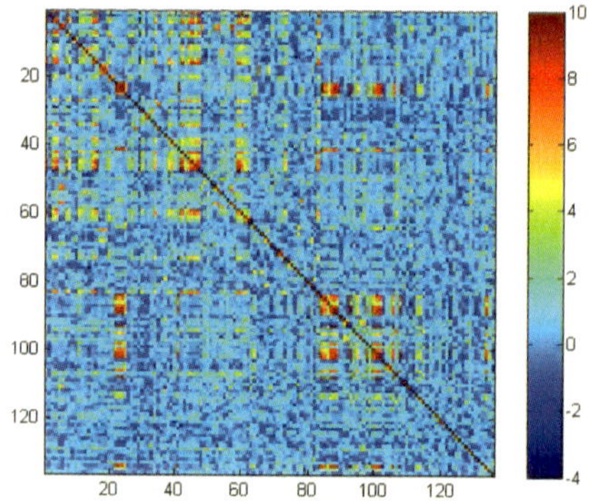

Fig. 16. Phone-call network: averaged pair-wise correlations (scaled by 10) between oscillators at t=10, coupling scenario $B.3$.

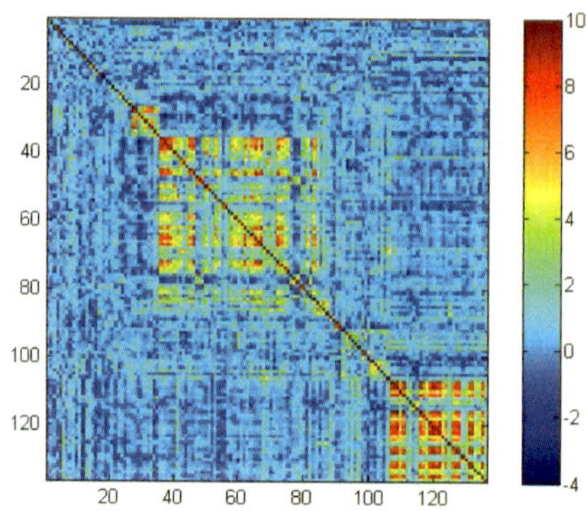

Fig. 17. Phone-call network: averaged pair-wise correlations re-ordered according to detected communities.

at Fig.12 are shown at Fig.16. Correlations between nodes, re-ordered according to one of the stable partitions detected by the random walk at $t=10$, reveal clearly the community structure (Fig.17). The phone-calls adjacency matrix (red circles) and all possible intra-community links (yellow squares) for the stable communities at $t = 10$ are depicted at Fig.18 (a). Links predicted by system dynamics (blue dots) inside and outside of yellow squares indicate predicted intra-community and inter-communities connections at different resolution levels and show the priority of the intra-community connections (Fig.18 (b) – Fig.18(c)). As the whole, the presented results for the coupled dynamical systems provide the formal basis for the recommendation rules formulated in Section 4.2.

As it is shown in Section 3, the dynamical process of opinions convergence may be seen as the first-order approximation of the network synchronization. At the same time, $S_{CC}^{(i,i)}(k,n)$

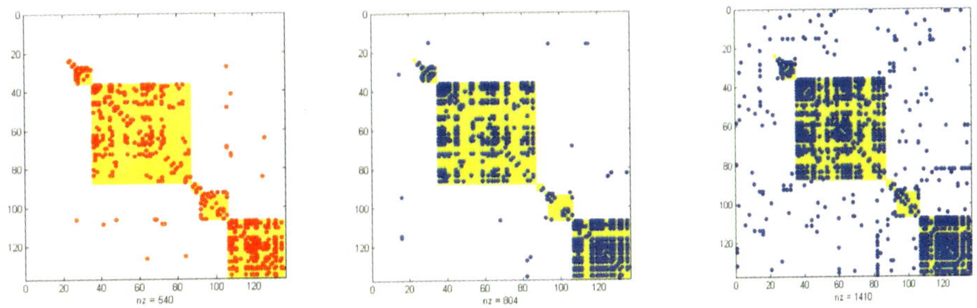

(a) (b) (c)

Fig. 18. Phone-call network: (a) adjacency matrix is marked by red dots, all possible intra-communities links are shown by yellow squares. Links predicted by dynamics (blue dots) tend to concentrate within communities: (b) $t = 10$; (c) $t = 15$.

has the lower computational complexity than $S^{(i,i)}_{\mathrm{DC}}(k,n)$, it makes $S^{(i,i)}_{\mathrm{CC}}(k,n)$ more suitable for large networks. Prediction scores $S_{\mathrm{CC}}(129,n)$ and $S_{\mathrm{KC}}(129,n)$ calculated according to (32) for cases with intra- and inter-communities links in the phone-call network are depicted at Fig.19. Here the scores are normalized as probabilities and sorted according to its priority; destination nodes n are listed along the x-axis; corresponding random-link probabilities, $p_{kn} = (d_k d_n)/2m$, are shown as the reference. Note that the link with the highest priority, {129,51} for $S^{(i,i)}_{\mathrm{CC}}(k,n)$, is the same as in the intra-community recommendation (cf. Table 2). However, the presence of inter-community links modifies priorities of other recommendations according to (30). To make verification we compare the predicted links at the phone-call network with links observed for the user "129" at the BT proximity layer. This comparison shows a good fit: 16 out of 18 predicted links are found at the BT proximity layer.
Results for the combined BT and phone-calls networks are presented at Fig.20. Pair-wise correlations between nodes obtained by dynamical systems approach are shown at Fig.20 (a). These correlations may be interpreted as probabilities for new links recommendations. Fig.20 (b) depicts recommended links based on the modified Katz predictor (blue circles) beyond the

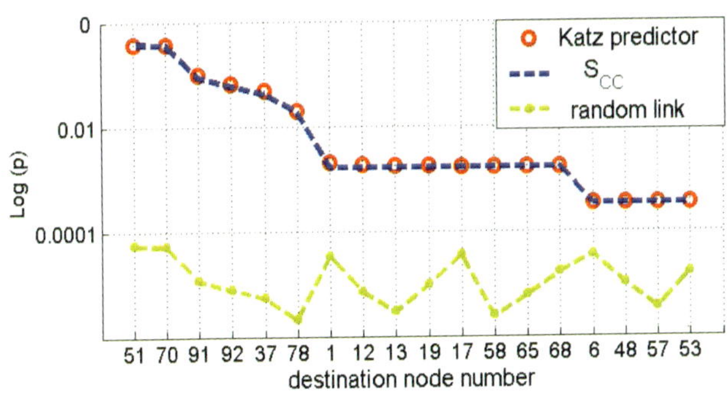

Fig. 19. Priorities of the FoF recommendations for the user 129 at Fig.12 to be connected to destination nodes shown along x-axis over all relevant communities.

given topology (red dots). We found that both recommenders mostly agree on the priority of intra-community links, but put different weights on inter-community predictions.
Depending on a purpose of recommendation we may select different prediction criteria. Since new links change topology, which in turn affects dynamical properties of the network, the recommendations may be seen as a distributed control driving the network evolution.
In general, the selection of topology-based recommendation criteria and their verifications are the open problems. Currently we are running experiments to evaluate different recommendation criteria and its acceptance rates.

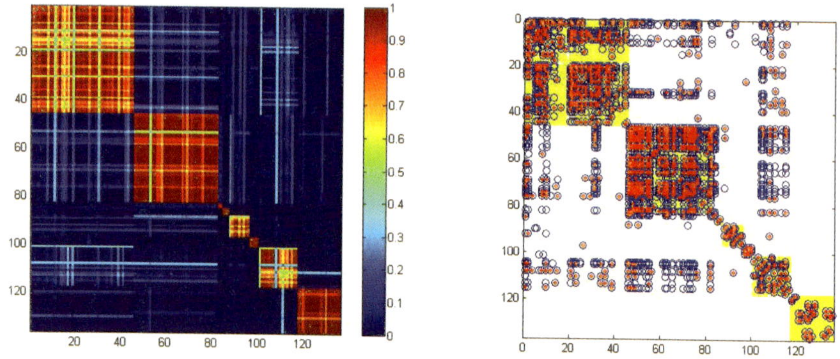

(a) (b)

Fig. 20. Combined BT and phone-call networks, nodes are ordered according to detected communities: (a) color-coded pair-wise correlations using dynamical systems; (b) links recommendations using modified Katz predictor (blue circles), adjacency matrix is marked by red dots, all possible intra-community links are shown by yellow squares.

8. Conclusions

In this chapter we present the framework for multi-membership communities detection in dynamical multi-layer graphs and its applications for links predictions/recommendations based on the network topology. The method is based on the dynamical formulation of modularity using a random walk and then extended to coupled dynamical systems to detect communities at different hierarchical levels. We introduce attractive and repulsive coupling and dynamical link updates that allow us to make predictions on a cooperative or a competing behavior of users in the network and analyze connectivity dynamics.
To address overlapping communities we suggest the method of soft community detection. This method may be used to improve marketing efficiency by identifying users which are strongly relevant to targeted groups, but are not detected by the standard community detection methods. Based on the soft community detection we suggest friend-recommendations in social networks, where new link recommendations are made as intra- and inter-clique communities completion and recommendations are prioritized according to similarity measures modified to include multiple-communities membership.
This developed methods are applied for analysis of datasets recorded during Nokia

mobile-data collection campaign to derive community structures in multi-layer graphs and to make new link recommendations.

9. Appendix: Clustering evaluation measures

Let's define $C = \{c_1, \ldots, c_M\}$ and $\Psi = \{\psi_1, \ldots, .\psi_M\}$ as partitions containing detected clusters c_i and the ground truth clusters ψ_i, respectively. Quality of clustering algorithms may be evaluated by different measures (Manning et al, 2008), in particular:

- Rand index:

$$RI = \frac{TruePositive + TrueNegative}{TruePositive + FalsePositive + FalseNegative + TrueNegative}; \tag{38}$$

- Purity test:

$$Purity(\Psi, C) = \frac{1}{n} \sum_{m=1}^{M} \max_j |\psi_m \cap c_j|; \tag{39}$$

- Normalized mutual information:

$$NMI(C, \Psi) = \frac{2\,I(\Psi, C)}{H(\Psi) + H(C))}, \tag{40}$$

where the mutual information $I(C_1, C_2)$ between the partitions C_1 and C_2 and their entropies $H(C_i)$ are

$$I(C_2, C_2) = \sum_{m_1}^{M} \sum_{m_2}^{M} \frac{c_{m_1,m_2}}{n} \log\left(\frac{n\, c_{m_1,m_2}}{n_{m_1} n_{m_2}}\right), \quad H(C_i) = -\sum_{m_i}^{M} \frac{n_{m_i}}{n} \log\left(\frac{n_{m_i}}{n}\right); \tag{41}$$

n is total number of data points; c_{m_1,m_2} is the number of common samples in the m_1-th cluster from C_1 and the m_2-th cluster in the partition C_2; n_{m_i} is the number of samples in the m_i-th cluster in the partition C_i . According to (41), $\max NMI(C_1, C_2) = 1$ if $C_1 = C_2$.

10. References

Acebrón, J., Bonilla, L., Pérez-Vicente, C., Ritort, F., Spigler, R. (2005). The Kuramoto model: A simple paradigm for synchronization phenomena. *Reviews of Modern Physics,* 77 (1), pp. 137–185.

Albert, R. & Barabási, A.-L. (2002). Statistical mechanics of complex networks. *Reviews of Modern Physics,* 74, pp. 47–97.

Arenas A., Díaz-Guilera, A., Pérez-Vicente, C. (2006). Synchronization reveals topological scales in complex networks. *Physical Review Letters,* 96, 114102.

Arenas, A., Diaz-Guilera, A., Kurths, J., Moreno, Y. and Zhou, C. (2008). Synchronization in complex networks, *Physics Reports,* 469, pp. 93–153.

Blondel, V., Guillaume, J.-L., Lambiotte, R. and Lefebvre, E. (2008). Fast unfolding of communites in large networks. *Journal of Statistical Mechanics: Theory and Experiment,* vol. 1742-5468, no. 10, pp. P10008+12.

Evans, T. S. and Lambiotte R. (2009). Line Graphs, Link Partitions and Overlapping Communities. *Physical Review,* E 80 016105.

Dong, X., Frossard, P., Vandergheynst, P. and Nefedov, N. (2011). Clustering with Multi-Layer Graphs: Spectral Perspective. *ArXiv, 1106.2233.*

Flake, G., Lawrence, S., Giles, C. and Coetzee, F. (2002). Self-organization and identification of Web communities. *IEEE Computer* 35, pp. 66–71.

Fortunato, S. (2011). Community detection in graphs. *Physics Reports,* 486, pp. 75–174.

Girvan, M. & Newman, M. E. J. (2002). Community structure in social and biological networks. *Proc. Natl. Acad. Sci. USA,* 99, pp. 7821–7826.

Newman, M.E.J. and Girvan, M. (2004). Finding and evaluating community structure in networks. *Physical Review,* E 69, 026113.

Kiukkonen, N., Blom, J., Dousse, O., Gatica-Perez, D. and Laurila, J. (2010). Towards Rich Mobile Phone Datasets: Lausanne Data Collection Campaign. *Proc. ACM Int. Conf. Pervasive Services,* Berlin.

Kolda, T. and Bader, B. (2009). Tensor decompositions and applications, *SIAM Review,* vol.51, pp. 455–500.

Kuramoto, Y. (1975). *Lectuer Notes in Physics,* 30, Springer NY.

Lambiotte, R., Delvenne, J.-C. and Barahona, M. (2009). Laplacian Dynamics and Multiscale Modular Structure in Networks. *ArXiv:0812.1770v3.*

Liben-Nowel, D. and Kleinberg, J. (2003). The Link Prediction Problem for Social Networks. *ACM Int. Conf. on Information and Knowledge Management.*

Manning, C., Raghava, P. and Schütze, H. (2008). *Introduction to Information Retrieval.* Cambridge University Press.

Newman, M. E. J. (2004). Fast algorithm for detecting community structure in networks. *Physical Review,* E 69, 066133.

Olfati-Saber, R. et al. (2007). Consensus and Cooperation in Networked Multi-Agent Systems. *IEEE Proceedings,* 95(1), pp. 215–233.

Strehl A. & Ghosh, J. (2002). Cluster Ensembles - A Knowledge Reuse Framework for Combining Multiple Partitions. *Journal of Machine Learning Research,* 3, pp. 583–617.

Tang, L., Wang, W. and Wang X. (2009). Uncovering Groups via Heterogeneous Interaction Analysis. *SDM workshop on Analysis of Dynamic Networks.*

Wasserman, S. & Faust, K. (1994). *Social Network Analysis,* Cambridge University Press, Cambridge.

Zachary, W. (1977). An information flow model for conflict and fission in small groups. *Journal of Anthropological Research,* 33, pp. 452–473.

3

Digital Backward Propagation: A Technique to Compensate Fiber Dispersion and Non-Linear Impairments

Rameez Asif, Chien-Yu Lin and Bernhard Schmauss
Chair of Microwave Engineering and High Frequency Technology (LHFT), Erlangen Graduate School in Advanced Optical Technologies (SAOT), Friedrich-Alexander University of Erlangen-Nuremberg (FAU), Cauerstr. 9, (91058) Erlangen Germany

1. Introduction

Recent numerical and experimental studies have shown that coherent optical QPSK (CO-QPSK) is the promising candidate for next-generation 100Gbit/s Ethernet (100 GbE) (Fludger et al., 2008). Coherent detection is considered efficient along with digital signal processing (DSP) to compensate many linear effects in fiber propagation i.e. chromatic dispersion (CD) and polarization-mode dispersion (PMD) and also offers low required optical signal-to-noise ratio (OSNR). Despite of fiber dispersion and non-linearities which are the major limiting factors, as illustrated in Fig. 1, optical transmission systems are employing higher order modulation formats in order to increase the spectral efficiency and thus fulfil the ever increasing demand of capacity requirements (Mitra et al., 2001). As a result of which compensation of dispersion and non-linearities (NL), i.e. self-phase modulation (SPM), cross-phase modulation (XPM) and four-wave mixing (FWM), is a point of high interest these days.

Various methods of compensating fiber transmission impairments have been proposed in recent era by implementing all-optical signal processing. It is demonstrated that the fiber dispersion can be compensated by using the mid-link spectral inversion method (MLSI) (Feiste et al., 1998; Jansen et al., 2005). MLSI method is based on the principle of optical phase conjugation (OPC). In a system based on MLSI, no in-line dispersion compensation is needed. Instead in the middle of the link, an optical phase conjugator inverts the frequency spectrum and phase of the distorted signals caused by chromatic dispersion. As the signals propagate to the end of the link, the accumulated spectral phase distortions are reverted back to the value at the beginning of the link if perfect symmetry of the link is assured. In (Marazzi et al., 2009), this technique is demonstrated for real-time implementation in 100Gbit/s POLMUX-DQPSK transmission.

Another all-optical method to compensate fiber transmission impairments is proposed in (Cvecek et al., 2008; Sponsel et al., 2008) by using the non-linear amplifying loop mirror (NALM). In this technique the incoming signal is split asymmetrically at the fiber coupler

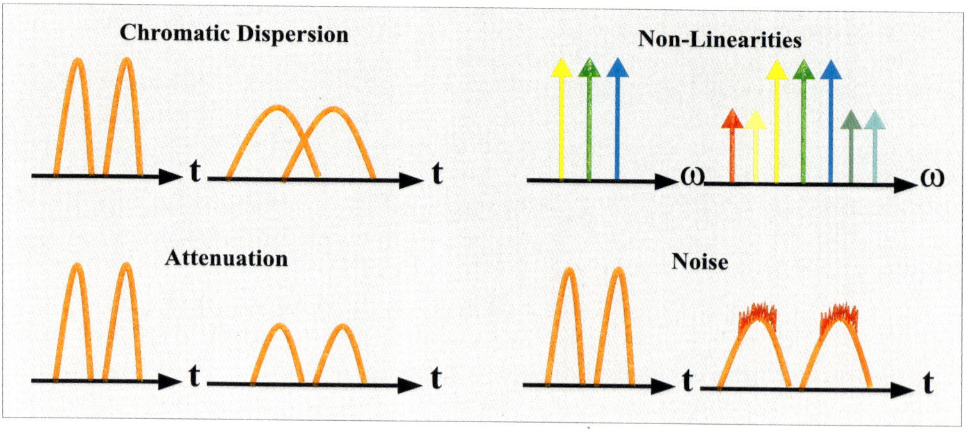

Fig. 1. Optical fiber transmission impairments.

into two counter-propagating signals. The weaker partial pulse passes first through the EDFA where it is amplified by about 20dB. It gains a significant phase shift due to self-phase modulation (Stephan et al., 2009) in the highly non-linear fiber (HNLF). The initially stronger pulse propagates through the fiber before it is amplified, so that the phase shift in the HNLF is marginal. At the output coupler the strong partial pulse with almost unchanged phase and the weak partial pulse with input-power-dependent phase shift interfere. The first, being much stronger, determines the phase of the output signal and therefore ensures negligible phase distortions.

Various investigations have been also been reported to examine the effect of optical link design (Lin et al., 2010a; Randhawa et al., 2010; Tonello et al., 2006) on the compensation of fiber impairments. However, the applications of all-optical methods are expensive, less flexible and less adaptive to different configurations of transmission. On the other hand with the development of proficient real time digital signal processing (DSP) techniques and coherent receivers, finite impulse response (FIR) filters become popular and have emerged as the promising techniques for long-haul optical data transmission. After coherent detection the signals, known in amplitude and phase, can be sampled and processed by DSP to compensate fiber transmission impairments.

DSP techniques are gaining increasing importance as they allow for robust long-haul transmission with compensation of fiber impairments at the receiver (Li, 2009; Savory et al., 2007). One major advantage of using DSP after sampling of the outputs from a phase-diversity receiver is that hardware optical phase locking can be avoided and only digital phase-tracking is needed (Noe, 2005; Taylor, 2004). DSP algorithms can also be used to compensate chromatic dispersion (CD) and polarization-mode dispersion (PMD) (Winters, 1990). It is depicted that for a symbol rate of τ, a $\frac{\tau}{2}$ tap delay finite impulse response (FIR) filter may be used to reverse the effect of fiber chromatic dispersion (Savory et al., 2006). The number of FIR taps increases linearly with increasing accumulated dispersion i.e the number of taps required to compensate 1280 ps/nm of dispersion is approximately 5.8 (Goldfarb et al., 2007). At long propagation distances, the extra power consumption required for this task becomes significant. Moreover, a longer FIR filter introduces a longer delay and requires more area on a DSP circuitry.

Alternatively, infinite impulse response (IIR) filters can used (Goldfarb et al., 2007) to reduce the complexity of the DSP circuit.
However, with the use of higher order modulation formats, i.e QPSK and QAM, to meet the capacity requirements, it becomes vital to compensate non-linearities along with the fiber dispersion. Due to this non-linear threshold point (NLT) of the transmission system can be improved and more signal power can be injected in the system to have longer transmission distances. In (Geyer et al., 2010) a low complexity non-linear compensator scheme with automatic control loop is introduced. The proposed simple non-linear compensator requires considerably lower implementation complexity and can blindly adapt the required coefficients. In uncompensated links, the simple scheme is not able to improve performance, as the non-linear distortions are distributed over different amounts of CD-impairment. Nevertheless the scheme might still be useful to compensate possible non-linear distortions of the transmitter. In transmission links with full in-line compensation the compensator provides 1dB additional noise tolerance. This makes it useful in 10Gbit/s upgrade scenarios where optical CD compensation is still present. Another promising electronic method, investigated in higher bit-rate transmissions and for diverse dispersion mapping, is the digital backward propagation (DBP), which can jointly mitigate dispersion and non-linearities. The DBP algorithm can be implemented numerically by solving the inverse non-linear Schrödinger equation (NLSE) using split-step Fourier method (SSFM) (Ip et al., 2008). This technique is an off-line signal processing method. The limitation so far for its real-time implementation is the complexity of the algorithm (Yamazaki et al., 2011). The performance of the algorithm is dependent on the calculation steps (h), to estimate the transmission link parameters with accuracy, and on the knowledge of transmission link design.
In this chapter we give a detailed overview on the advancements in DBP algorithm based on different types of mathematical models. We discuss the importance of optimized step-size selection for simplified and computationally efficient algorithm of DBP.

2. State of the art

Pioneering concepts on backward propagation have been reported in articles of (Pare et al., 1996; Tsang et al., 2003). In (Tsang et al., 2003) backward propagation is demonstrated as a numerical technique for reversing femtosecond pulse propagation in an optical fiber, such that given any output pulse it is possible to obtain the input pulse shape by numerically undoing all dispersion and non-linear effects. Whereas, in (Pare et al., 1996) a dispersive medium with a negative non-linear refractive-index coefficient is demonstrated to compensate the dispersion and the non-linearities. Based on the fact that signal propagation can be interpreted by the non-linear Schrödinger equation (NLSE) (Agrawal, 2001). The inverse solution i.e. backward propagation, of this equation can numerically be solved by using split-step Fourier method (SSFM). So backward propagation can be implemented digitally at the receiver (see section 3.2 of this chapter). In digital domain, first important investigations (Ip et al., 2008; Li et al., 2008) are reported on compensation of transmission impairments by DBP with modern-age optical communication systems and coherent receivers. Coherent detection plays a vital role for DBP algorithm as it provides necessary information about the signal phase. In (Ip et al., 2008) 21.4Gbit/s RZ-QPSK transmission model over 2000km single mode fiber (SMF) is used to investigate the role of dispersion mapping, sampling ratio and multi-channel transmission. DBP is implemented by using a asymmetric split-step Fourier method (A-SSFM). In A-SSFM method each calculation step is solved by linear operator ($\hat{D}$)

followed by a non-linear operator ($\hat{N}$) (see section 3.2.1 of this chapter). In this investigation the results depict that the efficient performance of DBP algorithm can be obtained if there is no dispersion compensating fiber (DCF) in the transmission link. This is due to the fact that in the fully compensated post-compensation link the pulse shape is restored completely at the input of the transmission fiber in each span. This reduces the system efficiency due to the maximized accumulation of non-linearities and the high signal-ASE (amplified spontaneous emission) interaction leading to non-linear phase noise (NLPN). So it is beneficial to fully compensate dispersion digitally at the receiver by DBP. The second observation in this article is about the oversampling rate which improves system performance by DBP.

A number of investigations with diverse transmission configurations have been done with coherent detection and split-step Fourier method (SSFM) (Asif et al., 2010; Mateo et al., 2011; Millar et al., 2010; Mussolin et al., 2010; Rafique et al., 2011a; Yaman et al., 2009). The results in these articles shows efficient mitigation of CD and NL. In (Asif et al., 2010) the performance of DBP is investigated for heterogeneous type transmission links which contain mixed spans of single mode fiber (SMF) and non-zero dispersion shifted fiber (NZDSF). The continuous growth of the next generation optical networks are expected to render telecommunication networks particularly heterogeneous in terms of fiber types. Efficient compensation of fiber transmission impairments is shown with different span configurations as well as with diverse dispersion mapping.

All the high capacity systems are realized with wavelength-division-multiplexed (WDM) to transmit multiple-channels on a single fiber with high spectral efficiency. The performance in these systems are limited by the inter-channel non-linearities (XPM,FWM) due to the interaction of neighbouring channels. The performance of DBP is evaluated for WDM systems in several articles (Gavioli et al., 2010; Li et al., 2008; Poggiolini et al., 2011; Savory et al., 2010). In (Savory et al., 2010) 112Gbit/s DP-QPSK transmission system is examined and investigations demonstrate that the non-linear compensation algorithm can increase the reach by 23% in a 100GHz spacing WDM link compared to 46% for the single-channel case. When the channel spacing is reduced to 50GHz, the reach improvement is minimal due to the uncompensated inter-channel non-linearities. Whereas, in (Gavioli et al., 2010; Poggiolini et al., 2011) the same-capacity and bandwidth-efficiency performance of DBP is demonstrated in a ultra-narrow-spaced 10 channel 1.12Tbit/s D-WDM long haul transmission. Investigations show that optimum system performance using DBP is obtained by using 2, 4 and 8 steps per fiber span for 14GBaud, 28GBuad and 56GBaud respectively. To overcome the limitations by inter-channel non-linearities on the performance of DBP (Mateo et al., 2010; 2011) proposed improved DBP method for WDM systems. This modification is based on including the effect of inter-channel walk-off in the non-linear step of SSFM. The algorithm is investigated in a 100Gbit/s per channel 16QAM transmission over 1000km of NZDSF type fiber. The results are compared for 12, 24 and 36 channels spaced at 50GHz to evaluate the impact of channel count on the DBP algorithm. While self-phase modulation (SPM) compensation is not sufficient in DWDM systems, XPM compensation is able to increase the transmission reach by a factor of 2.5 by using this DBP method. The results depicts efficient compensation of cross-phase modulation (XPM) and the performance of DBP is improved for WDM systems.

Polarization multiplexing (POLMUX) (Evangelides et al., 1992; Iwatsuki et al., 1993) opens a total new era in optical communication systems (Fludger et al., 2008) which doubles the capacity of a wavelength channel and the spectral efficiency by transmitting two signals via orthogonal states of polarization (SOPs). Although POLMUX is considered interesting for increasing the transmitted capacity, it suffers from decreased PMD tolerance (Nelson et al., 2000; 2001) and increased polarization induced cross-talk (X-Pol), due to the polarization-sensitive detection (Noe et al., 2001) used to separate the POLMUX channels. Previous investigations on DBP demonstrate the results for the WDM channels having the same polarization and solving the scaler NLSE equation is adequate. In (Yaman et al., 2009) it is depicted that the same principles can be applied to compensate fiber transmission impairments by using DBP but a much more advanced form of NLSE should be used which includes two orthogonal polarization states (E_x and E_y), i.e. Manakov equation. Polarization mode dispersion (PMD) is considered negligible during investigation. In this article the results depict that back-to-back performance for the central channel corresponds to a Q value of 20.6 dB. When only dispersion compensation is applied it results in a Q value of 3.9 dB. The eye-diagram is severely degraded and clearly dispersion is not the only source of impairment. Whereas, when DBP algorithm is applied the system observed a Q value of 12.6 dB. The results clearly shows efficient compensation of CD and NL by using the DBP algorithm. In (Mussolin et al., 2010; Rafique et al., 2011b) 100Gbit/s dual-polarization (DP) transmission systems are investigated with advanced modulation formats i.e. QPSK and QAM.

Another modification in recent times in conventional DBP algorithm is the optimization of non-linear operator calculation point (r). It is demonstrated that DBP in a single-channel transmission (Du et al., 2010; Lin et al., 2010b) can be improved by using modified split-step Fourier method (M-SSFM). Modification is done by shifting the non-linear operator calculation point $Nlpt$ (r) along with the optimization of dispersion D and non-linear coefficient γ to get the optimized system performance (see section 3.2.2 of this chapter). The modification in this non-linear operator calculation point is necessary due to the fact that non-linearities behave differently for diverse parameters of transmission, i.e. signal input launch power and modulation formats, and hence also due to precise estimation of non-linear phase shift ϕ_{NL} from span to span. The concept of filtered DBP (F-DBP) (Du et al., 2010) is also presented along with the optimization of non-linear point (see section 3.2.3 of this chapter). The system performance is improved through F-DBP by using a digital low-pass-filter (LPF) in each DBP step to limit the bandwidth of the compensating waveform. In this way we can optimize the compensation of low frequency intensity fluctuations without overcompensating for the high frequency intensity fluctuations. In (Du et al., 2010) the results depict that with four backward propagation steps operating at the same sampling rate as that required for linear equalizers, the Q at the optimal launch power was improved by 2 dB and 1.6 dB for single wavelength CO-OFDM and CO-QPSK systems, respectively, in a 3200 km (40x80km) single-mode fiber link, with no optical dispersion compensation.

Recent investigations (Ip et al., 2010; Rafique et al., 2011b) show the promising impact of DBP on OFDM transmission and higher order modulation formats, up to 256-QAM. However actual implementation of the DBP algorithm is now-a-days extremely challenging due to its complexity. The performance is mainly dependent on the computational step-size (h) (Poggiolini et al., 2011; Yamazaki et al., 2011) for WDM and higher baud-rate transmissions. In order to reduce the computational efforts of the algorithm by increasing the step-size (i.e. reducing the number of DBP calculation steps per fiber span), ultra-low-loss-fiber (ULF) is used (Pardo et al., 2011) and a promising method called correlated DBP (CBP) (Li et al., 2011; Rafique et al., 2011c) has been introduced (see section 4.1 of this chapter). This method takes into account the correlation between adjacent symbols at a given instant using a weighted-average approach, and an optimization of the position of non-linear compensator stage. In (Li et al., 2011) the investigations depict the results in 100GHz channel spaced DP-QPSK transmission and multi-span DBP shows a reduction of DBP stages upto 75%. While in (Rafique et al., 2011c) the algorithm is investigated for single channel DP-QPSK transmission. In this article upto 80% reduction in required back-propagation stages is shown to perform non-linear compensation in comparison to the standard back-propagation algorithm.

In the aforementioned investigations there is a trade-off relationship between achievable improvement and algorithm complexity in the DBP. Therefore DBP algorithms with higher improvement in system performance as compared to conventional methods are very attractive. Due to this fact simplification of the DBP model to efficiently describe fiber transmission especially for POLMUX signals and an estimation method to precisely optimize parameters are the keys for its future cost-effective implementation. By keeping in mind that existing DBP techniques are implemented with constant step-size SSFM methods. The use of these methods, however, need the optimization of D , γ and r for efficient mitigation of CD and NL. In (Asif et al., 2011) numerical investigation for the first time on logarithmic step-size distribution to explore the simplified and efficient implementation of DBP using SSFM is done (see section 3.2.4 of this chapter). The basic motivation of implementing logarithmic step-size relates to the fact of exponential decay of signal power and thus NL phase shift in the beginning sections of each fiber span. The algorithm is investigated in N-channel 112Gbit/s/ch DP-QPSK transmission (a total transmission capacity of 1.12Tbit/s) over 2000km SMF with no in-line optical dispersion compensation. The results depict enhanced system performance of DP-QPSK transmission, i.e. efficient mitigation of fiber transmission impairments, especially at higher baud rates. The benefit of the logarithmic step-size is the reduced complexity as the same forward propagation parameters can be used in DBP without optimization and computational time which is less than conventional M-SSFM based DBP.

The advancements in DBP algorithm till date are summarized in Appendix A. The detailed theory of split-step methods and the effect of step-size selection is explained in the following sections.

3. Non-linear Schrödinger equation (NLSE)

The propagation of optical signals in the single mode fiber (SMF) can be interpreted by the Maxwell's equations. It can mathematically be given as in the form of a wave equation as in Eq. 1 (Agrawal, 2001).

$$\nabla^2 E = \frac{1}{c^2}\frac{\partial^2 E}{\partial^2 t} - \mu_0 \frac{\partial^2 P(E)}{\partial^2 t} \tag{1}$$

Whereas, E is the electric field, μ_0 is the vacuum permeability, c is the speed of light and P is the polarization field. At very weak optical powers, the induced polarization has a linear relationship with E such that;

$$P_L(r,t) = \varepsilon_0 \int_{-\propto}^{\propto} x^{(1)}(t-\grave{t}) \cdot E(r,\grave{t}) d\grave{t} \tag{2}$$

Where ε_0 is the vacuum permittivity and $x^{(1)}$ is the first order susceptibility. To consider non-linearities in the system, the Eq. 2 can be re-written as illustrated in Eq. 3 (Agrawal, 2001).

$$P(r,t) = P_L(r,t) + P_{NL}(r,t) \tag{3}$$

Whereas, $P_{NL}(r,t)$ is the non-linear part of polarization. Eq. 3 can be used to solve Eq. 1 to derive the propagation equation in non-linear dispersive fibers with few simplifying assumptions. First, P_{NL} is treated as a small perturbation of P_L and the polarization field is maintained throughout the whole propagation path. Another assumption is that the index difference between the core and cladding is very small and the center frequency of the wave is assumed to be much greater than the spectral width of the wave which is also called as quasi-monochromatic assumption. The quasi-monochromatic assumption is the analogous to low-pass equivalent modelling of bandpass electrical systems and is equivalent to the slowly varying envelope approximation in the time domain. Finally, the propagation constant, $\beta(\omega)$, is approximated by a few first terms of Taylor series expansion about the carrier frequency, ω_0, that can be given as;

$$\beta(\omega) = \beta_0 + (\omega - \omega_0)\beta_1 + \frac{1}{2}(\omega - \omega_0)^2\beta_2 + \frac{1}{6}(\omega - \omega_0)^3\beta_3 + ... \tag{4}$$

Whereas;

$$\beta_n = \left[\frac{d^n\beta}{d\omega^n}\right]_{\omega=\omega_0} \tag{5}$$

The second order propagation constant $\beta_2[ps^2/km]$, accounts for the dispersion effects in the optical fibers communication systems. Depending on the sign of the β_2, the dispersion region can be classified into two parts as, normal($\beta_2 > 0$) and anomalous ($\beta_2 < 0$). Qualitatively, in the normal-dispersion region, the higher frequency components of an optical signal travel slower than the lower frequency components. In the anomalous dispersion region it occurs vice-versa. Fiber dispersion is often expressed by another parameter, $D[ps/(nm.km)]$, which is called as dispersion parameter. D is defined as $D = \frac{d}{d\lambda}\left[\frac{1}{v_g}\right]$ and the mathematical relationship between β_2 and D is given in (Agrawal, 2001), as;

$$\beta_2 = -\frac{\lambda^2}{2\pi c}D \tag{6}$$

Where λ is the wavelength of the propagating wave and v_g is the group velocity. The cubic and the higher order terms in Eq. 4 are generally negligible as long as the quasi-monochromatic assumption remains valid. However, when the center wavelength of an optical signal is near the zero-dispersion wavelength, as for broad spectrum of the signals, (that is $\beta \approx 0$) then the β_3 terms should be included.

If the input electric field is assumed to propagate in the $+z$ direction and is polarized in the x direction Eq. 1 can be re-written as;

$$\frac{\partial}{\partial z}E(z,t) = -\frac{\alpha}{2}E(z,t) \qquad \text{(linear attenuation)}$$

$$+j\frac{\beta_2}{2}\frac{\partial^2}{\partial^2 t}E(z,t) \qquad \text{(second order dispersion)}$$

$$+\frac{\beta_3}{6}\frac{\partial^3}{\partial^3 t}E(z,t) \qquad \text{(third order dispersion)}$$
$$-j\gamma|E(z,t)|^2E(z,t) \qquad \text{(Kerr effect)}$$
$$+j\gamma T_R\frac{\partial}{\partial_t}|E(z,t)|^2E(z,t) \qquad \text{(SRS)}$$
$$-\frac{\partial}{\omega_0}\frac{\partial}{\partial_t}|E(z,t)|^2E(z,t) \qquad \text{(self-steeping effect)} \tag{7}$$

Where $E(z,t)$ is the varying slowly envelope of the electric field, z is the propagation distance, t=t'-$\frac{z}{v_g}$ (t'= physical time, v_g=the group velocity at the center wavelength), α is the fiber loss coefficient $[1/km]$, β_2 is the second order propagation constant $[ps^2/km]$, β_3 is the third order propagation constant $[ps^3/km]$, γ=$\frac{2\pi n_2}{\lambda_0 A_{eff}}$ is the non-linear coefficient $[km^{-1}\cdot W^{-1}]$, n_2 is the non-linear index coefficient, A_{eff} is the effective core area of the fiber, λ_0 is the center wavelength and ω_0 is the central angular frequency. When the pulse width is greater than 1ps, Eq. 7 can further be simplified because the Raman effects and self-steepening effects are negligible compared to the Kerr effect (Agrawal, 2001). Mathematically the generalized form of non-linear Schrödinger equation suitable to describe the signal propagation in communication systems can be given as;

$$\frac{\partial E}{\partial z} = j\gamma|E|^2 + \left(-j\frac{\beta_2}{2}\frac{\partial^2}{\partial t^2} - \frac{\alpha}{2}\right)E = (\hat{N}+\hat{D})E \tag{8}$$

Also that $\hat{D}$ and $\hat{N}$ are termed as linear and non-linear operators as in Eq. 9.

$$\hat{N} = j\gamma|E|^2; \hat{D} = \left(-j\frac{\beta_2}{2}\frac{\partial^2}{\partial t^2} - \frac{\alpha}{2}\right) \tag{9}$$

3.1 Split-step Fourier method (SSFM)

As described in the previous section, it is desirable to solve the non-linear Schrödinger equation to estimate various fiber impairments occurring during signal transmission with high precision. The split-step Fourier method (SSFM) is the most popular algorithm because of its good accuracy and relatively modest computing cost.

As depicted in Eq. 8, the generalized form of NLSE contains the linear operator $\hat{D}$ and non-linear operators $\hat{N}$ and they can be expressed as in Eq. 9. When the electric field envelope, $E(z,t)$, has propagated from z to $z+h$, the analytical solution of Eq. 8 can be written as;

$$E(z+h,t) = exp\big(h(\hat{N}+\hat{D})\big)\cdot E(z,t) \tag{10}$$

In the above equation h is the propagation step length also called as step-size, through the fiber section. In the split-step Fourier method, it is assumed that the two operators commute with each other as in Eq. 11;

$$E(z+h,t) \approx exp\big(h(\hat{N})\big)exp\big(h(\hat{D})\big)\cdot E(z,t) \tag{11}$$

Eq.11 suggests that $E(z+h,t)$ can be estimated by applying the two operators independently. If h is small, Eq.11 can give high accuracy results. The value of h is usually chosen such that the maximum phase shift ($\phi_{max} = \gamma|Ep^2|h$, Ep=peak value of $E(z,t)$) due to the non-linear

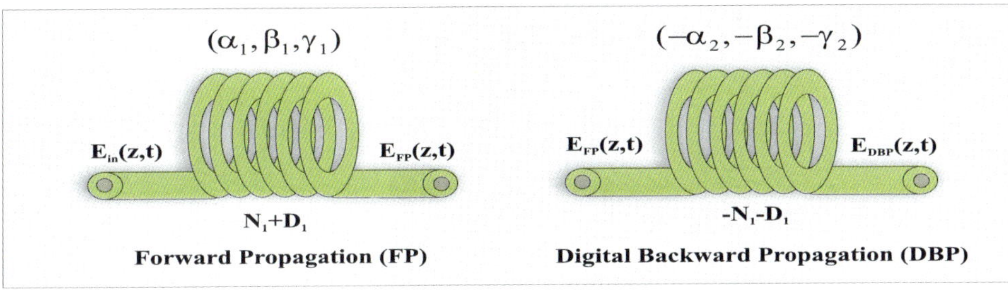

Fig. 2. Block diagram of forward propagation (FP) and digital backward propagation (DBP).

operator is below a certain value. It has been reported (Sinkin et al., 2003) that when ϕ_{max} is below 0.05 *rad*, the split-step Fourier method gives a good result for simulation of most optical communication systems. The simulation time of Eq.11 will greatly depend on the step-size of h. The block diagram of SSFM method is shown in Fig. 4.

3.2 Digital backward propagation (DBP)

The non-linear Schrödinger equation can be solved inversely to calculate the undistorted transmitted signal from the distorted received signal. The received signal at the receiver after transmission i.e. forward propagation (FP), is processed through a numerical model by using the negative sign with the propagation parameters i.e. dispersion D, non-linear coefficient γ. The method is termed as digital backward propagation (DBP) and is illustrated in Fig. 2. Mathematically inverse non-linear Schrödinger equation can be given as in Eq. 12;

$$\frac{\partial E}{\partial z} = (-\hat{N} - \hat{D})E \tag{12}$$

Whereas; the $\hat{D}$ and $\hat{N}$ are the linear and non-linear operators respectively.

The performance of DBP algorithm mainly depends on the estimation of propagating parameters of NLSE. To numerically solve NLSE with high accuracy, split-step Fourier method (SSFM) is used as discussed in the previous section. Both the operators i.e. linear $\hat{D}$ and non-linear $\hat{N}$ are solved separately and also that linear $\hat{D}$ part is solved in frequency domain whereas non-linear $\hat{N}$ is solved in time domain. This DBP model can be implemented both on the transmitter side as well as on the receiver side. When the signal is numerically distorted at the transmitter by DBP algorithm and then this pre-distorted signal is transmitted through fiber link it is termed as transmitter side DBP (Ip et al., 2008). While in majority of the cases DBP is implemented along with the coherent receiver, it is termed as receiver side DBP (Ip et al., 2008), and as an example QPSK receiver is illustrated as in Fig. 3. In the absence of noise in the transmission link both the schemes of DBP are equivalent. As the backward propagation operates on the complex-envelope of $E(z,t)$, this algorithm in principle is applicable with any modulation format of the transmission. It should be noted that the performance of DBP is limited by the amplified spontaneous emission (ASE) noise as it is a non-deterministic noise source and cannot be back propagated (Ip et al., 2008). DBP can only take into account the deterministic impairments. In terms of step-size h, DBP can be categorized in 3 types: (a) sub-span step size in which multiple calculation steps are processed over a single span of fiber; (b) per-span step size which is one calculation step per fiber span and (c) multi-span step size in which one calculation step is processed over several spans of

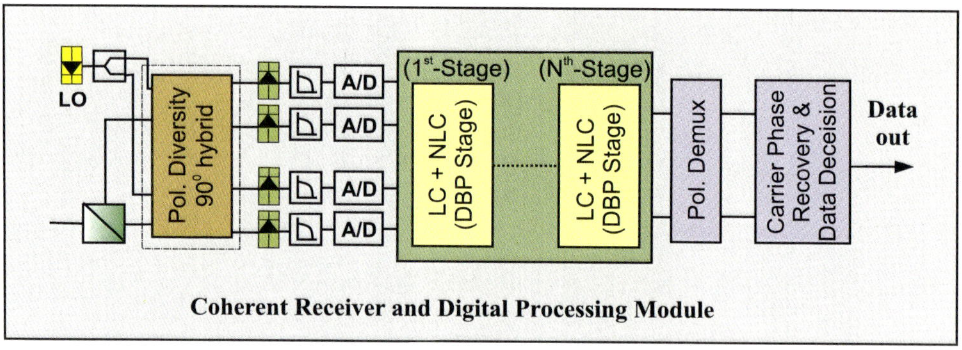

Fig. 3. Block diagram of coherent receiver with digital signal processing module of DBP (LC=linear compensation and NLC=non-linear compensation).

fiber. The SSFM methods which are used to implement the DBP algorithm are discussed in next sections.

3.2.1 Asymmetric and Symmetric SSFM (A-SSFM and S-SSFM)

SSFM can be implemented by using two conventional methods: asymmetric SSFM (A-SSFM) method where the linear operator ($\hat{D}$) is followed by a non-linear operator ($\hat{N}$) and symmetric SSFM (S-SSFM) method where the linear operator ($\hat{D}$) is split into two halves and is evaluated on both sides of non-linear operator ($\hat{N}$), as shown in Fig. 4. Mathematically S-SSFM can be given as in Eq. 13 and A-SSFM in Eq. 14.

$$E(z+h,t) = exp\left(\frac{h\hat{D}}{2}\right)exp\left(h\hat{N}\right)exp\left(\frac{h\hat{D}}{2}\right) \cdot E(z,t) \tag{13}$$

$$E(z+h,t) = exp\left(h\hat{D}\right)exp\left(h\hat{N}\right) \cdot E(z,t) \tag{14}$$

Two methods are adapted for computing parameters in S-SSFM (Asif et al., 2010; Ip et al., 2008). The method in which $\hat{N}(z+h)$ is calculated by initially assuming it as $\hat{N}(z)$ then estimating $E(z+h,t)$, which enables a new value of $\hat{N}_{new}(z+h)$ and subsequently estimating $E_{new}(z+h,t)$ is termed as iterative symmetric SSFM (IS-SSFM). The other method, which is less time consuming and has fewer computations, is based on the calculation of $\hat{N}(z+h)$ at the middle of propagation h is termed as non-iterative symmetric SSFM (NIS-SSFM). However computational efficiency of NIS-SSFM is better then IS-SSFM method (Asif et al., 2010).

3.2.2 Modified split-step Fourier method (M-SSFM)

For the modification of conventional SSFM method, (**?**) introduces a coefficient r which defines the position of non-linear operator calculation point ($Nlpt$), as illustrated in Fig. 4. Typically, r=0 for A-SSFM and r=0.5 for S-SSFM. Which means that with per-span DBP compensation A-SSFM models all the fiber non-linearities as a single lumped non-linearity calculation point which is at r=0 (at the end of DBP fiber span) and S-SSFM models all the fiber non-linearities as a single lumped non-linearity calculation point which is at r=0.5. This approximation becomes less accurate particularly in case of sub-span DBP or multi-span DBP due to inter-span non-linear phase shift estimation ϕ_{NL}, which may result in the over-compensation or under-compensation of the fiber non-linearity, reducing the mitigation of fiber impairments

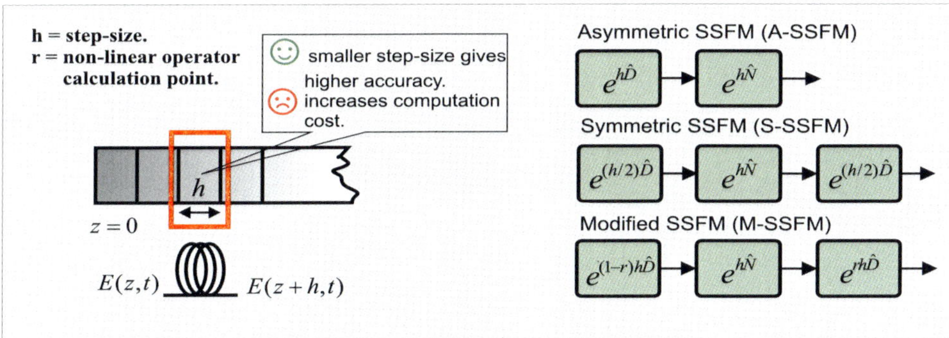

Fig. 4. Comparison of the split-step Fourier methods (SSFM).

(Du et al., 2010). Also that non-linearities behave differently for diverse input parameters of transmission i.e. input power and modulation formats. So we have to modify $Nlpt$ ($0 \leq r \leq 0.5$) along with the optimization of dispersion D and non-linear coefficient γ, used in the DBP, to get the optimum system performance. It is also well known in the SSFM literature that the linear section $\hat{D}$ of the two subsequent steps can be combined to reduce the number of Fourier transforms. This modified split-step Fourier method (M-SSFM) can mathematically be given as in Eq. 15.

$$E(z+h,t) = exp\left((1-r)h\hat{D}\right)exp(h\hat{N})exp\left((r)h\hat{D}\right) \cdot E(z,t) \tag{15}$$

3.2.3 Filtered split-step Fourier method (F-SSFM)

In (Du et al., 2010), the concept of filtered DBP (F-DBP) is introduced along with the optimization of non-linear operator calculation point. It is observed that during each DBP step intensity of the out-of-band distortion becomes higher. The distortion is produced by high-frequency intensity fluctuations modulating the outer sub-carriers in the non-linear sections of DBP. This limits the performance of DBP in the form of noise. To overcome this problem a low pass filter (LPF), as shown in Fig.5, is introduced in each DBP step. The digital LPF limits the bandwidth of the compensating waveform so we can optimize the compensation for the low frequency intensity fluctuations without overcompensating for the high-frequency intensity fluctuations. This filtering also reduces the required oversampling factor. The bandwidth of the LPF has to be optimized according to the DBP stages used to compensate fiber transmission impairments i.e bandwidth is very narrow when very few BP steps are used and bandwidth increases accordingly when more DBP stages are used. By using F-SSFM (Du et al., 2010), the results depict that with four backward propagation steps, the Q at the optimal launch power was improved by 2 dB and 1.6 dB for single wavelength CO-OFDM and CO-QPSK systems, respectively, in a 3200 km (40x80km) single-mode fiber link, with no optical dispersion compensation.

3.2.4 Logarithmic split-step Fourier method (L-SSFM)

As studies from (Asif et al., 2011) introduces the concept of logarithmic step-size based DBP (L-DBP) using split-step Fourier method. The basic motivation of implementing logarithmic step-size relates to the fact of exponential decay of signal power and thus NL phase shift in the beginning sections of each fiber span as shown in Fig 6. First SSFM methods were based

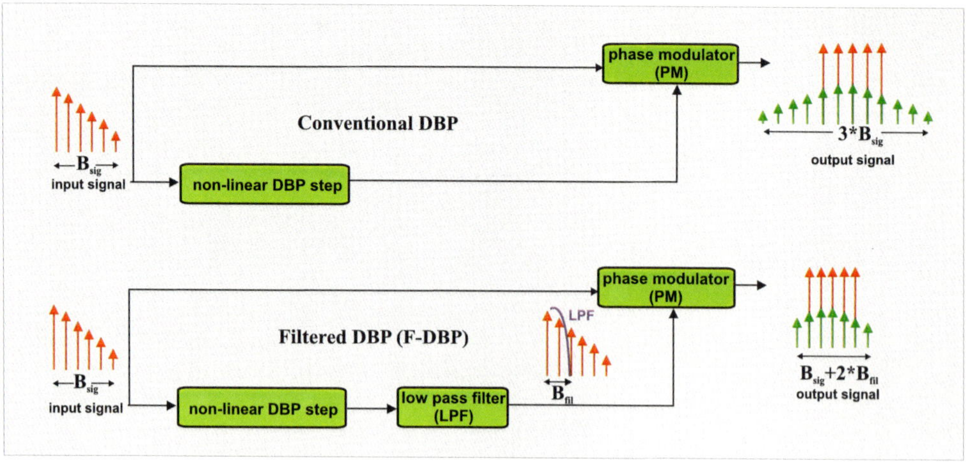

Fig. 5. Block diagram comparing the filtered DBP (F-DBP), conventional DBP schemes and also the bandwidth spectrum (B) at different locations of DBP steps (Du et al., 2010).

on the constant step-size methods. Numerical solution of NLSE using SSFM with constant step-size may cause the spurious spectral peaks due to fictitious four wave mixing (FWM). To avoid this numerical artifact and estimating the non-linear phase shift with high accuracy in fewer computations by SSFM, (Bosco et al., 2000; Sinkin et al., 2003) suggest a logarithmic step-size distribution for forward propagation simulations as given in Eq. 16.

$$h_n = -\frac{1}{A\Gamma} ln\left[\frac{1-n\sigma}{1-(n-1)\sigma}\right], \sigma = \left[1 - exp(-2\Gamma L)\right]/K \tag{16}$$

Whereas, L is the fiber span length, Γ is the loss coefficient and K is the number of steps per fiber span. So logarithmic step-size DBP based on the aforementioned equation is an obvious improvement of DBP. Note that the slope coefficient (A) for logarithmic distribution has been chosen as 1 to reduce the relative global error and also for L-DBP 2 minimum iterations are needed to evaluate the logarithmic step-size based DBP stage.

In (Asif et al., 2011), this L-DBP algorithm is evaluated for three different configurations: (a) 20 channel 56Gbit/s (14GBaud) with 25GHz channel spacing; (b) 10 channel 112Gbit/s (28GBaud) with 50GHz channel spacing and (c) 5 channel 224Gbit/s (56GBaud) with 100GHz channel spacing. So that each simulation configuration has the bandwidth occupancy of 500GHz. The DP-QPSK signals are transmitted over 2000km fiber. The algorithm shows efficient compensation of CD and NL especially at higher baud rates i.e. 56GBaud. For this baud rate the calculation steps per fiber span are also reduced from 8 to 4 as compared to the conventional DBP method. The non-linear threshold point (NLT) is improved by 4dB of signal power. One of the main strengths of the this algorithm is that L-DBP eliminates the optimization of DBP parameters, as the same forward propagation parameters can be used in L-DBP and calculation steps per fiber span are reduced up to 50%.

3.3 Future step-size distribution concepts

The global accuracy and computational efforts to evaluate the SSFM method mainly depends on the step-size (h) selection (Sinkin et al., 2003). In this article several step-size methods are

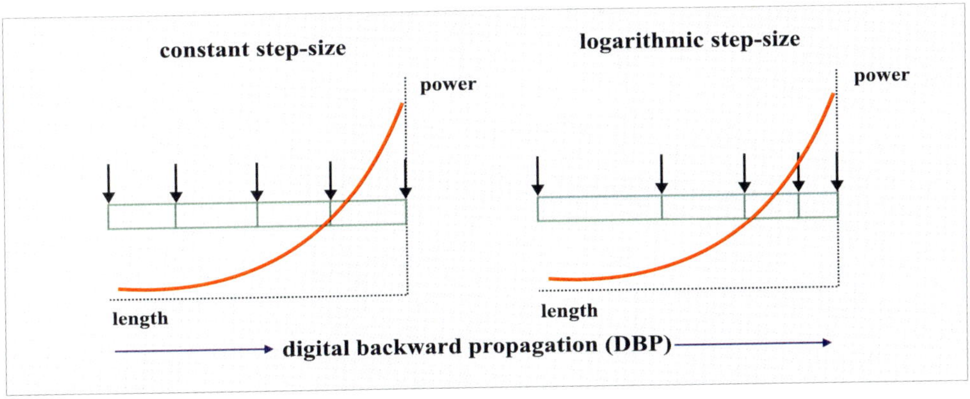

Fig. 6. Comparison of DBP algorithms based on constant step-size method and logarithmic step-size method. The red curves show the power dependence along per-span length.

discussed for forward simulation of optical communication systems. These techniques can be investigated to implement DBP in future. In this section we will discuss the figure of merit for different step-size distribution techniques.

3.3.1 Non-linear phase rotation method

In this method step-size is chosen so that the phase change due to non-linearities ϕ_{NL} does not exceed a certain limit (Sinkin et al., 2003). In Eq. 9 the effect of non-linear operator ($\hat{N}$) is to increase the non-linear phase shift ϕ_{NL} for a specific step-size (h) by an amount as given in Eq. 17.

$$\phi_{NL} = \gamma |E|^2 h \tag{17}$$

An upper-limit for the phase rotation ϕ_{NL}^{max} is ensured for this method is the step-size h fulfills Eq. 18.

$$h \leq \frac{\phi_{NL}^{max}}{\gamma |E|^2} \tag{18}$$

This step-size selection method is mainly used for soliton transmission.

3.3.2 Walk-off method

Walk-off method of implementing SSFM is suitable for investigating the WDM (Mateo et al., 2010) transmission systems. In these systems the wavelengths cover a braod spectrum due to which the interplay of chromatic dispersion and intra-channel cross phase modulation (XPM) plays dominant degradation role in system performance. In this method step-size is determined by the largest group velocity difference between channels. The basic intention is to choose the step size to be smaller than a characteristic walk-off length. The walk off length is the length of fiber required for the interacting channels to change their relative alignment by the time duration that characterizes the intensity changes in the optical signals. This length can be determined as: $L_{wo} \approx \triangle t / (D \triangle \lambda)$, where D is chromatic dispersion and $\triangle \lambda$ is the channel spacing between the interacting channels.

In a WDM transmission with large dispersion, pulses in different channels move through each other very rapidly. To resolve the collisions (Sinkin et al., 2003) between pulses in different channels the step-size in the walk-off method is chosen, so that in a single step two pulses in the two edge channels shift with respect to each other by a time that is a specified fraction of the pulse width. Mathematically it is depicted as in Eq. 19.

$$h = \frac{C}{\triangle v_g} \tag{19}$$

Whereas, C is a error bounding constant that can vary from system to system, $\triangle v_g$ is the largest group velocity difference between the channels. In any transmission model $\triangle v_g$=$|D|\Delta\lambda_{i,j}$. Where $\lambda_{i,j}$ is the wavelength difference between channels i and j. If the transmission link consists of same kind of fiber, step-size selection due to walk-off method is considered as constant (Sinkin et al., 2003).

3.3.3 Local error method

Local error method adaptively adjusts the step-size for required accuracy. In this method step-size is selected by calculating the relative local error δ_L of non-linear phase shift in each single step (Sinkin et al., 2003), taking into account the error estimation and linear extrapolation. The local error method provides higher accuracy than constant step-size SSFM method, since it is method of third order. On the other hand, the local error method needs additional 50% computational effort (Jaworski, 2008) comparing with the constant step-size SSFM. Simulations are carried out in parallel with coarse step-size ($2h$) and fine (h) steps. In each step the relative local error is being calculated: $\delta = \|u_f - u_c\| / \|u_c\|$. Whereas, u_f determines fine solution, u_c is the coarse solution and $\|u\| = \sqrt{|u(t)|^2 dt}$. The step size is chosen by keeping in each single step the relative local error δ within a specified range $(1/2\delta_G, \delta_G)$, where δ_G is the global local error. The main advantage of this algorithm is adaptively controlled step size (Jaworski, 2008) .

4. Recent developments in DBP

4.1 Correlated backward Propagation (CBP)

Recently a promising method to implement DBP is introduced by (Li et al., 2011; Rafique et al., 2011c) which is correlated backward propagation (CBP). The basic theme of implementing this scheme is to take into account the effect of neighbouring symbols in the calculation of non-linear phase shift ϕ_{NL} at a certain instant. The physical theory behind CBP is that the SPM imprinted on one symbol is not only related to the power of that symbol but also related to the powers of its neighbouring symbols because of the pulse broadening due to linear distortions. The schematic diagram of the CBP is as given in Fig. 7.

The correlation between neighbouring symbols is taken into account by applying a time-domain filter (Rafique et al., 2011c) corresponding to the weighted sum of neighbouring symbols. Non-linear phase shift on a given symbol by using CBP can be given as in Eq. 20 and 21.

$$E_x^{out} = E_x^{in} \cdot exp\left[-j \cdot \sum_{k=-(N-1)/2}^{+(N-1)/2} c_k \left\{ a \left| E_x^{in}\left(t - k\frac{T_s}{2}\right)\right|^2 + b \left| E_y^{in}\left(t - k\frac{T_s}{2}\right)\right|^2 \right\}\right] \tag{20}$$

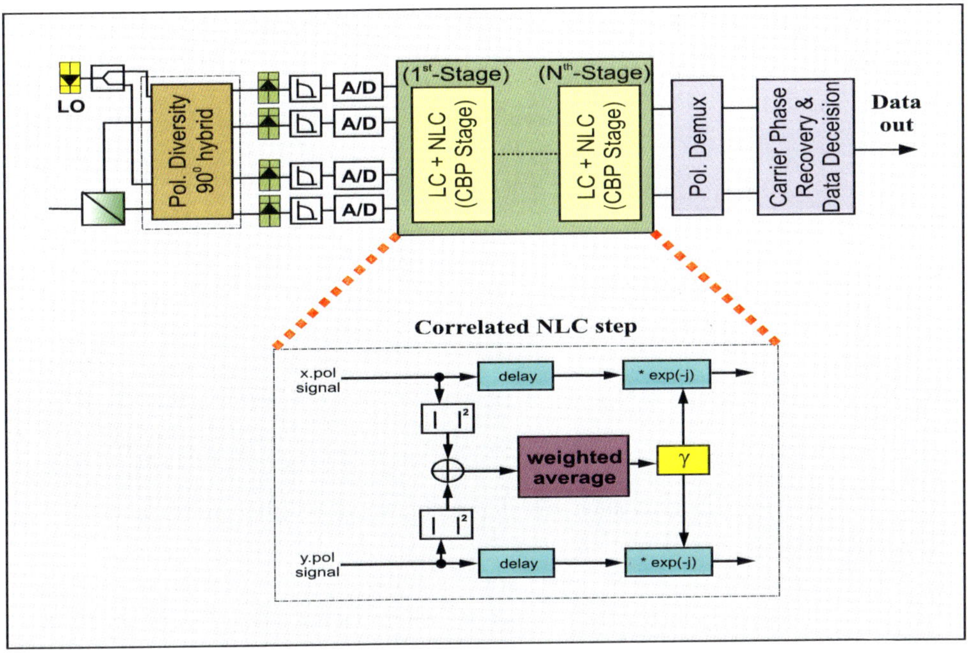

Fig. 7. Block diagram of coherent receiver with correlated backward propagation module (CBP) (Li et al., 2011; Rafique et al., 2011c).

$$E_y^{out} = E_y^{in} \cdot exp\left[-j \cdot \sum_{k=-(N-1)/2}^{+(N-1)/2} c_k \left\{ a \left| E_y^{in}\left(t - k\frac{T_s}{2}\right)\right|^2 + b\left|E_x^{in}\left(t - k\frac{T_s}{2}\right)\right|^2 \right\}\right] \quad (21)$$

Whereas, E is the electric field envelope of the orthogonal polarization states, a and b represent intra-polarization and inter-polarization parameters (Oda et al., 2009), N represents the number of symbols to be considered for a non-linear phase shift, c_k is the weighing vector, K is the delay order, and T_s is the symbol period. In (Li et al., 2011) the investigations depict the results in 100GHz channel spaced DP-QPSK transmission and multi-span DBP shows a reduction of DBP stages upto 75%. While in (Rafique et al., 2011c) the algorithm is investigated for single channel DP-QPSK transmission. In this article upto 80% reduction in required back-propagation stages is shown to perform non-linear compensation in comparison to the standard back-propagation algorithm. By using this method the number of DBP stages are significantly reduced.

4.2 Optical backward Propagation (OBP)

The DBP improves the transmission performance significantly by compensating dispersion and non-linearities. However, it requires a considerable amount of computational resources as described in previous sections thus upto now no real time experimental implementations are reported. In (Kumar et al., 2011) an alternative technique for real-time implementation is proposed in optical domain, realized by an effective non-linear coefficient using a pair of highly non-linear fibers (HNLFs). In this method the linear compensation is realized by using

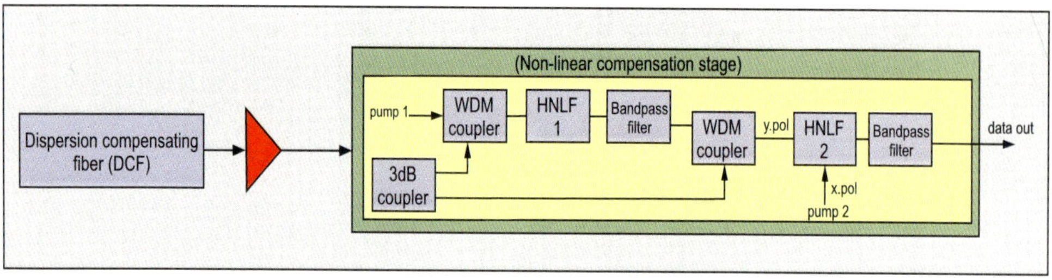

Fig. 8. Block diagram of optical backward propagation module (OBP) (Kumar et al., 2011).

dispersion compensation fibers (DCFs) and non-linear compensation by using HNLFs, as shown in Fig. 8. In this article the technique is evaluated for 32QAM modulation transmission with 25G-symbols/s over 800km fiber. The transmission reach without OBP (but with the DCF) is limited to 240km at the forward error correction limit of 2.1×10^{-3}. This is because the multilevel QAM signals are highly sensitive to fiber non-linear effects. The maximum reach can be increased to 640km and 1040km using two-span OBP (multi-span backward propagation) and one-span OBP (per-span backward propagation), respectively.
This technique is still in the early stages of development. As DCF in the OBP module can add additional losses and limit the performance of backward propagation algorithm, as a matter of fact we have to keep launch power to the DCF low so that the non-linear effects in the DCF can be ignored.

5. Analysis of step-size selection in 16-QAM transmission

In this section we numerically review the system performances of different step-size selection methods to implement DBP. We apply a logarithmic distribution of step sizes and numerically investigate the influence of varying step size on DBP performance. This algorithm is applied in a single-channel 16-QAM system with bit rate of 112Gbit/s over a 20x80km link of standard single mode fiber without in-line dispersion compensation. The results of calculating the non-linearity at different positions, including symmetric, asymmetric, and the modified (?) schemes, are compared. We also demonstrate the performance of using both logarithmic step sizes and constant step sizes, revealing that use of logarithmic step sizes performs better than constant step sizes in case of applying the same number of steps, especially at smaller numbers of steps. Therefore the logarithmic step-size method is still a potential option in terms of improving DBP performance although more calculation efforts are needed compared with the existing multi-span DBP techniques such as (Ip et al., 2010; Li et al., 2011). Similar to the constant step-size method, the logarithmic step-size methods is also applicable to any kind of modulation formats.

5.1 DBP algorithms and numerical model

Fig. 9, illustrates the different SSFM algorithms used in this study for a span compensated by 4 DBP-steps. The backward propagation direction is assumed from the left to the right, as the dashed arrows show. For the constant step-size scheme, step size remains the same for all steps, while for the logarithmic step-size scheme, step size increases with decreasing power. The basic principle is well known from the implementation of SSFM to calculate signal propagation in optical fibers, where adaptive step size methods are widely used. As signal

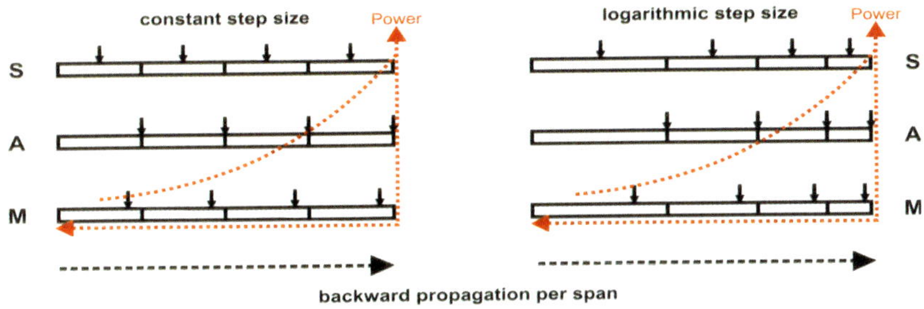

Fig. 9. Schemes of SSFM algorithms for DBP compensation. S: Symmetric-SSFM, A: Asymmetric-SSFM, and M: Modified-SSFM. The red-dotted curves show the power dependence along per-span length..

power exponentially decays along each fiber span, the step size is increased along the fiber. If backward propagation is regarded, the high power regime locates in the end of each span, illustrated in Fig. 1 by the red dotted curves and the step size has to be decreased along each backward propagation span.
Note that the slope coefficient for logarithmic step-size distribution (see section 3.2.4 of this chapter) has been chosen as 1 to reduce the relative global error according to (Jaworski, 2008). The solid arrows in Fig. 9 depict the positions for calculating the non-linear phase. For the symmetric scheme, the non-linearity calculating position (NLCP) is located in the middle of each step. For the asymmetric scheme, NLCP is located at the end of each step. For the modified scheme, NLCP is shifted between the middle and the end of each step and the position is optimized to achieve the best performance (**?**). In all schemes, the non-linear phase was calculated by $\phi_{NL} = \gamma_{DBP} \cdot P \cdot L_{eff}$, where the non-linear coefficient for DBP γ_{DBP} was optimized to obtain the best performance. All the algorithms were implemented for DBP compensation to recover the signal distortion in a single-channel 16-QAM transmission system with bit rate of 112Gbps (28Gbaud). In this simulation model, we used an 20x80km single mode fiber (SMF) link without any inline dispersion compensating fiber (DCF). SMF has the propagation parameters: attenuation α=0.2dB/km, dispersion coefficient D=16ps/nm-km and non-linear coefficient α=1.2 km^{-1}W^{-1}. The EDFA noise figure has been set to 4dB and PMD effect was neglected.

5.2 Simulation results

Fig. 10, compares the performance of all SSFM algorithms with varying number of steps per span. In our results, error vector magnitude (EVM) was used for performance evaluation of received 16-QAM signals. Also various launch powers are compared: 3dBm (Fig. 10(a)), 6dBm (Fig. 10(b)) and 9dBm (Fig. 10(c)). For all launch powers the logarithmic distribution of step sizes enables improved DBP compensation performance compared to using constant step sizes. This advantage arises especially at smaller number of steps (less than 8 steps per span). As the number of steps per span increases, reduction of EVM gets saturated and all the algorithms show the same performance. For both logarithmic and constant step sizes, the modified SSFM scheme, which optimizes the NLCP, shows better performance than symmetric SSFM and asymmetric SSFM, where the NLCP is fixed. This coincides with the results which have been presented in **?**. However, the improvement given from asymmetric to modified SSFM is almost negligible when logarithmic step sizes are used, which means the NLCP optimization reveals less importance and it is already sufficient to calculate the non-linearity at the end of each step if logarithmic step sizes are used. On the other hand, at higher launch powers, EVM increases and the saturation of EVM reduction happens toward larger number of steps. Note that with 9dBm launch power, the EVM cannot reach values below 0.15 (BER=10^{-3}) even if a large number of steps per span is applied.

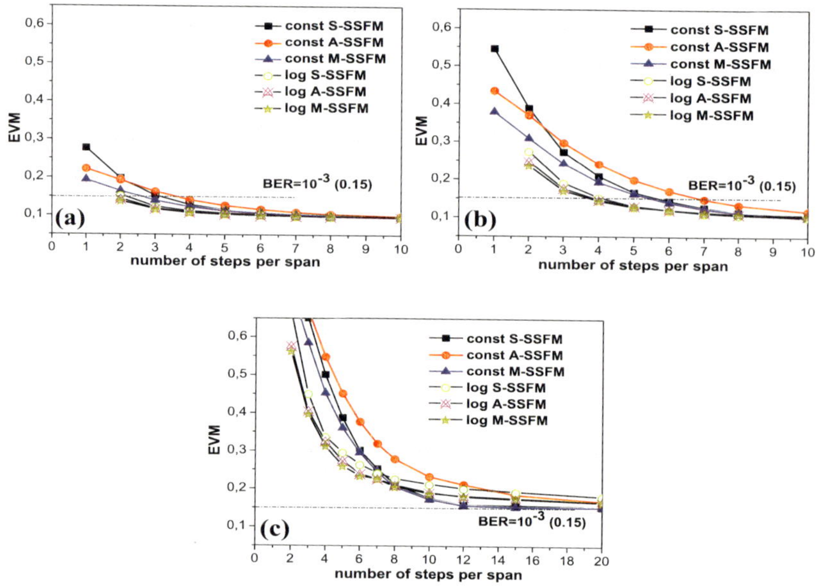

Fig. 10. EVM of all SSFM algorithms with varying number of steps per span for (a) 3dBm, (b) 6dBm and (c) 9dBm.

Fig. 11(a) shows the required number of steps per span to reach BER=10^{-3} at various launch powers for different SSFM algorithms. It is obvious that more steps are required for higher launch powers. Using logarithmic distribution of step sizes requires less steps to reach a certain BER than using uniform distribution of step sizes. At a launch power of 3dBm, the use of logarithmic step sizes reduces 50% in number of steps per span with respect to using the A-SSFM scheme with constant step sizes, and 33% in number of steps per span with respect to using the S-SSFM and M-SSFM schemes with constant step sizes. The advantage can be achieved because the calculated non-linear phase remains constant in every step along the complete. Fig. 11(b) shows an example of logarithmic step-size distribution using 8 steps per span. The non-linear step size determined by effective length of each step, L_{eff}, is represented as solid-square symbols and the average power in corresponding steps is represented as circle symbols. Uniformly-distributed non-linear phase for all successive steps can be verified by multiplication of L_{eff} and average power in each step resulting in a constant value. Throughout all simulations the non-linear coefficient for DBP γ_{DBP} was optimized to obtain the best performance. Fig. 12 shows constellation diagrams of received 16-QAM signals at 3dBm compensated by DBP with 2 steps per span. The upper diagrams show the results of using constant step sizes with non-optimized γ_{DBP} (Fig. 12(a)), and with optimized γ_{DBP} (Fig. 12(b)). The lower diagrams show the results of using logarithmic step sizes with non-optimized γ_{DBP} (Fig. 12(c)), and with optimized γ_{DBP} (Fig. 12(d)). The optimized value is 1.28($km^{-1}W^{-1}$). With optimization of γ_{DBP}, the constellation diagram can be rotated back completely.

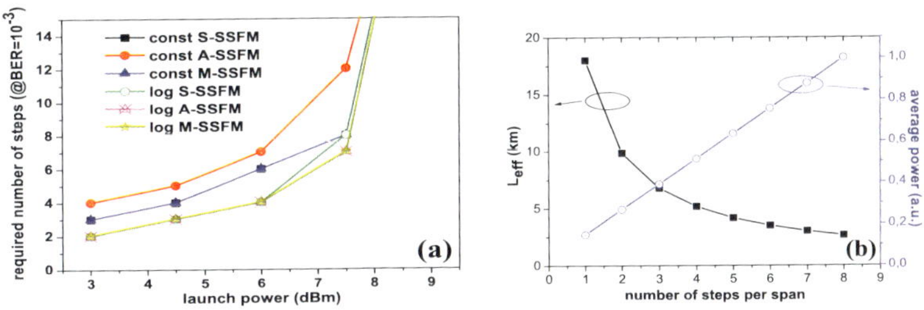

Fig. 11. (a) Required number of steps per span at various launch powers for different SSFM algorithms, and (b) Step-size distribution and average power in each step.

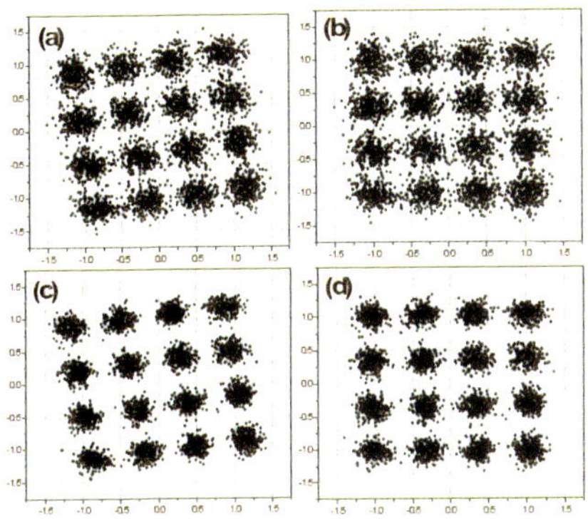

Fig. 12. Constellation diagrams of received 16-QAM signals. (a) constant step size with non-optimized γ_{DBP}, (b) constant step size with with optimized γ_{DBP}, (c) logarithmic step sizes with non-optimized γ_{DBP} and (d) logarithmic step sizes with optimized γ_{DBP}.

5.3 Conclusion

We studied logarithmic step sizes for DBP implementation and compared the performance with uniform step sizes in a single-channel 16-QAM transmission system over a length of 20x80km at a bit rate of 112Gbit/s. Symmetric, asymmetric and modified SSFM schemes have been applied for both logarithmic and constant step-size methods. Using logarithmic step sizes saves up to 50% in number of steps with respect to using constant step sizes. Besides, by using logarithmic step sizes, the asymmetric scheme already performs nicely and optimizing non-linear calculating position becomes less important in enhancing the DBP performance, which further reduces the computational efforts for DBP algorithms

6. Acknowledgement

The authors gratefully acknowledge funding of the Erlangen Graduate School in Advanced Optical Technologies (SAOT) by the German National Science Foundation (DFG) in the framework of the excellence initiative.

7. Appendix A

Method of Implementation	Literature
Symmetric split-step Fourier method (S-SSFM)	i) E. Ip et al.: IEEE JLT 2010. ii) C-Y Lin et al.: ECOC 2010. iii) E. Mateo et al.: Opt Express 2010.
Asymmetric split-step Fourier method (A-SSFM)	i) E. Ip et al.: IEEE JLT 2008. ii) C-Y Lin et al.: ECOC 2010. iii) D.S Millar et al.: ECOC 2009.
Modified split-step Fourier method (M-SSFM)	i) C.Y Lin et al.: ECOC 2010. ii) Du et al.: Opt Express 2010. iii) Asif et al.: Photonics North 2011.
Logarithmic split-step Fourier method (L-SSFM)	i) R. Asif et al.: ICTON Conference 2011.
Filtered split-step Fourier method (F-SSFM)	i) L. Du et al.: Opt Express 2010.
Correlated backward propagation (CBP)	i) L. Li et al.: OFC 2011. ii) Rafique et al.: Opt Express 2011.

Table 1. Summary of the literature of DBP based on implementation methods.

Modulation Formats	Literature
DPSK, DQPSK and QPSK	i) E. Ip et al.: IEEE JLT 2010. ii) C-Y Lin et al.: ECOC 2010. iii) E. Mateo et al.: App Optics 2009.
QAM (4,16,64,256)	i) D. Rafique et al.: Opt Express 2011. ii) S. Makovejs et al.: Opt Express 2010. iii) E. Mateo et al.: Opt Express 2011.
POLMUX and WDM (QPSK, QAM)	i) F. Yaman et al.: IEEE J.Phot 2010. ii) E. Mateo et al.: Opt Express 2010. iii) R. Asif et al.: Photonics North 2011.
OFDM	i) E. Ip et al.: IEEE JLT 2010. ii) E. Ip et al.: OFC 2011. iii) L. Du et al.: Opt Express 2010.

Table 2. Summary of the literature of DBP based on modulation formats.

System Configurations	Literature
10Gbit/s to 40Gbit/s	i) E. Ip et al.: IEEE JLT 2008. ii) C-Y Lin et al.: ECOC 2010. iii) L. Du et al.: Opt Express 2010.
$>$ 40Gbit/s till $<$ 100Gbit/s	i) D.S Millar et al.: ECOC 2009. ii) C-Y Lin et al.: ECOC 2010. iii) L. Du et al.: Opt Express 2010.
$>$ 100Gbit/s	i) O.S Tanimura et al.: OFC 2009. ii) E. Ip et al.: OFC 2011. iii) E. Mateo et al.: Opt Express 2011. iv) D. Rafique et al.: Opt Express 2011. v) R. Asif et al.: ICTON 2011.
WDM (25GHz channel spacing)	i) P. Poggiolini et al.: IEEE PTL 2011. ii) D. Rafique et al.: Opt Express 2011.
WDM (50GHz channel spacing)	i) P. Poggiolini et al.: IEEE PTL 2011. ii) R. Asif et al.: ICTON 2011. iii) S. Savory et al.: IEEE PTL 2010.
WDM (100GHz channel spacing)	i) P. Poggiolini et al.: IEEE PTL 2011. ii) R. Asif et al.: ICTON 2011. iii) S. Savory et al.: IEEE PTL 2010. iv) E. Mateo et al.: Opt Express 2011.

Table 3. Summary of the literature of DBP based on system configurations

Algorithm Complexity	Literature
Sub-span step size	i) E. Ip et al.: IEEE/LEOS 2008. ii) G. Li: Adv Opt Photon 2009.
Per-span step size	i) E. Ip et al.: IEEE JLT 2008. ii) E. Ip et al.: OFC 2011. iii) S. Savory et al.: IEEE PTL 2010.
Multi-span step size	i) L. Li et al.: OFC 2011. ii) D. Rafique et al.: Opt Express 2011. iii) L. Du et .: Opt Express 2011. iv) C-Y Lin et al.: ECOC 2010.

Table 4. Summary of the literature of DBP based on algorithm complexity

8. References

Agrawal, G. (2001). *Fiber-Optic Communication Systems*, John Wiley & Sons Inc, 2nd Edition, New York.

Asif, R., Lin, C.Y., Holtmannspoetter, M. & Schmauss, B. (2010). Optimized digital backward propagation for phase modulated signals in mixed-optical fiber transmission link. *Optics Express*, vol.18, (October 2010) pp.(22796-22807).

Asif, R., Lin, C.Y., Holtmannspoetter, M. & Schmauss, B. (2011). Logarithmic step-size distribution for implementing digital backward propagation in 112Gbit/s DP-QPSK transmission. *12th International Conference on Transparent Optical Networks (ICTON), 2011*, paper Tu.P.6, Stockholm Sweden, June 2011.

Bosco, G., Carena, A., Curri, V., Gaudino, R., Poggiolini, P. & Benedetto S. (2000). Suppression of spurious tones induced by the split-step method in fiber systems simulation. *IEEE Photonics Technology Letters*, vol.12, no.5, (May 2000), pp.(489-491).

Cvecek, K., Sponsel, K., Stephan, C., Onishchukov, G., Ludwig, R., Schubert, C., Schmauss, B., & Leuchs, G. (2008). Phase-preserving amplitude regeneration for a WDM RZ-DPSK signal using a nonlinear amplifying loop mirror. *Optics Express* vol.16, (January 2008), pp.(1923-19289).

Du, L. & Lowery, A. (2010). Improved single channel back-propagation for intra-channel fiber non-linearity compensation in long-haul optical communication systems. *Optics Express*, vol.18, (July 2010), pp.(17075-17088).

Evangelides, S.R., Mollenauer, L., Gordon, J., & Bergamo, N. (1992). Polarization multiplexing with solitons. *IEEE Journal of Lightwave Technology*, vol.10, no.1, (January 1992), pp.(28-35).

Feiste, U., Ludwig, R., Dietrich, E., Diez, S., Ehrke, H.J., Razic, Dz. & Weber, H.G. (1998). 40 Gbit/s transmission over 434 km standard fibre using polarisation independent mid-span spectral inversion. *IET Electronics Letters*, vol.34, no.21, (October 1998), pp.2044-2045.

Fludger, C.R.S., Duthel, T., van den Borne, D., Schulien, C., Schmidt, E., Wuth, T., Geyer, J., De Man, E., Khoe, G.D. & de Waardt, H. (2008). Coherent Equalization and POLMUX-RZ-DQPSK for Robust 100-GE Transmission. *IEEE Journal of Lightwave Technology*, vol.26, no.1, (January 2008) pp.(64-72).

Forghieri, F. (1997). Modeling of wavelength multiplexed lightwave systems. *Optical fiber communication conference (OFC 1997)*, Texas USA, February 1997.

Gavioli, G., Torrengo, E., Bosco, G., Carena, A., Savory, S., Forghieri, F. & Poggiolini, P. (2010). Ultra-narrow-spacing 10-Channel 1.12 Tb/s D-WDM long-haul transmission over uncompensated SMF and NZDSF. *IEEE Photonics Technology Letters*, vol.22, no.19, (October 2010) pp.(1419-1421).

Geyer, J.C., Fludger, C.R.S., Duthel, T., Schulien, C. & Schmauss, B. (2010). Simple automatic non-linear compensation with low complexity for implementation in coherent receivers, *36th European Conference Optical Communication (ECOC), 2010*, paper P3.02, Torino Italy, Sept 2010.

Goldfarb, G. & Li, G. (2007). Chromatic dispersion compensation using digital IIR filtering with coherent detection. *IEEE Photonics Technology Letters*, vol.19, no.13, (JULY 2007), pp.(969-971).

Ip, E. & Kahn, J.M. (2008). Compensation of dispersion and non-linear impairments using digital backpropagation. *IEEE Journal of Lightwave Technology*, vol.26, no.20, (October 2008), pp.(3416-3425).

Ip, E. & Kahn, J.M. (2010). Fiber impairment compensation using coherent detection and digital signal processing. *IEEE Journal of Lightwave Technology*, vol.28, no.4, (February 2010), pp.(502-519).

Iwatsuki, K., Suzuki, K., Nishi, S. & Saruwatari M. (1993). 80 Gb/s optical soliton transmission over 80 km with time/polarization division multiplexing. *IEEE Photonics Technology Letters*, vol.5, no.2, (Febuary 1993) pp. (245-248).

Jansen, S.L., van den Borne, D., Khoe, G., de Waardt, H., Monsalve, C., Splter, S. & Krummrich, P.M. (2005). Reduction of non-linear phase noise by mid-link spectral inversion in a DPSK based transmission system. *Conference on Optical Fiber communication/National Fiber Optic Engineers Conference (OFC/NFOEC) 2005*, paper OThO5, California USA, March 2005.

Jaworski, M. (2008). Step-size distribution strategies in SSFM simulation of DWDM links. *10th International Conference on Transparent Optical Networks (ICTON), 2008*, Athens Greece, June 2008.

Kumar, S. & Yang, D. (2011). Optical backpropagation for fiber-optic communications using highly non-linear fibers. *Optics Letters*, vol.36, (April 2011), pp.(1038-1040).

Li, G. (2009). Recent advances in coherent optical communication. *Advances in Optics and Photonics* vol.1, (February 2009), pp.(279-307).

Li, L., Tao, Z., Dou, L., Yan, W., Oda, S., Tanimura, T., Hoshida, T. & Rasmussen, J. (2011). Implementation efficient non-linear equalizer based on correlated digital back-propagation. *Conference on Optical Fiber communication/National Fiber Optic Engineers Conference (OFC/NFOEC) 2011*, paper OWW3, Los Angeles USA, March 2011.

Li, X., Chen, X., Goldfarb, G., Mateo, E., Kim, I., Yaman, F. & Li, G. (2008) Electronic post-compensation of WDM transmission impairments using coherent detection and digital signal processing. *Optics Express* vol.16, (January 2008), pp.(880-888).

Lin, C.Y., Asif, R., Holtmannspoetter, M. & Schmauss, B. (2010a). Evaluation of nonlinear phase noise in DPSK transmission for different link designs. *Physics Procedia*, vol.5, no.2, (August 2010), pp.(697-701).

Lin, C.Y., Holtmannspoetter, M. , Asif, R. & Schmauss, B. (2010b). Compensation of transmission impairments by digital backward propagation for different link designs, *36th European Conference Optical Communication (ECOC), 2010*, paper P3.16, Torino Italy, Sept 2010.

Marazzi, L., Parolari, P., Martelli, P., Siano, R., Boffi, P., Ferrario, M., Righetti, A., Martinelli, M., Pusino, V., Minzioni, P., Cristiani, I., Degiorgio, V., Langrock, C. & Fejer, M. (2009). Real-time 100-Gb/s polmux RZ-DQPSK transmission over uncompensated 500 km of SSMF by optical phase conjugation. *Conference on Optical Fiber communication/National Fiber Optic Engineers Conference (OFC/NFOEC) 2009*, paper JWA44, California USA, March 2009.

Mateo, E., Yaman, F. & Li, G. (2010). Efficient compensation of inter-channel non-linear effects via digital backward propagation in WDM optical transmission. *Optics Express*, vol.18, (June 2010), pp.(15144-15154).

Mateo, E., Zhou, X. & Li, G. (2011). Improved digital backward propagation for the compensation of inter-channel non-linear effects in polarization-multiplexed WDM systems. *Optics Express*, vol.19, (January 2011), pp.(570-583).

Millar, D.S., Makovejs, S., Behrens, C., Hellerbrand, S., Killey, R., Bayvel, P. & Savory, S.J. (2010). Mitigation of fiber non-linearity using a digital coherent receiver. *IEEE Journal of Selected Topics in Quantum Electronics*, vol.16, no.5, (September 2010) pp.(1217-1226).

Mitra P.P. & Stark J.B. (2001). Non-linear limits to the information capacity of optical fibre communications. *Nature* vol.411 no.6841, (April 2001) pp.(1027-1030).

Mussolin, M., Forzati, M,, Martensson, J., Carena, A. & Bosco, G. (2010). DSP-based compensation of non-linear impairments in 100 Gb/s polmux QPSK. *12th International Conference on Transparent Optical Networks (ICTON), 2010*, paper We.D1.2, Munich Germany, July 2010.

Nelson, L.E. & Kogelnik, H. (2000). Coherent crosstalk impairments in polarization multiplexed transmission due to polarization mode dispersion. *Optics Express*, vol.7, no.10, (November 2000) pp.(350-361).

Nelson, L.E., Nielsen, T. & Kogelnik, H. (2001). Observation of PMD-induced coherent crosstalk in polarization-multiplexed transmission. *IEEE Photonics Technology Letters*, vol.13, no.7, (July 2001), pp.(738-740).

Noe, R., Hinz, S., Sandel, D. & Wust, F. (2001). Crosstalk detection schemes for polarization division multiplexed transmission experiments. *IEEE Journal of Lightwave Technology*, vol.19, no.10, (October 2001), pp.(1469-1475).

Noe, R. (2005). PLL-free synchronous QPSK polarization multiplex/diversity receiver concept with digital I and Q baseband processing. *IEEE Photonics Technology Letters*, vol.17, no.4, (April 2005), pp.(887-889).

Oda, S., Tanimura, T., Hoshida, T., Ohshima, C., Nakashima, H., Zhenning, T. & Rasmussen, J. (2009). 112 Gb/s DP-QPSK transmission using a novel non-linear compensator in digital coherent receiver. *Conference on Optical Fiber communication/National Fiber Optic Engineers Conference (OFC/NFOEC) 2009*, paper OThR6, San-Diego USA, March 2009.

Pardo, O.B., Renaudier, J., Salsi, M., Tran, P., Mardoyan, H., Charlet, G. & Bigo, S. (2011). Linear and nonlinear impairment mitigation for enhanced transmission performance. *Conference on Optical Fiber communication/National Fiber Optic Engineers Conference (OFC/NFOEC) 2011*, paper OMR1, Los Angeles USA, March 2011.

Pare, C., Villeneuve, A., Blanger, P. & Doran, N. (1996). Compensating for dispersion and the nonlinear Kerr effect without phase conjugation. *Optics Letters* vol.21, (September 1996) pp.(459-461).

Poggiolini, P., Bosco, G., Carena, A., Curri, V., Miot, V. & Forghieri, F. (2011). Performance dependence on channel baud-rate of PM-QPSK systems over uncompensated links. *IEEE Photonics Technology Letters*, vol.23, no.1, (January 2011), pp.(15-17).

Rafique, D. & Ellis, A. (2011a). Impact of signal-ASE four-wave mixing on the effectiveness of digital back-propagation in 112 Gb/s PM-QPSK systems. *Optics Express*, vol.19, (February 2011) pp.(3449-3454).

Rafique, D., Zhao, J. & Ellis, A. (2011b). Digital back-propagation for spectrally efficient WDM 112 Gbit/s PM m-ary QAM transmission. *Optics Express*, vol.19, (March 2011), 5219-5224.

Rafique, D., Mussolin, M., Forzati, M., Martensson, J., Chugtai, M. & Ellis, A. (2011c). Compensation of intra-channel nonlinear fibre impairments using simplified digital back-propagation algorithm. *Optics Express*, vol.19, (April 2011), pp.(9453-9460).

Randhawa, R., Sohal, J. & Kaler, R. (2009). Pre-, post and hybrid dispersion mapping techniques for CSRZ optical networks with non-linearities. *Optik - International Journal for Light and Electron Optics*, vol.121, no.14, (August 2010), pp.(1274-1279).

Savory, S., Stewart, A.D., Wood, S., Gavioli, G., Taylor, M.G., Killey, R., & Bayvel, P. (2006). Digital equalisation of 40 Gbit/s per wavelength transmission over 2480 km of standard fibre without optical dispersion compensation. *32nd European Conference Optical Communication (ECOC), 2006*, paper Th2.5.5, Cannes France, September 2006.

Savory, S., Gavioli, G., Killey, R. & Bayvel P. (2007). Electronic compensation of chromatic dispersion using a digital coherent receiver. *Optics Express* vol.15, (March 2007) pp.(2120-2126).

Savory, S., Gavioli, G., Torrengo, E. & Poggiolini, P. (2010). Impact of inter-channel non-linearities on a split-step intra-channel non-linear equalizer, (IEEE Photonics Technology Letters), vol.22, no.10, (May 2010),pp.(673-675).

Sinkin, O.V., Holzlohner, R., Zweck, J. & Menyuk, C.R. (2003). Optimization of the split-step Fourier method in modelling optical-fiber communications systems. *IEEE Journal of Lightwave Technology*, vol.21, no.1, (january 2003), pp. (61-68).

Sponsel, K., Cvecek, K., Stephan, C., Onishchukov, G., Schmauss, B. & Leuchs, G. (2008). Effective negative non-linearity of a non-linear amplifying loop mirror for compensating non-linearity-induced signal distortions. *34th European Conference Optical Communication (ECOC), 2008*, paper Th.1.B5, Brussels Belgium, Sept 2008.

Stephan, C., Sponsel, K., Onishchukov, G., Schmauss, B. & Leuchs G. (2009). Suppression of non-linear phase noise in a DPSK transmission using a non-linear amplifying loop mirror. *Conference on Optical Fiber communication/National Fiber Optic Engineers Conference (OFC/NFOEC) 2009*, paper JthA60, San Diego USA, March 2009.

Taylor, M.G. (2004), Coherent detection method using DSP for demodulation of signal and subsequent equalization of propagation impairments. *IEEE Photonics Technology Letters*, vol.16, no.2, (February 2004), pp. (674676).

Tsang, M., Psaltis, D. & Omenetto, F. (2003). Reverse propagation of femtosecond pulses in optical fibers. *Optics Letters* vol.28, (March 2003), pp.(1873-1875).

Tonello, A., Wabnitz, S. & Boyraz, O. (2006). Duty-ratio control of nonlinear phase noise in dispersion managed WDM transmissions using RZ-DPSK modulation at 10 Gb/s. *IEEE Journal of Lightwave Technology*, vol.24, no.10, (October 2006), pp.(3719-3726).

Winters, J.H. (1990). Equalization in coherent lightwave systems using a fractionally spaced equalizer. *IEEE Journal of Lightwave Technology*, vol.8, no.10, (October 1990), pp.(1487-1491).

Yaman, F. & Li, G. (2009). Non-linear impairment compensation for polarization-division multiplexed WDM transmission using digital backward propagation. *IEEE Photonics Journal*, vol.2, no.5, (August 2009), pp.(144-152).

Yamazaki, E., Sano, A., Kobayashi, T., Yoshida, E. & Miyamoto, Y. (2011). Mitigation of non-linearities in optical transmission systems. *Conference on Optical Fiber communication/National Fiber Optic Engineers Conference (OFC/NFOEC) 2011*, paper OThF1, Los Angeles USA, March 2011.

Part 2

DSP in Monitoring, Sensing and Measurements

High-Precision Frequency Measurement Using Digital Signal Processing

Ya Liu[1,2], Xiao Hui Li[1] and Wen Li Wang[1]
[1]National Time Service Center, Chinese Academy Sciences, Xi'an, Shaanxi
[2]Key Laboratory of Time and Frequency Primary Standard, Institute of National Time Service Center Chinese Academy of Sciences, Xi'an, Shaanxi
China

1. Introduction

High-precision frequency measurement techniques are important in any branch of science and technology such as radio astronomy, high-speed digital communications, and high-precision time synchronization. At present, the frequency stability of some of atomic oscillators is approximately 1E-16 at 1 second and there is no sufficient instrument to measure it (C. A. Greenhall, 2007).

Kinds of oscillator having been developed, some of them have excellent long-term stability when the others are extremely stable frequency sources in the short term. Since direct frequency measurement methods is far away from the requirement of measurement high-precision oscillator, so the research of indirect frequency measurement methods are widely developed. Presently, common methods of measuring frequency include Dual-Mixer Time Difference (DMTD), Frequency Difference Multiplication (FDM), and Beat-Frequency (BF). DMTD is arguably one of the most precise ways of measuring an ensemble of clocks all having the same nominal frequency, because it can cancel out common error in the overall measurement process (D. A. Howe & DAVID A & D.B.Sulliivan, 1981). FDM is one of the methods of high-precision measurement by multiplying frequency difference to intermediate frequency. Comparing with forenamed methods, the BF has an advantage that there is the simplest structure, and then it leads to the lowest device noise. However, the lowest device noise doesn't means the highest accuracy, because it sacrifices accuracy to acquire simple configuration. Therefore, the BF method wasn't paid enough attention to measure precise oscillators.

With studying the BF methods of measuring frequency, we conclude that the abilities of measuring frequency rest with accuracy of counter and noise floor of beat-frequency device. So designing a scheme that it can reduce circuit noise of beat-frequency device is mainly mission as the model of counter has been determined. As all well known, reducing circuit noise need higher techniques to realize, and it is hardly and slowly, therefore, we need to look for another solution to improve the accuracy of BF method. In view of this reason, we design a set of algorithm to smooth circuit noise of beat-frequency device and realize the DFSA design goal of low noise floor (Ya Liu, 2008).

This paper describes a study undertaken at the National Time Service Center (NTSC) of combining dual-mixer and digital cross-correlation methods. The aim is to acquire high short-term stability, low cost, high reliability measurement system. A description of a classical DMTD method is given in Section 2. Some of the tests of the cross-correlation algorithm using simulated data are discussed in Section 3.2. The design of DFSA including hardware and software is proposed in Section 3.3-3.4. In section 4 the DFSA is applied to measure NTSC's cesium signal and the results of noise floor of DFSA is given. Future possible modifications to the DFSA and conclusions are discussed in Section 4.

2. Principle of DMTD method

The basic idea of the Dual Mixer Time Difference Method (DMTD) dates back to 1966 but was introduced in "precision" frequency sources measurement some 10 years later (S. STEIN, 1983). The DMTD method relies upon the phase measurement of two incoming signals versus an auxiliary one, called common offset oscillator. Phase comparisons are performed by means of double-balance mixers. It is based on the principle that phase information is preserved in a mixing process. A block diagram is shown in figure 1.

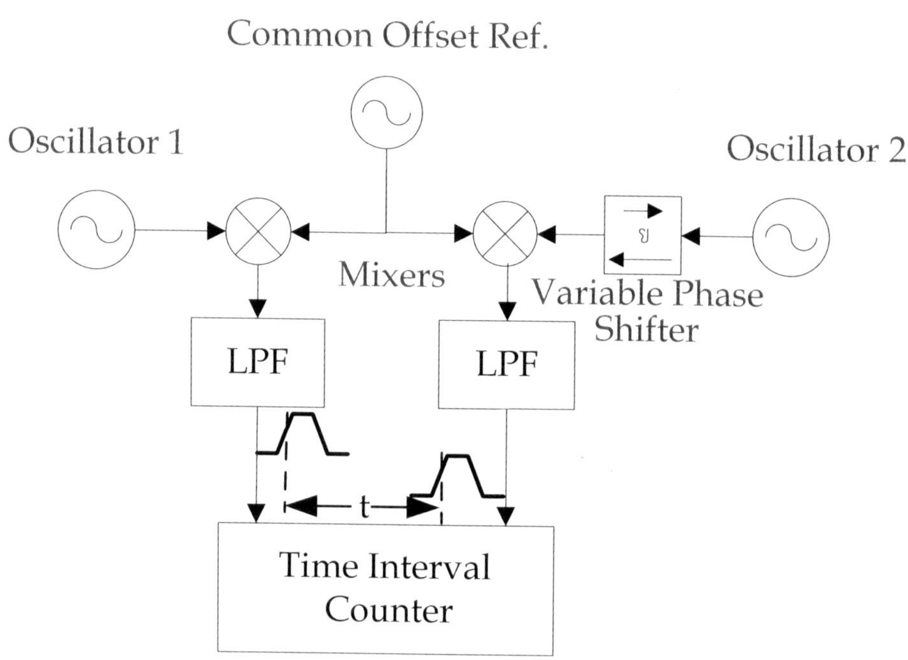

Fig. 1. Block diagram of a dual mixer time difference measuring system

DMTD combines the best features of Beat Method and Time Interval Counter Method, using a time interval counter to measure the relative phase of the beat signals. The measurement resolution is increased by the heterodyne factor (the ratio of the carrier to the beat frequency). For example, mixing a 10 MHz source against a 9.9999 MHz Hz offset reference will produce a 100 Hz beat signal whose period variations are enhanced by a factor of 10 MHz/100 Hz = 10^5. Thus, a period counter with 100 ns resolution (10 MHz clock) can resolve clock phase changes of 1 ps.

The DMTD setup is arguably the most precise way of measuring an ensemble of clocks all having the same nominal frequency. The usual idea thought that the noise of the common offset oscillator could be cancelled out in the overall measurement process. However, if the oscillator 1 and oscillator 2 are independent, then the beat signals of being fed into counter are not coherent. Figure 2 shows the beat signals that are fed into the time interval counter, thus, the beat signals of two test oscillators against the common offset oscillator are zero crossing at different sets of points on the time axis, such as t1 and t2. When time interval counter is used to measure the time difference of two beat signals, the time difference will be contaminated by short-term offset oscillator noise, here called common-source phase error (C. A. Greenhall, 2001, 2006). This DMTD method is inevitable common-source phase error when use counter to measure time difference. To remove the effect of common-source phase error need to propose other processing method.

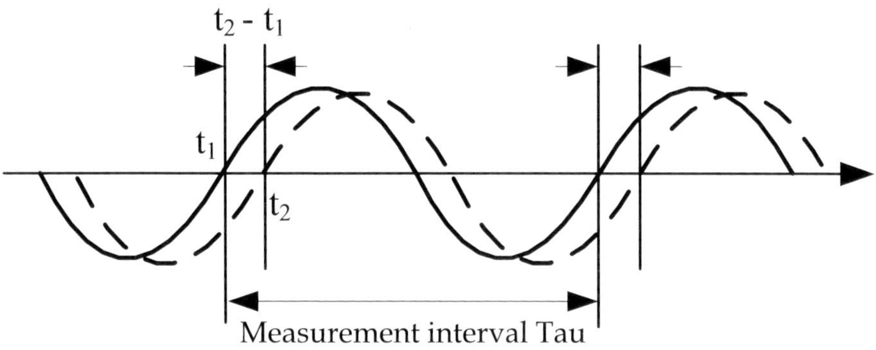

Fig. 2. Beat signals from double-balance mixers

3. Frequency measurement using digital signal processing

To remove the effect of common offset oscillator phase noise and improve the accuracy of measuring frequency, we proposed to make use of digital signal processing method measuring frequency. A Multi-Channel Digital Frequency Stability Analyzer has been developed in NTSC.

3.1 System configuration

This section will report on the Multi-Channel Digital Frequency Stability Analyzer (DFSA) based upon the reformed DMTD scheme working at 10MHz with 100Hz beat frequency. DFSA has eight parallel channels, and it can measure simultaneously seven oscillators. The block diagram of the DFSA that only includes two channels is reported in Fig. 3.
Common offset reference oscillator generates frequency signal, which has a constant frequency difference with reference oscillator. Reference oscillator and under test oscillator at the same nominal frequency are down-converted to beat signals of low frequency by mixing them with the common offset reference to beat frequency. A pair of analog-to-digital converters (ADC) simultaneously digitizes the beat signals output from the double-balance mixers. All sampling frequency of ADCs are droved by a reference oscillator to realize simultaneously sampling. The digital beat signals are fed into personal computer (PC) to computer the drift frequency or phase difference during measuring time interval.

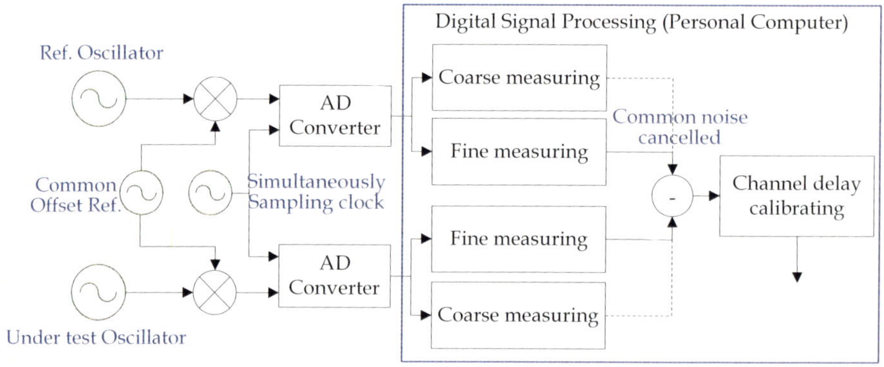

Fig. 3. Block diagram of the DFSA

3.2 Measurement methods

Digital beat signals processing is separated two steps that consist of coarse measuring and fine measuring. The two steps are parallel processed at every measurement period. The results of coarse measuring can be used to remove the integer ambiguity of fine measuring.

3.2.1 Coarse measurement

The coarse measurement of beat frequency is realized by analyzing the power spectrums beat signal. The auto power spectrums of the digital signals are calculated to find the frequency components of beat signal buried in a noisy time domain signal. Generating the auto power spectrum is by using a fast Fourier transform (FFT) method. The auto power spectrum is calculated as shown in the following formula:

$$S_x(f) = \frac{FFT(x)FFT^*(x)}{n^2} \tag{3.1}$$

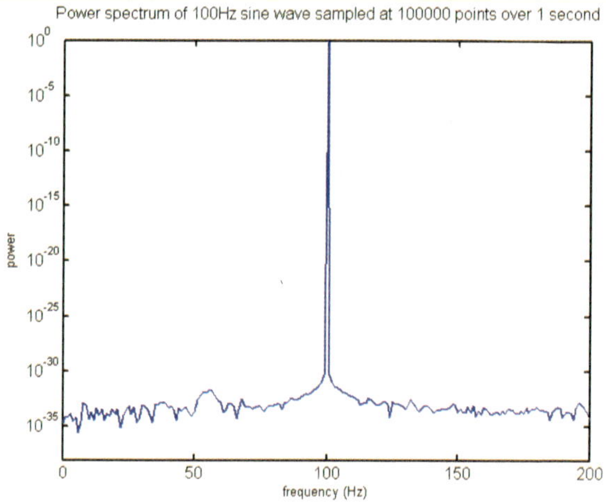

Fig. 4. The power vs. frequency in Hertz

Where x is the beat signals array; n is the number of points in the signal array x; * denotes a complex conjugate. According aforementioned formula, figure 4 plots power spectrum of a 100 Hz sine wave. As expected, we get a very strong peak at a frequency of 100 Hz. Therefore, we can acquire the frequency corresponding to the maximum power from the plot of auto power spectrum.

3.2.2 Fine measurement

The beat signals from the ADCs are fed into PC to realize fine measuring too. Fine measurement includes the cross-correlation and interpolation methods. To illuminate the cross-correlation method, figure 5 shows a group of simulation data. The simulation signals of 1.08Hz are digitized at the sampling frequency of 400Hz. The signal can be expressed by following formula.

$$x(n) = \sin(2\pi \frac{f}{f_s} n + \varphi_0) \tag{3.2}$$

Where f indicates the frequency of signal, the f_s is sampling frequency, n refers the number of sample, and φ_0 represents the initial phase. In the figure 5, the frequency of signal can be expressed:

$$f = f_N + \Delta f = (1 + 0.05)Hz \tag{3.3}$$

There the f_N refers the integer and Δf indicates decimal fraction. In addition, there is the initial phase $\varphi_0 = 0$ and $f_s = 400Hz$. There are sampled two seconds data in the figure 5, so we can divide it into data1 and data2 two groups. Data1 and data2 can be expressed respectively by following formulas:

$$x_1(n) = \sin(2\pi \frac{f_N + \Delta f}{f_s} n + \varphi_0), n \in [0,399] \tag{3.4}$$

$$\begin{aligned} x_2(n) &= \sin(2\pi \frac{f_N + \Delta f}{f_s} n + \varphi_0), n \in [400,799] \\ &= \sin(2\pi \frac{f_N + \Delta f}{f_s} n + \varphi_0 + 2\pi(f_N + \Delta f)), n \in [0,399] \end{aligned} \tag{3.5}$$

According the formula (3.5), the green line can be used to instead of the red one in the figure 5 to show the phase difference between data1 and data2. And then the phase difference is the result that the decimal frequency Δf of signal is less than 1Hz. Therefore, we can calculate the phase difference to get Δf. The cross-correlation method is used to calculate the phase difference of adjacent two groups data.
The cross-correlation function can be shown by following formula:

$$R_{x_1 x_2}(m) = \frac{1}{N} \sum_{n=0}^{N-1} x_1(n) x_2(n+m) = \frac{1}{2} \cos(2\pi \frac{f_N + \Delta f}{f_s} m + 2\pi(f_N + \Delta f)) \tag{3.6}$$

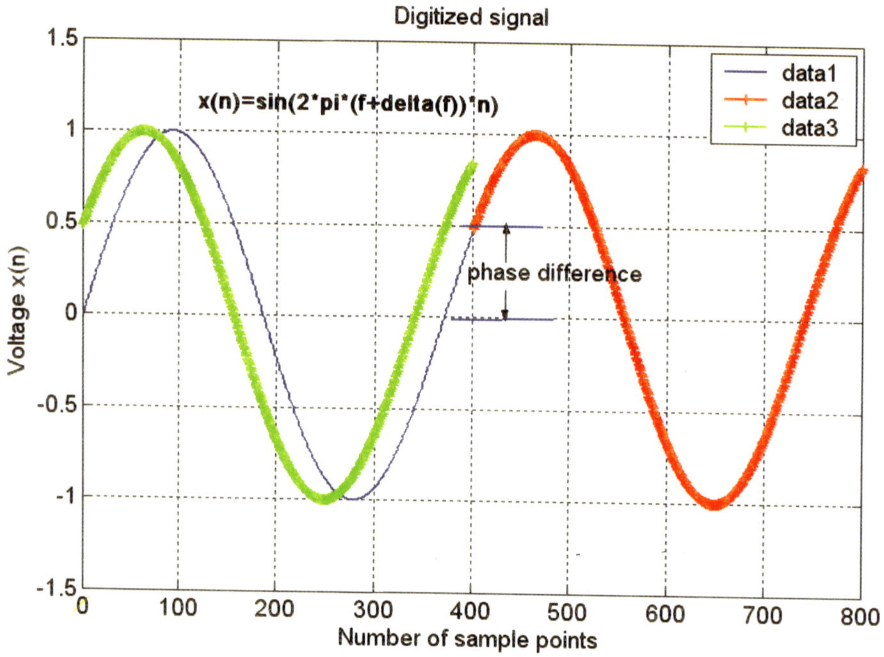

Fig. 5. Signals of 1.08Hz are digitized at the sampling frequency of 400Hz

Where m denotes the delay and m=0, 1, 2…N-1. To calculate the value of Δf , m is supposed to be zero. So we can get the formula (3.7):

$$R_{x_1x_2}(0) = \frac{1}{2}\cos(2\pi(f_N + \Delta f)) = \frac{1}{2}\cos(2\pi\Delta f)) \tag{3.7}$$

From the formula (3.7), the Δf that being mentioned in formula (3.3) means frequency drift of under test signal during the measurement interval can be acquired. On the other side, the f_N is measured by using the coarse measurement method. So combining coarse and fine measurement method, we can get the high-precision frequency of under test signals.

3.3 Hardware description

The Multi-Channel Digital Frequency Stability Analyzer consists of Multi-channel Beat-Frequency signal Generator (MBFG) and Digital Signal Processing (DSP) module. The multi-channel means seven test channels and one calibration channel with same physical structure. The system block diagram is shown in figure 6.

The MBFG is made up of Offset Generator (OG), Frequency Distribution Amplifier (FDA), and Mixer. There are eight input signals, and seven signals from under test sources when the other one is designed as the reference, generally the most reliable source to be chosen as reference. The reference signal f_0 is used to drive the OG. The OG is a special frequency synthesizer that can generate the frequency at $f_r = f_0 - f_b$. The output of OG drives FDA to acquire eight or more offset sources at frequency f_r. Seven under test signals, denoted frequency $f_{xi}, i = 1,2,3...$, are down-converted to sinusoidal beat-frequency signals at nominal frequency f_b by mixing them with the offset sources at frequency f_r. The signal flow graph is showed in figure 6.

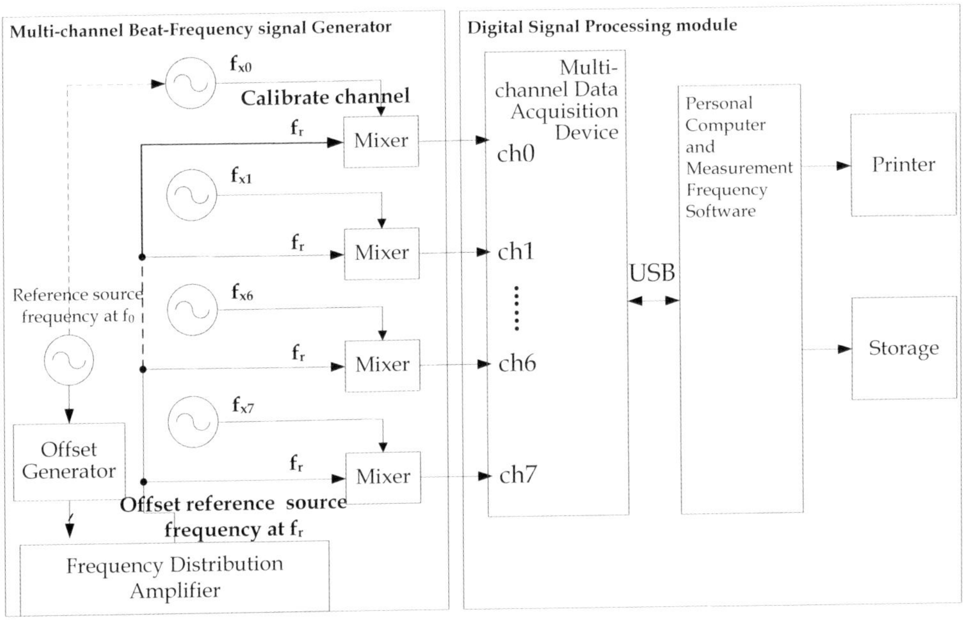

Fig. 6. Block Diagram of the Multi-Channel Digital Frequency Stability Analyzer

The channel zero is calibrating channel, which input the reference source running at frequency f_0 to test real time noise floor of the DFSA, and then can calibrate systematic errors of the other channels. The calibrating can be finished depending on the relativity between the input of channel zero and the output of OG. Because both signals come from one reference oscillator, they should have strong relativity that can cancel the effect of reference oscillator noise.

The Digital Signal Processing module consists of multi-channel Data Acquisition device (DAQ), personal computer (PC) and output devices. The Measurement Frequency (MF) software is installed in PC to analyze data from DAQ. The beat frequency signals, which are output from the MBFG that are connected to channels of analog-to-digital converter respectively, are digitized according to the same timing by the DAQ that are driven by a clock with sampling frequency N. Then, MF software retrieves the data from buffer of DAQ, maintains synchronization of the data stream, carries out processing of measurement (including frequency, phase difference, and analyzing stability), stores original data to disk, and manages the output devices.

The MBFG output must be sinusoidal beat frequency signals, because processing beat frequency signal make use of the property of trigonometric function. It has the obvious difference with traditional beat frequency method using square waveform and Zero Crosser Assembly.

3.4 Software description

The Measurement Frequency software (MF) of the Multi-Channel Digital Frequency Stability Analyzer is operated by the LabWindows/CVI applications. MF configures the parameters of DAQ, stores original data and results of measuring to disk, maintains synchronization of the data stream, carries out the algorithms of measuring frequency and phase difference, analyzes frequency stability, retrieves the stored data from disk and prepares plots of original data, frequency, phase difference, and Allan deviation. Figure 8 shows the main interface. To view interesting data, user can click corresponding control buttons to show beat signals graph, frequency values, phase difference and Allan deviation and so on.

MF consists of four applications, a virtual instrument panel that is the user interface to control the hardware and the others via DLL, a server program is used to manage data, processing program, and output program. Figure 7 shows the block diagram of MF software.

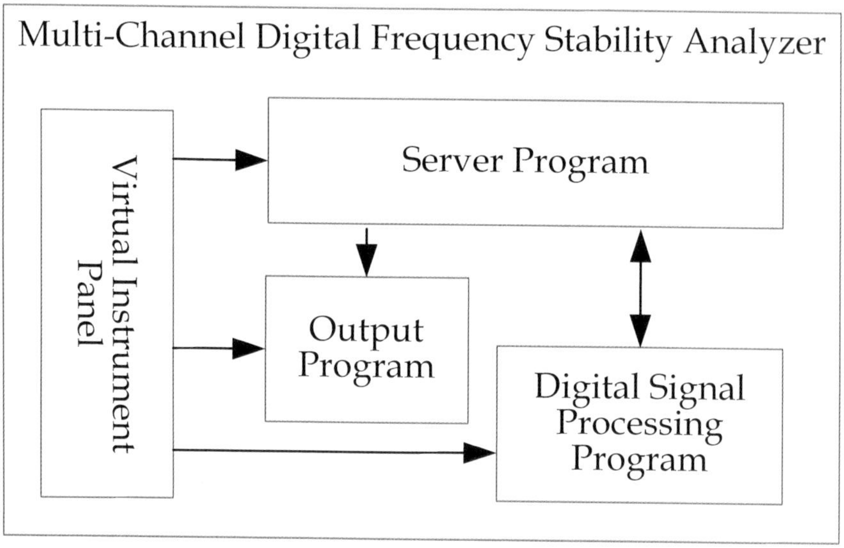

Fig. 7. Block Diagram of the Measurement Frequency Software

The virtual instrument panel have been developed what can be handled friendly by users. It looks like a real instrument. It consists of options pull-down menu, function buttons, choice menus. Figure 8 (a) shows the parameters setting child panel. Users can configure a set of parameters what involve DAQ, such as sampling frequency, amplitude value and time base of DAQ. Figure 8 (b) shows the screen shot of MF main interface. On the left of Fig. 8 (b), users can assign any measurement channel start or pause during measurement. On the right of Fig. 8 (b), strip chart is used to show the data of user interesting, such as real-time original data, measured frequency values, phase difference values and Allan deviation. To distinguish different curves, different coloured curves are used to represent different channels when every channel name has a specific colour. Figure 8 (c) shows the graph of the real-time results of frequency measurement when three channels are operated synchronously, and (d) shows the child panel what covers the original data, frequency values and Allan deviation information of one of channel.

Server program configures the parameters of each channel, maintains synchronization of the data stream, carries out the simple preprocessing (either ignore those points that are significantly less than or greater than the threshold or detect missing points and substitute extrapolated values to maintain data integrity), stores original data and results of measuring to disk.

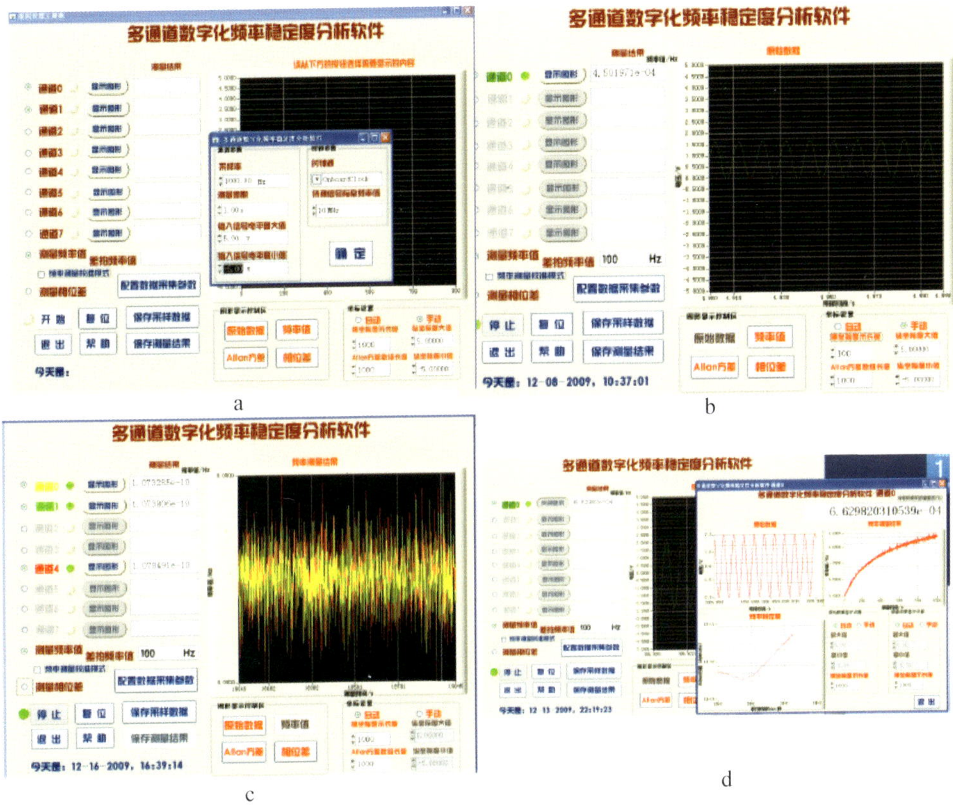

Fig. 8. MF software, (a) shows the window of configuring parameters and choosing channels, (b) shows the strip chart of real-time original data of one of channels, (c) shows the graph of the real-time results of frequency measurement, (d) shows the child panel that covers the original data, frequency values and Allan deviation information of one of channel.

Digital signal processing program retrieves the stored data from disk and carries out the processing. Frequency measurement includes dual-channel phase difference and single frequency measurement modes in the digital signal processing program. The program will run different functions according to the select mode of users. Single frequency measurement mode can acquire frequency values and the Allan deviation of every input signal source. In addition, the dual-channel phase difference mode can output the phase difference between two input signals.

The output program manages the interface that communicate with other instruments, exports the data of user interesting to disk or graph. Text files of these data are available if the user need to analyze data in the future.

3.5 Measurement precision

The dual-mixer and digital correlation algorithms are applied to DFSA. In this system, has symmetrical structure and simultaneously measurement to cancel out the noise of common offset reference source. (THOMAS E. PARKER, 2001) So the noise of common offset source can be ignored. The errors of the Multi-Channel Digital Frequency Stability Analyzer relate to thermal noise and quantization error (Ken Mochizuki, 2007 & Masaharu Uchino, 2004). The cross-correlation algorithm can reduce the effect of circuit noise floor and improve the measurement precision by averaging amount of sampling data during the measurement interval. In addition, this system is more reliability and maintainability because the structure of system is simpler than other high-precision frequency measurement system. This section will discuss the noise floor of the proposed system.

To evaluate the measurement precision of DFSA, we measured the frequency stability when the test signal and reference signal came from a single oscillator in phase (L.Sojdr, J. Cermak, 2003). Ideally, between the test channel and reference were operated symmetrically, so the error will be zero. However, since the beat signals output from MBFG include thermal noise, the error relate to white Gaussian noise with a mean value of zero.

Although random disturbance noise can be removed by running digital correlation algorithms in theory, we just have finite number of sampling data available in practice. So it will lead to the results that the cross-correlation between the signal and noise aren't completely uncorrelated. Then the effect of random noise and quantization noise can't be ignored. We will discuss the effect of ignored on measurement precision in following chapter.

According to above formula (3.7) introduction, the frequency drift Δf could be acquired by measuring the beat-frequency signal at frequency. But in the section 3.2.2, the beat signal is no noise, and that is inexistence in the real world. When the noises are added in the beat signal, it should be expressed like:

$$v_i(n) = V_i \sin(2\pi \frac{f_b + \Delta f_i}{N} n + \varphi_i) + g_i(n) + l_i(n), i = 1,2,3... \tag{3.8}$$

Where $v_i(n)$ represents beat-frequency signal, V_i indicates amplitude of channel i, f_b is the nominal frequency of beat-frequency signal, unknown frequency drift Δf_i of source under test in channel i, φ_i denotes the initial phase of channel i. Here N is sampling frequency of analog-to-digital converter (ADC), $g_i(n)$ denotes random noise of channel i, $l_i(n)$ is

quantization noise of channel i and generates by ADC, n is a positive integer and its value is in the range $1 \sim \infty$.
Formula (3.8) could be transformed into following normalized expression (3.9) to deduce conveniently.

$$v_i(n) = \sin(2\pi \frac{f_b + \Delta f_i}{N} n + \varphi_i) + g_i(n) + l_i(n) \tag{3.9}$$

To realize one time frequency measurement, sampling beat-frequency signal must be continuous operated at least two seconds. For example, the j-th measurement frequency of channel i will analyze the j second $v_{ij}(n)$ and j+1 second $v_{i(j+1)}(n)$ data from DAQ.
The cross-correlation between $v_{ij}(n)$ and $v_{i(j+1)}(n)$ have been used by following formula:

$$\begin{aligned} R_{ij}(m) &= \frac{1}{N}\sum_{n=0}^{N-1} v_{ij}(n)v_{i(j+1)}(n+m) \\ &= \frac{1}{N}\sum_{n=0}^{N-1}[x_{ij}(n)+g_{ij}(n)+l_{ij}(n)]\times[x_{i(j+1)}(n+m)+g_{i(j+1)}(n+m)+l_{i(j+1)}(n+m)] \\ &= \frac{1}{2}\cos(\omega_{ij}m+\Phi_{ij}) + R_{x_{ij}g_{i(j+1)}} + R_{x_{ij}l_{i(j+1)}} + R_{g_{ij}x_{i(j+1)}} + R_{g_{ij}g_{i(j+1)}} + R_{g_{ij}l_{i(j+1)}} + R_{l_{ij}x_{i(j+1)}} \\ &\quad + R_{l_{ij}g_{i(j+1)}} + R_{l_{ij}l_{i(j+1)}} \end{aligned} \tag{3.10}$$

Formula (3.10) could be split into three parts; with the first part is cross-correlation function between signals $x(n)$:

$$A = \frac{1}{2}\cos(\omega_{ij}m + \Phi_{ij}) \tag{3.11}$$

the second part is the cross-correlation function between noise and signal;

$$B = R_{x_{ij}g_{i(j+1)}} + R_{x_{ij}l_{i(j+1)}} + R_{g_{ij}x_{i(j+1)}} + R_{l_{ij}x_{i(j+1)}} \tag{3.12}$$

the third part is the cross-correlation function between noise and noise:

$$C = R_{g_{ij}g_{i(j+1)}} + R_{g_{ij}l_{i(j+1)}} + R_{l_{ij}g_{i(j+1)}} + R_{l_{ij}l_{i(j+1)}} \tag{3.13}$$

According to the property of correlation function, if two circular signals are correlated then it will result in a period signal with the same period as the original signal. Therefore, the C can be denoted average $R_{ij}(m)$ over m:

$$C = \frac{1}{N}\sum_{m=0}^{N-1} R_{ij}(m) \tag{3.14}$$

The term $B = R_{x_{ij}g_{i(j+1)}} + R_{x_{ij}l_{i(j+1)}} + R_{g_{ij}x_{i(j+1)}} + R_{l_{ij}x_{i(j+1)}}$ of cross-correlation can't be ignored. Because the term B isn't strictly zero. We will discuss the effect of ignoring B and C on measurement precision in following section.

According to the property of cross-correlation and sine function, we have

$$
\begin{aligned}
R_{x_{ij}g_{i(j+1)}}(m) &= R_{g_{i(j+1)}x_{ij}}(-m) = \frac{1}{N}\sum_{n=0}^{N-1} g_{i(j+1)}(n)x_{ij}(n-m) \\
&= \frac{1}{N}\sum_{n=0}^{N-1} g_{i(j+1)}(n)\sin(\varphi_{ij}+\omega_{ij}n-\omega_{ij}m) \\
&= \frac{1}{N}\sum_{n=0}^{N-1} g_{i(j+1)}(n)[\sin(\varphi_{ij}+\omega_{ij}n)\cos(\omega_{ij}m)-\cos(\varphi_{ij}+\omega_{ij}n)\sin(\omega_{ij}m)] \\
&= \frac{1}{N}\cos(\omega_{ij}m)\sum_{n=0}^{N-1} g_{i(j+1)}(n)\sin(\varphi_{ij}+\omega_{ij}n)-\frac{1}{N}\sin(\omega_{ij}m)\sum_{n=0}^{N-1} g_{i(j+1)}(n)\cos(\varphi_{ij}+\omega_{ij}n)
\end{aligned}
\tag{3.15}
$$

Similarly, for other cross-correlation, we have

$$
\begin{aligned}
R_{x_{ij}l_{i(j+1)}}(m) &= \frac{1}{N}\sum_{n=0}^{N-1} l_{i(j+1)}(n)x_{ij}(n-m) \\
&= \frac{1}{N}\cos(\omega_{ij}m)\sum_{n=0}^{N-1} l_{i(j+1)}(n)\sin(\varphi_{ij}+\omega_{ij}n)-\frac{1}{N}\sin(\omega_{ij}m)\sum_{n=0}^{N-1} l_{i(j+1)}(n)\cos(\varphi_{ij}+\omega_{ij}n)
\end{aligned}
\tag{3.16}
$$

$$
\begin{aligned}
R_{g_{ij}x_{i(j+1)}}(m) &= \frac{1}{N}\sum_{n=0}^{N-1} g_{ij}(n)x_{i(j+1)}(n+m) \\
&= \frac{1}{N}\cos(\omega_{ij}m)\sum_{n=0}^{N-1} g_{ij}(n)\sin(\varphi_{i(j+1)}+\omega_{ij}n)+\frac{1}{N}\sin(\omega_{ij}m)\sum_{n=0}^{N-1} g_{ij}(n)\cos(\varphi_{i(j+1)}+\omega_{ij}n)
\end{aligned}
\tag{3.17}
$$

$$
\begin{aligned}
R_{l_{ij}x_{i(j+1)}}(m) &= \frac{1}{N}\sum_{n=0}^{N-1} l_{ij}(n)x_{i(j+1)}(n+m) \\
&= \frac{1}{N}\cos(\omega_{ij}m)\sum_{n=0}^{N-1} l_{ij}(n)\sin(\varphi_{i(j+1)}+\omega_{ij}n)+\frac{1}{N}\sin(\omega_{ij}m)\sum_{n=0}^{N-1} l_{ij}(n)\cos(\varphi_{i(j+1)}+\omega_{ij}n)
\end{aligned}
\tag{3.18}
$$

Then, the B can be obtained as follows:

$$
\begin{aligned}
B = {} & \frac{1}{N}\cos(\omega_{ij}m)[\sum_{n=0}^{N-1} g_{i(j+1)}(n)\sin(\varphi_{ij}+\omega_{ij}n)+\sum_{n=0}^{N-1} l_{i(j+1)}(n)\sin(\varphi_{ij}+\omega_{ij}n) \\
& +\sum_{n=0}^{f_s-1} g_{ij}(n)\sin(\varphi_{i(j+1)}+\omega_{ij}n)+\sum_{n=0}^{f_s-1} l_{ij}(n)\sin(\varphi_{i(j+1)}+\omega_{ij}n)] \\
& +\frac{1}{N}\sin(\omega_{ij}m)[\sum_{n=0}^{N-1} g_{ij}(n)\cos(\varphi_{i(j+1)}+\omega_{ij}n)-\sum_{n=0}^{N-1} g_{i(j+1)}(n)\cos(\varphi_{ij}+\omega_{ij}n) \\
& +\sum_{n=0}^{N-1} l_{ij}(n)\cos(\varphi_{i(j+1)}+\omega_{ij}n)-\sum_{n=0}^{N-1} l_{i(j+1)}(n)\cos(\varphi_{ij}+\omega_{ij}n)]
\end{aligned}
\tag{3.19}
$$

The sum of formula (3.19) is equal to zero in the range [0,N-1].

$$\sum_{m=0}^{N-1} B = 0 \tag{3.20}$$

In view of the Eq. (3.20), although the B isn't strictly zero, their sum is equal to zero. We all known that on the right-hand side of Eq.(3.14) is the sum of cross-correlation function. Applying the Eq. (3.20) to (3.14) term by term, we obtain that the Eq.(3.14) strictly hold. Now we have the knowledge that the term C doesn't effect on the measurement results and we just need to discuss the term B as follows. Eq. (3.12) can be given by

$$R_{ij}(0) - \frac{1}{N}\sum_{m=0}^{N-1} R_{ij}(m) = \frac{1}{2}\cos(\Phi_{ij}) + B \tag{3.21}$$

Let the error terms that are caused by the white Gaussian noise and the quantization noise be represented by $B_1 = R_{x_{ij}g_{i(j+1)}} + R_{g_{ij}x_{i(j+1)}}$ and $B_2 = R_{x_{ij}l_{i(j+1)}} + R_{l_{ij}x_{i(j+1)}}$ respectively. So B can be expressed by $B = B_1 + B_2$.

Here, quantization noise is generally caused by the nonlinear transmission of AD converter. To analysis the noise, AD conversion usual is regarded as a nonlinear mapping from the continuous amplitude to quantization amplitude. The error that is caused by the nonlinear mapping can be calculated by using either the random statistical approach or nonlinear determinate approach. The random statistical approach means that the results of AD conversion are expressed with the sum of sampling amplitude and random noise, and it is the major approach to calculate the error at present.

We assume that $g(t)$ is Gaussian random variable of mean '0'and standard deviation 'σ_g^2'.

In the view of Eq.(3.15) and (3.17), we have obtained the standard deviation as follow:

$$\sigma_{B_1}^2 = \frac{2\sigma_g^2}{N} \tag{3.22}$$

Assume that the AD converter is round-off uniformly quantizer and using quantization step Δ . Then $l(t)$ is uniformly distributed in the range $\pm\Delta/2$ and its mean value is zero and standard deviation is $(\Delta^2/12)$. We have

$$\sigma_{B_2}^2 = \frac{2\Delta^2}{12N} \tag{3.23}$$

For B_1 and B_2 are uncorrelated, then

$$\sigma_B^2 = \sigma_{B_1}^2 + \sigma_{B_2}^2 = \frac{2\sigma_g^2}{N} + \frac{2\Delta^2}{12N} \tag{3.24}$$

The mean square value of $\frac{1}{2}\cos(\Phi_{ij}) + B$ on the right-hand side of formula (3.21) will be calculated by the following formula to evaluate the influence of noise on measurement initial phase difference.

$$
\begin{aligned}
&\frac{1}{N}\sum_{m=0}^{N-1}\left(\frac{1}{4}\cos^2(\Phi_{ij})+B\cos(\Phi_{ij})+B^2\right)\\
&=\frac{1}{4}\cos^2(\Phi_{ij})+\frac{1}{N}\sum_{m=0}^{N-1}(B\cos(\Phi_{ij})+B^2)\\
&=\frac{1}{4}\cos^2(\Phi_{ij})+(\frac{2\sigma_g^2}{N}+\frac{2\Delta^2}{12N})+\frac{1}{N}\sum_{m=0}^{N-1}B\cos(\Phi_{ij})\\
&\le\frac{1}{4}\cos^2(\Phi_{ij})+(\frac{2\sigma_g^2}{N}+\frac{2\Delta^2}{12N})+\frac{1}{N}\sum_{m=0}^{N-1}B\\
&=\frac{1}{4}\cos^2(\Phi_{ij})+(\frac{2\sigma_g^2}{N}+\frac{2\Delta^2}{12N})
\end{aligned}
\tag{3.25}
$$

Where σ_g^2 represent standard deviation of Gaussian random variable, Signal Noise Ratio $SN=\frac{V^2}{\sigma_g^2}$, and here the V is the amplitude of input signal, let amplitude resolution of a-bit digitize and quantization step be Δ, here variable 'a' can be 8~24. We have $\frac{\Delta}{V}=\frac{2}{2^a-1}$ (Ken Mochizuki, 2007). Applying this equation to formula (3.25) term by term, we obtain

$$
\sigma_e=\sqrt{\frac{1}{4}\cos^2(\Phi_{ij})+\frac{1}{N}(\frac{2V^2}{SN^2}+\frac{2\Delta^2}{12})}
\tag{3.26}
$$

Where σ_e is the standard deviation of measurement initial phase difference. The standard deviation of digital correlation algorithms depends on the sampling frequency N, SNR and amplitude resolution 'a', as understood from formula (3.26). Here the noise of amplitude resolution can be ignored if the 'a' is sufficiently bigger than 16-bit and the SNR is smaller than 100 dB. The measurement accuracy for this method is mostly related to SNR of signal. This method has been tested that has the strong anti-disturbance capability.

4. System noise floor and conclusion

To evaluate the noise floor, we designed the platform when the test signal and reference signal were distributed in phase from a single signal generator. The signal generator at 10MHz and the beat-frequency value of 100Hz were set. For this example obtained the Allan deviation (square root of the Allan variance (DAVID A, HOWE)) of $\sigma_y(\tau)=4.69E-14$ at $\tau=1$ second and $\sigma_y(\tau)=1.27E-15$ at $\tau=1000$ second.

The measurement ability could be optimized further by improving the performance of OG. Because the reference of the system is drove by the output of OG.

Since the digital correlation techniques can smooth the effects of random disturbance of the MBFG, it can achieve higher measurement accuracy than other methods even if on the same MBFG.

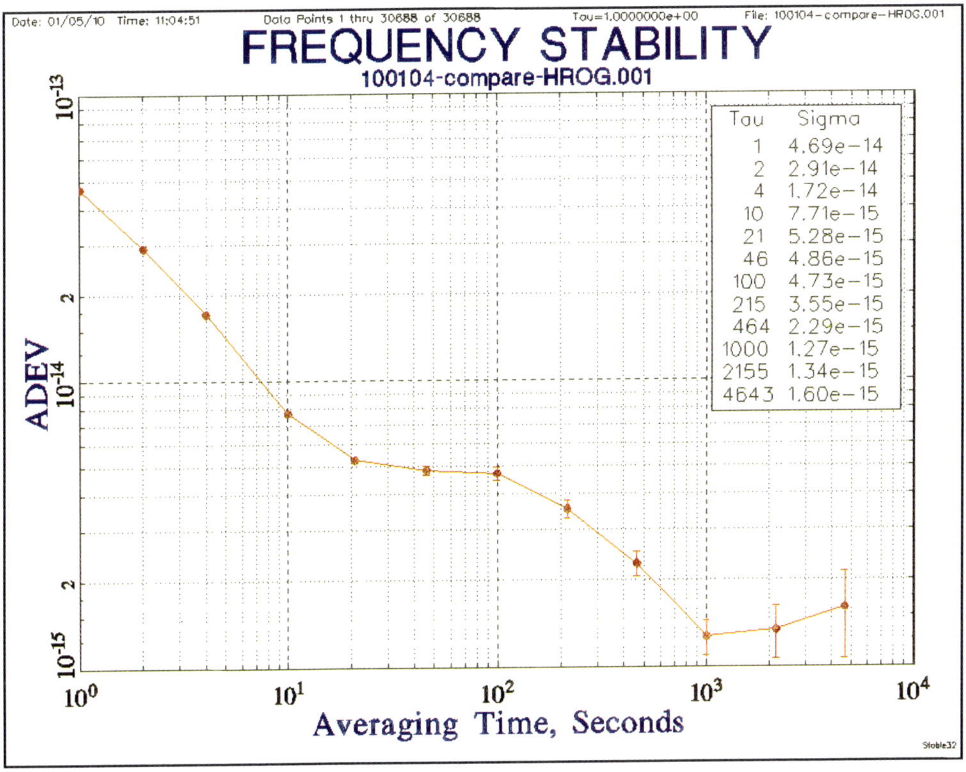

Fig. 9. An example of noise floor characteristics of the DFSA: Allan deviation

Additional, the design of calibration channel that is proposed to remove the systematic error is useful to acquire better performance for current application. A comprehensive set of noise floor tests under all conditions has not been carried out with the current system.
The system hardware consists only of MBFG, DAQ and PC. Compared with the conventional systems using counter and beat-frequency device, the system can be miniaturized and moved conveniently. As expected, system noise floor is good enough for current test requirement. The system will take measurement of wide range frequency into account in the future. Intuitive operator interface and command remotely will be design in following work.

5. Acknowledgment

The authors thank Bian Yujing and Wang Danni for instructing. I would like to thank the present of the Chinese Academy of sciences scholarship and Zhu Liyuehua Scholarship for the supporting. The work has been supported by the key program of West Light Foundation of The CAS under Grant 2007YB03 and the National Nature Science Funds 61001076 and 11033004.

6. References

Allan, D. W. - Daams, H.: *Picosecond Time Difference Measurement System* Proc. 29th Annual Symposium Frequency Control, Atlantic City, USA, 1975, 404–411.

C. A. Greenhall, A. Kirk, and G. L. Stevens, 2002, *A multi-channel dual-mixer stability analyzer: progress report,* in Proceedings of the 33rd Annual Precise Time and Time Interval (PTTI) Systems and Applications Meeting, 27-29 November 2001, Long Beach, California, USA, pp. 377-383.

C. A. Greenhall, A. Kirk, and R. L. Tjoelker. *A Multi-Channel Stability Analyzer for Frequency Standards in the Deep Space Network.* 38th Annual Precise Time and Time Interval (PTTI) Meeting.2006.105-115

C. A. Greenhall, A. Kirk, R. L. Tjoelker. *Frequency Standards Stability Analyzer for the Deep Space Network.* IPN Progress Report.2007.1-12

D. A. Howe, D. W. Allan, and J. A. Barnes, 1981, *Properties of signal sources and measurement methods,* in Proceedings of the 35th Annual Frequency Control Symposium, 27-29 May 1961, Philadelphia, Pennsylvania, USA (Electronic Industries Association, Washington, D.C.), 1–47

D.A.Howe,C.A.Greenhall.Total Variance: a Progress Report on a New Frequency Stabbility Characterization.1999.

David A, Howe. *Frequency Stability.*1703-1720, National Institute of Standards and Technology (NIST)

D.B.Sulliivan, D.W.Allan, D.A.Howe and F.L.Walls, *Characterization of Clocks and Oscillators,* NIST Technical Note 1337.

E. A. Burt, D. G. Enzer, R. T. Wang, W. A. Diener, and R. L. Tjoelker, 2006, *Sub-10-16 Frequency Stability for Continuously Running Clocks: JPL's Multipole LITS Frequency Standards*, in Proceedings of the 38th Annual Precise Time and Time Interval (PTTI)

Systems and Applications Meeting, 5-7 December 2006, Reston, Virginia, USA (U.S. Naval Observatory, Washington, D.C.), 271-292.

G. Brida, *High resolution frequency stability measurement system,* Review of Scientific Instruments, Vol., 73, NO. 5 May 2002, pp. 2171–2174.

J. Laufl M Calhoun, W. Diener, J. Gonzalez, A. Kirk, P. Kuhnle, B. Tucker, C. Kirby, R. Tjoelker. *Clocks and Timing in the NASA Deep Space Network.* 2005 Joint IEEE International Frequency Control Symposium and Precise Time and Time Interval (PTTI) Systems and Applications Meeting.2005.

Julian C. Breidenthal, Charles A. Greenhall, Robert L. Hamell, Paul F. Kuhnle. *The Deep Space Network Stability Analyzer.* The 26th Annual Precise Time and Time Interval (PTTI) Applications and Planning Meeting .1995.221-233

Ken Mochizuki, Masaharu Uchino, Takao Morikawa, *Frequency-Stability Measurement System Using High-Speed ADCs and Digital Signal Processing,* IEEE Transactions on Instrument, And Measurement, VOL. 56, NO. 5, Oct. 2007, pp. 1887–1893

L.Sojdr, J. Cermak, and G. Brida, *Comparison of High-Precision Frequency-Stability Measurement Systems,* Proceedings of the 2003 IEEE International Frequency Control Symposium, vol. A247, pp. 317–325, Sep. 2003

Masaharu Uchino, Ken Mochizuki, *Frequency Stability Measuring Technique Using Digital Signal Processing,* Electronics and Communications in Japan, Part 1, Vol. 87, No. 1, 2004, pp.21–33.

Richard Percival,Clive Green.*The Frequency Difference Between Two very Accurate and Stable Frequency Singnals.* 31st PTTI meeting.1999

R.T.Wang, M.D.Calhoun, A.Kirt, W. A. Diener, G. J. Dick, R.L. Tjoelker. *A High Performance Frequency Standard and Distribution System for Cassini Ka-Band Experiment.* 2005 Joint IEEE International Frequency Control Symposium (FCS) and Precise Time and Time Interval (PTTI) Systems and Applications Meeting.2005.919-924

S. Stein, D. Glaze, J. Levine, J. Gray, D. Hilliard, D. Howe, L. A. Erb, *Automated High-Accuracy Phase Measurement System. IEEE Transactions on Instrumentation and Measurement.* 1983.227-231

S. R. Stein, 1985, *Frequency and time their measurement and characterization,* in E. A. Gerber and A. Ballato, eds., Precision Frequency Control, Vol. 2 (Academic Press, New York), pp. 191–232, 399–416.

Thomas E. Parker. *Comparing High Performance Frequency Standards.* Frequency Control Symposium and PDA Exhibition, 2001. Proceedings of the 2001 IEEE International .2001.89-95

W. J. Riley, *Techniques for Frequency Stability Analysis,* IEEE International Frequency Control Symposium Tutorial Tampa, FL, May 4, 2003.

Ya Liu, Xiao-hui Li, Wen-Li Wang, Dan-Ni Wang, *Research and Realization of Portable High-Precision Frequency Set,* Computer Measurement & Control, vol.16, NO.1, 2008, pp21-23.

Ya Liu, Li Xiao-hui, Zhang Hui-jun, *Analysis and Comparison of Performance of Frequency Standard Measurement Systems Based on Beat-Frequency Method,* 2008 IEEE Frequency Control Symposium, 479-483

Ya Liu, Xiao-Hui Li, Yu-Lan Wang, *Multi-Channel Beat-Frequency Digital Measurement System for Frequency Standard*, 2009 IEEE International Frequency Control Symposium, 679-684

In-Situ Supply-Noise Measurement in LSIs with Millivolt Accuracy and Nanosecond-Order Time Resolution

Yusuke Kanno
Hitachi LTD.
Japan

1. Introduction

This chapter explores signal analysis of a circuit embedded in an LSI to probe the voltage fluctuation conditions, and is described as an example of digital signal processing[1]. As process scaling has continued steadily, the number of devices on a chip continues to grow according to Moore's Law and, subsequently, highly integrated LSIs such as multi-CPU-core processors and system-level integrated Systems-on-a-Chip (SoCs) have become available. This technology trend can also be applied to low-cost and low-power LSIs designed especially for mobile use. However, it is not the increase in device count alone that is making chip design difficult. Rather, it is the fact that parasitic effects of interconnects such as interconnect resistance now dominate the performance of the chip. Figure 1 shows the trends in sheet resistance and estimated power density of LSIs. These effects have greatly increased the design complexity and made power-distribution design a considerable challenge.

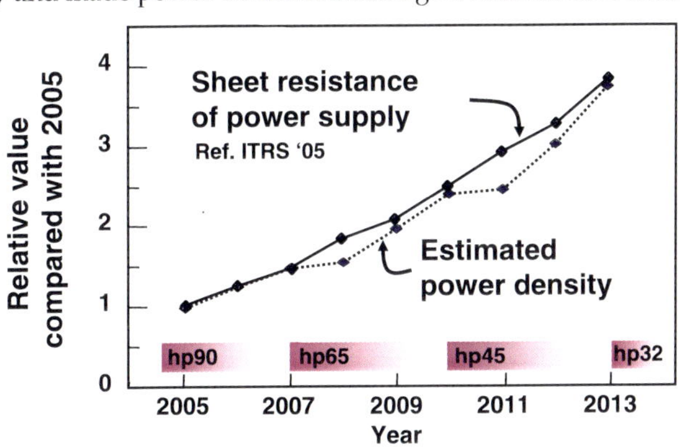

Fig. 1. Trends in sheet resistance and estimated power density.

Power supply integrity is thus a key for achieving higher performance of SoCs fabricated using an advanced process technology. This is because degradation of the power integrity causes a voltage drop across the power supply network, commonly referred to as the IR-drop, which, in turn, causes unpredictable timing violations or even logic failures (Saleh et al., 2000). To improve power integrity, highly accurate analysis of a power-supply network is required. However, sophisticated SoCs, such as those for mobile phones, have many IPs and many power domains to enable a partial-power-down mode in a single chip. Thus, many spots of concentrated power consumption, called "hot spots", appear at many places in the chip as shown in the Fig. 2. Analysis of the power-supply network is therefore becoming more difficult. To address these issues, it is necessary to understand the influence of supply noise in product-level LSIs, gain more knowledge of it, and improve evaluation accuracy in the design of power supply networks via this knowledge. Above all, this understanding is very important; therefore, in-situ measurement and analysis of supply-noise maps for product-level LSIs has become more important, and can provide valuable knowledge for establishing reliable design guidelines for power supplies.

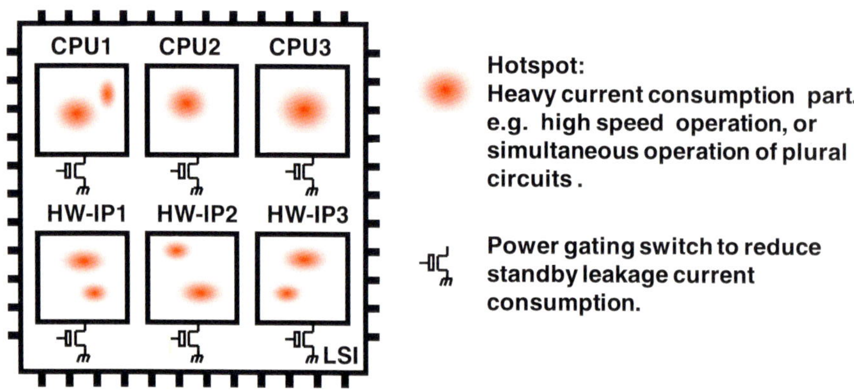

Fig. 2. Hotspots in the LSIs. The hotspots are defined as heavy current consumption parts in the LSIs. The sophisticated LSI has many CPUs and hardware Intellectual Properties (HW-IPs) in it, so the many hotspots become appearing.

In-depth analysis of the power supply network based on this in-situ power supply noise measurement can be helpful in designing the power supply network, which is becoming requisite for 65-nm process technology and beyond.

1.1 Related work

Several on-chip voltage measurement schemes have recently been reported (Okumoto et al., 2004; Takamiya et al., 2004), and the features are illustrated in Fig. 3.

One such scheme involves the use of an on-chip sampling oscilloscope (Takamiya et al., 2004). This function accurately measures high-speed signal waveforms such as the clock signal in a chip. Achieving such high measurement accuracy requires a sample/hold circuit which consist of an analog-to-digital converter (ADC) in the vicinity of the measurement point. This method can effectively avoid the influence of the noise on the measurement. Therefore, a large chip footprint is required for implementing measurement circuits such as a voltage noise filter, a reference-voltage generator and a timing controller.

	(a) On-chip sampling oscilloscope	(b) Simple analog measurement
Block Diagram	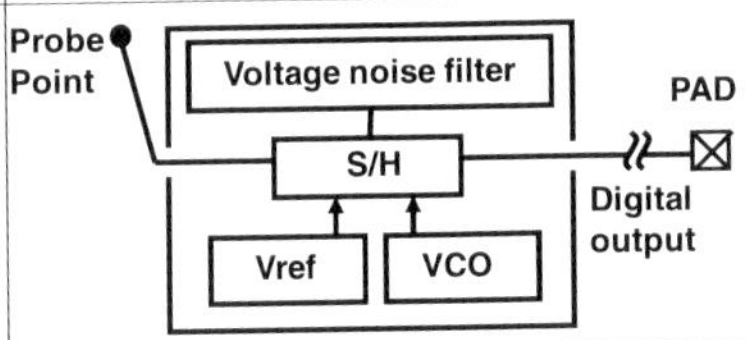	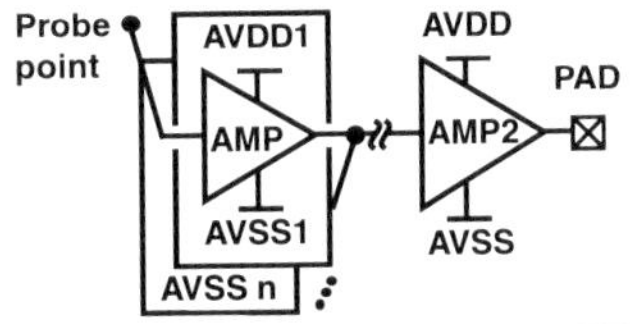
Feature	• Digitization in vicinity of measurement point	• Small footprint • Capability of many probes integration
Problem	•Large footprint	• Requires equality of local ground such as AVSS1 = ... = AVSSn = AVSS • Susceptible to noise

Fig. 3. Examples of on-chip voltage measurement scheme. (a) is an on-chip sampling oscilloscope (Takamiya et al., 2004) and (b) is a simple analog measurement (Okumoto et al., 2004).

A small, simple analog measurement was reported in (Okumoto et al., 2004). This probe consists of a small first amplifier, and the output signal of the probe is sent to a second amplifier and then transmitted to the external part of the chip. Because the probe is very small and has the same layout height as standard cells and needs only one second amplifier, many probes can be implemented in a single LSI with minimal area overhead. This method, however, requires dedicated power supplies for measuring voltages that are different from local power supplies V_{DD} and V_{SS}.

These measurements are therefore basically done under test-element-group (TEG) conditions, and they may find it difficult to capture supply noise at multiple points in product-level LSIs when actually running applications. To resolve this difficulty, an in-situ measurement scheme is proposed. This method requires only a CMOS digital process and can be applied to standard-cell based design. Thus, it is easy to apply to product-level LSIs. The effect was demonstrated on a 3G cellular phone processor (Hattori et al., 2006), and the measurement of power supply noise maps induced by running actual application programs was demonstrated.

1.2 Key points for an in-situ measurement

Three key points need to be considered in order to measure the power supply noise at multiple points on a chip: area overhead, transmission method, and dynamic range.

1. The first point is the area overhead of the measurement probes.
   Because the power-consumption sources are distributed over the chip and many independent power domains are integrated in an LSI, analyzing the power supply network for product-level LSIs is very complicated. To analyze these power-supply networks, many probes must be embedded in the LSI. Thus, the probes must be as small as possible. Minimal area overhead and high adaptability to process scaling and ready-made electrical design automation (EDA) tools are therefore very important factors regarding the probes.
2. The second point is the method used to transmit the measured signal.
   It is impossible to transmit the measured voltage by using a single-ended signal, because

there is no flat (global) reference voltage in an LSI. Dual-ended signal transmission is a promising technique to get around this problem; however, this method gives rise to another issue: the difficulty of routing by using a ready-made EDA tool. Noise immunity of the transmission is another concern, because analog signal transmission is still needed.

3. The third point is the dynamic range of the voltage measurement.
   To measure supply-voltage fluctuation, a dedicated supply voltage for the probes needs to have a greater range than that of the measured local supply voltage difference.

2. In-situ supply-noise map measurement

An in-situ power-supply-noise map measurement scheme was developed by considering the above key points. Figure 4 shows the overall configuration of our proposed measurement scheme. The key feature of this scheme is the minimal size of the on-chip measurement circuits and the support of off-chip high resolution digital signal processing with frequent calibration (Kanno, et al., 2006),(Kanno, et al., 2007). The on-chip measurement circuit therefore does not need to have a sample-and-hold circuit.

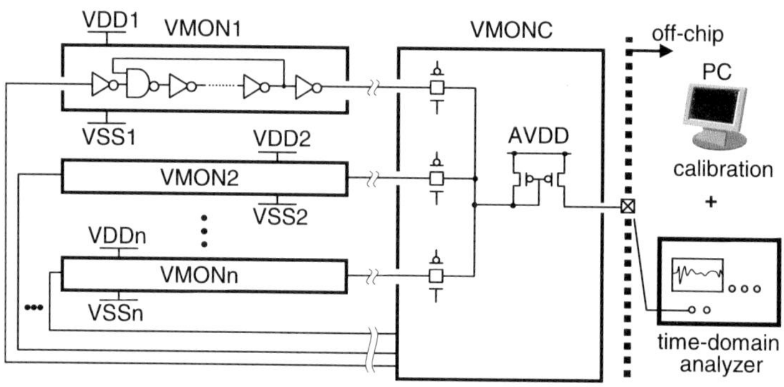

Fig. 4. In-situ supply-noise-map measurement scheme

The on-chip circuits consist of several voltage monitors (VMONs) and their controller (VMONC). The VMON is a ring oscillator that acts as a supply-voltage-controlled oscillator, so that the local supply difference (LSD) between V_{DD1} and V_{SS1} can be translated to a frequency-modulated signal (see Fig. 5). The VMONC activates only one of the VMONs and outputs the selected frequency-modulated signal to the external part of the chip. Every VMON can be turned off when measurement is not necessary.

The output signal is then demodulated in conjunction with time-domain analysis by an oscilloscope and calibrations by a PC. The frequency-modulated signal between the VMONs and VMONC is transmitted only via metal wires, so dozens of power-domain partitions can be easily implemented in an LSI (Kanno, et al., 2006). The frequency-modulated signal has high noise immunity for long-distance, wired signal transmission. Although the measurement results are averaged out in the nanoseconds of the VMON's sampling period, this method can analyze voltage fluctuation easily as the voltage fluctuation map in LSIs by using multi-point measurement.

The dynamic range of the measuring voltage is not limited despite requiring no additional dedicated supply voltage. This is because we measure a frequency fluctuation as a voltage

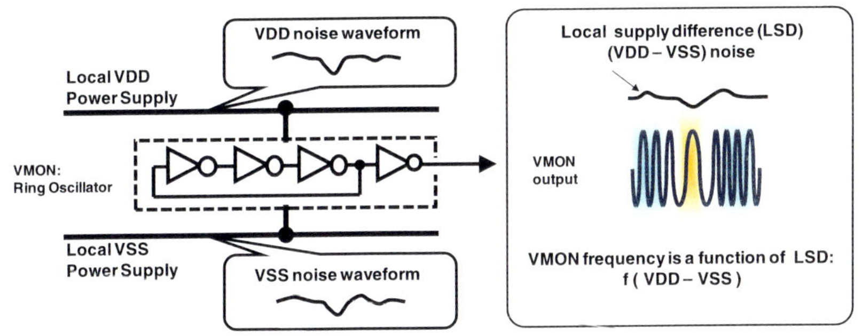

Fig. 5. Concept of the voltage-controlled oscillation. VMON is a ring oscillator whose frequency is modulated by the voltage fluctuation.

fluctuation, which is based on the fact that the oscillation frequency of a ring oscillator is a simple monotonic increasing function of the supply voltage(Chen, et al., 1996).

2.1 Time resolution and tracking of LSD

The ring oscillator's oscillation period consists of each inverter's delay, which depends on its LSD (Chen, et al., 1996). The voltage-measurement mechanism of the ring oscillator and the definition of our measured voltage are depicted in Fig. 6 in the simple case of a five-stage ring oscillator. The inverter circuit of each stage of the ring oscillator converts the LSD to

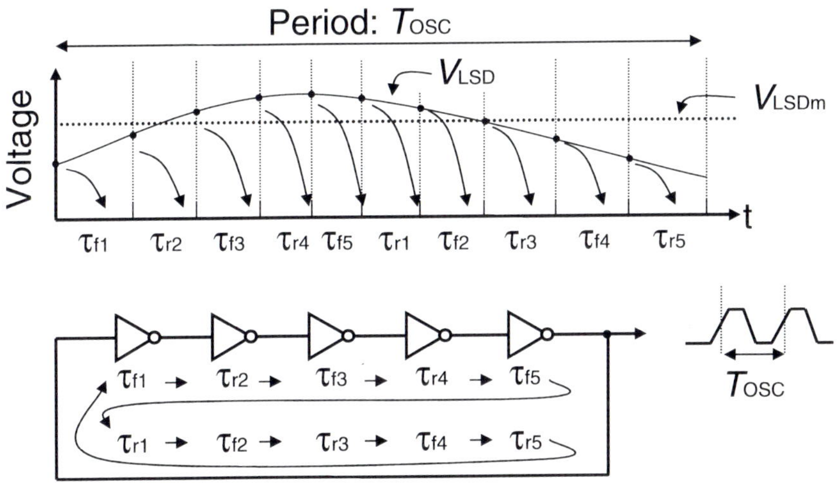

Fig. 6. Sampling of a ring oscillator

corresponding delay information. In the ring oscillator, since only one inverter in the ring is activated, each inverter converts the LSD voltages into delays one after another. This converted delay τ is a unique value based on the LSD,

$$\tau_{\mathrm{fi}} = f_{\mathrm{f}}(V_{\mathrm{LSDi}}), \tau_{\mathrm{ri}} = f_{\mathrm{r}}(V_{\mathrm{LSDi}}), \tag{1}$$

where τ_{ri} is the rise delay of the i-th stage, τ_{fi} is the fall delay of the i-th stage, and V_{LSDi} is the LSD supplying the i-th stage.

The output signal of the ring oscillator used to measure the external part of the chip has a period of T_{osc}, which is the sampling period of the ring oscillator. The T_{osc} is the total summation of all of the rise and fall delays of all the stages; that is,

$$T_{\text{osc}} = \sum_{i=1}^{5} \tau_{\text{ri}} + \sum_{i=1}^{5} \tau_{\text{fi}}, \tag{2}$$

$$= \sum_{i=1}^{5} f_{\text{f}}(V_{\text{LSDi}}) + \sum_{i=1}^{5} f_{\text{r}}(V_{\text{LSDi}}). \tag{3}$$

Since we can only measure the period of the ring oscillator T_{osc} and its inverse frequency(f_{osc}), we must calculate the voltage from (3) in order to determine the LSD. However, it is impossible to solve (3) because there are many combinations of V_{LSDi} that satisfy (3). Therefore, the measured LSD, V_{LSDm}, is defined as the constant voltage which provides the same period T_{osc},

$$T_{\text{osc}} = f(V_{\text{LSDm}}). \tag{4}$$

The period T_{osc} is thus the time resolution of the V_{LSDm}.

In this scheme, the LSD is calculated from a measured period T_{osc} or a measured frequency f_{osc}. The measured LSD denoted as V_{LSDm} is therefore an average value. Since the voltage fluctuation is integrated through the period T_{osc}, the time resolution is determined by the period T_{osc}.

Next the tracking of the LSD is discussed. There is a limitation in the tracking because the measurement of the voltage fluctuation is done by a ring oscillator as mentioned above, and the local voltage fluctuation is averaged out at the period of the ring oscillator. When the voltage fluctuation has a high-frequency element, the reproduction is difficult. In addition, a single measurement is too rough to track the target voltage fluctuation. However, although the voltage fluctuation is synchronized to the system clock, in general, since the ring oscillator oscillates asynchronously to the system frequency, the sampling points are staggered with each measurement. It is well known that averaging multiple low-resolution samples yields a higher resolution measurement if the samples have an appropriate dither signal added to them (Gray,et al., 1993).

For example, Fig. 7 (a) illustrates the case where the supply voltage fluctuation frequency is 150 MHz, which is about half the frequency of the ring oscillator. In this case, a single measurement cannot track the original fluctuation, but a composite of all measured voltages follows the power supply fluctuation. Another example is shown in Fig. 7 (b). In this case, since the frequency of the power supply fluctuation is similar to the frequency of the ring oscillator, the measured voltage V_{LSDm} is almost constant. These examples show that this scheme tracks the LSD as an averaged value during the period of T_{osc}. Therefore, as shown in these examples, a rounding error occurs even when the frequency of the LSD is the half that of the VMON frequency. Thus, for precise tracking, the frequency of the ring oscillator should be designed to be more than 10 times higher than that of the LSD. In general, the frequency of the power-supply voltage fluctuation can be classified into three domains; a low-frequency domain ($\sim$MHz), a middle-frequency domain ($\sim$100 MHz), and a high-frequency domain

(>GHz). Especially, the low-frequency domain is important in the case such as the operational mode switching and the power gating by on-chip power switches. Thus, in these cases, the accuracy of this method is sufficient to the tracking with high accuracy and the time resolution. Recently measurement of the influence of the on-chip power gating is reported (Fukuoka, 2007). Although the measured voltage is averaged out in the period of the VMON, however, the measurement of the voltage fluctuations at the actual operational mode in the product level LSI is innovative.

The higher the frequency of the ring oscillator, the higher the time resolution and improving the tracking accuracy; however, signal transmission at a higher frequency limits the length of the transmission line between the VMONs and VMONC due to the bandwidth limitation of the transmission line. There is therefore a trade-off between time resolution and transmission length. Although bandwidth can be widened by adding a repeater circuit, isolation cells, μI/O s (Kanno, et al., 2002), are needed when applying many power domains, and, thus, the design will be complicated.

2.2 Accuracy of waveform analysis

Accurate measurement of the VMON output frequency is also important in the in-situ measurement scheme. The accuracy also depends on the resolution of the oscilloscope

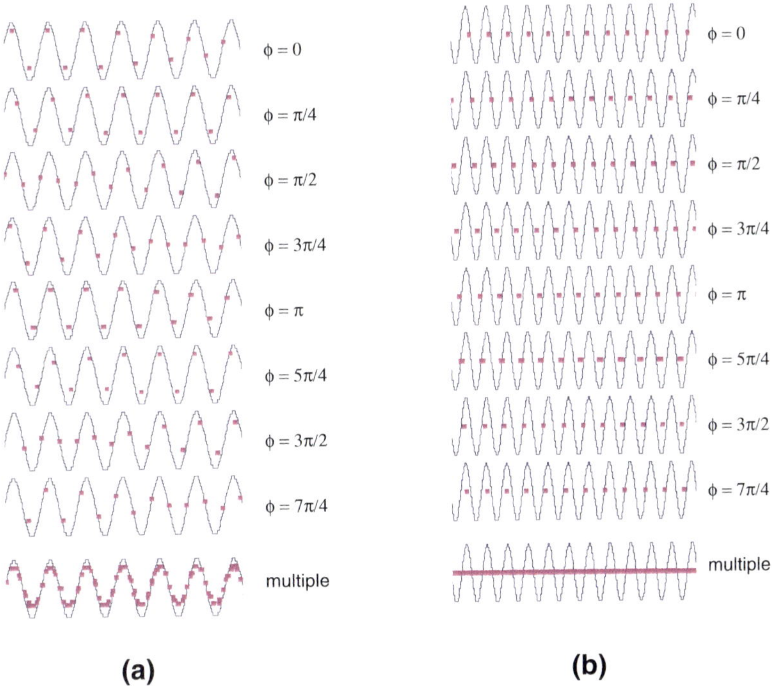

Fig. 7. Simulated results of voltage calculated by ring oscillator frequency: voltage fluctuation was (a) 150 MHz and (b) 300MHz. ϕ is the initial phase difference between voltage fluctuation and VMON output. The solid lines are voltage fluctuations and the dots are the calculated voltage from the VMON output.

used. Generally, frequency measurement is carried out by using a fast-Fourier-transform (FFT) based digital sampling oscilloscope. Sampling frequency and memory capacity of the oscilloscope are key for the FFT analysis.

First, the sampling frequency of the oscilloscope must be set in compliance with Shannon's sampling theorem. To satisfy this requirement, the sampling frequency must be set to at least double that of the VMONs. Second, the frequency resolution of the oscilloscope must be determined in order to obtain the necessary voltage resolution. Basically, the frequency resolution Δf of an FFT is equal to the inverse of the measurement period T_{meas}. If a 100-M word memory and a sampling speed of 40 GS/s are used, continuous measurement during a maximum measurement period of 25 ms can be carried out. If the frequency of the VMON output is several hundred megahertz and the coefficient of voltage-to-frequency conversion is about several millivolts per megahertz, highly accurate voltage measurement of the low-frequency LSD with an accuracy of about 1 mV can be achieved.

2.3 Support of off-chip digital signal processing

The proposed scheme has several drawbacks due to the simplicity of the ring-oscillator probe. One of the drawbacks is that the voltage-to-frequency dependence of the ring oscillator suffers from process and temperature variation. However, we can calibrate it by measuring the frequency-to-voltage dependence of each VMON before the in-situ measurement by setting the chip in standby mode. We can also compensate for temperature variation by doing this calibration frequently.

Figure 8 shows the measurement procedure of the proposed in-situ measurement scheme. First, the chip must be preheated in order to set the same condition for in-situ measurement, because the temperature is one of the key parameters for the measurement. This preheating is carried out by running a measuring program in the same condition as for the in-situ measurement. A test program is coded in order to execute an infinite loop because multiple measurements are necessary for improving the measurement accuracy. Because the measuring program is executed continuously, the temperature of the chip eventually reaches a state of thermal equilibrium. After the chip has reached this state, the calibration for the target VMON is executed just before the in-situ measurement. In the calibration, the frequency of the VMON output of a selected VMON is measured by varying the supply voltage while the chip is set in standby mode. Note that the calibration method can compensate for macroscopic temperature fluctuations, but not for microscopic fluctuations that occur in a short period of time that are much less than the calibration period. After the calibration, the in-situ measurement is executed by resetting the supply voltage being measured. In measuring the other VMONs continuously, the calibration step is repeated for each measurement. If other measurement conditions such as supply voltage, clock frequency, and the program being measured are changed, the chip must be preheated again.

Each VMON consumes a current of about 200 μA under the worst condition, and this current flows to and from the measurement points. This current itself also causes an IR drop; however, this current is almost constant, so the influence of this IR drop is also constant. In addition, the effect of the IR drop is assumed to obey a superposition principle, so the IR drop caused by the VMON can be separated from the IR drop caused by the chip operating current. Therefore, the IR drop caused by the VMON can be compensated for by the calibration.

Another drawback of our measurement scheme is that the simple ring-oscillator probe does not have any sample-and-hold circuits. This results in degradation of resolution. However,

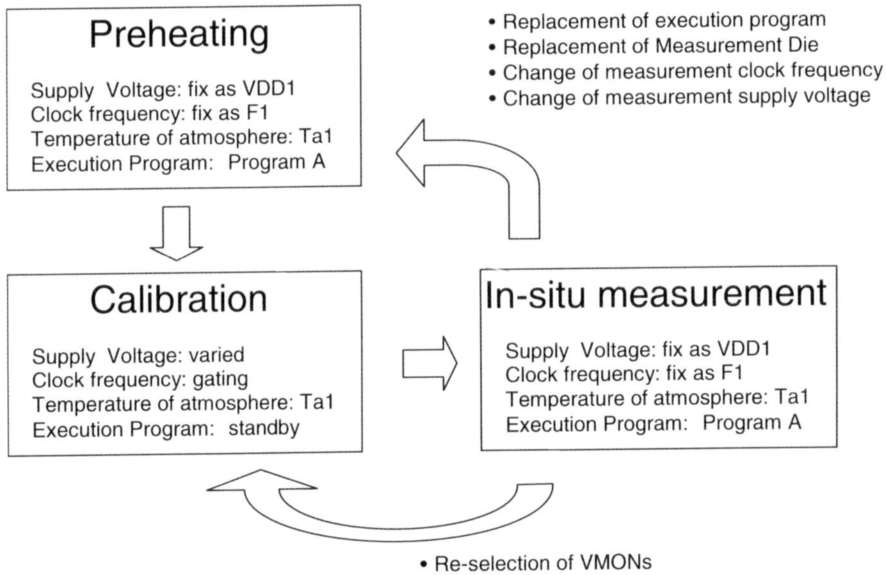

Fig. 8. Procedure for in-situ measurement

as described in section2.1, since the ring oscillator oscillates asynchronously with the chip operating frequency, high resolution can be achieved by averaging multiple low-resolution measurements using an oscilloscope (Abramzon, et al., 2004). This method is also effective for eliminating noise from measurements. If the wire length between VMON and VMONC is longer, the amplitude of the signal becomes small. This small amplitude suffers from the effect of noise. However, by using this averaging method, the influence of noise can be reduced, and signals can be measured clearly.

3. Measurement results

The in-situ measurement scheme was implemented in a 3G cellular phone processor (Hattori et al., 2006) as an example. Supply-noise maps for the processor were obtained while several actual applications were running. Figure 9 shows a chip photomicrograph. Three CPU cores and several IPs, such as an MPEG-4 accelerator, are implemented in the chip. A general-purpose OS runs on the AP-SYS CPU, and a real-time OS runs on the APL-RT CPU. The chip was fabricated using 90-nm, 8-Metal (7Cu+1Al), dual-Vth low-power CMOS process technology.

This chip has 20 power domains, and seven VMONs are implemented in several of the power domains (Kanno, et al., 2006). Five VMONs are implemented in the application part (AP-Part), and two VMONs are implemented in the baseband part (BB-Part). VMONs 1, 3, 4, and 5 are in the same power domain, whereas the others are in separate power domains. The reason these four VMONs were implemented in the same power domain is that this domain is the largest one, and many IPs are integrated in it.

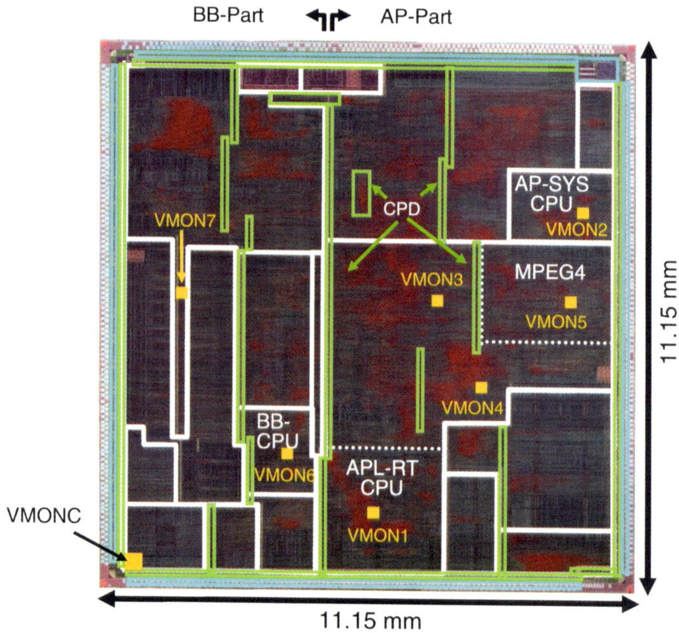

Fig. 9. Implementation example. This chip has three CPUs and several hardware accelerator such as a moving picture encoder (MPEG-4). The 20-power domains for partial power-shut down are implemented in a single LSI. This chip has a distributed common power domains (CPD) whose power-down opportunity is very rare. Seven VMONs and one VMOC are implemented in this chip.

Each VMON was only 2.52 μm $\times$ 25.76 μm, and they can be designed as a fundamental standard cell. Figure 10 shows the dependence of each VMON frequency on voltage, which were between 2.9 and 3.1 mV/MHz.

In Fig. 10, the frequency of the ring oscillators was designed to be about 200 MHz. Time resolution was about 5 ns. Note that we used LeCroy's SDA 11000 XXL oscilloscope with a 100-M-word-long time-interval recording memory and a maximum sampling speed of 40 GS/s.

3.1 Dhrystone measurement

We show the results of measurements taken while executing the Dhrystone benchmark program in the APL-RT CPU and a system control program in the AP-SYS CPU. The Dhrystone is known as a typical benchmark program for measuring performance per unit power, MIPS/mW, and the activation ratio of the circuit in the CPU core is thus high. Figure 11 shows the local supply noise from VMON1 embedded in the APL-RT CPU that was measured while executing the Dhrystone benchmark program. In these measurements, the cache of the APL-RT CPU was ON, and the hit ratio of the cache was 100%. This is the heaviest load for the APL-RT CPU executing the Dhrystone program. The measured maximum local supply noise was 69 mV under operation of the APL-RT CPU at 312 MHz and V_{DD}=1.25 V. In this measurement, the baseband part was powered on, but the clock distribution was stopped.

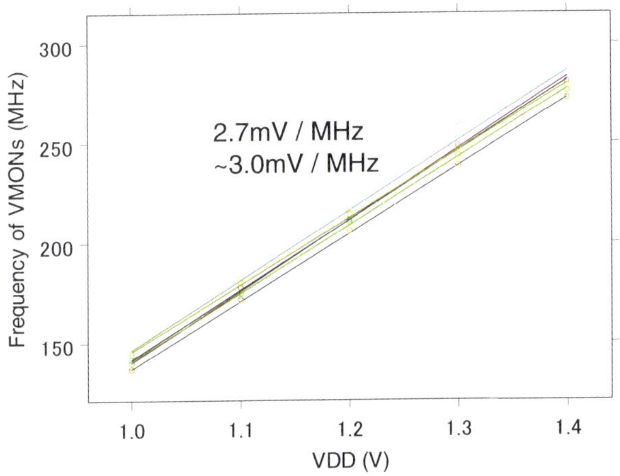

Fig. 10. Measured dependence of frequency of each VMON on voltage.

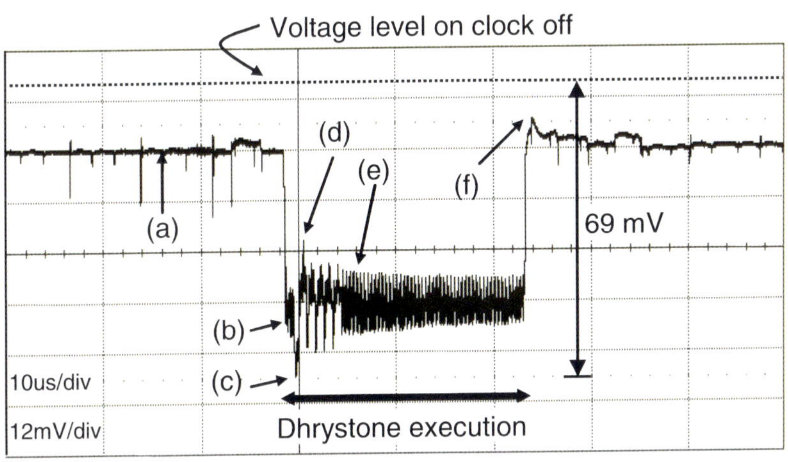

Fig. 11. Measured local supply noise by VMON1

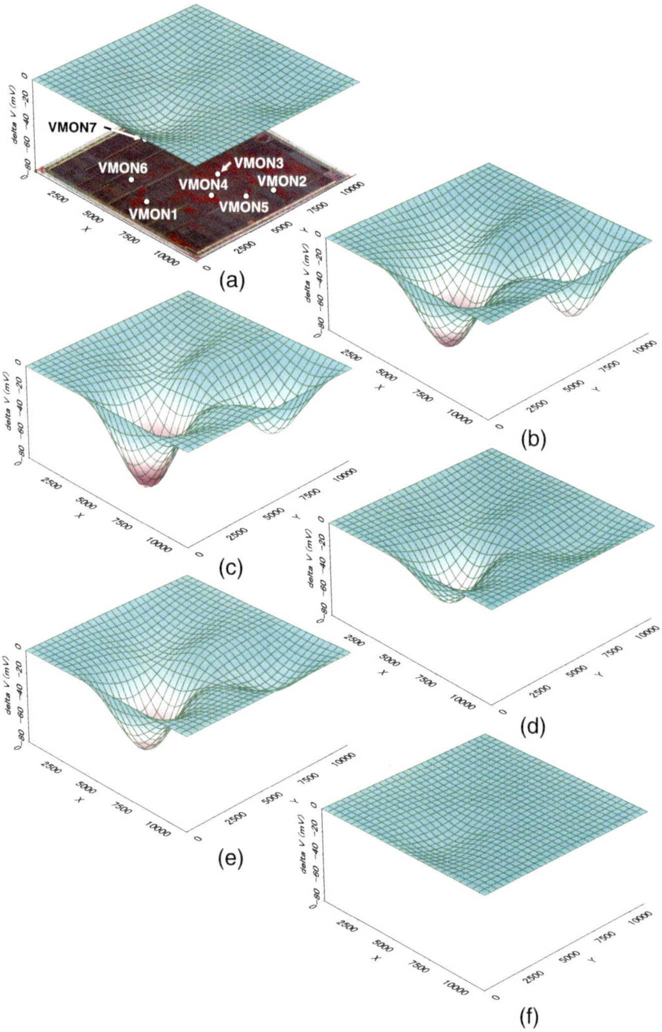

Fig. 12. Measured supply-noise maps for Dhrystone execution. (a) APL-RT CPU and AP-SYS CPU are consuming only clock power; (b) the Dhrystone program has just started in APL-RT CPU; (c) the local supply noise is at its maximum; (d) the AP-SYS CPU shows a supply "bounce" due to an inductive effect; (e) A typical situation where the APL-RT CPU executes Dhrystone and (f) both CPUs show a supply bounce due to an inductive effect. Although the seven measurement points are insufficient for showing in a 3D surface expression, this expression helps to understand the voltage relation between these points.

Figure 12 shows supply-noise maps obtained using these VMONs. Generally, although seven measurement points is insufficient for rendering in a 3D surface expression, this simple expression helps to understand the voltage relation between these points. This scheme can also produce a supply-noise-map animation, and Figs. 12(a) to (f) show snapshots of supply-noise maps corresponding to the timing points indicated in Fig. 11. Figure 12(a) is a snapshot when the CPUs are not operating but are consuming clock power. The location of each VMON is shown in Fig. 12(a). Note that the APL-RT CPU was running at 312 MHz, and the AP-SYS CPU was running at 52 MHz. Figure 12(b) is a snapshot taken when the Dhrystone program has just started. Two hot spots are clearly observed.

Figure 12(c) is a snapshot when the local supply noise is at its maximum. Figure 12(d) is an image taken when the AP-SYS CPU shows a supply "bounce" due to an inductive effect. A typical situation where the APL-RT CPU executes Dhrystone while the AP-SYS CPU is not operating but is consuming clock power is depicted in Fig. 12(e). Figure 12(f) is a snapshot when both CPUs show a supply bounce due to an inductive effect. At this time, the Dhrystone program was terminated, and both CPUs changed their operating modes, causing large current changes. It looks as if clock power consumption has vanished, although the clock remains active.

3.2 Measurement of moving picture encoding

Another measurement example involves moving picture encoding. A hardware accelerator that executes moving picture encoding and decoding (MPEG4) was implemented in this chip, as shown in Fig. 9, and VMON5 was embedded in it.

The waveform measured by VMON5 is shown in Fig. 13. In this MPEG4-encoding operation, a QCIF-size picture was encoded using the MPEG4 accelerator. In the measurement, the APL-RT CPU was running at 312 MHz, and the AP-SYS CPU was running at 208 MHz. The MPEG4 accelerator was running at 78 MHz, and V_{DD} was 1.25 V. The baseband part was powered on, but clock distribution was stopped.

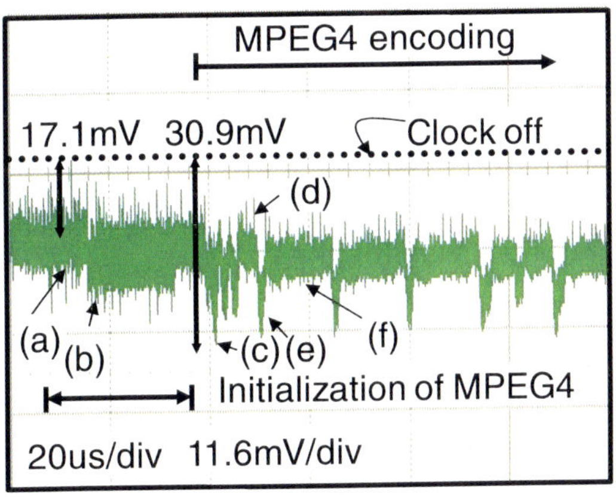

Fig. 13. Voltage noise measured while running MPEG encoding operation

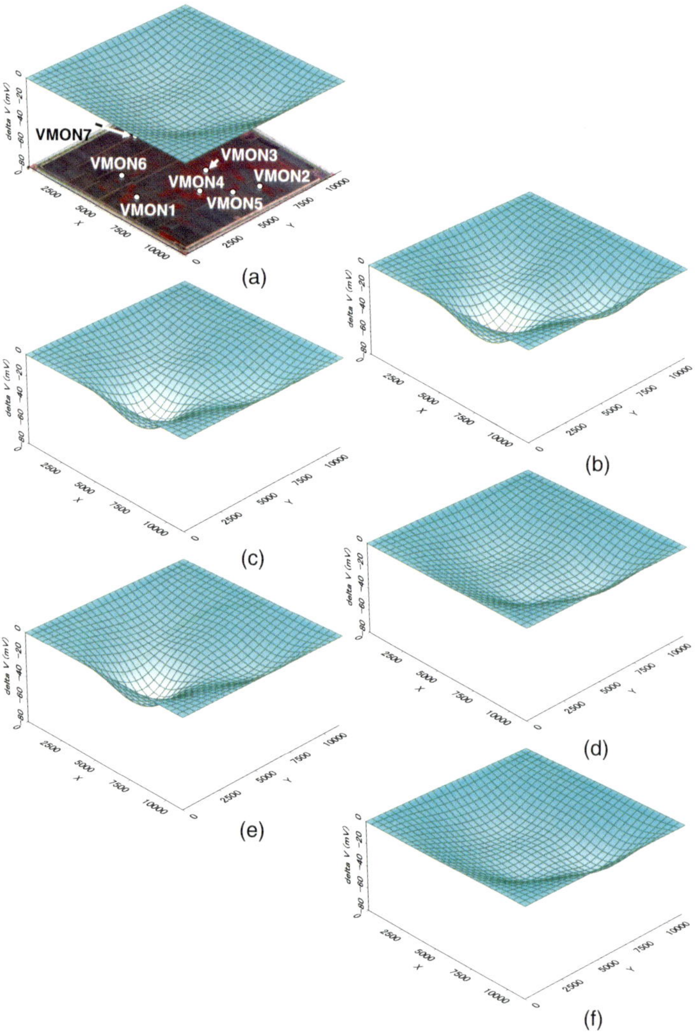

Fig. 14. Measured supply-noise maps for MPEG encoding operation. (a) neither CPU was operating but was consuming clock power; (b) the APL-RT CPU was initializing the MPEG4 accelerator; (c) the local supply noise was at its maximum; (d) the execution of the MPEG4 accelerator was dominant; (e) the APL-RT CPU was executing an interruption operation from the MPEG4 accelerator and (f) the MPEG4 accelerator was encoding a QCIF-size picture.

The maximum local supply noise measured by VMON5 was 30.9 mV, and the average voltage drop was smaller than that when executing the Dhrystone benchmark program. This result confirms that good power efficiency was attained using hardware accelerators. Measured maps of the typical situations are shown in Fig. 14. Figure 14(a) is a snapshot taken when neither CPU was operating but both were consuming clock power; it also shows the location of each VMON. Note that the APL-RT CPU was running at 312 MHz, and the AP-SYS CPU was running at 208 MHz. Figure 14(b) is a snapshot when the APL-RT CPU was initializing the MPEG4 accelerator. Figure 14(c) depicts the situation when the local supply noise was at its maximum. The image in Fig. 14(d) illustrates the period when the execution of the MPEG4 accelerator was dominant. Figure 14(e) is a snapshot when the APL-RT CPU was executing an interruption operation from the MPEG4 accelerator, and Fig. 14 (f) shows the typical situation where the MPEG4 accelerator was encoding a QCIF-size image.

This measurement was done using simple picture-encoding programs, so frequent interruptions were necessary to manage the execution of the program. However, in real situations, since operation would not be carried out with frequent interruptions , and the APL-RT CPU might be in the sleep mode, the power consumption of the APL-RT CPU would be reduced, and the map would show a calmer surface.

These results show that by using a hardware accelerator, the power consumption was also distributed over the chip, resulting in a reduction in the total power consumption. This voltage-drop map therefore visually presents the effectiveness of implementing a hardware accelerator.

4. Conclusion

An in-situ power supply noise measurement scheme for obtaining supply-noise maps was developed. The key features of this scheme are the minimal size of simple on-chip measurement circuits, which consist of a ring oscillator based probe circuit and analog amplifier, and the support of off-chip high resolution digital signal processing with frequent calibration. Although the probe circuit based on the ring oscillator does not require a sampling-and-hold circuit, high accuracy measurements were achieved by off-chip digital signal processing and frequent calibrations. The frequent calibrations can compensate for process and temperature variations. This scheme enables voltage measurement with millivolt accuracy and nanosecond-order time resolution, which is the period of the ring oscillator.

Using the scheme, we demonstrated the world's first measured animation of a supply-noise map in product-level LSIs, that is, 69-mV local supply noise with 5-ns time resolution in a 3G-cellular-phone processor.

5. Acknowledgment

This work was done in cooperation with H. Mizuno, S. Komatsu, and Y. Kondoh of the Hitachi, Ltd., and T. Irita, K. Hirose, R. Mori, and Y. Yasu of the Renesas Electronics Corporation. We thank T. Yamada and N. Irie of Hitachi Ltd., and T. Hattori, T. Takeda of Renesas Electronics Corporation, and K. Ishibashi of The University of Electro-Communications, for their support and helpful comments. We also express our gratitude to Y. Tsuchihashi, G. Tanaka, Y. Miyairi, T. Ajioka, and N. Morino of Renesas Electronics Corporation for their valuable advice and assistance.

6. References

Abramzon, V.; Alon, E.; Nezamfar, B. & Horowitz, M., *"Scalable Circuit for Supply Noise Measurement, "* in ESSCIRC Dig. Tech. Papers, Sept. 2005, pp. 463-466.

Chen, K.; Wann, H. C.; KO, P. K. & Hu, C., *"The Impact of Device Scaling and Power Supply Change on CMOS Gate Performance, "* IEEE Electron Device Letters, Vol. 17, No. 5, pp. 202 - 204, May 1996

Fukuoka, K.; Ozawa, O.; Mori, R.; Igarashi, Y.; Sasaki, T.; Kuraishi, T.; Yasu, Y. & Ishibashi, K.; "A 1.92 μs-wake-up time thick-gate-oxide power switch technique for ultra low-power single-chip mobile processors," in Symp. VLSI Circuits Dig. Tech. Papers, pp. 128-129, Jun. 2007.

Gray, R. M. & Stockham Jr. T. G.; "Dithered quantizers," IEEE Transactions on Information Theory, Vol. 39, No. 3, May 1993, pp. 805-812.

Hattori, T.; Ito, M.; Irita, T.; Tamaki, S.; Yamamoto, E.; Nishiyama, K.; Yagi, H.; Higashida, M.; Asano, H.; Hayashibara, I.; Tatezawa, K.; Hirose, K.; Yoshioka, S.; Tsuchihashi, R.; Arai, N.; Akiyama, T. & Ohno, K., *"A power management scheme controlling 20 power domains for a single chip mobile processor,"* ISSCC Dig. Tech. Papers, Feb. 2006, pp. 542-543.

Kanno, Y.; Mizuno, H.; Oodaira, N.; Yasu, Y. & Yanagisawa, K., *"µI/O Architecture for 0.13-um Wide-Voltage-Range System-on-a-Package (SoP) Designs"*, Symp. on VLSI Circuit Dig. Tech. Papers, pp. 168-169, June 2002.

Kanno, Y.; Mizuno, H.; Yasu, Y.; Hirose, K.; Shimazaki, Y.; Hoshi, T.; Miyairi, Y.; Ishii, T.; Yamada, T.; Irita, T.; Hattori, T.; Yanagisawa, K. & Irie, N., *"Hierarchical power distribution with 20 power domains in 90-nm low-power multi-CPU Processor,"* ISSCC Dig. Tech. Papers, Feb. 2006, pp. 540-541.

Kanno, Y.; Kondoh, Y.; Irita, T.; Hirose, K.; Mori, R.; Yasu, Y.; Komatsu, S.; Mizuno, H.; *"In-Situ Measurement of Supply-Noise Maps With Millivolt Accuracy and Nanosecond-Order Time Resolution,"* Symposium on VLSI Circuits 2006, Digest of Technical Papers, June, 2006, pp. 63-64.

Kanno, Y.; Kondoh, Y.; Irita, T.; Hirose, K.; Mori, R.; Yasu, Y.; Komatsu, S.; Mizuno, H.; *"In-Situ Measurement of Supply-Noise Maps With Millivolt Accuracy and Nanosecond-Order Time Resolution,"* IEEE Journal of Solid-State Circuits, Volume: 42, April, 2007, pp. 784-789.

Okumoto, T.; Nagata, M. & Taki, K., *"A built-in technique for probing power-supply noise distribution within large-scale digital integrated circuits, "* in Symp. VLSI Circuits Dig. Tech. Papers, Jun. 2004, pp. 98-101.

Saleh, R.; Hussain, S. Z. ; Rochel, S. & Overhauser, D., *"Clock skew verification in the presence of IR-drop in the power distribution network, "* IEEE Trans. Comput.-Aided Des. Integrat. Circuits Syst., vol. 19, no. 6, pp. 635-644, Jun. 2000.

Takamiya, M. & Mizuno, M., *"A Sampling Oscilloscope Macro toward Feedback Physical Design Methodology, "* in Symp. VLSI Circuits Dig. Tech. Papers , pp. 240-243, Jun. 2004.

6

Comparative Analysis of Three Digital Signal Processing Techniques for 2D Combination of Echographic Traces Obtained from Ultrasonic Transducers Located at Perpendicular Planes

Miguel A. Rodríguez-Hernández[1], Antonio Ramos[2] and J. L. San Emeterio[2]
[1]*ITACA. Universitat Politècnica de Valencia*
[2]*Lab. Ultrasonic Signal, Systems and Technologies, CSIC. Madrid*
Spain

1. Introduction

In certain practical cases of quality control in the manufacturing industry, by means of ultrasonic non-destructive evaluation (NDE), it is very difficult to detect certain types of internal flaw using conventional instrumentation based in ultrasonic transducers located on a unique external surface of the piece under inspection. In these cases, the detection problems are due to the especial flaws orientation or their spatial location, and some technological solutions for it are still pendent to be proposed.
In addition, it is convenient, in a more general scope, to improve the flaw-location in two dimensions, by using several ultrasonic transducers emitting beams from distinct places. In fact, the utilization of more than one detection transducer provides complementary information in the NDE of many pieces. These transducers can be located at the same or at different planes depending on the piece shape and the detection necessities. In any case, the result of such arrangement is a set of ultrasonic traces, which have to be carefully fussed using digital signal processing techniques in order to extract more accurate and more complete detection results.
The usual trend for reducing the mentioned limitations in flaw detection is to increase the number of ultrasonic channels involved in the testing. On the other hand, it is important to reduce this ultrasonic channels number in order to minimize technological costs. In addition, it should be noted that the detection capability also depends on other important factors, because, from a more general point of view, still some physical limitations of the ultrasonic beams remain for a) certain angles of the scanning (Chang and Hsieh 2002), b) for certain complex geometries of the industrial components to be tested (Roy et al 1999) or c) for biological elements in medical diagnosis (Defontaine et al 2004, Reguieg et al 2006).
Schemes have been preliminarily proposed in order to improve flaw detection in difficult conditions, trying to resolve these type of aspects well with two transducers and additional digital signal processing of echoes (Chang and Hsieh 2002), or well with several arrays of few elements (Engl and Meier 2002). Other posterior alternative proposals, based on perpendicular scanning from two planes with a reduced transducers number and ultrasonic

beams overlapping, were reported (Meyer and Candy 2002, Rodríguez et al 2004). But an extensive research in order to find simple and complete solutions to these problems is still needed. In particular, the authors are currently investigating techniques for ultrasonic radiation from perpendicular planes using arrays of few radiators working in near field conditions. In parallel, we are developing digital signal processing tools for improving the signal to noise ratio (SNR) in the echoes acquired in NDE of media with complex internal structure (Lázaro et al 2002, Rodríguez et al 2004a, Pardo et al 2008).

In this technological context, a set of novel ultrasonic signal combination techniques have been developed to be applied in flaw detection ultrasonic systems based on multiple transducers. These combination techniques use a spatial-combination approach from the echographic traces acquired by several transducers located at different external planes of the piece under testing. In all these alternative techniques, the A-scan echo-information, received from the different transducers involved, is fused in a common integrated two-dimensional (2D) pattern, in which, each spot displayed incorporates distinct grades of SNR improvement, depending on particular processing parameters.

In this chapter, some linear and non-linear digital processing techniques to fuse echo-traces coming from several NDE ultrasonic transducers distributed on two perpendicular scanning planes are described. These techniques are also applied to the flaw detection by using a 2D combination of the ultrasonic traces acquired from the different transducers. The final objective is to increase the detection capabilities of unfavorable-orientation flaws and also to achieve a good 2D spatial location of them.

Individual ultrasonic echo-signals are measured by sucesively exciting several transducers located at two perpendicular planes with electrical short-time pulses. Each transducer acquires a one-dimensional (1D) trace, thus it becomes necessary to fuse all the measured 1D signals with the purpose of obtaining an useful 2D representation of the material under inspection. Three combination techniques will be presented in this chapter; they are based on different processing tools: Hilbert, Wavelets and Wigner-Vile transforms. For each case, the algorithms are presented and the mathematical expressions of the resulting 2D SNRs are deduced and evaluated by means of controlled experiments.

Simulated and experimental results show certain combinations of simple A-scans registers providing relatively high detection capacities for single flaws. These good results are obtained in spite that the very-reduced number of ultrasonic channels involved and confirm the accuracy of the theoretical expressions deduced for 2D-SNR of the combined registers.

2. Some antecedents of ultrasonic evaluation from perpendicular planes

Techniques for combining ultrasonic signal traces coming from perpendicular planes have few antecedents. As a precedent of this type of scanning performed from two distinct planes, the inspection of a high-power laser with critical optic components using ultrasonic transducers situated in perpendicular planes is mentioned in (Meyer and Candy 2002). In this particular case, the backscattering noise is valueless and the method seems centred in the combination from the arrival time of the ultrasonic echoes, and thus the combination is made with a time domain technique.

In (Rodríguez et al 2004), a testing piece containing a flaw was evaluated by using transducers located at two scanning planes. In this case, the receiving ultrasonic traces contain backscattering noise and the combination was performed in the time domain. Two combination options were there presented: one based on a 2D sum operator and the other

using a 2D product operator. The SNR was used as a quality index to evaluate both methods; and the resulting evaluation data showed a better performance of the product operator. Nevertheless, their performances were limited in both cases by the time representation of the signals.
A technique in this same line that introduces the combination in the time-frequency domain, based on the Wigner-Ville transform (WVT), was preliminary applied in (Rodríguez 2003). This technique took into account the temporal and the frequency information of the ultrasonic traces. A better SNR result than with the time domain method (Rodríguez et al 2004) was obtained. But this option presented two drawbacks: a lost of linearity of the processed signals and a high computational cost.
In (Rodríguez et al 2004b) a new method was presented, performing the combination in the time-frequency domain with a low computational cost by the use of a linear transform (based on the wavelet transform (Daubechies 1992); its 2D SNR performance seemed to be closed to that obtained in (Rodríguez 2003) with Wigner-Ville transforms.
The present chapter summarizes these three combination techniques previously proposed by the authors for flaw detection from perpendicular transducers. A comparative analysis (based on theoretic and experimental results) of their performances over a common set of specific experiments is made. The objective is to establish the respective advantages and inconveniences of each technique in a rather rigorous frame. For experimental evaluations, we have arranged an ultrasonic prototype to generate (from 2 planes) ultrasonic near-field beams collimated along the inspected piece, and to acquire the echoes from the transducers involved in our experiments. The different combination results calculated in each case, from the measured echo-responses, will be discussed.

3. Description of processing techniques for combination. Expressions of SNR

A number of distinct combination techniques to fuse several ultrasonic traces, coming from perpendicular transducers, have been proposed by the authors. There are two important parameters that define all these techniques: a) the initial type of the traces representation, and b) the particular operator utilized in their combination process.
To choose the best representation for the processing of signals is an open general problem with multiples solutions; the two most popular representations are in time or in frequency domains: a) the direct time domain is very useful for NDE problems because the spatial localization of possible defects or flaws (in the material under testing) is closely related with the apparition time of the echoes; b) the frequency domain is less used in this type of ultrasound based applications because does not permit a spatial localization; in addition, the spectrum of the ultrasonic information with interest for testing in some industrial applications, is almost coincident with the mean spectrum of the "grain" noise originated from the material texture, which some times appears corrupting the signals waveforms associated to the investigated reflectors.
An interesting possibility for introducing spectral information in these applications is the use of time-frequency representations (Cohen 1995) for the echo-graphic signals. This option shows in a 2D format the time information for the different frequency bands in which the received ultrasonic signals range. Therefore, each point of a 2D time-frequency representation corresponds with one spectral frequency and with one time instant. Two different time-frequency techniques, the wavelet transform (Daubechies 1992, Shensa 1992)

and the Wigner-Vile transform (Claasen and Mecklenbrauker 1980), will be applied in the following as complementary tools during the combination procedure.
In relation to the other abovementioned parameter defining the combination techniques, several operators to make the trace combination have been used in previous author's works: maximum, minimum, mean, median, sum and product. Theoretical and experimental results obtained by applying these operators indicate that the best performances obtained, for all the combination methods, were produced when a product operator was employed.
For this reason, we have selected (among all the possible operators) the 2D product between echo-traces, in order to properly perform the comparison among all the methods considered in this paper. In the following, the three alternative processing techniques proposed for trace combination are described, showing their performance in relation to the resultant SNR.

3.1 Time-domain combination technique

This first technique performs the combination using the envelope of the ultrasonic traces. The first step in this method is the acquisition of the traces from the ultrasonic transducers involved, which are located over two perpendicular planes in the external part of the inspected piece. The following step is the matching in time of all the different pairs of traces, each one with echo-information corresponding to precisely the same volumetric elemental area, i.e. coming from the two specific transducers which projections define such area. To reduce problems due to no perfect synchronization of the two matched traces in those pairs, the signal envelopes are utilized instead of the original signals, because this option is less sensitive to little time-matching errors. These envelopes are obtained by means of applying them the Hilbert transform. The final step is the trace combination process, by using the mentioned 2D product operator.
Briefly, the method can be resumed in four successive steps: first, the collection of the traces from the different transducers; second, the traces envelope calculation; third, the matching between the information segments of each perpendicular transducers specifically related to the same inspection area; and fourth, the combination among all the segment couples. The corresponding functional scheme is presented in Figure 1 for the particular case of four

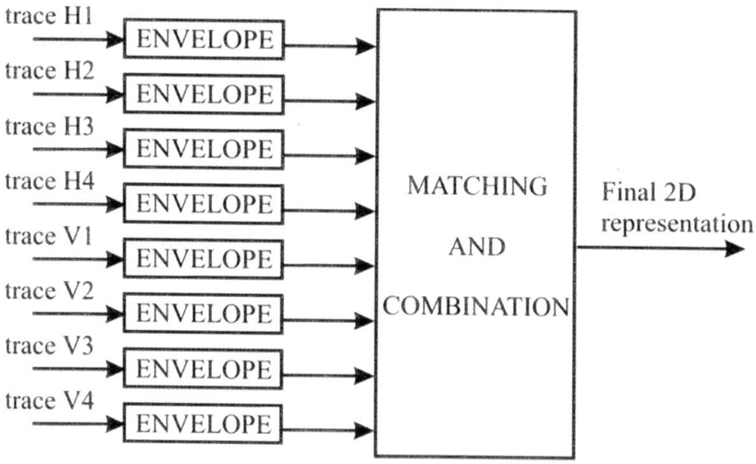

Fig. 1. Functional scheme of the time-domain echo-traces combination technique.

ultrasonic transducers (H1, H2, H3 and H4) with horizontal propagation beams and four transducers (V1, V2, V3 and V4) with vertical propagation beams.
Some theoretical characterizations of this method, including statistical distributions of the combined noise and some results about SRN enhancements were presented in (Rodríguez et al 2004). The more important result of that work is the expression of the resulting SNR for the 2D ultrasonic representation after the combination process.
The SNR of the initial traces, SNR_{ini}, containing an echo-pulse and noise, is defined as:

$$SNR_{ini}(dB) = 10\log\frac{\frac{1}{M}\sum_{i=1}^{M}(p(i))^2}{\frac{1}{L}\sum_{i=1}^{L}(n(i))^2} \tag{1}$$

where, p denotes the echo-pulse and n represents the noise; M is the length of the pulse and L is the length of the whole ultrasonic trace.
The SNR of the final 2D representation is:

$$SNR_{2D}(dB) = 10\log\frac{\frac{1}{M^2}\sum_{i=1}^{M}\sum_{j=1}^{M}(p_{2D}(i,j))^2}{\frac{1}{L^2}\sum_{i=1}^{L}\sum_{j=1}^{L}(n_{2D}(i,j))^2} \tag{2}$$

where, p_{2D} and n_{2D} denotes the 2D representation of the echo-pulse and of the grain noise; M and L are the dimensions of the 2D rectangular representations of the echo-pulse and of the ultrasonic trace, respectively.
The SRN of the 2D representation obtained by using this time-domain combination method, SNR_{2Dtime}, can be expressed as a function of SNR_{ini} :

$$SNR_{2Dtime}(dB) = 2 \cdot SNR_{ini}(dB) \tag{3}$$

In consequence, the resulting SNR with this method, SNR_{2Dtime} , expressed in dB, is the double of the initial SNR of the A-scans before combination (SNR_{ini}).

3.2 Linear time-frequency combination technique

The time-domain traces combination, previously described, works without any frequency consideration. In order to obtain a further improving of SNR, it would be necessary to use some type of processing in the frequency domain. Nevertheless, the ultrasonic echoes coming from flaws in some NDE applications, and the grain noise produced by the own material structure, have similar global mean spectra, which difficult the flaw discrimination in the frequency domain. But if these spectra are instantaneously analyzed, it can be observed that the instantaneous spectrum is more regular for echo-signal than for grain noise. The tools that permit the analysis of these differences between signal and noise are the time-frequency representations, which can be obtained by using a linear or also a non-linear transformation.
In this section, we will deal with the application of linear time-frequency representations to improve our signal-combination purpose. The two most popular linear time-frequency

representations are the Short-Time Fourier Transform and the Wavelet transform (Hlawatsch and Boudreaux-Barlets 1992). Both types of transforms can be implemented in an easy way by means of linear filter banks.

In the present linear technique, the combination process begins with the time-frequency representation of the all the acquired ultrasonic traces. A linear time-frequency transform is applied and the frequency bands with maximum ultrasonic energy are selected in each trace. The number of selected bands will be denoted as *L*. At this point, we have to resolve *L* problems similar to that presented in the previous time-domain combination method. In this way, *L* separate 2D displays are produced, one for each frequency band. The final step is the combination of these 2D displays by using a point-to-point product of them. A simple case, where combination is performed by selecting the same frequency bands for all the transducers, was presented in (Rodríguez et al 2004b), but also it could be possible to make the combination by using different bands for each transducer. The method scheme is presented in the Figure 2 for 4 horizontal and 4 vertical transducers.

Here, the combination for each frequency band is similar to the case of the time-domain technique. Then, it will be necessary to make the following steps: a) to match in time the common information of the different transducer pairs (for each frequency band), b) to calculate the time-envelope for all the bands selected in each trace, c) to perform the combinations obtaining several one-band 2D representations, and d) to fuse all these 2D displays, so resulting the final 2D representation.

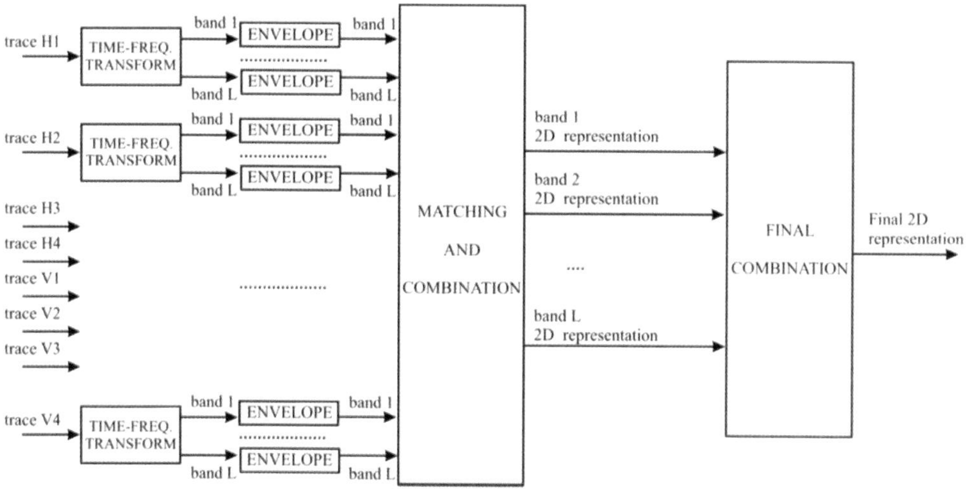

Fig. 2. Functional scheme of the linear time-frequency traces combination technique

The matching process can be common for all the frequency bands if the point number of the initial traces is maintained and if the delays of the filtering process are compensated in each band. The SNR of the 2D representation of each individual band, $SNR_{2DTFlinear}^{(band-i)}$ is obtained from expression (3).

$$SNR_{2DTFlinear}^{(band-i)}(dB) = 2 \cdot SNR_{ini}(dB) \tag{4}$$

The final global SNR, after the combination of all the 2D displays belonging to the different frequency bands, $SNR_{2DTFlinear}$, can be obtained supposing that the 2D representations for each band are independent and perfectly synchronized (Rodríguez et al 2004b):

$$SNR_{2DTFlinear}(dB) = 2 \cdot L \cdot SNR_{ini} \quad (dB) \tag{5}$$

being, L, the number of the selected frequency bands.

Consequently, in this case, the resulting $SNR_{2DTFlinear}$ presents an important factor of improvement over the SNR_{ini} . This factor is the double of the number of frequency bands used in the combination. It must be noted that comparing expressions (5) and (3), the SNR improvements is multiplied by L, but the computational complexity of the algorithm is also multiplied by the same factor L. In the experimental results section of this chapter, the accuracy of this expression will be confirmed comparing (5) with simulations using as linear time-frequency tool the undecimated wavelet packet transform (Shensa 1992, Coifman and Wickerhauser 1992). In any case, it must be noted that this expression is also valid for any linear time-frequency transform.

3.3 Wigner-Ville Transform (WVT) combination technique

The non-linear time-frequency distributions present some advantages over linear transforms, but some non-linear terms ("cross-terms") appear degrading the quality of the distributions and usually the computational cost is incremented. One of the most popular non-linear time-frequency representations is the Wigner-Ville transform (WVT) (Claasen and Mecklenbrauker 1980), which has been previously utilized in ultrasonic applications with good results (Chen and Guey 1992, Malik and Saniie 1996, Rodríguez et al 2004a).

The WVT presents an useful property for dealing with ultrasonic traces: its positivity for some kind of signals (Cohen 1995). In order to illustrate the suitability of this transform for the processing of the ultrasonic pulses typical in NDE applications, we will show that they fulfil that property. For it, an ultrasonic pulse-echo like to those acquired in such NDE equipment can be approximately modelled by the following expression:

$$p(t) = A \cdot e^{-(at^2/2)} \cos(\omega_0 t) \tag{6}$$

where A is the pulse amplitude, a is a constant related to the duration and bandwidth of the pulse (a>0), and ω_0 is the central frequency of its spectrum.

The WVT of the ultrasonic pulse modelled by (6) is (Rodríguez 2003):

$$WVT_p(t,\omega) = \frac{A^2}{(a\pi)^{\frac{1}{2}}} \cdot e^{-(at^2/2)-(\omega-\omega_0)^2/a} \tag{7}$$

The expression (7) shows that the WVT of an ultrasonic pulse with Gaussian envelope has only positive values. The chirp with Gaussian envelope is the most general signal for which the WVT is positive through-out the time-frequency plane (Cohen 1995). The ultrasonic grain noise does not carry out this property, so resulting that the sign of the WVT values represents a useful option to discriminate this type of difficult-to-eliminate noise of the echo pulses coming from real flaws.

The combination method begins in this case by calculating the WVT in all the ultrasonic traces. After the band selection is performed, the negative values (that correspond mainly with noise) are set to cero. For each frequency band, a combination is made by using the 2D product operator, like as it was used in the time-domain combination technique. The final 2D representation is obtained with a point to point product of all the 2D displays related to the different frequency bands. A functional scheme of this WVT based combination method is presented in the Figure 3, for the case of eight transducers considered in this section.

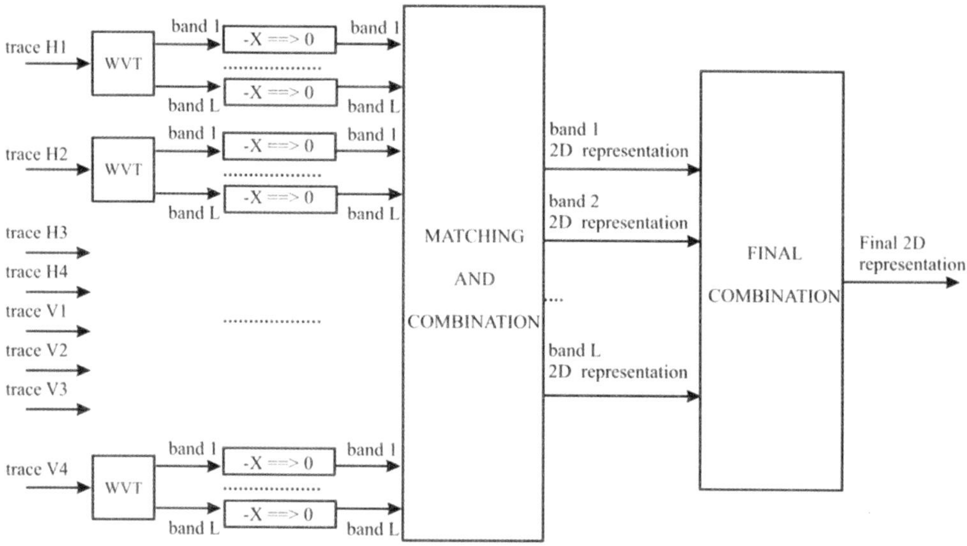

Fig. 3. Functional scheme of the WVT traces combination method.

A good estimation of the resulting SNR for the 2D representation in this WVT case, SNR_{2DWVT}, can be obtained from the results presented in (Rodríguez 2003):

$$SNR_{2DWVT}(dB) \cong 3 \cdot L \cdot SNR_{ini} \ (dB) \tag{8}$$

Therefore, the improvement factor of the SNR, expressed in dB, which can be obtained by this WVT method, is the triple of the number of frequency bands that had been selected.

In consequence, the theoretic improvement levels in the SNR provided by the three alternative techniques for combining ultrasonic traces coming from two perpendicular transducers, (i.e., the basic option using traces envelope product, and the others two options based on linear time-frequency and WVT trace decompositions), are quite different.

So, the quality of the resulting 2D combinations, in a SNR sense, is predicted to be quite better when time-frequency decompositions are chosen, and the best results must be expected for the Wigner-Ville option, which in general seems to be potentially the more effective processing. Nevertheless, in spite of these good estimated results for the WVT case, it must be noted that in general this option supposes higher computational cost. Therefore, the more effective practical option should be decided in each NDE situation depending on the particular requirements and limitations in performance and cost being needed. In the following sections, the confirmation of these predictions will be carried out, by means of several experiments from simulated and measured ultrasonic traces.

4. Protocols used in the different testing experiments

Two types of experiments (I and II) have been designed with the purpose of evaluating and comparing the three trace combination methods presented in the previous section. The comparison will be performed over the same set of ultrasonic traces for the three cases. The type-I experiments are based on simulated noisy ultrasonic traces and those of type-II use experimentally acquired echo-traces. The protocols used in these experiments are an extension of those we have planned in references (Rodríguez et al 2004a, Rodríguez 2003, Rodríguez et al 2004b).

4.1 Experiments type-I based on simulated noisy traces

Type-I experiments were carried out with simulated signal registers. They provide adequate calculation results to confirm the accuracy of the expressions estimated from the theoretical models of the processing techniques proposed in the equations (3), (5) and (8) to predict the distinct SNRs (SNR_{2Dtime}, $SNR_{2DTFlinear}$ and SNR_{2DWVT}). So, those expressions could be validated for an ample range of values in SNR_{ini} with perfectly controlled characteristics in echo-signals and their associated grain noises. Some results, in a similar context, using these same rather simple simulated registers, have been compared in a previous work (Rodríguez et al 2004a) with the obtained results when a more accurate ultrasonic trace generator was used. A very close agreement between them was observed, which confirms the suitability of these registers to evaluate those expressions.

The testing case proposed to attain this objective is the location of a punctual reflector into a rectangular parallelepiped from 2 external surfaces, perpendicular between them, and using 4 transducers by surface. The general scheme of these experiments, with 4 horizontal (H1, H2, H3, H4) and 4 vertical (V1, V2, V3, V4) transducers is depicted in the Figure 4. Transducers H3 and V2 receive echoes from the reflector whereas the other transducers (H1, H2, H4, V1, V3 and V4) only receive grain noise. To assure compatibility of experiments type-I with experiments type-II, ultrasonic propagation in a piece of 24x24 mm has been simulated assuming for calculations a propagation velocity 2670 m/s very close to that corresponding to methacrylate material. The sampling frequency was 128 MHz.

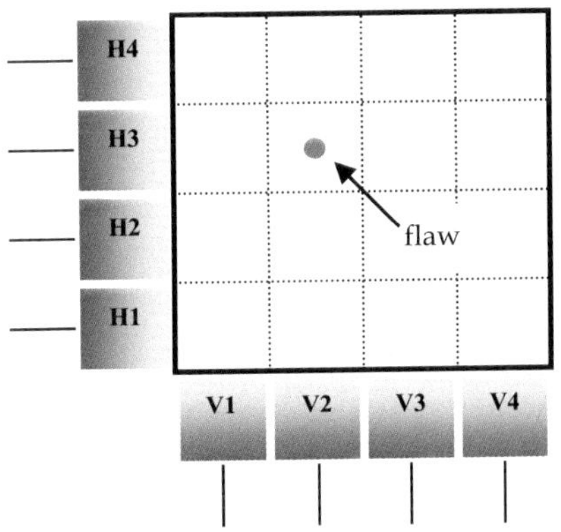

Fig. 4. Geometry of the inspection case planned to evaluate the different combination methods: detection of a single-flaw in a 2D arrangement with 16 elemental-cells.

The simulation of the echo-traces produced by the reflector was made by integrating a real echographic signal with a synthetic noise-component similar to the grain reflections registered in some industrial inspections, and that are quite difficult to be cleaned. The echographic echo was acquired from one of the 4 MHz transducers of the perpendicular array used for experiments type-II. The sampling frequency was 128 MHz. The echo is shown in figure 5. The "coherent" grain noise, to be associated with the basic echo-signal, was obtained by means of a synthetic white gaussian noise generator. To assure the frequency coherence with the main reflector echo-pulse (simulating an unfavourable case), this initial noise register was passed thought a digital filter just having a frequency response as the ultrasonic echo-pulse spectrum. Finally, the composed traces containing noisy echoes are obtained by the addition of the real echo-signals with the synthetic noise register. Previously, the noise had been unit power normalized and the echo-signal had been multiplied by a constant with the finality of obtaining the desired SNR_{ini}.

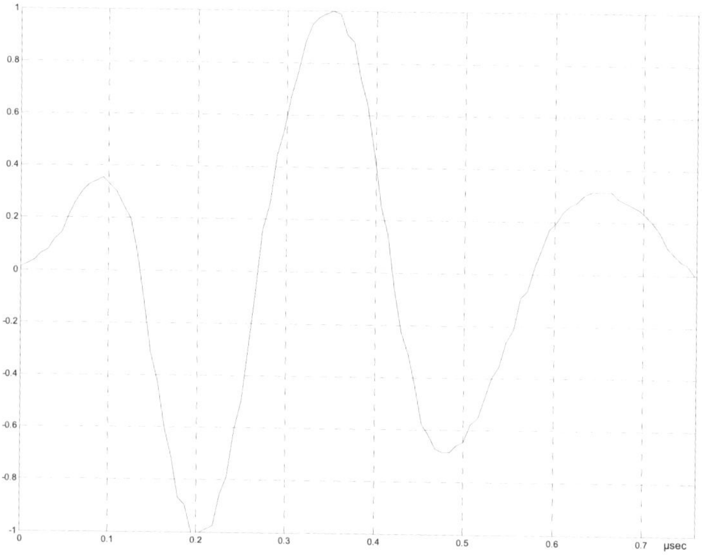

Fig. 5. Ultrasonic echo utilised in type-I experiments.

Several sets of tests were prepared with 11 different SNR_{ini} (0, 1, 2, 3, 4, 5, 6, 7, 8, 9 and 10 dB). For each SNR_{ini}, 10.000 tests were performed using the three combination methods described in section 3, and their respective results were compared. The length of the each individual ultrasonic trace was of 2304 points (corresponding to 18 microseconds with a sampling frequency of 128 MHz). 18 microseconds is the time of flight of 48 (24 +24) mm with a propagation velocity of 2670 m/s, very close to the total echo length from the methacrylate piece considered in experiments. The length of the echo-signals contained in these traces was of 98 samples. The size of the final 2D representation is 2304x2304 (5308416) points (corresponding with an inspected area of 24x24 mm). Thus, from 18432 initial points (2304 by transducer), a 2D display with 5308416 points was obtained for the whole piece. To measure the different SNR's, the echo-signal power was measured over its associated area 98x98 points in the 2D display, whereas for the noise power, the rest of the 2D display points were used.

4.2 Experiments type-II with echographic traces measured from an ultrasonic prototype

The type-II experiments are based on real ultrasonic echoes measured from an isolated-flaw (hole drilled in a plastic piece) with a multi-channel ultrasonic prototype designed for this kind of tests in laboratory. The two array transducers are disposed in a perpendicular angle and the square plastic piece with the hole are inside and in contact with the radiation area of arrays. There are 4 broadband elemental transducers in each perpendicular array, 8 in the whole system. Transducers work in the 4 MHz frequency band range. The dimensions of the emitting surface of each individual transducer are 6x6 mm, being 24 mm the total length of both arrays. Then, the area of the methacrylate piece to be inspected by the ultrasonic system is 24x24 mm. Arrays manufacturing was ordered to the Krautkramer company. The methacrylate piece has a drilled cylindrical hole in a position similar as used in experiment type I. Then, simulations of experiment type-I are almost coincident with real measurements of experiment type-II. The main difference is that methacrylate generates a very low level of ultrasonic grain noise. Figure 6 shows the disposition of transducers and inspected piece.

Fig. 6. Perpendicular array transducers and the inspected plastic piece with the hole.

In all the measurement cases, the transducers are driven for transmission and selected for echoe reception in a sequential way. We deal with near field radiations and only one transducer emits and receives at the same time, in our eight-shots successive measurement process. Thus, among all the echoes produced by the isolated reflector in each transducer shot, only those received in the two transducers located in front of the reflector (at the perpendicular projections of the flaw on the horizontal and vertical apertures) will be captured, because, in each shot, the echoes acquisitions are cancelled in the other seven transducers. Additionally, these two transducers in front of the reflector could receive certain amount of noise. And under these conditions, the rest of transducers of the two array apertures, in each plane, only could eventually acquire some noise signal during its shot, but not echoes from the reflector hole. Concretely, in the flaw scheme of the figure 4 (before shown for the simulated type-I experiments), the pulsed-echoes from the discontinuity of the reflector will be received by transducers H3 and V2 (with the apparition time of these echoes being determined by the distance to each transducer and the sound propagation velocity in the piece), and the traces in H1, H2, H4, V1, V3 and V4, will not contain flaw reflections.

For measurements, an experimental prototype, with eight ultrasonic transceivers, has been arranged for the validation and comparative assessment of the three flaw localization techniques by 2D traces combination in a real NDE context. It includes as emitter-receiver probes two 4 MHz piezoelectric linear arrays of 4 elements each one (as it is shown in figure 6), which are controlled by a Krautkramer NDE system model USPC-2100, disposed in the pulse-echo mode. The main characteristics of this NDE system in the signal receiving stage are the following: a dynamic range of 110 dB; a maximum effective sampling of 200 MHz in the digitalizing section. A signal gain of 44 dB and a sampling rate of 128 MHz were selected in reception for all the signal acquisitions performed in this work. Other general characteristics of this system are: pulse repetition rate of up to 10 KHz per channel, and 15 MHz of effective bandwidth in emission-reception. The high-voltage pulser sections of this commercial system were programmed in order to work with the highest electric excitation disposable for the driven transducers, which is about 400 Volts (measured across a nominal load of 100 Ohm). A relatively low value for the E/R damping resistance of 75 Ohm was selected looking for the assurance of a favourable SNR and a good bandwidth in the received echoes. Finally, the maximum value offered by this equipment for the energy level, contained into the driving spike, was selected.

It must be noted that in the experimental ultrasonic evaluations performed with the two arrays, their elemental transducers were operated with the restriction of that only one transducer was emitting and receiving at the same time. So, the two transducers located in front of the flaw (in this case: transducers H3 & V2) were operated separately as receivers in order to obtain useful information from the artificially created flaw (by drilling the plastic piece), which is clearly smaller than transducer apertures. Thus, only ultrasonic beams of H3 & V2 transducers (which remain collimated into a 6 mm width due to the imposed near-field conditions) attain the hole, whereas the other six elemental transducers radiate theirs beams far away of that hole, and therefore, in any case, they are not covering the artificial flaw and are not receiving echoes reflected from this flaw during their acquisition turns.

5. Simulated and experimental flaw detection results for the three combination techniques. Discussion of their performance

Three sets of experiments are shown in this section. First, the results related to the final SNR calculated for seven type-I simulated experiments using different combination options will be presented in the first section part. Second, 2D displays about the location of an isolated reflector, calculated for a particular combination case and a small SNR_{ini} are also shown. Third, as results illustrating the type-II experiments, 3 pairs of representations of a real flaw obtained by means of the 3 different combination techniques of section 3 will be shown and commented, analyzing the respective performances of the three techniques. The initial data for these type-II experiments were a set of measured ultrasonic traces acquired with the ultrasonic set-up of section 4.

The first tasks in type-I experiments (with simulated traces) were performed to confirm the accuracy of expressions (3), (5) and (8). In these experiments, 11 SNR_{ini} were selected (0, 1, 2, 3, 4, 5, 6, 7, 8, 9, 10). 10.000 sets of measures were generated using a real 4 MHz echo response sampled at 128 MHz and synthetic noise, composed in this case by 66.66% of white Gaussian noise (accounting by the "thermic" noise induced by the usual electronic instrumentation) and 33.34% of coherent noise (accounting by "grain" noise tied to material texture). Seven experiments were realized: 1 with time domain technique, 3 based on linear time-frequency decomposition using 2, 3 and 4 bands, and finally 3 utilising WVT with 2, 3 and 4 band again. The SNR after the 7 experiments were measured. The results are exposed in Tables 1 and 2, together with the values expected from expressions (3), (5) and (8).

In the first column of Tables 1 and 2, the initial SNR, SNR_{ini} of the ultrasonic traces are shown. The experiment 1 in the Table 1 was planned in order to measure the behaviour of the 2D time-combination method in terms of SNR_{2Dtime} improvement. The experiments number 2, 3 and 4 had as objective to evaluate the accuracy of the expression $SNR_{2DTFlinear}$ corresponding with the linear time-frequency combination. The difference among these 3 cases is the number of bands utilized [parameter L in expression (5)]; thus, the experiments 2, 3 and 4 were performed with 2, 3 and 4 bands respectively. The particular linear time-frequency transform used in these latter experiments was the undecimated wavelet packet transform, (Mallat 1989, Shensa 1992, Coifman and Wickerhauser 1992), with Daubechies 4 as mother wavelet, as it was used in the work (Rodríguez et al 2004b) but with some new adjusts included in this case, which provide a better agreement (as it can be seen in Table 1) between estimated and measured expressions of $SNR_{2DTFlinear}$ that in the mentioned work.

Finally, experiments 5 to 7 in Table 2 show the improvements obtained by using the WVT transform in the combination. The differences among these 3 WVT experiments are again the number of bands being involved: 2, 3 or 4, respectively. The SNR related to these 7 experiments are presented in Table 1 and Table 2. The expected SNRs estimated from their theoretic expressions, together with the measured SNRs, are detailed for each case. The measured SNR values, which are shown in these tables, were calculated as the mean of different 10.000 SNRs obtained for each set of simulated traces.

SNR_{ini} (dB)	SNR_{2Dtime}(dB) experiment 1		$SNR_{2DTFlinear}$(dB) 2 bands experiment 2		3 bands experiment 3		4 bands experiment 4	
	Est.	Meas.	Est.	Meas.	Est.	Meas.	Est.	Meas.
0	0	0.11	0	0.34	0	0.05	0	0.75
1	2	2.08	4	3.53	6	5.72	8	8.81
2	4	4.07	8	7.62	12	11.54	16	16.63
3	6	6.06	12	11.46	18	17.53	24	24.57
4	8	8.11	16	15.42	24	23.41	32	32.26
5	10	9.97	20	19.39	30	29.34	40	40.44
6	12	12.01	24	23.43	36	35.28	48	48.42
7	14	14.11	28	27.38	42	41.23	56	56.24
8	16	16.13	32	31.34	48	47.31	64	64.25
9	18	18.16	36	35.32	54	53.24	72	72.17
10	20	20.08	40	39.33	60	59.27	80	80.43

Table 1. SNRs of the 2D representations obtained by means of the experiments 1 to 4.

SNR_{ini} (dB)	SNR_{2DWVT}(dB)					
	2 bands experiment 5		3 bands experiment 6		4 bands experiment 7	
	Est.	Meas.	Est.	Meas.	Est.	Meas.
0	0	4.93	0	8.64	0	12.88
1	6	8.90	9	12.81	12	18.08
2	12	11.91	18	19.01	24	25.31
3	18	16.76	27	28.02	36	38.92
4	24	21.63	36	35.70	48	50.45
5	30	27.65	45	45.32	60	64.33
6	36	34.63	54	56.13	72	80.90
7	42	41.53	63	63.17	84	94.67
8	48	48.91	72	78.46	96	111.31
9	54	56.88	81	90.69	108	127.91
10	60	64.24	90	101.73	120	142.04

Table 2. SNRs of the 2D representations obtained by means of the experiments 5 to 7.

The estimated and measured values of the SNR_{2Dtime} (Table 1, columns 2 and 3) and $SNR_{2DTFlinear}$ ratios, obtained for 2 bands (Table 1, columns 4 and 5), 3 bands (Table 1, columns 6 and 7) and 4 bands (Table 1, columns 8 and 9), present a very good agreement. Finally, the SNR_{2DWVT} (Table 2) for different bands number show a high correlation between estimated and measured values, but in some cases small differences appear. These are due to the fact that the estimated expression for SNR_{2DWVT} was obtained by means of approximations, but in any case, the global correspondence between estimated and measured values is also reasonably good.

Apart from SNR improvements, the three techniques described in this chapter allow the accurate detection of flaws inside pieces.

A second type-I experiment was realised to show this good accuracy in the defect detection capability inside the pieces. A new set of ultrasonic traces was generated, simulating again a hole in a rectangular piece as it is depicted in figure 4. In this case, the selected SNR_{ini} of the initial A-scan was 3 dB.

The echo is the real 4 MHz trace sampled at 128 MHz, and the noise contained in the initial eight traces was composed by white noise and coherent noise with amplitudes of 50% each one. This set of simulated measures is displayed in figure 7, being the units shown in horizontal axis micro-seconds. In these graphics, it can be appreciated that noise and echo amplitudes are similar, thus it is very difficult to distinguish the reflector echo from the noise. In fact, the echo only appears in graphics corresponding to transducers H3 and V2. The real echo-pulse of H3 transducer is located in the middle of the noise beginning approximately at 5.5 microseconds whereas the echo-pulse of V2 transducer begins around 10.75 microseconds.

Using the ultrasonic registers of figure 7, the three combinations of the traces by applying the different techniques exposed in the chapter were performed. The first combination was done using the time domain method and the resulting 2D representation is shown in figure 8.a., where the 24x24 mm inspected area is displayed (the axis units are in mm). The searched hole location is around 8 mm in horizontal axis and 15 mm in vertical axis. It can

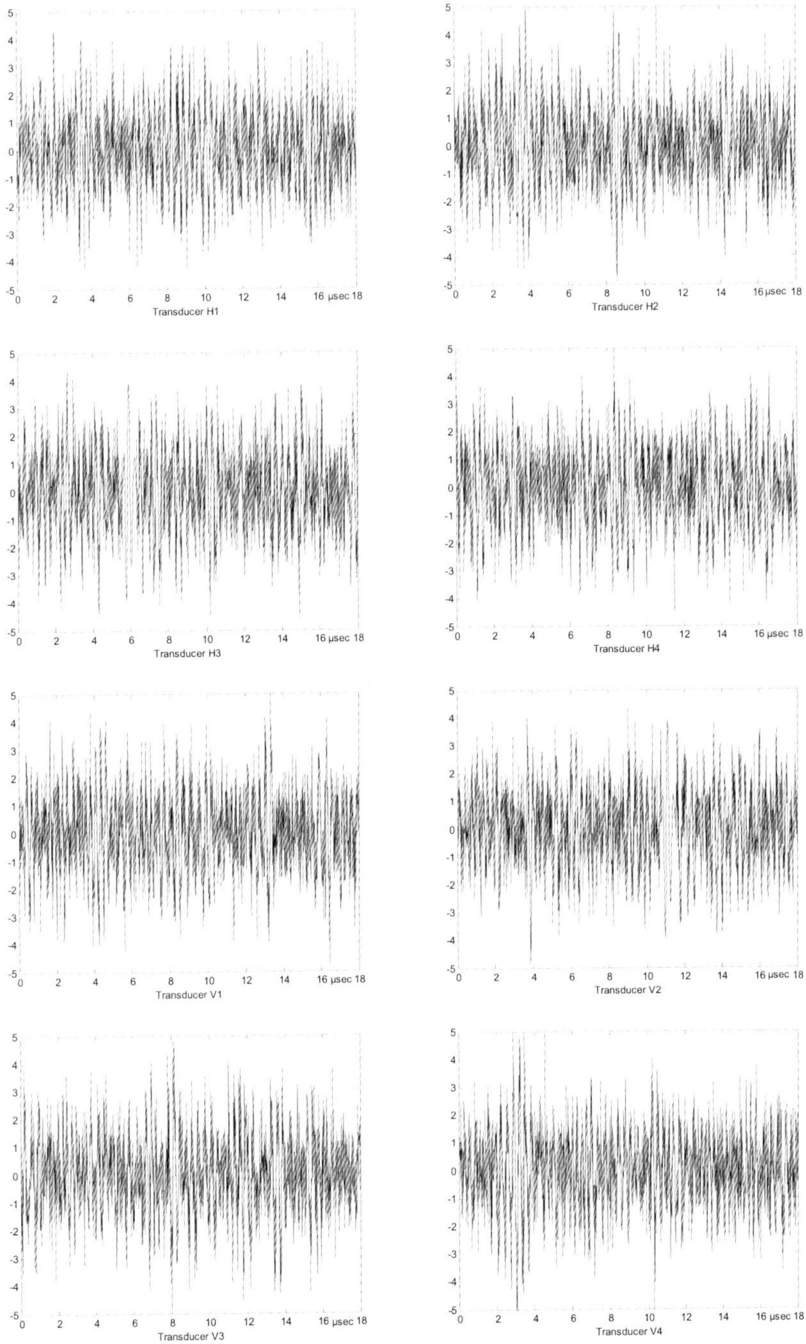

Fig. 7. Ultrasonic traces from the 8 transducers of figure 4 with a simulated SNR_{ini} = 3 dB.

be deduced that by using this time domain technique, the flaw is not very well marked and a lot of noise appear, but it is must taken into account that, in the initial traces shown in figure 7, the echo level was under noise level, in some cases.

The linear time-frequency transform used for second combination in this comparative analysis was the undecimated wavelet packet transform with Daubechies 4 as mother wavelet, as in the previous set of experiments. Figures 8.b, 8.d and 8.e show the 2D representations obtained using wavelets with 2, 3 and 4 bands. In these graphics, which amplitudes are in linear scale, it can be clearly distinguished the mark corresponding to the hole. Figure 8.f represents the same result than 8.e, but with the gray scale of amplitudes measured in dB, in order to appreciate with more detail the low levels of noise.

Finally figures 8.c, 8.g and 8.h show the 2D representations obtained using WVT with 2, 3 and 4 bands and using a linear scale for amplitudes. Figure 8.h and 8.i correspond to the same results, but figure 8.i is displayed with its amplitude scale expressed in dB. Thus, in figure 8.h, the noise has disappeared but in figure 8.i the low level noise can still be observed. It must be noted that, for all the cases, the 2D representations of figure 8 mark the flaw that we are looking for, although in the initial traces, shown in figure 7, the echoes coming from the flaw were very difficult to see.

Additionally, in the first strip of the figure 8, the 2D graphic resulting when time domain method is used, is shown. It can be seen its performance in contrast with the wavelet method with minimum quality (L=2) and WVT option with minimum quality (L=2), in such a way that a quick comparison can be made among improvements applying the different methods.

In that concerning to results of type-II experiments, displays of 2D representations, obtained by combination of experimental traces acquired from the ultrasonic prototype described in section 4 are presented in figure 9. Two scales have been used for each 2D result: linear and logarithmic scales. With the logarithmic scale, the small flaw distortions and secondary detection indications, produced by each combination method, can be more easily observed and quantified. It must be noted that the logarithmic scales have an ample resolution of 60 dB, giving a better indication of techniques performance.

In all these cases, the initial traces had a low level of grain noise because these echo-signals correspond to reflections from the small cylindrical hole drilled in a plastic piece made of a rather homogeneous material without internal grains. The patterns of figure 9 were obtained using similar processing parameters than those used with the simulated traces in the type-I experiments, and only two bands were considered for frequency decomposition. The results of the figure 9, using the time-combination method, present clear flaw distortions (more clearly visible in 9.b) with shadow zones in form of a cross, but even in this unfavourable case, a good spatial flaw location is achieved.

The mentioned crossing distortions appear already very attenuated in the results shown in figures 9.c and 9.d, corresponding to the linear time-frequency combination technique (wavelet using 2 bands), and practically disappear in the results of figures 9.e and 9.f obtained by using to the WVT combination technique.

Similar good results could be also achieved in many practical NDE cases with isolated-flaws patterns, but this performance could be not extended to other more complicated testing situations whit flaws very close among them, i.e. with two or more flaws located into a same elemental cell and thus being insonifyed by the same two perpendicular beams. Under these more severe conditions, some ambiguity situations, with apparition of "phantom" flaws, could be produced [Rodríguez et al 2005]. We are working order to propose the extension of this type of ultrasonic traces combination methods (using perpendicular NDE transducers) from echoes coming from two ultrasonic imaging array apertures, where this particular restriction (for only isolated reflectors) will be solved, by means of an improved procedure, that includes an additional processing step involving additional echographic information acquired not only from the emitting transducers.

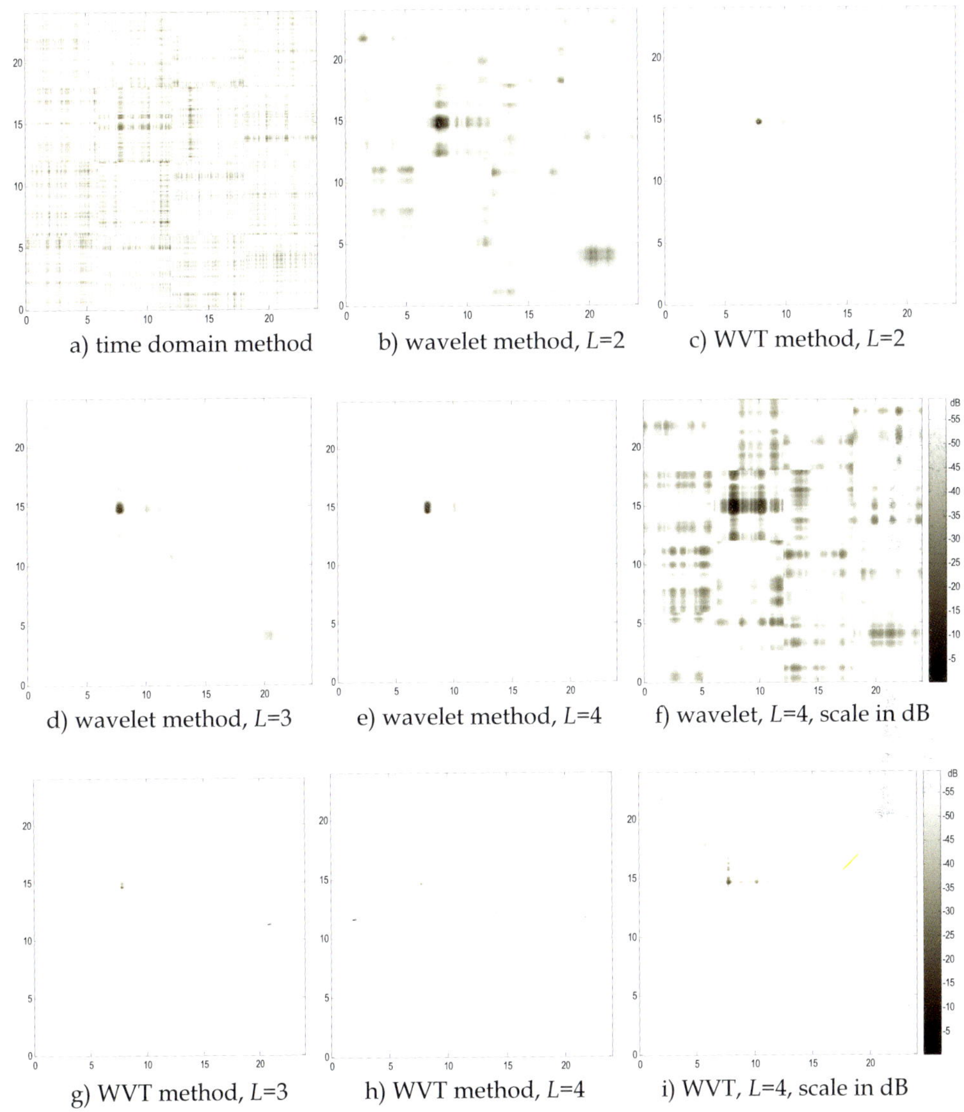

Fig. 8. Different 2D representations, after the combination of the traces shown in the figure 7; different methods and L values were used.

6. Conclusion

Three variants of a recent digital signal processing procedure for ultrasonic NDE, based on the scanning with a small number of transducers sized to work in near field conditions (located at two perpendicular planes to obtain different ultrasonic perspectives), are evaluated. They originate distinct techniques to fuse echo information coming from two planes: time-domain, linear time-frequency, and WVT based, 2D combination methods.

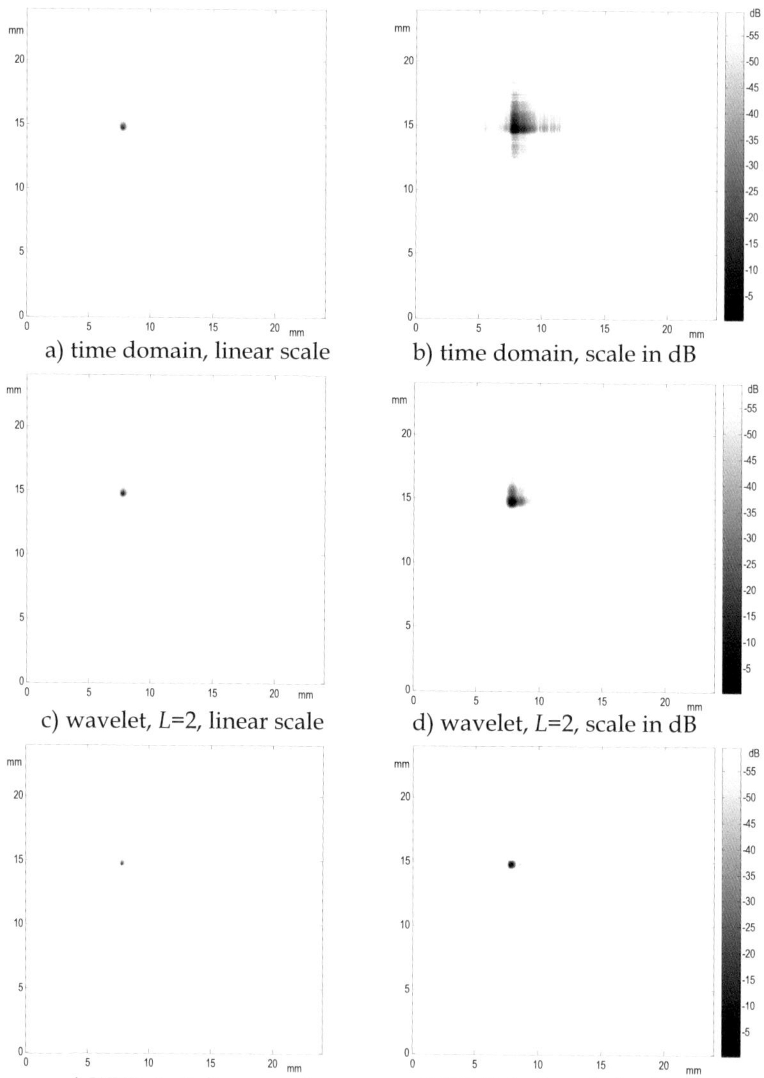

a) time domain, linear scale   b) time domain, scale in dB

c) wavelet, L=2, linear scale   d) wavelet, L=2, scale in dB

e) WVT, L=2, linear scale   f) WVT, L=2, scale in dB

Fig. 9. Different 2D representations after combination of real traces in experiments type-II, with linear scale (a, c, e) and logarithmic scale (b, d, f).

Two types of experiments have been performed to evaluate these techniques. Results of the first type, involving simulated noisy signal traces, have confirmed the accuracy of our theoretical SNR expressions proposed for the three combination variants. The first type experiments also demonstrate a great capability for accuracy detection of internal flaws.

Results from the second type, using an experimental ultrasonic prototype, permit to validate the proposed methods in a real NDE context.

More concretely, the three combination methods described and applied in this chapter, based on different processing tools (the Hilbert, Wigner-Ville, and Undecimated Wavelet packet Transforms) produce accurate 2D displays for isolated-flaws location. Additionally, these methods drastically improve the SNR of these 2D displays in relation to the initially acquired traces, very especially with the two latter processing cases, being the best flaw discrimination results obtained with the WVT option, but with a mayor computational cost than the wavelet technique, which also offers a good performance.

These good results for isolated-flaws patterns could be not directly extended to other more complicated testing situations with flaws very close among them, because some ambiguous flaw indications could be produced. In a future work, this particular restriction will be addressed by means of a specifically extended imaging procedure.

7. Acknowledgment

This work was supported by the National Plan of the Spanish Ministry of Science & Innovation (R&D Project DPI2008-05213).

8. References

Chang Y F and Hsieh C I 2002 Time of flight diffraction imaging for double-probe technique *IEEE Trans. Ultrason. Ferroel, Freq. Cont.* vol 49(6), pp 776-783.

Chen C.H. and Guey J.C. 1992 On the use of Wigner distribution in Ultrasonic NDE *Rev. of Progress in Quantitative Nondestructive Evaluation*, vol. 11A, pp. 967-974,.

Claasen T.A.C.M. and Mecklenbrauker W.F.G. 1980 The Wigner Distribution - A tool for time-frequency signal analysis *Philips J. Res.*, vol. 35, pp. 217-250, 276-300, 372-389.

Cohen L 1995 *Time-Frequency Analysis* Prentice Hall PTR Englewood Cliffs New Jersey.

Coifman R. and Wickerhauser M.V. 1992 Entropy-based algorithms for best basis selection *IEEE Trans. on Information Theory*, vol. 38, pp. 713-718.

Daubechies I 1992 *Ten Lectures on Wavelets* Society for Industrial and Applied Mathematics PhiladelphiaPA

Defontaine M, Bonneau S, Padilla F, Gomez M.A, Nasser Eddin M, Laugier P and Patat F 2004 2D array device for calcaneus bone transmission: an alternative technological solution using crossed beam forming *Ultrasonics* vol 42, pp 745-752.

Engl G and Meier R 2002 Testing large aerospace CFRP components by ultrasonic multichannel conventional and phased array pulse-echo techniques *NDT.net* vol. 7 (10).

Hlawatsch F and Boudreaux-Barlets G 1992 Linear and Quadratic Time-Frequency Signal Representations *IEEE Signal Processing Magazine* vol 9(2), pp. 21-67.

Lazaro J C, San Emeterio J L, Ramos A and Fernandez-Marron J L 2002 Influence of thresholding procedures in ultrasonic grain noise reduction using wavelets *Ultrasonics* vol. 40, pp 263-267.

Malik M.A.and Saniie J. 1996 Performance comparison of time-frequency distributions for ultrasonic non-destructive testing *Proc .IEEE Ultrasonic Symposium*, pp. 701-704.

Mallat S 1989 A theory for multiresolution signal decomposition: the wavelet representation *IEEE Transaction on Pattern Analysis and Machine Intelligence* vol 11, pp 674-693.

Meyer A W and Candy J V 2002 Iterative Processing of Ultrasonic Measurements to Characterize Flaws in Critical Optical Components *IEEE Trans. on Ultrason. Ferroel. and Freq. Cont.* vol 8, pp 1124-1138.

Pardo E, San Emeterio J L, Rodríguez M A and Ramos A 2008 Shift Invariant Wavelet Denoising of Ultrasonic Traces *Acta Acustica United with Acustica* vol 94 (5), pp 685-693.

Reguieg D, Padilla F, Defontaine M, Patat F and Laugier P 2006 Ultrasonic transmission device based on crossed beam forming *Proc. of the 2006 IEEE Ultrasonic Symposium*, pp. 2108-2111

Roy O, Mahaut S and Serre M 1999 Application of ultrasonic beam modeling to phased array testing of complex geometry components. *Review of Progress in Quantitative Non destructive Evaluation* Kluwer Acad. Plenum Publ. NewYork vol 18, pp. 2017-2024.

Rodríguez M A 2003 Ultrasonic non-destructive evaluation with spatial combination of Wigner-Ville transforms *ndt&e international* vol 36 pp. 441-445.

Rodríguez M A, Ramos A and San Emeterio J L 2004 Localization of isolated flaws by combination of noised signals detected from perpendicular transducers *NDT&E International* 37, pp. 345-352.

Rodríguez M A, San Emeterio J L, Lázaro J C and Ramos A 2004a Ultrasonic Flaw Detection in NDE of Highly Scattering Materials using Wavelet and Wigner-Ville Transform Processing *Ultrasonics* vol 42, pp 847-851.

Rodríguez M A, Ramos A, San Emeterio J L and Pérez J J 2004b Flaw location from perpendicular NDE transducers using the Wavelet packet transform *Proc. IEEE International Ultrasonics Symposium 2004 (IEEE Catalog 05CH37716C)*, pp 2318-2232.

Rodríguez M A, Ramos A and San Emeterio J L 2005 Multiple flaws location by means of NDE ultrasonic arrays placed at perpendicular planes *Proc. IEEE International Ultrasonics Symposium 2005 (IEEE Catalog 0-7803-9383-X/05)*, pp. 2074-2077.

Shensa M, 1992, The discrete wavelet transform: wedding the trous and Mallat algorithms, *IEEE Trans. Signal Process*, vol. 40, pp. 2464-2482.

7

High-Speed VLSI Architecture Based on Massively Parallel Processor Arrays for Real-Time Remote Sensing Applications

A. Castillo Atoche[1], J. Estrada Lopez[2],
P. Perez Muñoz[1] and S. Soto Aguilar[2]
[1]*Mechatronic Department, Engineering School, Autonomous University of Yucatan*
[2]*Computer Engineering Dept., Mathematics School, Autonomous University of Yucatan*
Mexico

1. Introduction

Developing computationally efficient processing techniques for massive volumes of hyperspectral data is critical for space-based Earth science and planetary exploration (see for example, (Plaza & Chang, 2008), (Henderson & Lewis, 1998) and the references therein). With the availability of remotely sensed data from different sensors of various platforms with a wide range of spatiotemporal, radiometric and spectral resolutions has made remote sensing as, perhaps, the best source of data for large scale applications and study. Applications of Remote Sensing (RS) in hydrological modelling, watershed mapping, energy and water flux estimation, fractional vegetation cover, impervious surface area mapping, urban modelling and drought predictions based on soil water index derived from remotely-sensed data have been reported (Melesse et al., 2007). Also, many RS imaging applications require a response in (near) real time in areas such as target detection for military and homeland defence/security purposes, and risk prevention and response. Hyperspectral imaging is a new technique in remote sensing that generates images with hundreds of spectral bands, at different wavelength channels, for the same area on the surface of the Earth. Although in recent years several efforts have been directed toward the incorporation of parallel and distributed computing in hyperspectral image analysis, there are no standardized architectures or Very Large Scale Integration (VLSI) circuits for this purpose in remote sensing applications.

Additionally, although the existing theory offers a manifold of statistical and descriptive regularization techniques for image enhancement/reconstruction, in many RS application areas there also remain some unsolved crucial theoretical and processing problems related to the computational cost due to the recently developed complex techniques (Melesse et al., 2007), (Shkvarko, 2010), (Yang et al., 2001). These descriptive-regularization techniques are associated with the unknown statistics of random perturbations of the signals in turbulent medium, imperfect array calibration, finite dimensionality of measurements, multiplicative signal-dependent speckle noise, uncontrolled antenna vibrations and random carrier trajectory deviations in the case of Synthetic Aperture Radar (SAR) systems (Henderson & Lewis, 1998), (Barrett & Myers, 2004). Furthermore, these techniques are not suitable for

(near) real time implementation with existing Digital Signal Processors (DSP) or Personal Computers (PC).
To treat such class of real time implementation, the use of specialized arrays of processors in VLSI architectures as coprocessors or stand alone chips in aggregation with Field Programmable Gate Array (FPGA) devices via the hardware/software (HW/SW) co-design, will become a real possibility for high-speed Signal Processing (SP) in order to achieve the expected data processing performance (Plaza, A. & Chang, 2008), (Castillo Atoche et al., 2010a, 2010b). Also, it is important to mention that cluster-based computing is the most widely used platform on ground stations, however several factors, like space, cost and power make them impractical for on-board processing. FPGA-based reconfigurable systems in aggregation with custom VLSI architectures are emerging as newer solutions which offer enormous computation potential in both cluster-based systems and embedded systems area.
In this work, we address two particular contributions related to the substantial reduction of the computational load of the Descriptive-Regularized RS image reconstruction technique based on its implementation with massively processor arrays via the aggregation of high-speed low-power VLSI architectures with a FPGA platform.
First, at the algorithmic-level, we address the design of a family of Descriptive-Regularization techniques over the range and azimuth coordinates in the uncertain RS environment, and provide the relevant computational recipes for their application to imaging array radars and fractional imaging SAR operating in different uncertain scenarios. Such descriptive-regularized family algorithms are computationally adapted for their HW-level implementation in an efficient mode using parallel computing techniques in order to achieve the maximum possible parallelism.
Second, at the systematic-level, the family of Descriptive-Regularization techniques based on reconstructive digital SP operations are conceptualized and employed with massively parallel processor arrays (MPPAs) in context of the real time SP requirements. Next, the array of processors of the selected reconstructive SP operations are efficiently optimized in fixed-point bit-level architectures for their implementation in a high-speed low-power VLSI architecture using 0.5um CMOS technology with low power standard cells libraries. The achieved VLSI accelerator is aggregated with a FPGA platform via HW/SW co-design paradigm.
Alternatives propositions related to parallel computing, systolic arrays and HW/SW co-design techniques in order to achieve the near real time implementation of the regularized-based procedures for the reconstruction of RS applications have been previously developed in (Plaza, A. & Chang, 2008), (Castillo Atoche et al., 2010a, 2010b). However, it should be noted that the design in hardware (HW) of a family of reconstructive signal processing operations have never been implemented in a high-speed low-power VLSI architecture based on massively parallel processor arrays in the past.
Finally, it is reported and discussed the implementation and performance issues related to real time enhancement of large-scale real-world RS imagery indicative of the significantly increased processing efficiency gained with the proposed implementation of high-speed low-power VLSI architectures of the descriptive-regularized algorithms.

2. Remote sensing background

The general formalism of the RS imaging problem presented in this study is a brief presentation of the problem considered in (Shkvarko, 2006, 2008), hence some crucial model elements are repeated for convenience to the reader.

The problem of enhanced remote sensing (RS) imaging is stated and treated as an ill-posed nonlinear inverse problem with model uncertainties. The challenge is to perform high-resolution reconstruction of the power spatial spectrum pattern (SSP) of the wavefield scattered from the extended remotely sensed scene via space-time processing of finite recordings of the RS data distorted in a stochastic uncertain measurement channel. The SSP is defined as a spatial distribution of the power (i.e. the second-order statistics) of the random wavefield backscattered from the remotely sensed scene observed through the integral transform operator (Henderson & Lewis, 1998), (Shkvarko, 2008). Such an operator is explicitly specified by the employed radar signal modulation and is traditionally referred to as the signal formation operator (SFO) (Shkvarko, 2006). The classical imaging with an array radar or SAR implies application of the method called "matched spatial filtering" to process the recorded data signals (Franceschetti et al., 2006), (Shkvarko, 2008), (Greco & Gini, 2007). A number of approaches had been proposed to design the constrained regularization techniques for improving the resolution in the SSP obtained by ways different from the matched spatial filtering, e.g., (Franceschetti et al., 2006), (Shkvarko, 2006, 2008), (Greco & Gini, 2007), (Plaza, A. & Chang, 2008), (Castillo Atoche et al., 2010a, 2010b) but without aggregating the minimum risk descriptive estimation strategies and specialized hardware architectures via FPGA structures and VLSI components as accelerators units. In this study, we address a extended descriptive experiment design regularization (DEDR) approach to treat such uncertain SSP reconstruction problems that unifies the paradigms of minimum risk nonparametric spectral estimation, descriptive experiment design and worst-case statistical performance optimization-based regularization.

2.1 Problem statement

Consider a coherent RS experiment in a random medium and the narrowband assumption (Henderson & Lewis, 1998), (Shkvarko, 2006) that enables us to model the extended object backscattered field by imposing its time invariant complex scattering (backscattering) function $e(\mathbf{x})$ in the scene domain (scattering surface) $X \ni \mathbf{x}$. The measurement data wavefield $u(\mathbf{y}) = s(\mathbf{y}) + n(\mathbf{y})$ consists of the echo signals s and additive noise n and is available for observations and recordings within the prescribed time-space observation domain $Y = T \times P$, where $\mathbf{y} = (t, \mathbf{p})^{\mathrm{T}}$ defines the time-space points in Y. The model of the observation wavefield u is defined by specifying the stochastic equation of observation (EO) of an operator form (Shkvarko, 2008):

$$u = \tilde{S}e + n; \; e \in \mathrm{E}; \; u, n \in \mathrm{U}; \; \tilde{S} : \mathrm{E} \to \mathrm{U}, \tag{1}$$

in the Hilbert signal spaces E and U with the metric structures induced by the inner products, $[u_1, u_2]_{\mathrm{U}} = \int_Y u_1(\mathbf{y}) u_2^*(\mathbf{y}) d\mathbf{y}$, and $[e_1, e_2]_{\mathrm{E}} = \int_X e_1(\mathbf{x}) e_2^*(\mathbf{x}) d\mathbf{x}$, respectively. The operator model of the stochastic EO in the conventional integral form (Henderson & Lewis, 1998), (Shkvarko, 2008) may be rewritten as

$$u(\mathbf{y}) = (\tilde{S}e(\mathbf{x}))(\mathbf{y}) = \int_X \tilde{S}(\mathbf{y},\mathbf{x})\, e(\mathbf{x}) d\mathbf{x} + 4\, n(\mathbf{y}) = \int_X S(\mathbf{y},\mathbf{x})\, e(\mathbf{x}) d\mathbf{x} + \int_X \delta S(\mathbf{y},\mathbf{x})\, e(\mathbf{x}) d\mathbf{x} + n(\mathbf{y}) . \tag{2}$$

The random functional kernel $\tilde{S}(\mathbf{y},\mathbf{x}) = S(\mathbf{y},\mathbf{x})+\delta S(\mathbf{y},\mathbf{x})$ of the stochastic signal formation operator (SFO) $\tilde{S}$ given by (2) defines the signal wavefield formation model. Its mean, $<\tilde{S}(\mathbf{y},\mathbf{x})> = S(\mathbf{y},\mathbf{x})$, is referred to as the nominal SFO in the RS measurement channel specified by the time-space modulation of signals employed in a particular radar system/SAR (Henderson & Lewis, 1998), and the variation about the mean $\delta S(\mathbf{y},\mathbf{x}) = \mu(\mathbf{y},\mathbf{x})S(\mathbf{y},\mathbf{x})$ models the stochastic perturbations of the wavefield at different propagation paths, where $\mu(\mathbf{y},\mathbf{x})$ is associated with zero-mean multiplicative noise (so-called Rytov perturbation model). All the fields e, n, u in (2) are assumed to be zero-mean complex valued Gaussian random fields. Next, we adopt an incoherent model (Henderson & Lewis, 1998), (Shkvarko, 2006) of the backscattered field $e(\mathbf{x})$ that leads to the δ-form of its correlation function, $R_e(\mathbf{x}_1,\mathbf{x}_2) = b(\mathbf{x}_1)\delta(\mathbf{x}_1 - \mathbf{x}_2)$. Here, $e(\mathbf{x})$ and $b(\mathbf{x}) = <|e(\mathbf{x})|^2>$ are referred to as the scene random complex scattering function and its average power scattering function or spatial spectrum pattern (SSP), respectively. The problem at hand is to derive an estimate $\hat{b}(\mathbf{x})$ of the SSP $b(\mathbf{x})$ (referred to as the desired RS image) by processing the available finite dimensional array radar/SAR measurements of the data wavefield $u(\mathbf{y})$ specified by (2).

2.2 Discrete-form uncertain problem model

The stochastic integral-form EO (2) to its finite-dimensional approximation (vector) form (Shkvarko, 2008) is now presented.

$$\mathbf{u} = \tilde{\mathbf{S}}\mathbf{e} + \mathbf{n} = \mathbf{S}\mathbf{e} + \boldsymbol{\Delta}\mathbf{e} + \mathbf{n}\,, \tag{3}$$

in which the perturbed SFO matrix

$$\tilde{\mathbf{S}} = \mathbf{S} + \boldsymbol{\Delta}\,, \tag{4}$$

represents the discrete-form approximation of the integral SFO defined for the uncertain operational scenario by the EO (2), and $\mathbf{e}$, $\mathbf{n}$, $\mathbf{u}$ are zero-mean vectors composed of the decomposition coefficients $\{e_k\}_{k=1}^{K}$, $\{n_m\}_{m=1}^{M}$, and $\{u_m\}_{m=1}^{M}$, respectively. These vectors are characterized by the correlation matrices: $\mathbf{R_e} = \mathbf{D} = \mathbf{D}(\mathbf{b}) = \mathrm{diag}(\mathbf{b})$ (a diagonal matrix with vector $\mathbf{b}$ at its principal diagonal), $\mathbf{R_n}$, and $\mathbf{R_u} = <\tilde{\mathbf{S}}\mathbf{R_e}\tilde{\mathbf{S}}^+>_{p(\boldsymbol{\Delta})} + \mathbf{R_n}$, respectively, where $<\cdot>_{p(\boldsymbol{\Delta})}$ defines the averaging performed over the randomness of $\boldsymbol{\Delta}$ characterized by the *unknown* probability density function $p(\boldsymbol{\Delta})$, and superscript $^+$ stands for Hermitian conjugate. Following (Shkvarko, 2008), the distortion term $\boldsymbol{\Delta}$ in (4) is considered as a random zero mean matrix with the bounded second-order moment $\eta \geq \langle \|\boldsymbol{\Delta}\|^2 \rangle$. Vector $\mathbf{b}$ is composed of the elements, $b_k = \mathcal{B}(e_k) = <e_k e_k^*> = <|e_k|^2>$; $k = 1, \ldots, K$, and is referred to as a K-D vector-form approximation of the SSP, where $\mathcal{B}$ represents the second-order statistical ensemble averaging operator (Barrett & Myers, 2004). The SSP vector $\mathbf{b}$ is associated with the so-called lexicographically ordered image pixels (Barrett & Myers, 2004). The corresponding conventional $K_y{\times}K_x$ rectangular frame ordered scene image $\mathbf{B} = \{b(k_x, k_x);\ k_x, = 1,\ldots,K_x;\ k_v, = 1,\ldots,K_y\}$ relates to its lexicographically ordered vector-form representation $\mathbf{b} = \{b(k);\ k = 1,\ldots,K = K_y{\times}\ K_x\}$ via the standard row by row concatenation (so-called lexicographical reordering) procedure, $\mathbf{B} = L\{\mathbf{b}\}$ (Barrett & Myers, 2004). Note that in the simple case of certain operational scenario (Henderson & Lewis, 1998), (Shkvarko, 2008), the discrete-form (i.e. matrix-form) SFO **S** is assumed to be deterministic, i.e. the random perturbation term in (4) is irrelevant, $\mathbf{\Delta} = \mathbf{0}$.

The digital enhanced RS imaging problem is formally stated as follows (Shkvarko, 2008): to map the scene pixel frame image $\hat{\mathbf{B}}$ via lexicographical reordering $\hat{\mathbf{B}} = L\{\hat{\mathbf{b}}\}$ of the SSP vector estimate $\hat{\mathbf{b}}$ reconstructed from whatever available measurements of independent realizations of the recorded data vector **u.** The reconstructed SSP vector $\hat{\mathbf{b}}$ is an estimate of the second-order statistics of the scattering vector **e** observed through the perturbed SFO (4) and contaminated with noise **n**; hence, the RS imaging problem at hand must be qualified and treated as a statistical nonlinear inverse problem with the uncertain operator. The high-resolution imaging implies solution of such an inverse problem in some optimal way. Recall that in this paper we intend to follow the unified descriptive experiment design regularized (DEDR) method proposed originally in (Shkvarko, 2008).

2.3 DEDR method

2.3.1 DEDR strategy for certain operational scenario

In the descriptive statistical formalism, the desired SSP vector $\hat{\mathbf{b}}$ is recognized to be the vector of a principal diagonal of the estimate of the correlation matrix $\mathbf{R_e}(\mathbf{b})$, i.e. $\hat{\mathbf{b}} = \{\hat{\mathbf{R}}_{\mathbf{e}}\}_{\text{diag}}$. Thus one can seek to estimate $\hat{\mathbf{b}} = \{\hat{\mathbf{R}}_{\mathbf{e}}\}_{\text{diag}}$ given the data correlation matrix $\mathbf{R_u}$ pre-estimated empirically via averaging $J \geq 1$ recorded data vector snapshots $\{\mathbf{u}_{(j)}\}$

$$\mathbf{Y} = \hat{\mathbf{R}}_{\mathbf{u}} = \underset{j \in J}{\text{aver}}\{\mathbf{u}_{(j)}\mathbf{u}_{(j)}^{+}\} = \frac{1}{J}\sum_{j=1}^{J}\mathbf{u}_{(j)}\mathbf{u}_{(j)}^{+}, \tag{5}$$

by determining the solution operator (SO) **F** such that

$$\hat{\mathbf{b}} = \{\hat{\mathbf{R}}_{\mathbf{e}}\}_{\text{diag}} = \{\mathbf{FYF}^{+}\}_{\text{diag}} \tag{6}$$

where $\{\cdot\}_{\text{diag}}$ defines the vector composed of the principal diagonal of the embraced matrix.

To optimize the search for **F** in the *certain* operational scenario the DEDR strategy was proposed in (Shkvarko, 2006)

$$\mathbf{F} \rightarrow \min_{\mathbf{F}}\{\mathfrak{R}(\mathbf{F})\}, \tag{7}$$

$$\mathfrak{R}(\mathbf{F}) = \text{trace}\{(\mathbf{FS} - \mathbf{I})\mathbf{A}(\mathbf{FS} - \mathbf{I})^{+}\} + \alpha\, \text{trace}\{\mathbf{F}\mathbf{R_n}\mathbf{F}^{+}\} \tag{8}$$

that implies the minimization of the weighted sum of the systematic and fluctuation errors in the desired estimate $\hat{\mathbf{b}}$ where the selection (adjustment) of the regularization parameter α and the weight matrix **A** provide the additional experiment design degrees of freedom incorporating any descriptive properties of a solution if those are known a priori (Shkvarko, 2006). It is easy to recognize that the strategy (7) is a structural extension of the statistical minimum risk estimation strategy for the nonlinear spectral estimation problem at hand because in both cases the balance between the gained spatial resolution and the noise energy in the resulting estimate is to be optimized.

From the presented above DEDR strategie, one can deduce that the solution to the optimization problem found in the previous study (Shkvarko, 2006) results in

$$F = \mathbf{K}\mathbf{S}^{+}\mathbf{R}_{\mathbf{n}}^{-1}, \tag{9}$$

where
$$\mathbf{K} = (\mathbf{S}^{+}\mathbf{R}_{\mathbf{n}}^{-1}\mathbf{S} + \alpha\mathbf{A}^{-1})^{-1} \tag{10}$$

represents the so-called regularized reconstruction operator; $\mathbf{R}_{\mathbf{n}}^{-1}$ is the noise whitening filter, and the adjoint (i.e. Hermitian transpose) SFO $\mathbf{S}^{+}$ defines the matched spatial filter in the conventional signal processing terminology.

2.3.2 DEDR strategy for uncertain operational scenario

To optimize the search for the desired SO **F** in the *uncertain* operational scenario with the randomly perturbed SFO (4), the *extended DEDR* strategy was proposed in (Shkvarko, 2006)

$$\mathbf{F} = \arg\min_{\mathbf{F}} \max_{<||\Delta||^2>_{p(\Delta)} \leq \delta} \{\mathfrak{R}_{ext}(\mathbf{F})\} \tag{11}$$

$$\text{subject to } <||\mathbf{\Delta}||^2>_{p(\mathbf{\Delta})} \leq \delta \tag{12}$$

where the conditioning term (12) represents the worst-case statistical performance (WCSP) regularizing constraint imposed on the unknown second-order statistics $<||\mathbf{\Delta}||^2>_{p(\mathbf{\Delta})}$ of the random distortion component $\mathbf{\Delta}$ of the SFO matrix (4), and the DEDR "extended risk" is defined by

$$\mathfrak{R}_{ext}(\mathbf{F}) = \mathrm{tr}\{<(\mathbf{F}\tilde{\mathbf{S}} - \mathbf{I})\mathbf{A}(\mathbf{F}\tilde{\mathbf{S}} - \mathbf{I})^{+}>_{p(\mathbf{\Delta})}\} + \alpha\,\mathrm{tr}\{\mathbf{F}\mathbf{R}_{\mathbf{n}}\mathbf{F}^{+}\} \tag{13}$$

where the regularization parameter α and the metrics inducing weight matrix **A** compose the processing level "degrees of freedom" of the DEDR method.
To proceed with the derivation of the robust SFO (11), the risk function (13) was next decomposed and evaluated for its the maximum value applying the Cauchy-Schwarz inequality and Loewner ordering (Greco & F. Gini, 2007) of the weight matrix $\mathbf{A} \leq \gamma\mathbf{I}$ with the scaled Loewner ordering factor $\gamma = \min\{\tilde{\gamma} : \mathbf{A} \leq \tilde{\gamma}\mathbf{I}\} = 1$. With these robustifications, the extended DEDR strategy (11) is transformed into the following optimization problem

$$\mathbf{F} = \rightarrow \min_{\mathbf{F}} \{\mathfrak{R}_{\Sigma}(\mathbf{F})\} \tag{14}$$

with the *aggregated* DEDR risk function

$$\mathfrak{R}_{\Sigma}(\mathrm{F})\} = \mathrm{tr}\{(\mathrm{FS} - \mathrm{I})\mathrm{A}(\mathrm{FS} - \mathrm{I})^{+}\} + \alpha\mathrm{tr}\{\mathrm{F}\,\mathbf{R}_{\Sigma}\,\mathrm{F}^{+}\}, \tag{15}$$

Where
$$\mathbf{R}_{\Sigma} = \mathbf{R}_{\Sigma}(\beta) = (\mathrm{R_n} + \beta\mathrm{I});\ \beta = \delta/\alpha \geq 0. \tag{16}$$

The optimization solution of (14) follows a structural extension of (9) for the augmented (diagonal loaded) $\mathbf{R}_{\Sigma}$ that yields

$$F = \mathbf{K}_{\Sigma}\mathbf{S}^{+}\mathbf{R}_{\Sigma}^{-1}, \tag{17}$$

Where $$\mathbf{K}_{\Sigma} = (\mathbf{S}^{+}\mathbf{R}_{\Sigma}^{-1}\mathbf{S} + \alpha\mathbf{A}^{-1})^{-1} \tag{18}$$

represents the robustified reconstruction operator for the uncertain scenario.

2.3.3 DEDR imaging techniques

In this sub-section, three practically motivated DEDR-related imaging techniques (Shkvarko, 2008) are presented that will be used at the HW co-design stage, namely, the conventional matched spatial filtering (MSF) method, and two high-resolution reconstructive imaging techniques: (i) the robust spatial filtering (RSF), and (ii) the robust adaptive spatial filtering (RASF) methods.

1. *MSF*: The MSF algorithm is a member of the DEDR-related family specified for $\alpha >> ||\mathbf{S}^{+}\mathbf{S}||$, i.e. the case of a dominating priority of suppression of noise over the systematic error in the optimization problem (7). In this case, the SO (9) is approximated by the matched spatial filter (MSF):

$$\mathbf{F}_{MSF} = \mathbf{F}^{(1)} \approx \mathbf{S}^{+}. \tag{19}$$

2. *RSF*: The RSF method implies no preference to any prior model information (i.e., $\mathbf{A} = \mathbf{I}$) and balanced minimization of the systematic and noise error measures in (14) by adjusting the regularization parameter to the inverse of the signal-to-noise ratio (SNR), e.g. $\alpha = N_0/B_0$, where B_0 is the prior average gray level of the image. In that case the SO $\mathbf{F}$ becomes the Tikhonov-type robust spatial filter

$$\mathbf{F}_{RSF} = \mathbf{F}^{(2)} = (\mathbf{S}^{+}\mathbf{S} + \alpha_{RSF}\mathbf{I})^{-1}\mathbf{S}^{+}. \tag{20}$$

   in which the RSF regularization parameter α_{RSF} is adjusted to a particular operational scenario model, namely, $\alpha_{RSF} = (N_0/b_0)$ for the case of a certain operational scenario, and $\alpha_{RSF} = (N_{\Sigma}/b_0)$ in the uncertain operational scenario case, respectively, where N_0 represents the white observation noise power density, b_0 is the average a priori SSP value, and $N_{\Sigma} = N_0 + \beta$ corresponds to the augmented noise power density in the correlation matrix specified by (16).

3. *RASF*: In the statistically optimal problem treatment, α and $\mathbf{A}$ are adjusted in an adaptive fashion following the minimum risk strategy, i.e. $\alpha\,\mathbf{A}^{-1} = \hat{\mathbf{D}} = \mathrm{diag}(\hat{\mathbf{b}})$, the diagonal matrix with the estimate $\hat{\mathbf{b}}$ at its principal diagonal, in which case the SOs (9), (17) become itself solution-dependent operators that result in the following robust adaptive spatial filters (RASFs):

$$\mathbf{F}_{RASF} = \mathbf{F}^{(3)} = (\mathbf{S}^{+}\mathbf{R}_{\mathbf{n}}^{-1}\mathbf{S} + \hat{\mathbf{D}}^{-1})^{-1}\,\mathbf{S}^{+}\mathbf{R}_{\mathbf{n}}^{-1} \tag{21}$$

for the certain operational scenario, and

$$\mathrm{F}_{RASF\Sigma} = \mathrm{F}^{(4)} = (\mathbf{S}^{+}\mathbf{R}_{\Sigma}^{-1}\mathbf{S} + \hat{\mathbf{D}}^{-1})^{-1}\,\mathbf{S}^{+}\mathbf{R}_{\Sigma}^{-1} \tag{22}$$

for the uncertain operational scenario, respectively.

Using the defined above SOs, the DEDR-related data processing techniques in the conventional pixel-frame format can be unified now as follows

$$\hat{\mathbf{B}} = L\{\hat{\mathbf{b}}\} = L\{\{\mathbf{F}^{(p)}\mathbf{Y}\mathbf{F}^{(p)+}\}_{\mathrm{diag}}\};\ ); \quad p = 1, 2, 3, 4 \tag{23}$$

with $\mathbf{F}^{(1)} = \mathbf{F}_{MSF}$; $\mathbf{F}^{(2)} = \mathbf{F}_{RSF}$, and $\mathbf{F}^{(3)} = \mathbf{F}_{RASF}$, $\mathbf{F}^{(4)} = \mathbf{F}_{RASF\Sigma}$, respectively.
Any other feasible adjustments of the DEDR degrees of freedom (the regularization parameters α, β, and the weight matrix $\mathbf{A}$) provide other possible DEDR-related SSP reconstruction techniques, that we do not consider in this study.

3. VLSI architecture based on Massively Parallel Processor Arrays

In this section, we present the design methodology for real time implementation of specialized arrays of processors in VLSI architectures based on massively parallel processor arrays (MPPAs) as coprocessors units that are integrated with a FPGA platform via the HW/SW co-design paradigm. This approach represents a real possibility for low-power high-speed reconstructive signal processing (SP) for the enhancement/reconstruction of RS imagery. In addition, the authors believe that FPGA-based reconfigurable systems in aggregation with custom VLSI architectures are emerging as newer solutions which offer enormous computation potential in RS systems.
A brief perspective on the state-of-the-art of high-performance computing (HPC) techniques in the context of remote sensing problems is provided. The wide range of computer architectures (including homogeneous and heterogeneous clusters and groups of clusters, large-scale distributed platforms and grid computing environments, specialized architectures based on reconfigurable computing, and commodity graphic hardware) and data processing techniques exemplifies a subject area that has drawn at the cutting edge of science and technology. The utilization of parallel and distributed computing paradigms anticipates ground-breaking perspectives for the exploitation of high-dimensional data processing sets in many RS applications. Parallel computing architectures made up of homogeneous and heterogeneous commodity computing resources have gained popularity in the last few years due to the chance of building a high-performance system at a reasonable cost. The scalability, code reusability, and load balance achieved by the proposed implementation in such low-cost systems offer an unprecedented opportunity to explore methodologies in other fields (e.g. data mining) that previously looked to be too computationally intensive for practical applications due to the immense files common to remote sensing problems (Plaza & Chang, 2008).
To address the required near-real-time computational mode by many RS applications, we propose a high-speed low-power VLSI co-processor architecture based on MPPAs that is aggregated with a FPGA via the HW/SW co-design paradigm. Experimental results demonstrate that the hardware VLSI-FPGA platform of the presented DEDR algorithms makes appropriate use of resources in the FPGA and provides a response in near-real-time that is acceptable for newer RS applications.

3.1 Design flow

The all-software execution of the prescribed RS image formation and reconstructive signal processing (SP) operations in modern high-speed personal computers (PC) or any digital signal processors (DSP) platform may be intensively time consuming. These high computational complexities of the general-form DEDR-POCS algorithms make them definitely unacceptable for real time PC-aided implementation.
In this section, we describe a specific design flow of the proposed VLSI-FPGA architecture for the implementation of the DEDR method via the HW/SW co-design paradigm. The

HW/SW co-design is a hybrid method aimed at increasing the flexibility of the implementation and improvement of the overall design process (Castillo Atoche et al., 2010a). When a co-processor-based solution is employed in the HW/SW co-design architecture, the computational time can be drastically reduced. Two opposite alternatives can be considered when exploring the HW/SW co-design of a complex SP system. One of them is the use of standard components whose functionality can be defined by means of programming. The other one is the implementation of this functionality via a microelectronic circuit specifically tailored for that application. It is well known that the first alternative (the software alternative) provides solutions that present a great flexibility in spite of high area requirements and long execution times, while the second one (the hardware alternative) optimizes the size aspects and the operation speed but limits the flexibility of the solution. Halfway between both, hardware/software co-design techniques try to obtain an appropriate trade-off between the advantages and drawbacks of these two approaches.

In (Castillo Atoche et al., 2010a), an initial version of the HW/SW- architecture was presented for implementing the digital processing of a large-scale RS imagery in the operational context. The architecture developed in (Castillo Atoche et al., 2010a) did not involve MPPAs and is considered here as a simply reference for the new pursued HW/SW co-design paradigm, where the corresponding blocks are to be designed to speed-up the digital SP operations of the DEDR-POCS-related algorithms developed at the previous SW stage of the overall HW/SW co-design to meet the real time imaging system requirements.

The proposed co-design flow encompasses the following general stages:

i. Algorithmic implementation (reference simulation in MATLAB and C++ platforms);
ii. Partitioning process of the computational tasks;
iii. Aggregation of parallel computing techniques;
iv. Architecture design procedure of the addressed reconstructive SP computational tasks onto HW blocks (MPPAs);

3.1.1 Algorithmic implementation

In this sub-section, the procedures for computational implementation of the DEDR-related robust space filter (RSF) and robust adaptive space filter (RASF) algorithms in the MATLAB and C++ platforms are developed. This reference implementation scheme will be next compared with the proposed architecture based on the use of a VLSI-FPGA platform.

Having established the optimal RSF/RASF estimator (20) and (21), let us now consider the way in which the processing of the data vector $\mathbf{u}$ that results in the optimum estimate $\hat{\mathbf{b}}$ can be computationally performed. For this purpose, we refer to the estimator (20) as a multi-stage computational procedure. We part the overall computations prescribed by the estimator (16) into four following steps.

a. First Step: Data Innovations

At this stage the a priori known value of the data mean $\langle \mathbf{u} \rangle = \mathbf{S}\mathbf{m}_b$ is subtracted from the data vector $\mathbf{u}$. The innovations vector $\mathbf{u}^{\circ} = \mathbf{u} - \mathbf{S}\mathbf{m}_b$ contains all new information regarding the unknown deviations $\mathbf{b}^{\circ}$ = ($\mathbf{b}$ – $\mathbf{m}_b$) of the vector $\mathbf{b}$ from its prescribed (known) mean value $\mathbf{m}_b$.

b. Second Step: Rough Signal Estimation

At this stage we obtain the vector $\mathbf{q} = \mathbf{S}^{+}\mathbf{u}^{\circ}$. The operator $\mathbf{S}^{+}$ operating on $\mathbf{u}^{\circ}$ is mapped. Thus, the result, $\mathbf{q}$, can be interpreted as a rough estimate of $\mathbf{b}^{\circ} = (\mathbf{b} - \mathbf{m}_b)$ referred to as a degraded image.

c. Third Step: Signal Reconstruction

At this stage we obtain the estimate $\hat{\mathbf{b}}^{\circ} = \mathbf{A}_{\alpha}^{-1}\mathbf{q} = (\mathbf{S}^{+}\mathbf{S} + \alpha_{RSF}\mathbf{I})^{-1}\mathbf{q}$ of the unknown signal referred to as the reconstructed image frame. The matrix $\mathbf{A}_{\alpha}^{-1} = (\mathbf{S}^{+}\mathbf{S} + \alpha_{RSF}\mathbf{I})^{-1}$ operating on $\mathbf{q}$ produces some form of inversion of the degradations embedded in the operator $\mathbf{S}^{+}\mathbf{S}$. It is important to note that in the case $\alpha = 0$, we have $\hat{\mathbf{b}}^{\circ} = \mathbf{A}_{(\alpha=0)}^{-1}\mathbf{q} = \mathbf{S}^{\#}\mathbf{u}$, where matrix $\mathbf{S}^{\#} = (\mathbf{S}^{+}\mathbf{S})^{-1}\mathbf{S}^{+}$ is recognized to be the pseudoinverse (i.e., the well known Moore-Penrouse pseudoinverse) of the SFO matrix $\mathbf{S}$.

d. Fourth Step: Restoration of the Trend

Having obtained the estimate $\hat{\mathbf{b}}^{\circ}$ and known the mean value $\mathbf{m}_b$, we can obtain the optimum RSF estimate (20) simply by adding the prescribed mean value $\mathbf{m}_b$ (referred to as the non-zero trend) to the reconstructed image frame as $\hat{\mathbf{b}} = \mathbf{m}_b + \hat{\mathbf{b}}^{\circ}$.

3.1.2 (ii) Partitioning process of the computational tasks

One of the challenging problems of the HW/SW co-design is to perform an efficient HW/SW partitioning of the computational tasks. The aim of the partitioning problem is to find which computational tasks can be implemented in an efficient hardware architecture looking for the best trade-offs among the different solutions. The solution to the problem requires, first, the definition of a partitioning model that meets all the specification requirements (i.e., functionality, goals and constraints).

Note that from the formal SW-level co-design point of view, such DEDR techniques (20), (21), (22) can be considered as a properly ordered sequence of the vector-matrix multiplication procedure that one can next perform in an efficient high performance computational fashion following the proposed bit-level high-speed VLSI co-processor architecture. In particular, for implementing the fixed-point DEDR RSF and RASF algorithms, we consider in this partitioning stage to develop a high-speed VLSI co-processor for the computationally complex matrix-vector SP operation in aggregation with a powerful FPGA reconfigurable architecture via the HW/SW co-design technique. The rest of the reconstructive SP operations are employed in SW with a 32 bits embedded processor (MicroBlaze).

This novel VLSI-FPGA platform represents a new paradigm for real time processing of newer RS applications. Fig. 1 illustrates the proposed VLSI-FPGA architecture for the implementation of the RSF/RASF algorithms.

Once the partitioning stage has been defined, the selected reconstructive SP sub-task is to be mapped into the corresponding high-speed VLSI co-processor. In the HW design, the precision of 32 bits for performing all fixed-point operations is used, in particular, 9-bit integer and 23-bits decimal for the implementation of the co-processor. Such precision guarantees numerical computational errors less than 10^{-5} referring to the MATLAB Fixed Point Toolbox (Matlab, 2011).

3.1.3 Aggregation of parallel computing techniques

This sub-section is focused in how to improve the performance of the complex RS algorithms with the aggregation of parallel computing and mapping techniques onto HW-level massively parallel processor arrays (MPPAs).

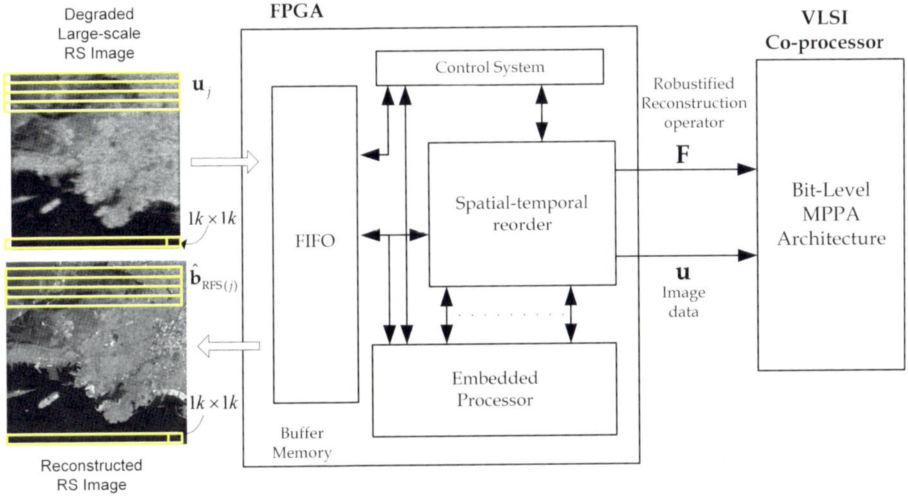

Fig. 1. VLSI-FPGA platform of the RSF/RASF algorithms via the HW/SW co-design paradigm.

The basic algebraic matrix operation (i.e., the selected matrix-vector multiplication) that constitutes the base of the most computationally consuming applications in the reconstructive SP applications is transformed into the required parallel algorithmic representation format. A manifold of different approaches can be used to represent parallel algorithms, e.g. (Moldovan & Fortes, 1986), (Kung, 1988). In this study, we consider a number of different loop optimization techniques used in high performance computing (HPC) in order to exploit the maximum possible parallelism in the design:

- Loop unrolling,
- Nested loop optimization,
- Loop interchange.

In addition, to achieve such maximum possible parallelism in an algorithm, the so-called data dependencies in the computations must be analyzed (Moldovan & Fortes, 1986), (Kung, 1988). Formally, these dependencies are to be expressed via the corresponding dependence graph (DG). Following (Kung, 1988), we define the dependence graph **G=[P, E]** as a composite set where **P** represents the nodes and **E** represents the arcs or edges in which each $e \in \mathbf{E}$ connects $p_1, p_2 \in \mathbf{P}$ that is represented as $e = p_1 \rightarrow p_2$. Next, the data dependencies analysis of the matrix-vector multiplication algorithms should be performed aimed at their efficient parallelization.

For example, the matrix-vector multiplication of an $n \times m$ matrix **A** with a vector **x** of dimension m, given by **y=Ax**, can be algorithmically computed as $y_j = \sum_{i=1}^{n} a_{ji} x_i$, *for* $j = 1,...,m$, where **y** and a_{ji} represents an n-dimensional (n-D) output vector and the corresponding element of **A**, respectively. The first SW-level transformation is the so-called single assignment algorithm (Kung, 1988), (Castillo Atoche et al., 2010b) that performs the computing of the matrix-vector product. Such single assignment algorithm corresponds to a loop unrolling method in which the primary benefit in loop unrolling is to perform more computations per iteration. Unrolling also reduces the overall number of branches significantly and gives the processor more instructions between branches (i.e., it increases the size of basic blocks).

Next, we examine the computation-related optimizations followed by the memory optimizations. Typically, when we are working with nests of loops, we are working with multidimensional arrays. Computing in multidimensional arrays can lead to non-unit-stride memory access. Many of the optimizations can be perform on loop nests to improve the memory access patterns. The second SW-level transformation consists in to transform the matrix-vector single assignment algorithm in the locally recursive algorithm representation without global data dependencies (i.e. in term of a recursive form). At this stage, nested-loop optimizations are employed in order to avoid large routing resources that are translated into the large amount of buffers in the final processor array architecture. The variable being broadcasted in single assignment algorithms is removed by passing the variable through each of the neighbour processing elements (PEs) in a DG representation.

Additionally, loop interchange techniques for rearranging a loop nest are also applied. For performance, the loop interchange of inner and outer loops is performed to pull the computations into the center loop, where the unrolling is implemented.

3.1.4 Architecture design onto MPPAs

Massively parallel co-processors are typically part of a heterogeneous hardware/software-system. Each processor is a massive parallel system consisting of an array of PEs. In this study, we propose the MPPA architecture for the selected reconstructive SP matrix-vector operation. This architecture is first modelled in a processor Array (PA) and next, each processor is implemented also with an array of PEs (i.e., in a highly-pipelined bit-level representation). Thus, we achieved the pursued MPPAs architecture following the space-time mapping procedures.

First, some fundamental proved propositions are given in order to clarify the mapping procedure onto PAs.

Proposition 1. There are types of algorithms that are expressed in terms of regular and localized DG. For example, basic algebraic matrix-form operations, discrete inertial transforms like convolution, correlation techniques, digital filtering, etc. that also can be represented in matrix formats (Moldovan & Fortes, 1986), (Kung, 1988).

Proposition 2. As the DEDR algorithms can be considered as properly ordered sequences vector-matrix multiplication procedures, then, they can be performed in an efficient computational fashion following the PA-oriented HW/SW co-design paradigm (Kung, 1988).

Following the presented above *propositions*, we are ready to derive the proper PA architectures. (Moldovan & Fortes, 1986) proved the mapping theory for the transformation $\mathbf{T}$. The transformation $\mathbf{T}':\mathbf{G}^N \rightarrow \hat{\mathbf{G}}^{N-1}$ maps the *N*-dimensional DG ($\mathbf{G}^N$) onto the (*N*–1)-dimensional PA ($\hat{\mathbf{G}}^{N-1}$), where *N* represents the dimension of the DG (see proofs in (Kung, 1988) and details in (CastilloAtoche et al., 2010b). Second, the desired linear transformation matrix operator $\mathbf{T}$ can be segmented in two blocks as follows

$$\mathbf{T} = \begin{bmatrix} \mathbf{\Pi} \\ \mathbf{\Sigma} \end{bmatrix}, \quad (24)$$

where $\mathbf{\Pi}$ is a $(1{\times}N)$-D vector (composed of the first row of $\mathbf{T}$) which (in the segmenting terms) determines the time scheduling, and the $(N-1){\times}N$ sub-matrix $\mathbf{\Sigma}$ in (24) is composed of the rest rows of $\mathbf{T}$ that determine the space processor specified by the so-called projection vector $\mathbf{d}$ (Kung, 1988).Next, such segmentation (24) yields the regular PA of $(N-1)$-D specified by the mapping

$$\mathbf{T\Phi} = \mathbf{K}, \tag{25}$$

where $\mathbf{K}$ is composed of the new revised vector schedule (represented by the first row of the PA) and the inter-processor communications (represented by the rest rows of the PA), and the matrix $\mathbf{\Phi}$ specifies the data dependencies of the parallel representation algorithm.

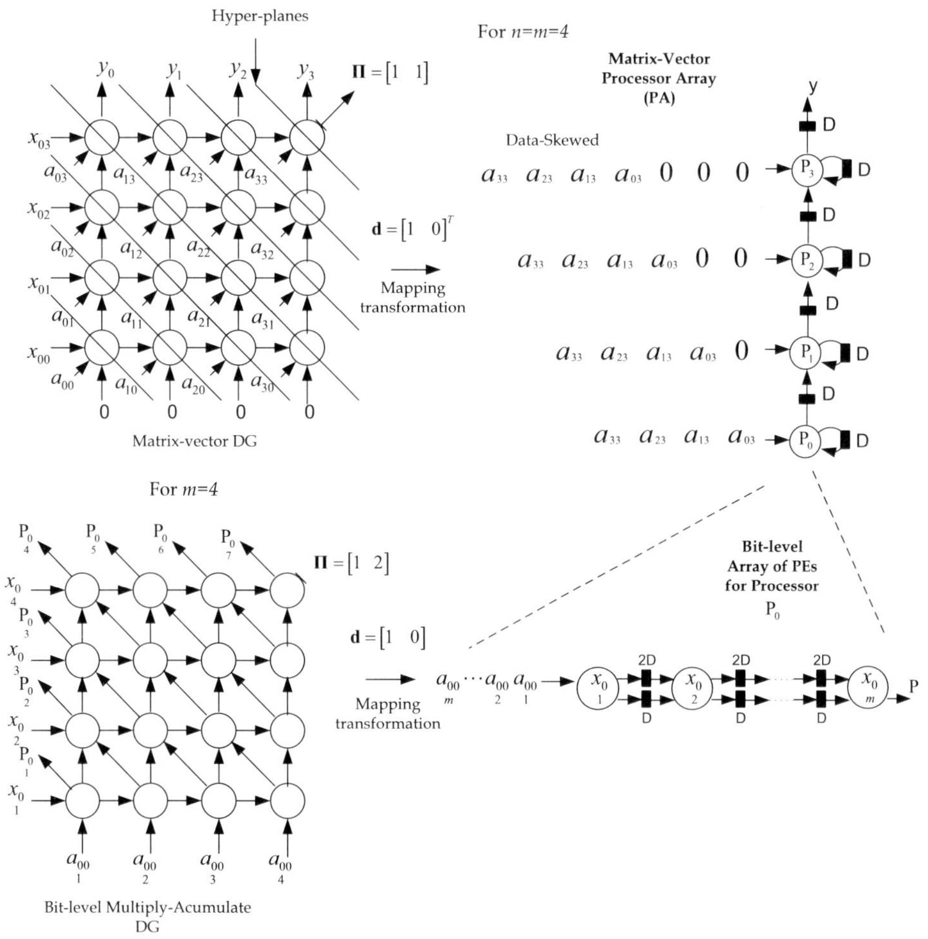

Fig. 2. High-Speed MPPA approach for the reconstructive matrix-vector SP operation

For a more detailed explanation of this theory, see (Kung, 1988), (CastilloAtoche et al., 2010b). In this study, the following specifications for the matrix-vector algorithm onto PAs are employed: $\mathbf{\Pi}=\begin{bmatrix}1 & 1\end{bmatrix}$ for the vector schedule, $\mathbf{d}=\begin{bmatrix}1 & 0\end{bmatrix}$ for the projection vector and, $\mathbf{\Sigma}=\begin{bmatrix}0 & 1\end{bmatrix}$ for the space processor, respectively. With these specifications the transformation matrix becomes $\mathbf{T}=\begin{bmatrix}\mathbf{\Pi}\\ \mathbf{\Sigma}\end{bmatrix}=\begin{bmatrix}1 & 1\\ 0 & 1\end{bmatrix}$. Now, for a simplified test-case, we specify the following operational parameters: $m = n = 4$, the period of clock of 10 ns and 32 bits data-word length. Now, we are ready to derive the specialized bit-level matrix-format MPPAs-based architecture. Each processor of the vector-matrix PA is next derived in an array of processing elements (PEs) at bit-level scale. Once again, the space-time transformation is employed to design the bit-level architecture of each processor unit of the matrix-vector PA. The following specifications were considered for the bit-level multiply-accumulate architecture: $\mathbf{\Pi}=\begin{bmatrix}1 & 2\end{bmatrix}$ for the vector schedule, $\mathbf{d}=\begin{bmatrix}1 & 0\end{bmatrix}$ for the projection vector and, $\mathbf{\Sigma}=\begin{bmatrix}0 & 1\end{bmatrix}$ for the space processor, respectively. With these specifications the transformation matrix becomes $\mathbf{T}=\begin{bmatrix}\mathbf{\Pi}\\ \mathbf{\Sigma}\end{bmatrix}=\begin{bmatrix}1 & 2\\ 0 & 1\end{bmatrix}$. The specified operational parameters are the following: l=32 (i.e., which represents the dimension of the word-length) and the period of clock of 10 ns. The developed architecture is next illustrated in Fig. 2.

From the analysis of Fig. 2, one can deduce that with the MPPA approach, the real time implementation of computationally complex RS operations can be achieved due the highly-pipelined MPPA structure.

3.2 Bit-level design based on MPPAS of the high-speed VLSI accelerator

As described above, the proposed partitioning of the VLSI-FPGA platform considers the design and fabrication of a low-power high-speed co-processor integrated circuit for the implementation of complex matrix-vector SP operation. Fig. 3 shows the Full Adder (FA) circuit that was constantly used through all the design.

An extensive design analysis was carried out in bit-level matrix-format of the MPPAs-based architecture and the achieved hardware was studied comprehensively. In order to generate an efficient architecture for the application, various issues were taken into account. The main one considered was to reduce the gate count, because it determines the number of transistors (i.e., silicon area) to be used for the development of the VLSI accelerator. Power consumption is also determined by it to some extent. The design has also to be scalable to other technologies. The VLSI co-processor integrated circuit was designed using a Low-Power Standard Cell library in a 0.6μm double-poly triple-metal (DPTM) CMOS process using the Tanner Tools® software. Each logic cell from the library is designed at a transistor level. Additionally, S-Edit® was used for the schematic capture of the integrated circuit using a hierarchical approach and the layout was automatically done through the Standard Cell Place and Route (SPR) utility of L-Edit from Tanner Tools®.

4. Performance analysis

4.1 Metrics

In the evaluation of the proposed VLSI-FPGA architectue, it is considered a conventional side-looking synthethic aperture radar (SAR) with the fractionally synthesized aperture as an RS imaging system (Shlvarko et al., 2008), (Wehner, 1994). The regular SFO of such SAR

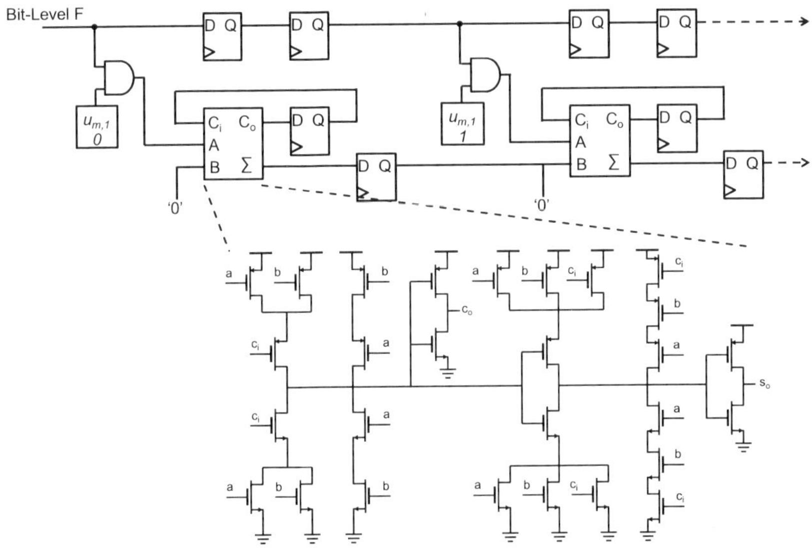

Fig. 3. Transistor-level implementation of the Full Adder Cell.

is factored along two axes in the image plane: the azimuth or cross-range coordinate (horizontal axis, x) and the slant range (vertical axis, y), respectively. The conventional triangular, $\Psi_r(y)$, and Gaussian approximation, $\Psi_a(x)=\exp(-(x)^2/a^2)$ with the adjustable fractional parameter a, are considered for the SAR range and azimuth ambiguity function (AF), (Wehner, 1994). In analogy to the image reconstruction, we employed the quality metric defined as an improvement in the output signal-to-noise ratio (IOSNR)

$$\text{IOSNR} = 10\log_{10}\frac{\sum_{k=1}^{K}\left(\hat{b}_k^{(MSF)}-b_k\right)^2}{\sum_{k=1}^{K}\left(\hat{b}_k^{(p)}-b_k\right)^2}\ ;\ p = 1, 2 \qquad (26)$$

where b_k represents the value of the k*th* element (pixel) of the original image **B**, $\hat{b}_k^{(MSF)}$ represents the value of the k*th* element (pixel) of the degraded image formed applying the MSF technique (19), and $\hat{b}_k^{(p)}$ represents a value of the k*th* pixel of the image reconstructed with two developed methods, $p = 1, 2$, where $p = 1$ corresponds to the RSF algorithm and $p = 2$ corresponds to the RASF algorithm, respectively.

The quality metrics defined by (26) allows to quantify the performance of different image enhancement/reconstruction algorithms in a variety of aspects. According to these quality metrics, the higher is the *IOSNR*, the better is the improvement of the image enhancement/reconstruction with the particular employed algorithm.

4.2 RS implementation results

The reported RS implementation results are achieved with the VLSI-FPGA architecture based on MPPAs, for the enhancement/reconstruction of RS images acquired with different fractional SAR systems characterized by the PSF of a Gaussian "bell" shape in both directions of the 2-D scene (in particular, of 16 pixel width at 0.5 from its maximum for the 1K-by-1K BMP pixel-formatted scene). The images are stored and loaded from a compact flash device for the image enhancement process, i.e., particularly for the RSF and RASF techniques. The initial test scene is displayed in Fig. 4(a). Fig. 4(b) presents the same original image but degraded with the matched space filter (MSF) method. The qualitative HW results for the RSF and RASF enhancement/reconstruction procedures are shown in Figs. 4(c) and 4(d) with the corresponding IOSNR quantitative performance enhancement metrics reported in the figure captions (in the [dB] scale).

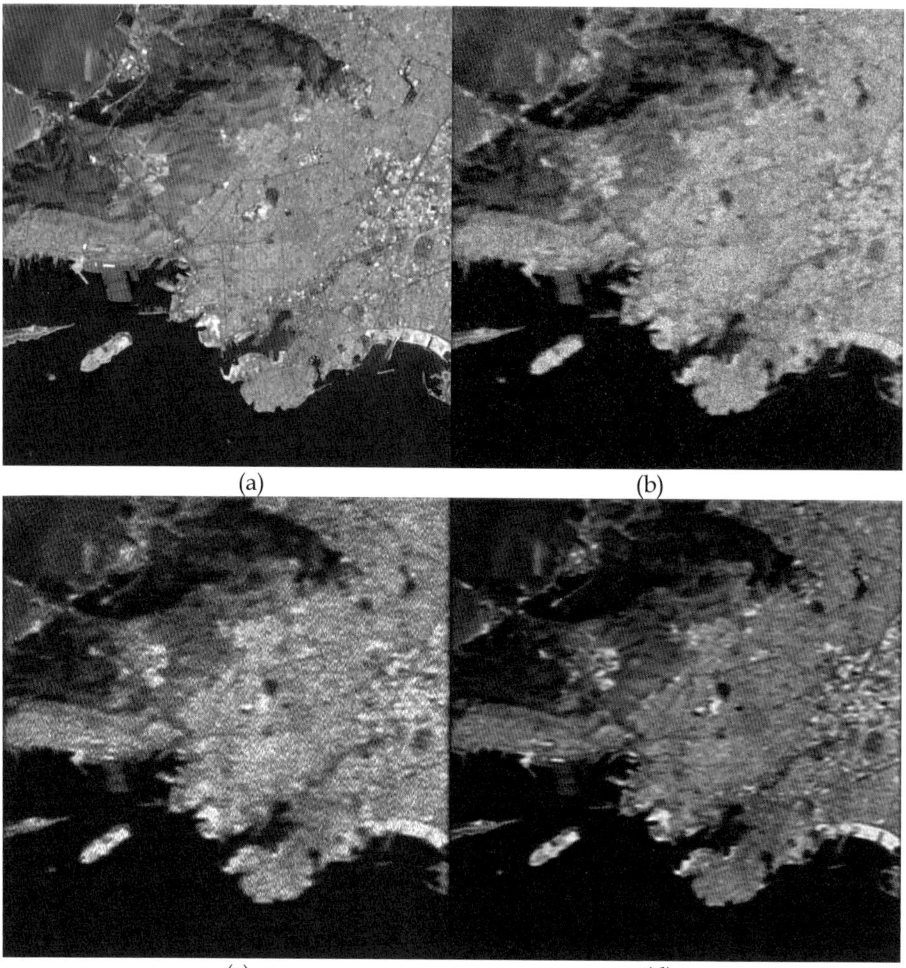

(a) (b) (c) (d)

Fig. 4. VLSI-FPGA results for SAR images with 15dB of SNR: (a) Original test scene; (b) degraded MSF-formed SAR image; (c) RSF reconstructed image (IOSNR = 7.67 dB); (d) RASF reconstructed image (IOSNR = 11.36 dB).

The quantitative measures of the image enhancement/reconstruction performance achieved with the particular employed DEDR-RSF and DEDR-RASF techniques, evaluated via IOSNR metric (26), are reported in Table 1 and Fig. 4.

SNR [dB]	RSF Method *IOSNR* [dB]	RASF Method *IOSNR* [dB]
5	4.36	7.94
10	6.92	9.75
15	7.67	11.36
20	9.48	12.72

Table 1. Comparative table of image enhancenment with DEDR-related RSF and RASF algorithms

From the RS performance analysis with the VLSI-FPGA platform of Fig.4 and Table 1, one may deduce that the RASF method over-performs the robust non-adaptive RSF in all simulated scenarios.

4.3 MPPA analysis

The matrix-vector multiplier chip and all of modules of the MPPA co-processor architecture were designed by gate-level description. As already mentioned, the chip was designed using a Standard Cell library in a 0.6μm CMOS process (Weste & D. Harris, 2004), (Rabaey et al., 2003). The resulting integrated circuit core has dimensions of *7.4 mm* x *3.5 mm*. The total gate count is about 32K using approximately 185K transistors. The 72-pin chip will be packaged in an 80 LD CQFP package and can operate both at 5 V and 3 V. The chip is illustrated in Fig. 5.

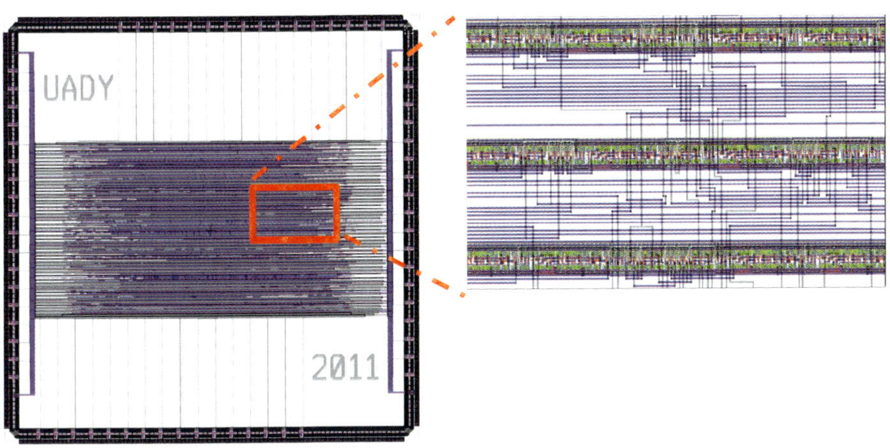

Fig. 5. Layout scheme of the proposed MPPA architecture

Next, Table 2 shows a summary of hardware resources used by the MPPA architecture in the VLSI chip.

Function	Complexity	For $m = 32$
AND	$m \times m$	1024
Adder	$(m + 1) \times m$.	1056
Mux	M	32
Flip-Flop	$[(4m + 2) \times m] + m$	4160
Demux	M	32

Table 2. Summary of hardware resource utilization for the proposed MPPA architecture

Having analyzed Table 2, Fig. 4 and 5, one can deduce that the VLSI-FPGA platform based on MPPAs via the HW/SW co-design reveals a novel high-speed SP system for the real time enhacement/reconstruction of highly-computationally demanded RS systems. On one hand, the reconfigurable nature of FPGAs gives an increased flexibility to the design allowing an extra degree of freedoom in the partitioning stage of the pursued HW/SW co-design technique. On the other side, the use of VLSI co-processors introduces a low power, high-speed option for the implementation of computationally complex SP operations. The high-level integration of modern ASIC technologies is a key factor in the design of bit-level MPPAs. Considering these factors, the VLSI/ASIC approach results in an attractive option for the fabrication of high-speed co-processors that perform complex operations that are constantly demanded by many applications, such as real-time RS, where the high-speed low-power computations exceeds the FPGAs capabilities.

5. Conclusions

The principal result of the reported study is the addressed VLSI-FPGA platform using MPPAs via the HW/SW co-design paradigm for the digital implementation of the RSF/RASF DEDR RS algorithms.

First, we algorithmically adapted the RSF/RASF DEDR-related techniques over the range and azimuth coordinates of the uncertain RS environment for their application to imaging array radars and fractional imaging SAR. Such descriptive-regularized RSF/RASF algorithms were computationally transformed for their HW-level implementation in an efficient mode using parallel computing techniques in order to achieve the maximum possible parallelism in the design.

Second, the RSF/RASF algorithms based on reconstructive digital SP operations were conceptualized and employed with MPPAs in context of the real time RS requirements. Next, the bit-level array of processors elements of the selected reconstructive SP operation was efficiently optimized in a high-speed VLSI architecture using 0.6um CMOS technology with low-power standard cells libraries. The achieved VLSI accelerator was aggregated with a reconfigurable FPGA device via HW/SW co-design paradigm.

Finally, the authors consider that with the bit-level implementation of specialized arrays of processors in VLSI-FPGA platforms represents an emerging research field for the real-time RS data processing for newer Geospatial applications.

6. References

Barrett, H.H. & Myers, K.J. (2004). *Foundations of Image Science*, Willey, New York, NY.

Castillo Atoche A., Torres, D. & Shkvarko, Y. V. (2010). Descriptive Regularization-Based Hardware/Software Co-Design for Real-Time Enhanced Imaging in Uncertain Remote Sensing Environment, *EURASIP Journal on Advances in Signal Processing*, Vol. 2010, pp. 1-31.

Castillo Atoche A., Torres D. & Shkvarko, Y. V. (2010). Towards Real Time Implementation of Reconstructive Signal Processing Algorithms Using Systolic Arrays Coprocessors, *Journal of Systems Architecture*, Vol. 56, No. 8, pp. 327-339.

Franceschetti, G., Iodice, A., Perna, S. & Riccio, D. (2006). Efficient simulation of airborne SAR raw data of extended scenes, *IEEE Trans. Geoscience and Remote Sensing*, Vol. 44, No. 10, pp. 2851-2860.

Greco, M.S. & Gini, F. (2007). Statistical analysis of high-resolution SAR ground clutter data, *IEEE Trans. Geoscience and Remote Sensing*, Vol. 45, No. 3, pp. 566-575.

Henderson, F.M. & Lewis, A.V. (1998). *Principles and Applications of Imaging Radar : Manual of Remote Sensing*, 3rd ed., John Willey and Sons Inc., New York, NY.

Kung, S.Y. (1988). *VLSI Array Processors*, Prentice Hall, Englewood Cliffs, NJ.

Matlab, (2011). Fixed-Point Toolbox™ User's Guide. Available from http://www.mathworks.com

Melesse, A. M., Weng, Q., Thenkabail, P. S. & Senay, G. B. (2007). Remote Sensing Sensors and Applications in Environmental Resources Mapping and Modelling. *Journal Sensors*, Vol. 7, No. 12, pp. 3209-3241, ISSN 1424-8220.

Moldovan, D.I. & Fortes, J.A.B. (1986). Partitioning and Mapping Algorithms into Fixed Size Systolic Arrays, *IEEE Trans. On Computers*, Vol. C-35, No. 1, pp. 1-12, ISSN: 0018-9340.

Plaza, A. & Chang, C. (2008). High-Performance Computer Architectures for Remote Sensing Data Analysis: Overview and Case Study, In: *High Performance Computing in Remote Sensing*, Plaza A., Chang C., (Ed.), 9-42, Chapman & Hall/CRC, ISBN 978-1-58488-662-4, Boca Raton, Fl., USA.

Rabaey, J. M., Chandrakasan, A., Nikolic, B. (2003). *Digital Integrated Circuits: A Design Perspective*, 2nd Ed., Prentice-Hall.

Shkvarko, Y.V. (2006). From matched spatial filtering towards the fused statistical descriptive regularization method for enhanced radar imaging, *EURASIP J. Applied Signal Processing*, Vol. 2006, pp. 1-9.

Shkvarko, Y.V., Perez Meana, H.M., & Castillo Atoche, A. (2008). Enhanced radar imaging in uncertain environment: A descriptive experiment design regularization paradigm, *Intern. Journal of Navigation and Observation*, Vol. 2008, pp. 1-11.

Shkvarko, Y.V. (2010). Unifying Experiment Design and Convex Regularization Techniques for Enhanced Imaging With Uncertain Remote Sensing Data—Part I: Theory. *IEEE Transactions on Geoscience and Remote Sensing*, Vol. 48, No. 1, pp. 82-95, ISSN: 0196-2892.

Wehner, D.R. (1994). *High-Resolution Radar*, 2nd ed., Artech House, Boston, MS.

Weste, N. & D. Harris. (2004). *CMOS VLSI Design: A Circuits and Systems Perspective*, Third Ed., Addison-Wesley.

Yang, C. T., Chang, C. L., Hung C.C. & Wu F. (2001). Using a Beowulf cluster for a remote sensing application, *Proceedings of 22nd Asian Conference on Remote Sensing*, Singapore, Nov. 5-9, 2001.

8

A DSP Practical Application: Working on ECG Signal

Cristian Vidal Silva[1], Andrew Philominraj[2] and Carolina del Río[3]
[1]*University of Talca, Business Informatics Administration*
[2]*University of Talca, Language Program*
[3]*University of Talca, Business Administration*
Chile

1. Introduction

An electrocardiogram (ECG) is a graphical record of bioelectrical signal generated by the human body during cardiac cycle (Goldschlager, 1989). ECG graphically gives useful information that relates to the heart functioning (Dubis, 1976) by means of a base line and waves representing the heart voltage changes during a period of time, usually a short period (Cuesta, 2001).

Putting leads on specific part of the human body, it is possible to get changes of the bioelectrical heart signal (Goldschlager, 1989) where one of the most basic forms of organizing them is known as Einthoven lead system which is shown in Figure 1 (Vidal & Pavesi, 2004; Vidal et al., 2008).

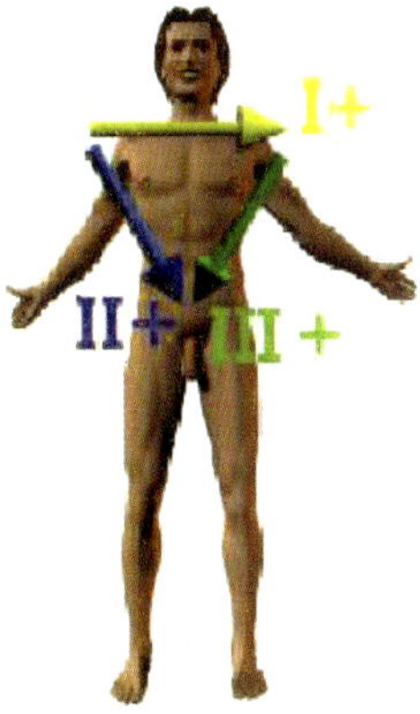

Fig. 1. Einthoven lead system

1.1 ECG usefulness

The ECG has a special value in the following clinical situations (Goldschlager, 1989):

- Auricular and ventricular hypertrophy.
- Myocardial Infarction (heart attack).

- Arrhythmias.
- Pericarditis.
- Generalized suffering affecting heart and blood pressure.
- Cardiac medicine effects, especially digital and quinidine.
- Electrolytic transformations.

In spite of the special value, the ECG is considered only a laboratory test. It is not an absolute truth concerning the cardiac pathologies diagnosis. There are examples of patients presenting string heart diseases which present a normal ECG, and also perfectly normal patients getting an abnormal ECG (Goldschlager, 1989). Therefore, an ECG must always be interpreted with the patient clinical information.

2. Electrocardiographic signal

According to (Proakis & Manolakis, 2007) a signal can be analyzed and processed in two domains, time and frequency. ECG signal is one of the human body signals which can be analyzed and worked in these two domains.

2.1 Time domain of an ECG signal

P, Q, R, S, T and U are specific wave forms identified in the time domain of an ECG signal. The QRS complex, formed by Q, R and S waves, represents a relevant wave form because the heart rate can be identified locating two successive QRS complex. Figure 2 presents typical waves in an ECG signal.

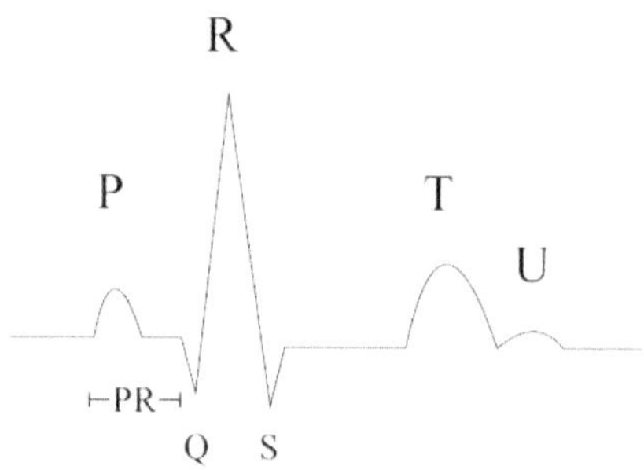

Fig. 2. Typical wave forms of an ECG signal record

2.2 Frequency domain of an ECG signal

Frequency values of an ECG signal vary from 0 Hz to 100 Hz (Cuesta, 2001; Vidal & Pavesi, 2004; Vidal et al., 2008; Vidal & Gatica, 2010) whereas the associated amplitude values vary from 0.02 mV to 5 mV. Table 1 describes the frequency and amplitude values of ECG, EMG (electromiogram), and EEG (electroencephalogram) signals.

Signal	Amplitude (mV)	Frequency range (Hz)
ECG	0.02 - 5.0	0.05 - 100
EEG	0.0002 - 0.3	DC - 150
EMG	0.1 - 5.0	DC - 10000

Table 1. Amplitude and Frequency Range of Basic Bioelectrical Signals of the Human Being

As it is appreciated, the amplitude values of human body bioelectrical signals are measured in micro volts (mV). Furthermore, the amplitude values of these signals are small voltage values and are being caught using traditional electronic devices. This is an important characteristic which must be considered to implement an electronic device in order to obtain bioelectrical signals.
There are different sources of noise at the moment of getting a human body signal. The frequency domain helps us to know of how additional sources affect the important signal in the time domain.
Figure 3 shows frequency range of QRS complex of an ECG signal next to the frequency range of common noise sources.

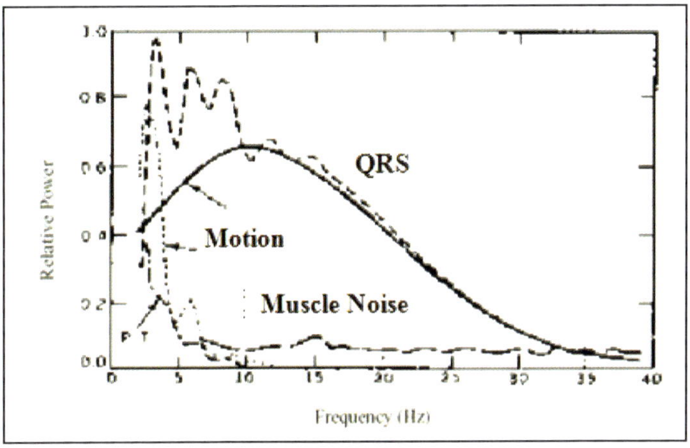

Fig. 3. Frequency range of QRS complex on an ECG signal next to noise sources (Vidal et al., 2008)

3. Digital ECG

Building a device to get and process the ECG signal must consider the signal characteristics. According to (Cuesta, 2001; Vidal & Pavesi, 2004), facing individually each part of the global problems is a technique applicable in order to get good practical results.
Figure 4 presents each part or block of a basic digital ECG according to reviewed literature

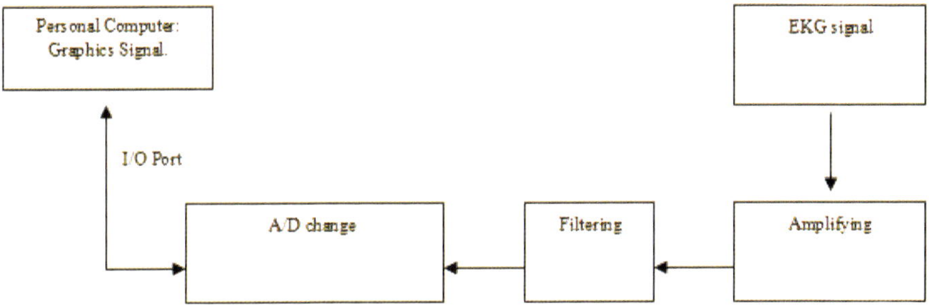

Fig. 4. Blocks Diagram of a Basic Digital ECG.

(Cuesta, 2001; Vidal et al., 2008; Vidal & Gatica, 2010) where the most important part corresponds to the amplifying module because of a bioelectrical signal that represents a low potential, and sophisticated amplifiers are required for obtaining and recording it (Vidal & Pavesi, 2004; Vidal et al., 2008; Vidal & Gatica, 2010).
The following sections present experiences building a device for getting the ECG signal, and works related to processing ECG signal.

3.1 Digital ECG design

Signals produced by bioelectric phenomenon are small potential values and due to this, sophisticated amplifiers are required so as to easily obtain signal values (Vidal & Pavesi, 2004).
Against a physiologic backdrop, these ionic signals are transmitted at a fast-rate without synaptic delay in both direction directed by the electric synapse transmission model. This electric potential is later transformed in a mechanical signal as of using calcium ion that comes from extracellular condition which is also useful for cooking calcium that is released from the internal section of cardiac cells provoking a massive cardiac muscle like a sincitio or functional unit (Clusin, 2008). In this sense, the main finality of an amplifier is to increment the measurable level of the gotten signal by electrodes, avoiding any kind of interference. The capacitive interference of the patient body, electrical fields of electric installations, and other environment electronic devices are examples of interference or noise. (Proakis & Manolakis, 2007) indicate that the quantification can be done using single pole configurations or bipolar. In the single pole quantification, difference between a signal and a common base is measured whereas the bipolar mode measures the difference of two voltage sources (two electrodes) with respect to a common base where any interference voltage generated at the quantification point appears at the amplifier input as common-mode interference signals. Figure 5 illustrates this phenomenon in a bipolar quantification.

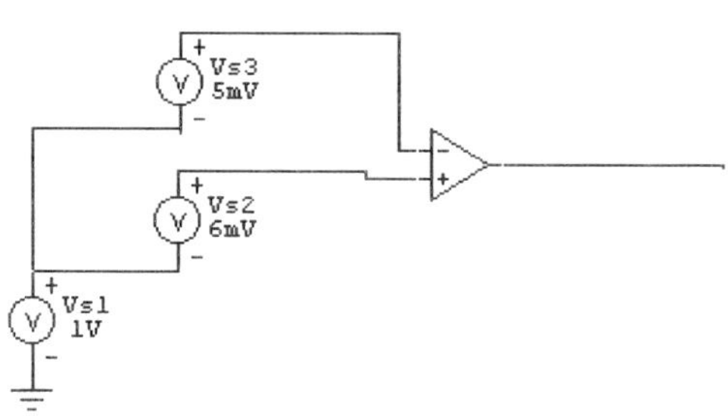

Fig. 5. Common-Mode Interferences in a bipolar quantification

A strong source noise which interferes on the ECG signal is the capacitive interference of the patient body. This interference voltage is coupled to the ECG signal reaching values of 2.4 V approximately. A value which is very higher than the ECG signals value range (0.02 mV to 5 mV). In addition to this interference, the capacitive interference due to the equipment or device used to measure the ECG signal which is produced by the equipment power supply. Another noise source is the denominated inductive interference that is caused by the electric net which produces variable in time magnetic fields inducing extra voltages on the next of patient electrodes (Townsend, 2001).

For these reasons, common mode rejection ratio (CMRR) rate is a desirable characteristic of an amplifier working on differential mode. On a day today practice, a problem denominated contact impedance disbalance appears (Townsend, 2001) that is produced when there are different interfaces impedances between the skin and electrodes in a form that the common-mode potential is higher in one of the two voltage sources. Therefore, part of the common-mode voltage is worked as differential voltage and amplified according to the amplifier gain. This occasionally produces saturation on the next amplifying module stage, if the amplification module were composed by more stages. This voltage, which is generally continuous, can be eliminated using a simple high-pass filter. Hence, the output voltage of the differential amplifier would consist of 3 components (Townsend, 2001; Vidal & Pavesi, 2004):

- Wished output due to the differential amplification on the ECG signal.
- Common-mode signal not wished due to the CMRR is not infinite.
- Common-mode signal not wished due to the disbalance on the impedance contact.

(Wells & Crampton, 2006) indicate that weak signals require an amplification of 1000 at least to produce adequate signal levels for future works on it. (Vidal & Pavesi, 2004) used an instrument amplifier model INA131 which presents a fixed CMRR of 100, and according to the associated datasheet it is adequate for biomedical instrumentation. The analog to digital conversion stage (A/D conversion) is always done when the signal is amplified. The electronic schemes of a digital electrocardiographic device according to (Vidal & Gatica, 2010) are presented on figures 6 and 7, respectively. (Vidal & Pavesi, 2004; Vidal & Gatica, 2010) use the TLC1541 A/D converter. It is necessary to indicate that both electronic items, INA131 and TLC1541, are less expensive.

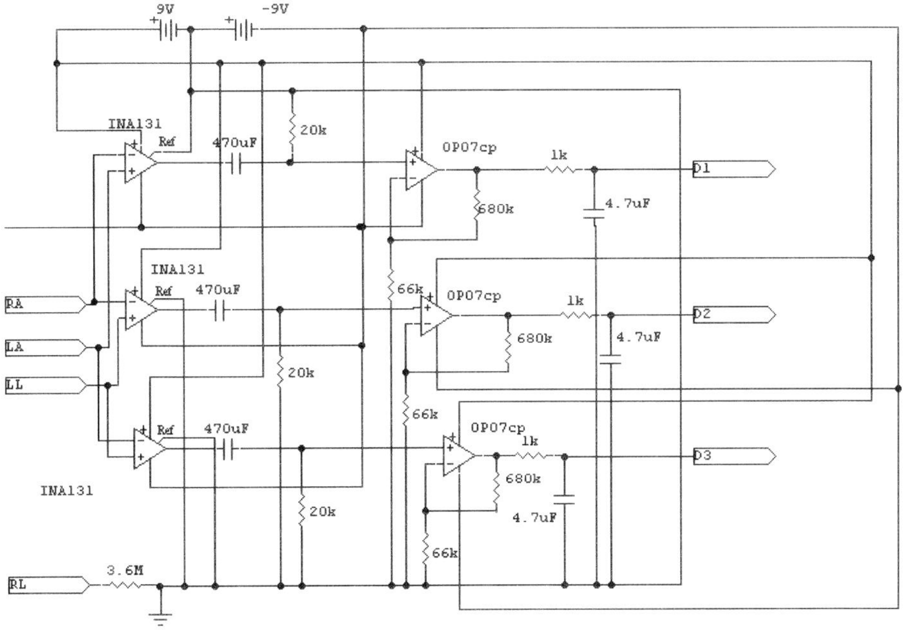

Fig. 6. ECG Signal Amplifying Module Circuit

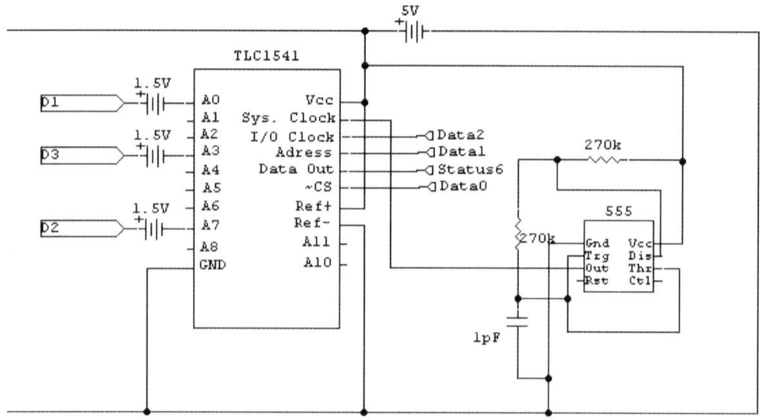

Fig. 7. Data Acquisition Module Circuit

3.2 Acquiring and processing ECG signal

The acquisition data stage has a hardware part composed by the A/D converter, and a software part which is in charge of directing the A/D converter work. Any programming language allowing low level hardware instruction is usable. (Vidal & Pavesi, 2004) and (Vidal & Gatica, 2010) describe the use of C and Visual Basic programming languages for getting and processing the ECG signal. According to these works, the routine written in C language is used to direct the A/D converter functioning using non-standard functions to access the personal computer ports. The obtained quantity of samples is stored in a binary file which is rescued by the Visual Basic programming language routine to processing (applying filters and QRS detection algorithms) and showing the signal. Showing the signal at the computer is done "off-line" from the generated file with the ECG signal samples. As (Vidal & Gatica, 2010) highlights using current high level programming languages would be possible to build a showing graphics routine. Using lineal interpolation it is possible to get high level graphic results. Even though the Nyquist's sample theorem indicates that a signal can be rebuild using an ideal interpolation method (Lindner, 2009; Proakis & Manolakis 2007), by means of lineal interpolation, and through this it is possible to get good results for low frequency signals like ECG. It is possible to build a universal graphics generator for getting signals (Vidal & Pavesi, 2004; Vidal & Gatica, 2010). Figures 8 and 9 present a universal graphics generator for a sine curve signal and a triangle signal, respectively. These signals are low frequency signals (2 Hz) generated by a function or electrical waves generator with some acquisition deformities (high negative values are not considered). Figure 10 shows a pure ECG signal got by means of an implemented ECG system (Vidal & Gatica, 2010).

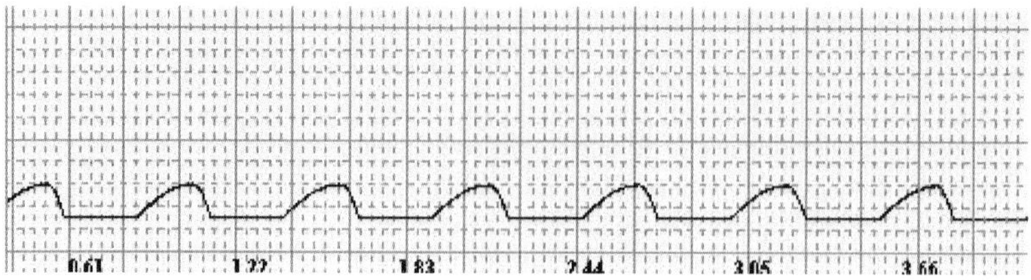

Fig. 8. Sine Signal obtained by the A/D Change Module

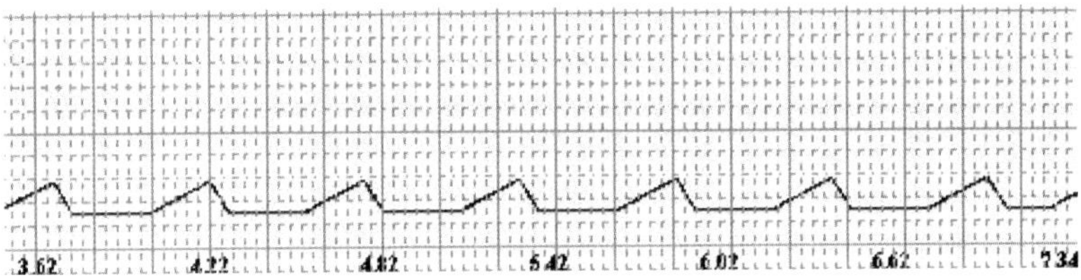
Fig. 9. Triangle Signal obtained by the A/D Change Module

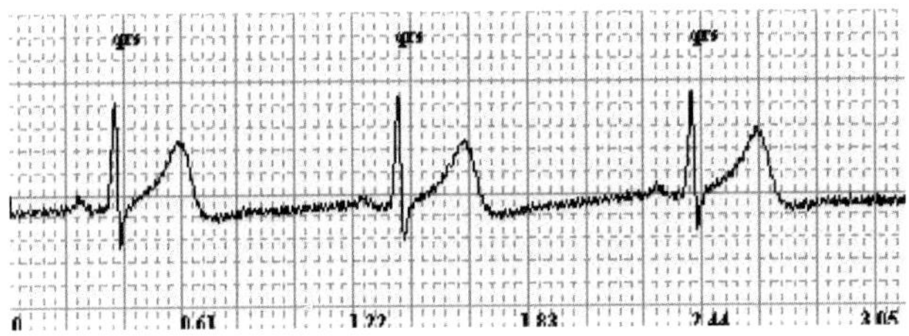
Fig. 10. ECG Signal obtained by the A/D Change Module

4. ECG signal processing

(Vidal & Pavesi, 2004; Vidal & Gatica, 2010) worked on the digital filters application to eliminate noise on an ECG signal, and the use of algorithms for QRS complex detecting. Following subsections describe digital filters to work on the ECG signal, and present the main principles of a QRS detector algorithm (Vidal et al., 2008).

4.1 Digital filters for ECG signal

To work the ECG signal it is necessary to apply digital filters which helps to diminish the noise present on it. One of the most useful filters is Lynn's filters (Goldschlager, 1989) and there are previous works where Lynn's filters are successfully applied to processing ECG signal (Thakor et al., 1984; Kohler et al., 2002; Ahlstrom & Tompkins, 1985). These filters present desirable properties of real-time filters like lineal phase and integer coefficients. There are low-pass and high-pass Lynn's filters versions which are described as follows.

4.1.1 Low-pass filter

Lynn's filters described in (Ahlstrom & Tompkins, 1985) and used on ECG signal processing in (Pan & Tompkins, 1985; Hamilton & Tompkins, 1986), represent a simple and effective form of applying low-pass filter on ECG signals. These filters obey the next transfer function:

$$H(z) = \frac{(1 - z^{-\alpha})^2}{(1 - z^{-1})^2} = \frac{(1 - 2z^{-\alpha} + z^{-2\alpha})}{(1 - 2z^{-1} + z^{-2})} \tag{1}$$

This filter can be implemented by means of the following differences equation:

$$y[n] = 2y[n-1] - y[n-2] + x[n] - 2x[n-\alpha] + x[n-2\alpha] \tag{2}$$

The amplitude answer of this filter is calculated as follows:

$$H(\omega)=\frac{1-2\cos\alpha\omega+\cos 2\alpha\omega+j(2sen\alpha\omega-sen2\alpha\omega)}{1-2\cos\omega+\cos 2\omega+j(2sen\omega-sen2\omega)}=$$

$$\frac{|\cos\alpha\omega-1|}{|\cos\omega-1|}=\frac{sen^2\left(\frac{\alpha}{2}\omega\right)}{sen^2\left(\frac{\omega}{2}\right)} \quad (3)$$

For a sample frequency of 430 Hz, possible α values and associated cut frequency (-3 dB.) are shown in Table 2. Figures 11, 12, and 13 show associated amplitude response for these filters.

α Value	Cut Frequency
3	48 Hz
4	35 Hz
12	11.46 Hz

Table 2. Cut Frequencies of Low-Pass Lynn Filter

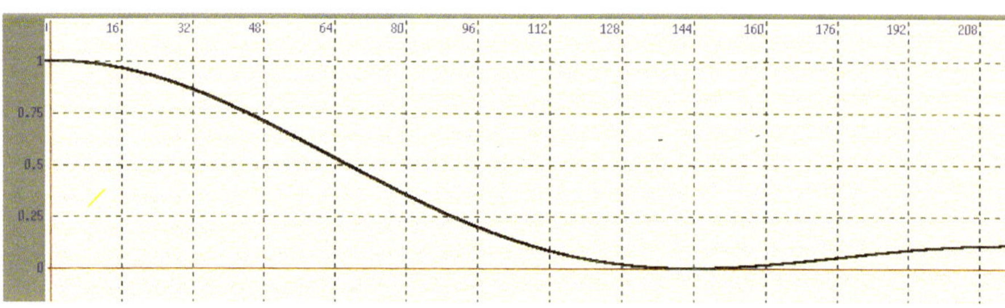

Fig. 11. Amplitude Response of Low-Pass Lynn's Filter for α=3

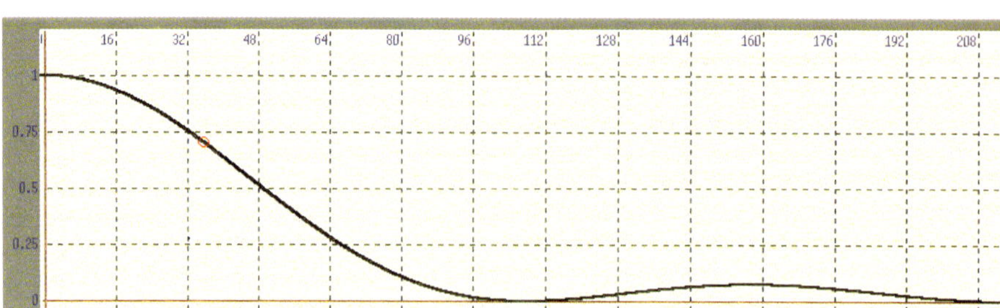

Fig. 12. Amplitude Response of Low-Pass Lynn's Filter for α=4

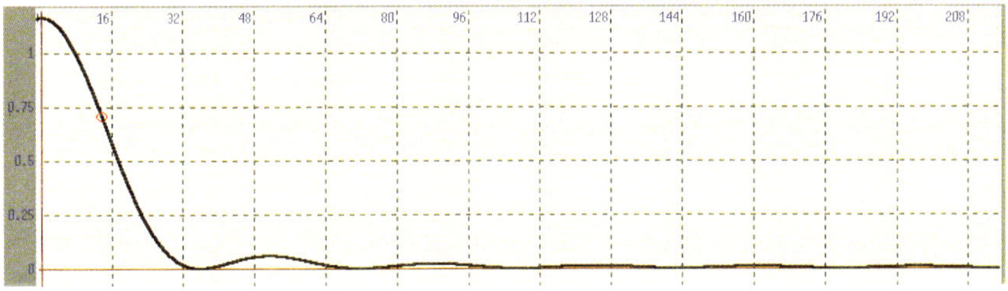

Fig. 13. Amplitude Response of Low-Pass Lynn's Filter for α=5

4.1.2 High pass filters

Like a low-pass Lynn's filters, there are high-pass Lynn's filters which are described in (Ahlstrom & Tompkins, 1985) and applied to ECG signal processing on (Pan & Tompkins, 1985; Hamilton & Tompkins, 1986). These filters are designed using an all-pass filter and resting over it a low-pass filter, and the result is a high-pass filter (Vidal & Pavesi, 2004). However for an effective design, low-pass filter and all-pass filter must be in phase (Smith, 1999).
The High-Pass Lynn's filter starts using the following low-pass filter transfer equation:

$$H(z) = \frac{\left(1 - z^{-\alpha}\right)}{\left(1 - z^{-1}\right)} \tag{4}$$

Amplitude and phase responses are got by:

$$\begin{aligned}
H(\omega) &= \frac{\left(1 - e^{-j\alpha\omega}\right)}{\left(1 - \omega^{-j\omega}\right)} = \frac{1 - \cos\alpha\omega + jsen\alpha\omega}{1 - \cos\omega + jsen\omega} = \\
&\frac{2sen^2\frac{\alpha\omega}{2} + j2sen\frac{\alpha\omega}{2}\cos\frac{\alpha\omega}{2}}{2sen^2\frac{\omega}{2} + j2sen\frac{\omega}{2}\cos\frac{\omega}{2}} = \\
&\frac{sen\frac{\alpha\omega}{2}\left(sen\frac{\alpha\omega}{2} + j\cos\frac{\alpha\omega}{2}\right)}{sen\frac{\omega}{2}\left(sen\frac{\omega}{2} + j\cos\frac{\omega}{2}\right)} = \\
&\frac{sen\frac{\alpha\omega}{2}\left(sen\frac{\alpha\omega}{2} + j\cos\frac{\alpha\omega}{2}\right)\left(sen\frac{\omega}{2} - j\cos\frac{\omega}{2}\right)}{sen\frac{\omega}{2}\left(sen\frac{\omega}{2} + j\cos\frac{\omega}{2}\right)\left(sen\frac{\omega}{2} - j\cos\frac{\omega}{2}\right)} = \\
&\frac{sen\frac{\alpha\omega}{2}\left[\left(sen\frac{\alpha\omega}{2}sen\frac{\omega}{2} + \cos\frac{\alpha\omega}{2}\cos\frac{\omega}{2}\right) - j\left(sen\frac{\alpha\omega}{2}\cos\frac{\omega}{2} - sen\frac{\omega}{2}\cos\frac{\alpha\omega}{2}\right)\right]}{sen\frac{\omega}{2}} = \\
&\frac{sen\frac{\alpha\omega}{2}\left(\cos\left(\frac{\omega}{2}(\alpha-1)\right) - jsen\left(\frac{\omega}{2}(\alpha-1)\right)\right)}{sen\frac{\omega}{2}}
\end{aligned} \tag{5}$$

Finally, amplitude and phase responses are showed on Eq. 6 and Eq. 7, respectively.

$$|H(\omega)| = \left| \frac{sen\frac{\alpha\omega}{2}}{sen\frac{\omega}{2}} \right| \tag{6}$$

$$\Theta(\omega) = -\frac{\omega}{2}(\alpha - 1) \tag{7}$$

The filter's group delay is $(\alpha-1)/2$, and the associated gain for ω=0 is α determined evaluating | H (ω=0) |.

Once completely characterized the low-pass filter, designing the high-pass filter is an easy task using the following transfer function:

$$H(z) = z^{\frac{-(\alpha-1)}{2}} - \left(\frac{1-z^{-\alpha}}{1-z^{-1}}\right)/\alpha = \frac{-1/\alpha + z^{\frac{-(\alpha-1)}{2}} - z^{\frac{-(\alpha-1)}{2}-1} + z^{-\alpha}/\alpha}{1-z^{-1}} \tag{8}$$

This filter can be implemented directly by the following difference equation:

$$y[n] = y[n-1] - x[n]/\alpha + x\left[n - \frac{(\alpha-1)}{2}\right] - x\left[n - \frac{(\alpha-1)}{2} - 1\right] + x[n-\alpha]/\alpha \tag{9}$$

Getting amplitude response for this filter is mathematically complex. Nevertheless, theoretically this filter must have the same cut frequency of the subjacent low-pass filter in inverse order. Furthermore, the values of phase response and group delay of the high-pass filter are the equal to the same parameters for the low-pass filter (Smith, 1999).

For a cut frequency of 430 Hz, α values and associated cut frequency (-3 dB.) are shown on Table 3.

Valor de α	Frecuencia de Corte
850	0.2 Hz.
320	0.5 Hz.
35	5 Hz.

Table 3. Cut Frequencies of High-Pass Lynn Filter

Figures 14, 15 and 16 show the low-pass filter amplitude response which give an idea of the amplitude response of the associated high-pass filter because the cut frequencies are the same.

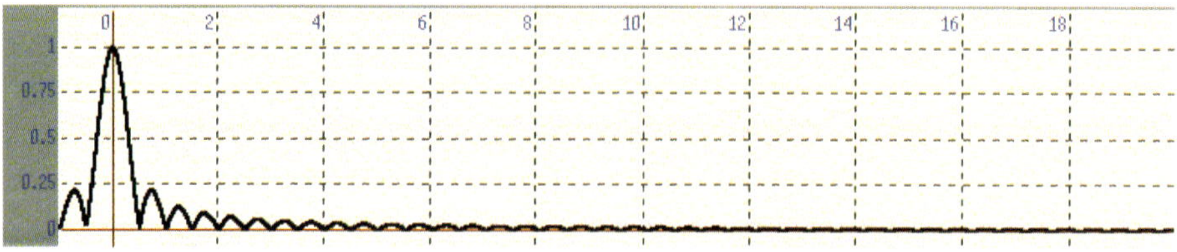

Fig. 14. Low-Pass / High-Pass Lynn's Filter Amplitude Response - Cut Frequency 0.2 Hz

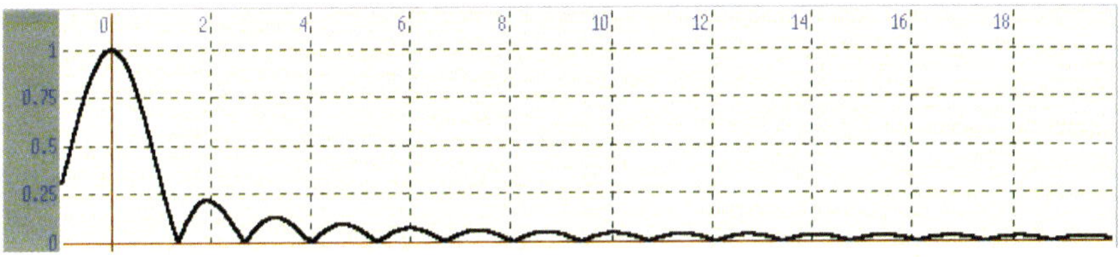

Fig. 15. Low-Pass / High-Pass Lynn's Filter Amplitude Response - Cut Frequency 0.5 Hz

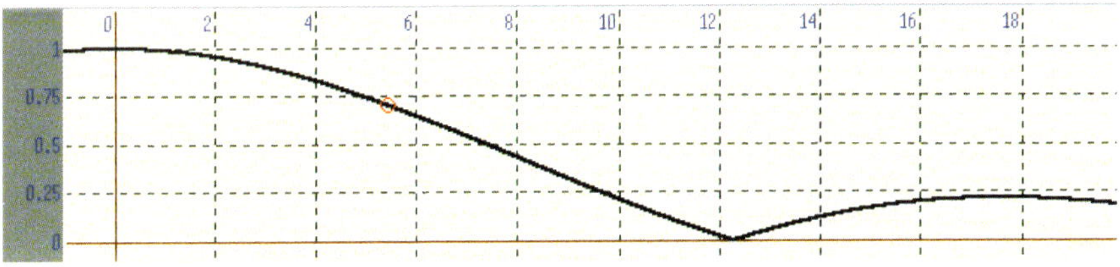

Fig. 16. Low-Pass / High-Pass Lynn's Filter Amplitude Response - Cut Frequency 5 Hz

Figures 17, 18, 19, 20 and 21 present signals registered by an implement ECG device using Figure 4 and 5 circuits (Vidal & Gatica, 2010). Figure 15 shows a pure signal ECG without applying filters to delete noise. Figure 18 shows the 35 Hz low-pass Lynn's filter application on the Figure 17 signal. Figure 18 presents the application of a 48 Hz low-pass filter application over the Figure 17 signal. In Figures 20 and 21 the application of 0.2 and 0.5 high-pass Lynn's filters respectively on the Figure 17 signal is shown. It is important to be aware of the group delay effect on the ECG signal after the 0.2 Hz high-pass Lynn's filter application, 423 samples in this case (around 1 second). Likewise, for the 0.5 Hz high-pass Lynn's filter application there is a group delay of 160 samples.

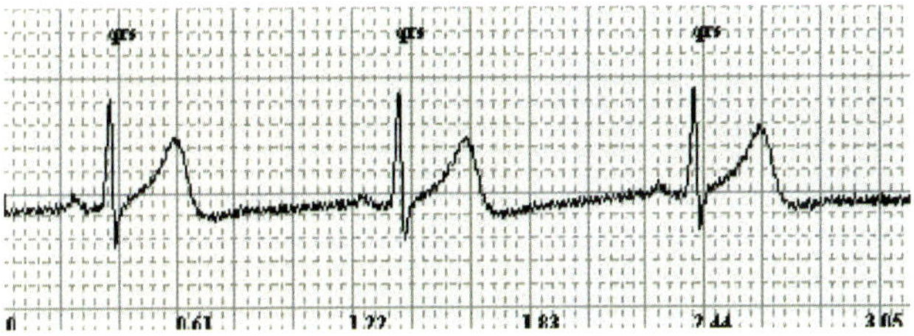

Fig. 17. Pure ECG Signal

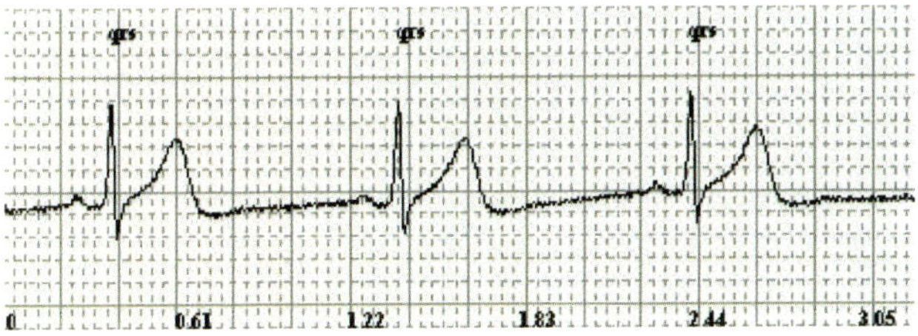

Fig. 18. Filtered ECG Signal Using Low-Pass 35 Hz Lynn's Filter

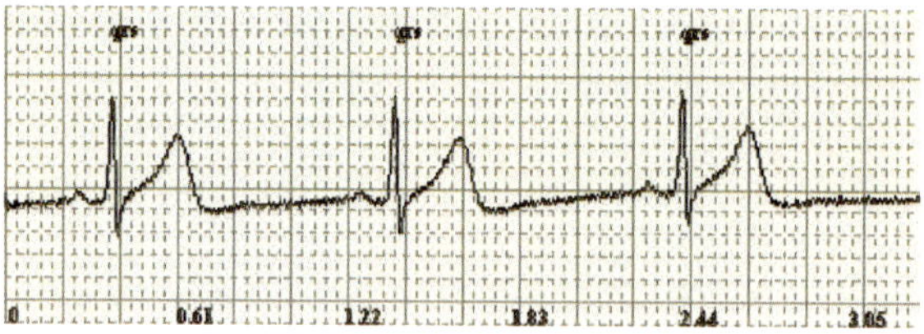

Fig. 19. Filtered ECG Signal Using Low-Pass 48 Hz Lynn's Filter

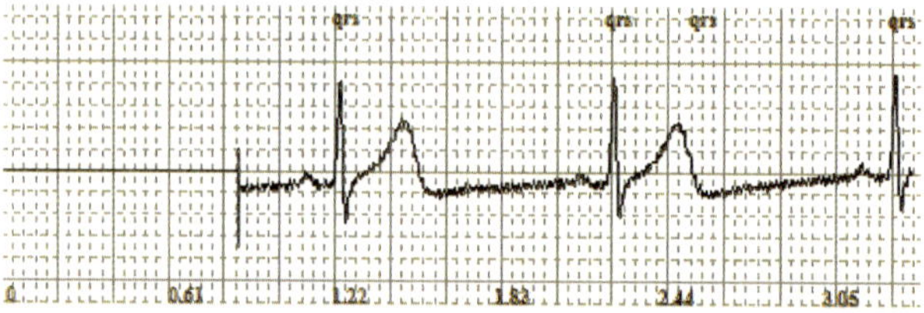

Fig. 20. Filtered ECG Signal Using High-Pass 0.2 Hz Lynn's Filter

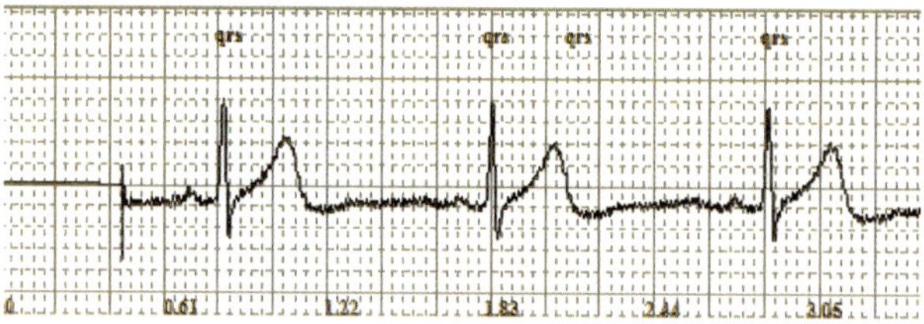

Fig. 21. Filtered ECG Signal Using High-Pass 0.5 Hz Lynn's Filter

The filters application allows improving the ECG signal quality in a remarkable manner. Figure 22 shows the application of a low-pass Lynn's filter of 48 Hz and a high-pass Lynn's filter of 0.5 Hz.

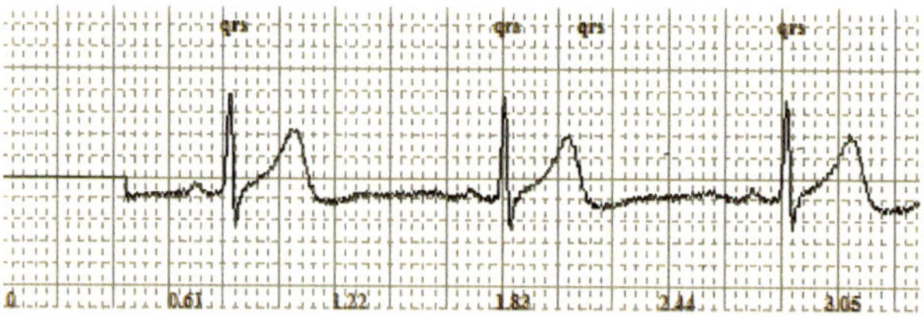

Fig. 22. Filtered ECG Signal Using a Low-Pass 48 Hz Lynn's Filter and a High-Pass 0.5 Hz Lynn's Filter

4.2 QRS detection algorithm on ECG signal

Within the automatic detection waveform of the ECG signal, it is important to detect QRS complex (Cuesta, 2001; Vidal & Pavesi, 2004). This is the dominant feature of the ECG signal. The QRS complex marks the beginning of the contraction of the left ventricle, so the detection of this event has many clinical applications (Vidal et al., 2008; Townsend, 2001).

In the literature there are several algorithmic approaches for detecting QRS complexes of ECG signal with pre-filtering of the signal (Thakor et al., 1984)
The implementation of incremental improvements to a classical algorithm to detect QRS complexes was realized in an experiment as mentioned in (Vidal et al., 2008; Vidal & Gatica, 2010) which in its original form do not have a great performance. The first improvement based on the first derivative is proposed and analyzed in (Friese at al., 1990). The second improvement is based on the use of nonlinear transformations proposed in (Pan & Tompkins, 1985) and analyzed in (Suppappola & Ying, 1994; Hamilton & Tompkins, 1986). The third is proposed and analyzed in (Vidal & Pavesi, 2004; Vidal et al., 2008), as an extension and improvement of that is presented in (Friesen et al., 1994) using characteristics of the algorithm proposed in (Pan & Tompkins, 1985). It should be noted that the three algorithmic improvements recently mentioned, used classical techniques of DSP (Digital Signal Processing). It is noteworthy to indicate that the second improvement proposed in (Pan & Tompkins, 1985) is of great performance in the accurate detection of QRS complexes, for even the modern technology are not able to provide better results.
To test the algorithms that work on ECG signal, it is not necessary to implement a data acquisition system. There are specialized databases with ECG records for analyzing the performance of any algorithm to work with ECG signals (Cuesta, 2001; Vidal & Pavesi, 2004). One of the most important is the MIT DB BIH (database of arrhythmias at Massachusetts Institute of Technology,) (MIT DB, 2008).
In Tables 4, 5, 6 and 7, respectively, are the results obtained with the application of incremental improvements made to the first algorithm for detecting QRS complexes in some records at MIT DB BIH. A good level of performance reached in the final version of algorithm of detection of QRS complexes implemented in this work could be appreciated, (Table 7), compared to its original version (Table 4)

Signal	Pulses Heart (NL)	True Positives (PV)	False Positives (PF)	False Negatives (NF)	(PF + NF) / NL
R. 1118 - S. 1	2278	2278	79676	0	3497,63%
R. 118 - S. 2	2278	2278	77216	0	3389,64%
R. 108 – S. 1	562	562	8933	0	1589,50%
R. 108 – S. 2	562	562	17299	0	3078,11%

Table 4. Results obtained with the Holsinger Algorithm in its Original version, for some of the MIT Database records.

Signal	Pulses Heart (NL)	True Positives (PV)	False Positives (PF)	False Negatives (NF)	(PF + NF) / NL
R. 1118 - S. 1	2278	1558	874	720	69,97%
R. 118 - S. 2	2278	1650	798	628	62,60%
R. 108 – S. 1	562	346	246	216	82,20%
R. 108 – S. 2	562	490	182	72	45,20%

Table 5. Results obtained with the Holsinger Algorithm in its Modified version 1, for some of the MIT Database records.

Signal	Pulses Heart (NL)	True Positives (PV)	False Positives (PF)	False Negatives (NF)	(PF + NF) / NL
R. 1118 - S. 1	2278	2265	4	13	0,5%
R. 118 - S. 2	2278	2263	11	15	1,80%
R. 108 – S. 1	562	538	35	24	10,49%
R. 108 – S. 2	562	524	76	38	20,28%

Table 6. Results obtained with the Holsinger Algorithm Modified Version 2, for some of the MIT Database records

Signal	Pulses Heart (NL)	True Positives (PV)	False Positives (PF)	False Negatives (NF)	(PF + NF) / NL
R. 1118 - S. 1	2278	2265	1	1	0,08%
R. 118 - S. 2	2278	2263	1	2	0,13%
R. 108 – S. 1	562	542	1	15	2,84%
R. 108 – S. 2	562	538	23	21	7,82%

Table 7. Results obtained with the Holsinger Algorithm Modified Version 3, for some of the MIT Database records

5. Conclusion

The implementation of equipments for the acquisition and processing of bioelectrical human signals such as the ECG signal is currently a viable task. This chapter is a summary of previous works with simple equipment to work with the ECG signal. Currently the authors are working on:

- Improvements to the work done:
- Increase the number of leads purchased. The A/D converter allows up to 11 simultaneous inputs and supports a sampling rate of 32 KHz. Under certain conditions. 12 simultaneous leads are required for a professional team.
- Modify RC filters in the filter stage for more elaborate filters to ensure a better discrimination of the frequencies that are outside the pass-band.
- Include isolation amplifiers to increase levels for the security of patients, isolating the direct loop with the computer, which is generated with the design proposed in this chapter. Even with the probability of a catastrophe to occur which are low, but the possibility exists and such massive use should be avoided, before including these amplifiers.
- Unifying routine readings of A/D converter and display of results.
- Certify the technical characteristics of the circuits mounted in order to validate its massive use.

Future works:

- Increase the use of this equipment for capturing other bioelectrical signals such as electroencephalographic and electromygraphic.
- Implement a tool to validate algorithms of detection QRS, based on the MIT DB.

- Apply wavelets in the design and implementation of filtering algorithms and detector of waveforms.
- Analyze other techniques for detection of parameters like, fuzzy logic, genetic approaches and neural networks.
- Make use of information technologies, such as a database in order to obtain relevant information of the patients and their pathologies.

Finally, this work is a good demonstration of the potential applications of Hardware - Software, especially in the field of biotechnology. The quantity and quality of the possible future works show the validity of the affirmation in academic and professional aspects. In addition to the likely use of this work in medical settings, it also gives account of the scope of works such as ECG digital, which are practically limitless.

6. Acknowledgment

To Dr. David Cuesta of the Universidad Politécnica de Valencia for his valuable contributions and excellent disposition to the authors of this work; to cardiologist Dr. Patricio Maragaño, director of the Regional Hospital of Talca's Cardiology department, for his clinical assessment and technical recommendations for the development of the algorithmic procedures undertaken.

7. References

Ahlstrom, M. L.; Tompkins, W. J. (1985). Digital Filters for Real-Time ECG Signal Processing Using Microprocessors, *IEEE Transaction on Biomedical Engineering,* Vol.32, No.9, (March 2007), pp. 708-713, ISSN 0018-9294

Clusin, W. T. (2008). Mechanisms of calcium transient and action potential alternans in cardiac cells and tissues. *American Journal of Physiology, Heart and Circle Physiology, Volume 294, No 1,* (October 2007), H1-H10, Maryland, USA.

Cuesta, D. (September 2001). Estudio de Métodos para Procesamiento y Agrupación de Señales Electrocardiográficas. *Doctoral Thesis, Department of Systems Data Processing and Computers (DISCA) , Polytechnic University of Valencia,* Valencia, Spain.

Dubin, D. (August 1976). *Electrocardiografía Práctica : Lesión, Trasado e Interpretación,* McGraw Hill Interamericana, 3rd edition, ISBN 978-968-2500-824, Madrid, Spain

Goldschlager, N. (June 1989). *Principles of Clinical Electrocardiographic,* Appleton & Lange, 13th edition, ISBN 978-083-8579-510, Connecticut, USA

Friesen, G. M.; Janett, T.C.; Jadallah, M.A.; Yates, S.L.; Quint, S. R.; Nagle, H. T. (1990). A Comparison of the Noise Sensitivity of Nine QRS Detection Algorithms, *IEEE Transactions on Biomedical Engineering,* Vol.31, No.1, (January 1990), pp. 85-98., ISSN 0018-9294

Hamilton, P. S.; Tompkins, W. J. (1986). Quantitative Investigation of QRS Detection Rules Using MIT/BIH Arrhythmia Database, *IEEE Transactions on Biomedical Engineering,* Vol.31, No.3, (March 2007), pp. 1157-1165, ISSN 0018-9294

Kohler, B. –U.; Henning, C.; Orglmeister, R. (2002). The Principles of Software QRS Detection, *IEEE Engineering in Medicine and Biology,* Vol.21, No.1, (January-February 2002), pp. 42-57, ISSN 0739-5175

IEEE Transactions on Biomedical Engineering, Vol.31, No.11, (November 1984), pp. 702-706, ISSN 0018-9294

Lindner, D. (January 2009). *Introduction to Signals and Systems,* Mc Graw Hill Company, First Edition, ISBN 978-025-6252-590, USA

MIT DB. (2008). , MIT-BIH Arrhytmia Database, 20.06.2011, Avalaible from http://www.physionet.org/physiobank/database/mitdb/

Pan, J.; Tompkins, W. J. (1985). A Real-Time QRS Detection Algorithm, *IEEE Transactions on Biomedical Engineering,* Vol.32, No.3, (March 2007), pp. 230-236, ISSN 0018-9294

Proakis, J. ; Manolakis, D. (2007). *Digital Signal Processing : Principles, Algorithms, and Applications,* Prentice Hall, 3rd edition, ISBN 978-013-3737-622, New Jersey, USA

Smith, S. W. (1999). *The Scientist and Engineer's Guide to Digital Signal Processing,* Second Edition, California Technical Publishing, 1999, ISBN 978-096-6017-632, California, USA

Suppappola, S; Ying, S. (1994). Nonlinear Transform of ECG Signals for Digital QRS Detection: A Quantitative Analysis, *IEEE Transactions on Biomedical Engineering,* Vol.41, No. 4, (April 1994), pp. 397-400, ISSN: 0018-9294

Thakor, N. V.; Webster, J.; Tompkins, W. J. (1984). Estimation of QRS Spectra for Design of a QRS Filter, *IEEE Transactions on Biomedical Engineering,* Vol.31, No.11, (2007), pp. 702-706, ISSN 0018-9294

Townsend, N. (2001). Medical Electronics, *Signal Processing & Neural Networks Group, Dept. of Engineering Science, University of Oxford,* 21.06.2011, Available from http://www.robots.ox.ac.uk/~neil/teaching/lectures/med_elec/

Vidal, C.; Pavesi, L. (January 2004). Implementación de un Electrocardiográfo Digital y Desarrollo de Algoritmos Relevantes al Diagnóstico Médico. *Bacherlor Thesis, Computer Engineering, Catholic University of Maule,* Talca, Chile

Vidal, C.; Charnay, P.; Arce, P. (2008). Enhancement of a QRS Detection Algorithm Based on the First Derivative Using Techniques of a QRS Detector Algorithm Based on Non-Linear Transformation, *Proceedings of IFMBE 2008 4th European Conference of the International Federation for Medical and Biological Engineering,* Volume 22, Part 6, pp. 393-396, ISBN 978-354-0892-076, Antwerp, Belgium, December 2009

Vidal, C.; Gatica, V. (2010). Design and Implementation of a Digital Electrocardiographic System, *University of Antioquia Engineering Faculty Scientific Magazine,* No. 55, (September 2010), pp. 99-107, ISSN 0120-0230, Antioquia, Colombia

Wells, J. K.; Crampton, W. G. R. (2006). A Portable Bioamplifier for Electric Fish Research: Design and Construction, *Neotropical Ichthyology,* Volume 4, (2006), pp. 295-299, ISSN 1679-6225, Porto Alegre, Brazil

Applications of the Orthogonal Matching Pursuit/ Nonlinear Least Squares Algorithm to Compressive Sensing Recovery

George C. Valley and T. Justin Shaw
The Aerospace Corporation
United States

1. Introduction

Compressive sensing (CS) has been widely investigated as a method to reduce the sampling rate needed to obtain accurate measurements of sparse signals (Donoho, 2006; Candes & Tao, 2006; Baraniuk, 2007; Candes & Wakin, 2008; Loris, 2008; Candes et al., 2011; Duarte & Baraniuk, 2011). CS depends on mixing a sparse input signal (or image) down in dimension, digitizing the reduced dimension signal, and recovering the input signal through optimization algorithms. Two classes of recovery algorithms have been extensively used. One class is based on finding the sparse target vector with the minimum ell-1 norm that satisfies the measurement constraint: that is, when the vector is transformed back to the input signal domain and multiplied by the mixing matrix, it satisfies the reduced dimension measurement. In the presence of noise, recovery proceeds by minimizing the ell-1 norm plus a term proportional to ell-2 norm of the measurement constraint (Candes and Wakin, 2008; Loris, 2008). The second class is based on „greedy" algorithms such as orthogonal matching pursuit (Tropp and Gilbert, 2007) and iteratively, finds and removes elements of a discrete dictionary that are maximally correlated with the measurement.

There is, however, a difficulty in applying these algorithms to CS recovery for a signal that consists of a few sinusoids of arbitrary frequency (Duarte & Baraniuk, 2010). The standard discrete Fourier transform (DFT), which one expects to sparsify a time series for the input signal, yields a sparse result only if the duration of the time series is an integer number of periods of each of the sinusoids. If there are N time steps in the time window, there are just N frequencies that are sparse under the DFT; we will refer to these frequencies as being on the frequency grid for the DFT just as the time samples are on the time grid. To recover signals that consist of frequencies off the grid, there are several alternative approaches: 1) decreasing the grid spacing so that more signal frequencies are on the grid by using an overcomplete dictionary, 2) windowing or apodization to improve sparsity by reducing the size of the sidelobes in the DFT of a time series for a frequency off the grid, and 3) scanning the DFT off integer values to find the frequency (Shaw & Valley, 2010). However, none of these approaches is really practical for obtaining high precision values of the frequency and amplitude of arbitrary sinusoids. As shown below in Section 6, calculations with time windows of more than 10,000 time samples become prohibatively slow; windowing distorts the signal and in many cases, does not improve sparsity enough for CS recovery algorithms to work; scanning the DFT off integer values requires performing the CS recovery algorithm over and over again with an unknown sparse transform and becomes prohibitively expensive when the number of sinusoids in the signal exceeds 1.

Here we present a new approach to recovering sparse signals to arbitrary accuracy when the parameters of the signal do not lie on a grid and the sparsifying transform is unknown. Our approach is based on orthogonal matching pursuit (OMP), which has been applied to recovering CS signals by many authors (Donoho et al., 2006; Tropp and Gilbert, 2007; Liu and Temlyakov, 2010; Huang and Zhu, 2011). The major difference between our work and previous work is that we add a nonlinear least squares (NLS) step to each OMP iteration. In the first iteration of conventional OMP applied to finding sinusoids, one finds the frequency that maximizes the correlation between the measurement matrix evaluated on an overcomplete dictionary and the CS measurement, solves a linear least squares problem to find the best estimate of the amplitude of the sinusoid at this frequency, and subtracts this sinusoid multiplied by the measurement matrix from the CS measurement. In the second iteration, one finds the frequency that maximizes the correlation between the measurement matrix and the residual measurement, solves a least squares problem for both frequencies to get new estimates of both amplitudes, and subtracts the sum of the two sinusoids multiplied by the measurement matrix from the previous residual. This process is described in detail in „Algorithm 3 (OMP for Signal Recovery)" in the paper by Tropp and Gilbert (2007) and in our Table 1 in Section 3. Our approach proceeds in the same way as conventional OMP but we substitute a **Nonlinear Least Squares** step for the linear least squares step. In the NLS step, we use a minimizer to find better values for the frequencies without reference to a discrete grid. Because the amplitudes are extremely sensitive to the accuracy of the frequencies, this leads to a much better value for the amplitudes and thus to a much more accurate expansion of the input signal. Just as in conventional OMP, we continue our algorithm until a system level threshold in the residual is reached or until a known number of sinusoids is extracted. A related procedure for matching pursuit but not yet applied to compressive sensing or orthogonal matching pursuit is described by Jacques & De Vleeschouwer (2008). What we refer to as the NLS step appears in their Section V, eq. (P.2).

Our approach to CS recovery differs from most methods presented to date in that we assume our signal (or image) is sparse in some model as opposed to sparse under some transform. Of course, for every sparse model there is some sparsifying transform, but it may be easier in some problems to find the model as opposed to the transform. Models inevitably involve parameters, and in most cases of practical interest, these parameters do not lie on a discrete grid or lie on a grid that is too large for efficient discrete processing techniques (see the discussion in Section 1 of Jacques & De Vleeschouwer, 2008). For instance, to recover the frequency of a sinusoid between 0 and 1 to precision of 10^{-16} would require 10^{16} grid points. While we first developed our method to find the frequency and amplitude of sinusoids, like OMP it is readily adaptable to signals that are the superposition of a wide range of other models. In Section 2, we present background material on the OMP, NLS and CS methods on which our method is based. In Section 3, we develop the model-based OMP/NLS formulation. Sections 4 and 5 contains the application to signals that consist of a sum of a small number of sinusoids. Section 6 compares performance of our algorithm to conventional OMP using a linear least square step and to penalized ell-1 norm methods.

2. Background

Our method and results rely heavily on work in three well-known areas: orthogonal matching pursuit, nonlinear least squares and compressive sensing.

2.1 Compressive sensing

In compressive sensing (Donoho, 2006; Candes & Tao, 2006; Baraniuk, 2007), a sparse vector **s** of dimension *N* can be recovered from a measured vector *y* of dimension ***M*** **(*M* << *N*)** after transformation by a sensing matrix Θ as shown in eq. (1)

$$y = \Theta s + n \quad (1)$$

where n is a noise vector. Often, Θ is factored into two matrices, $\Theta = \Phi\Psi$ where Φ is a „random" mixing matrix and Ψ is a Hermitian matrix with columns that form a basis in which the input vector is sparse. A canonical example is the case in which the input is a time series with samples taken from a single sinusoid with an integer number of periods in the time window. These data are not sparse but are transformed into a sparse vector by the discrete Fourier transform (DFT). Note that although Θ is not square and hence not invertible, Ψ is both square and invertible. Work in compressive sensing has shown (Donoho, 2006; Candes & Tao, 2006; Baraniuk, 2007) that under quite general conditions, all *N* components of **s** may be recovered from the much smaller number of measurements of ***y***. With no noise ($\boldsymbol{n} = 0$) recovery proceeds by minimizing the ell-1 norm of a test vector **s′** (the ell-1 norm of **s′** is given by the sum of the absolute values of the elements of **s′**) subject to the constraint $y = \Theta s'$. In the presence of noise, recovery proceeds by minimizing a linear combination of the ell-1 norm of the target vector and the ell-2 norm of the residual vector given by $y - \Theta$ **s**

$$s'(\lambda) = \text{argmin}_s(\lambda\,||s||_1 + ||y - \Theta s||^2) \quad (2)$$

where the parameter λ is chosen such that the signal is optimally recovered (Baraniuk, 2007; Loris, 2008).

2.2 Orthogonal Matching Pursuit method

Orthogonal matching pursuit (OMP) is an alternative method that can be used to find the target vector **s** from the measurement vector ***y***. Matching pursuit has a rich history in signal processing long before CS and has appeared under many names (Mallat & Zhang, 1993; Pati et al., 1993; Davis et al., 1997). With the advent of CS, many variants of OMP have been applied to recovery including methods called MOMP, ROMP, CoSaMP, etc. (Needell and Tropp, 2008; Needell and Vershynin, 2009; Huang and Zhu, 2011) but with one exception (Jacques and De Vleeschouwer, 2008) discussed below, all of these methods recover frequencies (or other parameters) from discrete grids.

The basic idea of all matching pursuit algorithms is to minimize a cost function to obtain frequencies of sinusoids present in the signal. First, take the frequency corresponding to the smallest value of the cost function and calculate the linear least squares estimate for the complex amplitude at this frequency. Second, mix this sinusoid with the known CS mixing matrix Φ and remove this mixed approximation to the first sinusoid from the CS measurement vector [***y*** in eq. (1)]. This process is repeated until a known number of sinusoids is found or a system-defined threshold is reached. For frequencies not on the DFT grid of the time series, OMP can be improved by evaluating the cost function on an overcomplete dictionary (Candes et al. 2011), but as in the ell-1 estimates discussed above, this step becomes computationally intractable long before machine precision can be obtained for arbitrary frequencies.

2.3 Nonlinear Least Squares method

Here we follow the development of nonlinear least squares (NLS) given by Stoica and Moses (1997). Their eq. (4.3.1) defines a cost function to be minimized as a function of the vectors ω,α,φ

$$f(\omega,\alpha,\varphi) = \Sigma_t \, | y(t) - \Sigma_k \, \alpha_k \exp[i(\omega_k t+\varphi_k)] |^2 \tag{3}$$

where the sums are over the number of sinusoids present in the signal, $\boldsymbol{k}$ = 1 to $\boldsymbol{K}$ and the time points run from $\boldsymbol{t}$ = 0 to N-1. Stoica and Moses also show (see their eqs. 4.3.2-4.3.8), that the frequency vector ω is the critical unknown and the amplitude and phase (or complex amplitude) are simply „nuisance" parameters that are obtained from ω. While eq. (3) appears to require simultaneous solution for three real vectors, each of length K, Stoica and Moses (eqs. 4.3.2-4.3.8) show that the problem can be reduced to solving for just the frequency vector ω and that the complex amplitude vector can be calculated directly from the frequency vector. We use a version of these equations below in eqs. (8) and (13).

In principle, solution of the CS analog of eq. (3) could be performed to directly obtain the parameters of a sparse signal, but in practice, direct solution of eq. (3) is not computationally practical (Stoica and Moses, 1997). The difficulty is that even for a small $\boldsymbol{K}$, eq. (3) is highly multimodal (see for example, Fig. 1 in Li et al., 2000) and the solution requires extremely good first guesses for the vector ω. Even with good initial values for ω, performance guarantees are difficult to find and continue to be the subject of intense investigation (Salzo and Villa, 2011 and references therein).

Similar two-step model-based approaches to estimating the frequency and amplitude of real and complex sinusoids have been discussed previously in the literature (Stoica et al., 2000: Li et al., 2000; Chan and So, 2004; Christensen and Jensen, 2006). Stoica et al. discuss the use of NLS to obtain the amplitude for complex sinusoidal signals given the frequency; Li et al. and Chan and So discuss a combined matching pursuit NLS approach similar to ours for obtaining the frequencies of complex and real harmonic sinusoidal signals, respectively; and Christensen and Jensen use matching pursuit plus NLS to estimate frequencies in a signal that is the sum of arbitrary frequency real sinusoids. To the best of our knowledge, our paper is the first application of an OMP/NLS algorithm to estimate the frequency and amplitude from CS measurements.

3. Formulation of OMP with an NLS step for CS

3.1 Mathematical development

Consider a continuous time signal $\boldsymbol{X}(\boldsymbol{t})$ consisting of $\boldsymbol{K}$ complex sinusoids of the form

$$X(t) = \sum_{k=1}^{K} a_k \exp(2\,\pi \mathrm{i}\, f_k\, t) \tag{4}$$

where $\boldsymbol{a}_k$ is the complex valued amplitude and f_k is the real valued frequency of the kth sinusoid. This signal model is broadly applicable [see Duarte and Baraniuk (2010) and references therein]. We take f_{min} = 0 and f_{max} = 1 to set up our test problem; we sample $X(t)$ at the Nyquist rate for complex signals, $\Delta t = 1/f_{max}=1$, to obtain the sampled time series X_S of length N from t = 0 to t = N-1 where N is the number of time samples. As in all applications of compressive sensing, we make a sparsity assumption, $K << N$, and mix the input signal vector X_S plus sampled noise n down in dimension to the measured vector y of dimension M:

$$y = \Phi(X_S + n), \tag{5}$$

where Φ is an M x N mixing matrix and $K<M << N$. Note that in eq. (5) n is added to X_S prior to mixing. In other models noise is added to the mixed version of X_S, ΦX_S, or even to Φ itself. We generate the elements of Φ using the pseudo-random number functions in our software (*Mathematica* and *Python*) such that they are taken uniformly from the set of nine complex numbers: {-1 – i, -1, -1 + i, -i, 0, i, 1 – i, 1, 1 + i} or equivalently, the elements are taken from the sum of random integers drawn from {-1,0,1} plus i times different random integers drawn from {-1,0,1}. We use a complex mixing matrix because our signal model is complex. The noise is assumed to be independent and identically distributed (i.i.d.) Gaussian noise with standard deviation $\sigma/2^{1/2}$ and is added to the real and the imaginary part of each element of X_S, so that the covariance of n is $\sigma^2 I$, where I is the N x N identity matrix. If the frequencies lie on the DFT frequency grid associated with the time series t = 0 to t = N-1, eq. (5) can be solved for the frequencies by writing s = **DFT** X_S, substituting X_S = **IDFT** s (**IDFT** = Inverse DFT) in eq. (5), and solving $y = \Phi(\mathbf{IDFT}\ s+n)$ by minimizing the ell-1 norm of s subject to the measurement constraint eq. (5) if n = 0 or by minimizing the ell-1 norm penalized by an arbitrary fraction of the constraint (LASSO) in the presence of noise (Candes & Wakin, 2008; Loris, 2008).

Although the noise is assumed to be i.i.d., the mixing matrix Φ colors the noise in the observation vector y. The covariance of y is given by

$$\mathrm{Cov}[y] = \sigma^2 \Phi\, \Phi^H, \tag{6}$$

and the standard maximum likelihood estimator requires definition of a weighting matrix W,

$$W = (\Phi\, \Phi^H)^{-1}, \tag{7}$$

where the superscript H indicates the Hermitian conjugate (see Stoica et al., 2000 and Chan and So, 2004, for a discussion of weighted estimators in NLS). If the inverse in eq. (7) is ill-conditioned or does not exist, this indicates a poor choice of mixing matrix Φ and another one should be chosen. The maximum likelihood estimator (MLE) for X_S, $C(Z)$ is solved by finding the vector Z that minimizes the weighted square of the residual given by

$$C(Z) = (\Phi Z - y)^H\, W\, (\Phi Z - y): \tag{8}$$

Z is a vector taken from the linear subspace spanned by at most K complex sinusoids sampled over t = 0 to N-1 (see the corresponding equation for determining the amplitude and frequency of a sum of complex sinusoids in a system that does not have compressive sensing, Stoica and Moses, 1997, eq. 4.3.6). CS recovery is equivalent to determining the spectral support (that is, the K unknown frequencies) of the input signal X_S, or equivalently determining the vector Z that minimizes eq. (8) (Duarte & Baraniuk, 2010). In the absence of noise, weighting with W is unnecessary because the solution is exact and both the weighted and un-weighted residuals are zero. Finding the K sinusoids that solve eq. (8) is the standard NLS problem and if this were computationally tractable, the problem would be solved. But as pointed out by Li et al. (2000) [see also the discussion in Stoica & Moses (1997)], "the NLS cost function in (3) is usually multimodal with many local minima," and "the minimization of the NLS cost function requires the use of a very fine searching algorithm and may be computationally prohibitive."

One way out of this dilemma is to use NLS in place of least squares within OMP. This has two advantages over using NLS by itself. First, the frequency band over which one has to search in NLS is reduced from the entire frequency band to the frequency grid spacing in the over-complete dictionary used in OMP. Second, the estimates of the frequencies at any given iteration are improved from the values on the grid by using NLS in the previous iteration (see the discussion of a similar continuous version of matching pursuit by Jacques & De Vleeschouwer, 2008).

The first step in our formulation is to define the vector function of frequency, $x(f)$, as the time series for a unity amplitude complex sinusoid at frequency f evaluated at integral sampling times $\boldsymbol{t} = 0$ to $\boldsymbol{t} = \boldsymbol{N} - 1$,

$$x(f) = [1, e^{i2\pi f}, e^{i4\pi f}, \ldots, e^{i2(N-1)\pi f}]. \tag{9}$$

Note that the solution for $\boldsymbol{X}_S$ in eq. (5) is a linear combination of $\boldsymbol{K}$ vectors $x(f_i)$, $(\boldsymbol{i} = 1, \boldsymbol{K})$. To use OMP, we need an over-complete dictionary (Candes et al., 2010) which means that $\boldsymbol{x(f)}$ is evaluated on a fine grid oversampled by the factor $\boldsymbol{N_f}$ from the DFT grid. The second step is to define a function that can be evaluated on the fine grid to find a grid frequency close to one of the true frequencies in the input signal $\boldsymbol{X_s}$. Here we use the function $\boldsymbol{G(f, r)}$ given by

$$G(f, r) = 1 \Bigg/ \left| r^* W r - \frac{[\Phi x(f)]^* W r}{\sqrt{[\Phi x(f)]^* W [\Phi x(f)]}} \right| \tag{10}$$

where initially $\boldsymbol{r} = \boldsymbol{y}$ and subsequently, $\boldsymbol{r}$ equals the residual of $\boldsymbol{y}$ with 1 to $\boldsymbol{K}$ components removed as discussed below. We calculate the argmax of $\boldsymbol{G(f,y)}$ over $\boldsymbol{f}$ in the dictionary frequencies to make a first estimate of one of the frequencies present in the input signal $\boldsymbol{X}(\boldsymbol{t})$. If there is no noise and if there is only one sinusoid, this procedure provides the dictionary vector whose frequency is nearest that of the input sinusoid. If multiple sinusoids are present, the maximum of $\boldsymbol{G(f,y)}$ occurs at one of the dictionary vectors whose frequency is near one of the input sinusoids provided that the dictionary is sufficiently over-complete and that $\boldsymbol{\Phi}$ possesses the restricted isometry property (Duarte and Baraniuk, 2010). Note that $\boldsymbol{G(f,r)}$ is the inverse square of the distance between $\boldsymbol{r}$ and the linear span of $\boldsymbol{x(f)}$ in the $\boldsymbol{W}$-normed inner product space (defined by $< \boldsymbol{a}, \boldsymbol{b} > = \boldsymbol{a}^H \boldsymbol{W} \boldsymbol{b}$). Thus finding the argmax of $\boldsymbol{G(f,r)}$ is equivalent to finding the argmax of the inner product of the residual with the product of $\boldsymbol{\Phi}$ times the dictionary vectors $\boldsymbol{x(f_j)}$ for all f_j on the over-complete frequency grid (see Tropp and Gilbert, 2007, Algorithm 3, Step 2).

Given estimates of the frequencies $\{\boldsymbol{f_1, f_2, \ldots f_K}\}$ present in the input signal, we can find estimates of the amplitudes of each sinusoid by using the least squares estimator $\boldsymbol{A(U)}$ for the amplitude vector $\{\boldsymbol{a_1, a_2, \ldots a_K}\}$ (see Stoica and Moses, 1997, eq. 4.3.8 and Stoica et al., 2000)

$$A(U) = (U^H W U)^{-1} U^H W y \tag{11}$$

where $\boldsymbol{U}$ is the spectral support matrix given that depends on $\{f_1, f_2, \ldots f_K\}$ through

$$\boldsymbol{U} = \Phi\,\{\, x(f_1),\ x(f_2), \ldots x(f_K)\} \tag{12}$$

Note that if there is no noise and if all frequencies are known exactly, eq. (11) can be verified by substituting $\boldsymbol{y} = \boldsymbol{U}\,\boldsymbol{A}(\boldsymbol{U})$, which is equivalent to eq. (5), on the right hand side of eq. (11).

Finally, starting from estimates of the frequencies and amplitudes from OMP as described above, apply weighted NLS to get better values. This is done by finding the frequency or set of frequencies $f = \{\, f_1, f_2, \ldots f_K \}$ that minimize the functional $\boldsymbol{R(f)}$ given by

$$R(f) = |\{A[U(f)]U(f) - y\}^{H}\, W\,\{A[U(f)]U(f) - y\}\,|, \tag{13}$$

which is the same as the weighted least squares estimator given by eq. (8) with the substitution of $A[\boldsymbol{U(f)}]$ defined by eq. (11) for the amplitude vector and $\boldsymbol{U(f)}$ defined by eq. (12) for the mixed sinusoids (see the analogous equations in Stoica and Moses, 1997, eqs. 4.3.7 and 4.3.8). The product $A[\boldsymbol{U(f)}]\ \boldsymbol{U(f)}$ in eq. (13) is the same as $\boldsymbol{\Phi Z}$ in eq. (8).

3.2 Algorithm description

As described in Table 1, the first step in the first iteration of the Do loop is estimation of the first frequency in the spectral support of the signal X_s. This is given by the frequency of the sinusoid whose image after multiplication by $\boldsymbol{\Phi}$ has the maximum correlation with the observation vector y (see, for example, Tropp and Gilbert, 2007 Algorithm 3, step 2). Here we use the equivalent form, the argmax of $\boldsymbol{G(f, y)}$ with respect to f to obtain the first estimate of the frequency of the first sinusoid f_1 in eq. (4). At this point previous implementations of discrete OMP use the amplitude estimator eq. (11) to estimate the amplitude of the first sinusoid $a_1 = A[\boldsymbol{\Phi}\ x(f_1)]$, multiply this amplitude estimate times $x(f_1)$, given by eq. (9), and by the measurement or mixing matrix $\boldsymbol{\Phi}$ and subtract this vector from the measurement vector y to form the first residual r_1.

In our algorithm, we proceed differently by improving the precision of the frequency estimates using NLS before finding the amplitude estimate. We take the frequency f_1 from the argmax of $\boldsymbol{G(f, y)}$ evaluated on a discrete set of frequencies and use that as the starting value to solve the NLS problem given by eq. (13). We have used several methods and several different software packages to solve the NLS problem. A simple decimation routine [i.e., tabulating $\boldsymbol{R(f)}$ from f_1-Δf to f_1+Δf (Δf is the over-complete grid spacing) in 10 steps, finding the argmin, decreasing Δf by a factor of 10, tabulating and finding the argmin of $\boldsymbol{R(f)}$ again until the specified precision is reached] works well but is not very efficient. Powell's method in *Python* ("scipy.optimize.fmin_powell") and one of the Newton methods, the PrincipalAxis method, and the Conjugate Gradient method in *Mathematica*'s minimizer "FindMinimum" all work and take less time than the decimation routine. A detailed investigation of minimizers for the NLS step in our version of OMP is beyond the scope of this chapter. The oversampling N_f required for our method and that required for conventional OMP are nearly identical as discussed below in Section 6.

Given the better value of f_1, we compute a_1 from eq. (11) and a new value of the residual $\boldsymbol{r}$ with the NLS estimate of the first signal removed from y as in OMP

$$r_1 = y - A(U_1)U_1 = y - a_1\,\boldsymbol{\Phi}\,x(f_1). \tag{14}$$

where $U_1 = \Phi x(f_1)$. The argmax of $G(f, r_1)$ now yields a first estimate of the frequency of the second sinusoid, f_2. Next improve the estimates of both f_1 and f_2 by again solving the NLS problem by minimizing the functional $R(f)$ over $f = \{f_1, f_2\}$. Note that this overwrites the previous estimate of the first frequency f_1. The amplitudes a_1 and a_2 are recalculated using (8) with U_2 given by

$$U_2 = [\ \Phi x(f_1), \Phi x(f_2)\ ] \tag{15}$$

for the latest values of f_1 and f_2, Finally, in this iteration estimates of the first two sinusoids are removed from y:

$$r_2 = y - A(U_2)\, U_2. \tag{16}$$

If K, the total number of sinusoids present in the signal, is known, this process is repeated K times until f_K and a_K are obtained. In the absence of noise, the sum of these sinusoids solves (5) exactly and $r_K = 0$.

```
Inputs:
        CS Mixing Matrix Φ
        Measured data y
        Maximum number of sinusoids K or threshold T
        f_min, f_max
        Oversampling ratio for dictionary N_f
Initialize
        U = [ ]
        r_0 = y
        K_T = K
        W = (Φ Φ^H)^-1
        Δf = (f_max-f_min)/(N N_f)
Do i = 1 to K
        f_i = Argmax G(f, r_i-1) over {f_min, f_min+Δf,...f_max-Δf,f_max}
        {f_1,f_2,...f_i} = Argmin[R(f) with initial value f = {f_1, f_2, ...,f_i} ]
        U = {Φx(f_1), Φx(f_2)... Φx(f_i)}
        r_i = y- A(U) U
        If r_i^H W r_i < T:
                K_T = i
                Break
        End If
End Do
Output of Do:
        K_T
        {f_1, f_2, ...f_KT}
        {a_1,a_2,...a_KT}= A [[{ Φx (f_1), Φx (f_2)... Φx (f_K) }]
```

Table 1. OMP/NLS Algorithm.

There are two methods to handle the case where the actual number of sinusoids present is unknown, yet still smaller than K. The simpler method, applicable for high SNR (σ small compared to the smallest signal amplitude), is to perform K iterations of the OMP/NLS algorithm, which will incur an additional noise folding penalty, by projecting the additional noise dimensions onto the solution. The second method is to stop when the residual can be explained by noise alone though hypothesis testing. At the solution, the weighted squared residual $r^H_k W r_k$ will display a χ-squared statistic with $2k$ degrees of freedom, where k is the actual number of sinusoids present in the signal. The hypothesis that the residual is caused by noise alone, is accepted when $r_k^H W r_k < \sigma^2 T$ for some threshold T and rejected otherwise. The value for T is dependent on a user selectable significance level, the probability of incorrectly rejecting the given hypothesis. For a significance level of α, T = CDF^{-1}(1-α), where CDF is the cumulative distribution function of the chi-squared distribution with $2k$ degrees of freedom. We used $\alpha = 0.05$ in our simulations, but α is an application-specific parameter.

4. Results for sparse sinusoids without noise

4.1 Signal composed of a single sinusoid

Consider first a signal of the form given by eq. (4) with $K = 1$, $f_1 = 0.44681715350529533$ and $a_1 = 0.8018794857541801$. Fig. 1 shows a plot of $G(f, y)$ for $N = 1024$, $M = 4$. Finding the argmax of $G(f, y)$ evaluated for 32,768 frequencies between f_{min} and f_{max} ($N_f = 32$) yields an initial value for the frequency and amplitude of $f_1 = 0.446808$ and $a_1 = 0.801070+0.026421i$. Minimization of $R[\{\Phi x(f_1)\}]$ starting at $f_1 = 0.446808$ yields a final value f_1 equal to the input frequency f_1 with error less than $1\text{x}10^{-16}$ (machine precision) and the amplitude a_1 through $A[\{\Phi\, x(f_1)\}]$ equal to the input amplitude with error less than $4\text{x}10^{-16}$.

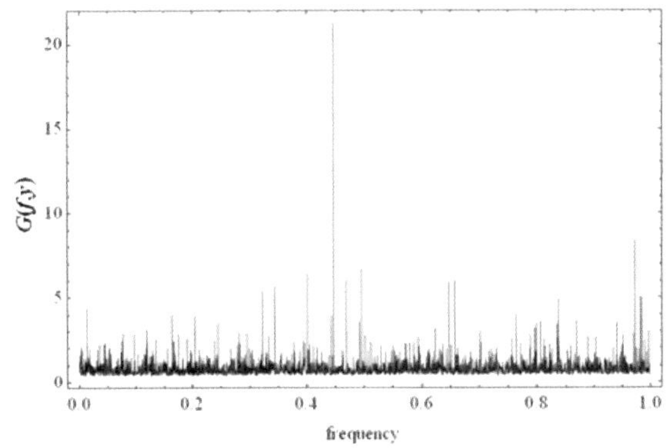

Fig. 1. $G(f, y)$ as a function of frequency for a signal composed of a single sinusoid mixed with N = 1024x4. Note the appearance of a single strong peak in the estimator that serves as an excellent starting value for minimizing the functional $R(f)$ given in eq. (13).

4.2 Signal composed of a 20 sinusoids

The algorithm also works for multiple frequencies. More than 20 independent tests were performed for an input signal composed of 20 independent frequencies randomly chosen between 0 and 1; all frequency components have amplitude of 1. In all tests our algorithm recovered the 20 frequencies to machine precision with a 128x1024 mixing matrix. For test 1, shown in detail here, the closest frequency pairs in the signal are {0.2663, 0.2689} and {0.7715, 0.7736}, but while signals with nearly the same frequency are difficult cases, here the combined OMP/NLS recovers all the sinusoids to machine precision. Fig. 2 shows the initial calculation of $\boldsymbol{G(f, y)}$ for a 128x1024 mixing matrix and 8192 frequency points (N_f = 8). Note that most, but not all of the frequencies have peaks in the initial scan of $\boldsymbol{G(f, y)}$. Fig. 3 shows $\boldsymbol{G(f, r_{19})}$ during the 20th iteration of the Do loop in the algorithm shown in Table 1. After refining the frequencies by finding the minimum of $\boldsymbol{R(f)}$ in (10), the frequency errors are reduced to less than 10^{-16} and the amplitude errors are reduced to 4×10^{-14}. Our results compare favorably to those obtained using other recovery methods for a test problem with 20 arbitrary frequency complex sinusoids, $\boldsymbol{N}$ = 1024, and variable numbers of measurements $\boldsymbol{M}$ (Duarte and Baraniuk, 2010).

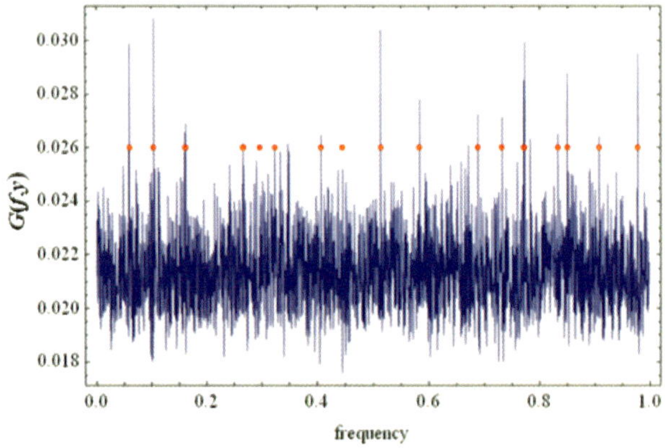

Fig. 2. The initial calculation of $G(f, y)$ for a signal with 20 input frequencies mixed with a 128 x 1024 matrix. The red dots indicate the input frequencies.

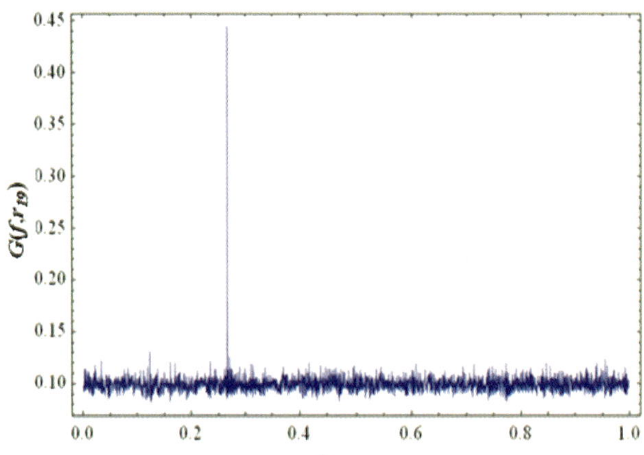

Fig. 3. The next to last calculation of $G(f, r_{19})$ for a signal with 20 input frequencies mixed with a 128x1024 matrix showing a large peak near the frequency of the only remaining sinusoid.

4.3 Signal composed of 2 sinusoids with large dynamic range

For signals composed of 2 well separated frequencies but widely different amplitudes in the absence of noise, we recover the amplitude and frequency of the 2 sinusoids when $a_1 = 1$ and a_2 is as small as 10^{-14} with an 8x1024 mixing matrix. For this case the amplitude and frequency of the large signal are recovered to machine precision while the frequency and amplitude error of the weak signal are 3×10^{-4} and 1%, respectively. Naturally, such performance is not found in the presence of noise as discussed below.

4.4 Signal composed of 2 sinusoids with closely spaced frequencies

We have also input a signal with 2 very closely spaced frequencies and unity amplitudes. For frequencies {0.3389, 0.3390} we recover the frequencies to machine precision with a 16x1024 mixing matrix. Smaller values of M for the mixing matrix yield one root half way between the two frequencies. For frequencies {0.33899,0.33900} mixed with 16x1024 and 32x1024 matrices the OMP part of our algorithm yields a signal with one frequency at 0.338995 and an amplitude of 1.9996. Attempts to find a second frequency yield a badly conditioned matrix for $U^H W U$ and the inversion required to find the 2nd amplitude in eq. (11) fails. For a 64x1024 mixing matrix OMP finds two separated estimates of the frequencies and this allows NLS determination of both frequencies to an accuracy of a few parts in 10^5. These results are in contrast to those obtained using the "spectral compressive sensing" algorithms that use "a signal model that inhibits closely spaced sinusoids" (Duarte and Baraniuk, 2010).

4.5 Dependence on dimensions of the mixing matrix

We have investigated the requirements on M, the small dimension of the measurement matrix, to recover a signal composed of a small number of sinusoids using the OMP-NLS algorithm. Fig. 4 shows the fraction of failed recoveries as a function of M for a problem in which the signal is composed of 1,3,5, or 7 sinusoids and $N = 128$. For each value of K we performed 1000 trials so a failure fraction of 0.1 corresponds to 100 failures. The conventional relation between K, M, and N for recovery is given by $M = C\ K\ \log(N/K)$ (Baraniuk, 2007; Candes and Wakin, 2008). From Fig. 4 we see that the curves for K = 3,5 and 7 are equispaced and correspond to $C \sim 1.5$.

We have also investigated several different types of the measurement matrix as displayed in Fig. 5. The three curves correspond to three different measurement matrices. For the blue curve the mixing matrix is generated from the sum of random integers drawn from {-1,0,1} plus i times different random integers drawn from {-1,0,1}; for the red curve, complex numbers with the real and imaginary parts given by reals uniformly distributed between -1 and 1 and i times uniformly distributed reals; for the magenta curve, the mixing matrix is generated from randomly chosen -1's and 1's. The magenta curve for a real mixing matrix made from 1's and -1's is inferior to the blue and red curves for the two complex mixing matrices. The differences between the red and blue curves in Fig. 5 appear to be random fluctuations and are in agreement with other CS results that Gaussian and Bernoulli measurement matrices perform equally well (Baraniuk, 2007; Candes and Wakin, 2008). Fig. 6 compares calculations with the weighting matrix given by eq. (7) to calculations with the weighting matrix set to the identity matrix. One can see that the green curve with the weighting matrix set to the identity matrix is significantly worse in the important region of less than 1% failure.

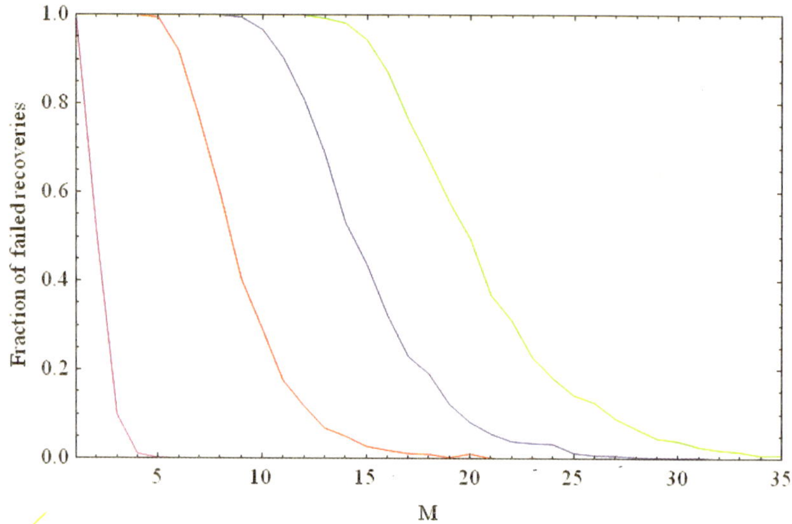

Fig. 4. Fraction of failed recoveries as a function of the small dimension of the mixing matrix M for signals consisting of 1 (magenta), 3 (red), 5 (blue) and 7 (green) sinusoids. The large dimension of the mixing matrix is $N = 128$ and 1000 trials were performed for each value of M.

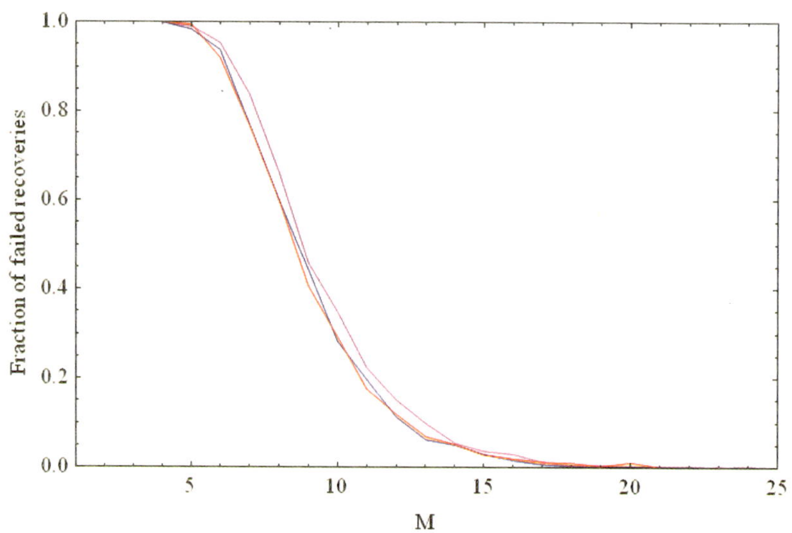

Fig. 5. Fraction of failed recoveries as a function of the small dimension of the mixing matrix M. For the blue curve the mixing matrix is generated from the sum of random integers drawn from {-1,0,1} plus i times different random integers drawn from {-1,0,1}; for the red curve, complex numbers with the real and imaginary parts given by reals uniformly distributed between -1 and 1; for the magenta curve, the entries of the mixing matrix are randomly chosen from -1 and 1.

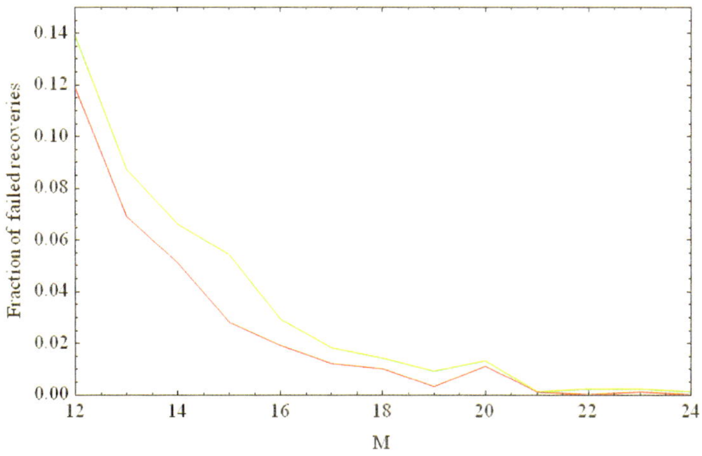

Fig. 6. Fraction of failed recoveries as a function of the small dimension of the mixing matrix M. The red curve is with the weighting matrix defined by eq. (7). The green curve has the weighting matrix set to the identity matrix.

5. Results for sparse sinusoids with noise

5.1 Signal composed of a single sinusoid with noise

Figs. 7 (a) and (b) show the error in the recovery of a single-frequency, unity amplitude signal as a function of the small dimension M of an Mx1024 mixing matrix $\boldsymbol{\Phi}$ with $\sigma = 10^{-2}$ for 100 realizations of the noise. As M increases the standard deviations of the errors in both frequency and amplitude, σ_f and σ_a, decrease as expected since more measurements are made to average a given noise level. The decrease of about a factor of 3 in σ_f and σ_a for a factor of 10 increase in M is consistent with estimates based on SNR (Shaw and Valley, 2010; Davenport et al., 2006). Fig. 8 shows σ_f and σ_a as a function of s averaged over 20 different 4x1024 mixing matrices. Both σ_f and σ_a are proportional to σ with σ_a about 2 to 3 orders of magnitude larger than σ_f.

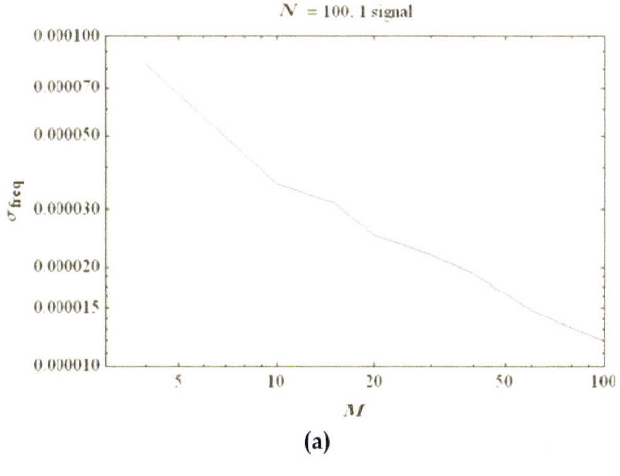

(a)

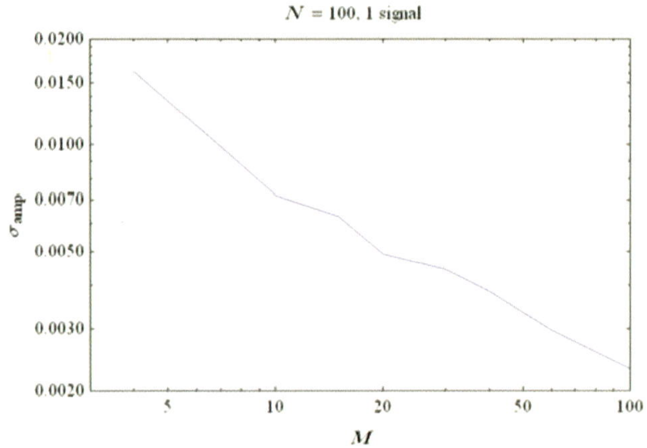

(b)

Fig. 7. Standard deviation of the errors in frequency and amplitude of sinusoids mixed by a mixing matrix Φ with dimensions *M* x 1024 recovered using OMP/NLS as a function of the small dimension *M* of the mixing matrix Φ for $\sigma = 10^{-2}$. The results are obtained from the average of 100 independent calculations. (a) Frequency, (b) amplitude error.

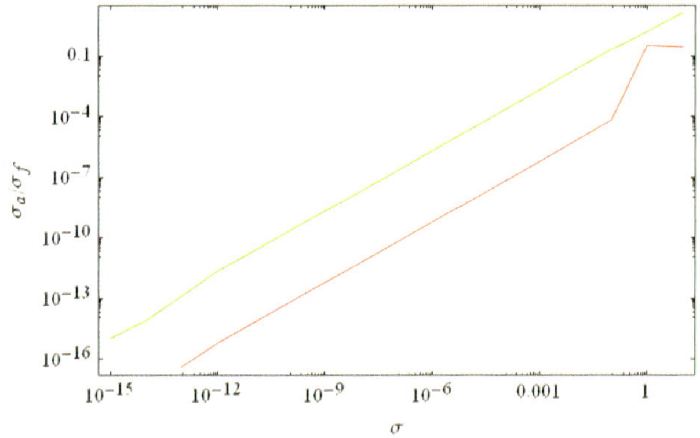

Fig. 8. Standard deviation of the frequency and amplitude errors, σ_f (lower red curve) and σ_a (upper green curve), as a function of σ averaged over 20 different 4x1024 mixing matrices.

5.2 Signal composed of 2 sinusoids with 100:1 dynamic range

Noise also affects the ability of our algorithm to recover a small signal in the presence of a large signal. Figs. 9 and 10 show σ_f and σ_a for a test case in which the amplitudes are given by {1.0, 0.01}, $\boldsymbol{M}$ = 10, $\boldsymbol{N}$ =1024 and the frequencies are well separated. These results are for a single realization of the mixing matrix and averaged over 20 realizations of the noise. Note that as expected, the frequency and amplitude of the large-amplitude component are much better recovered than those of the small-amplitude component. Knowledge of the parameters of the small component essentially disappears for σ greater than about 0.005. Tests with the small amplitude equal to 0.001 and 0.0001 suggest that this threshold scales with the amplitude of the small signal.

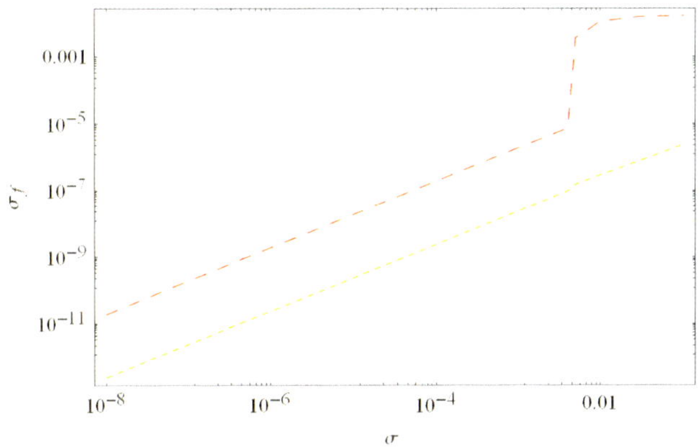

Fig. 9. Standard deviation of the error in the recovered frequency σ_f as a function of noise standard deviation σ for an input signal that consists of two complex sinusoids with amplitudes 1 and 0.01. The green, short dashed curve corresponds to the strong signal; the red, long dashed, to the weak signal. Each curve is averaged over 20 realizations of the noise.

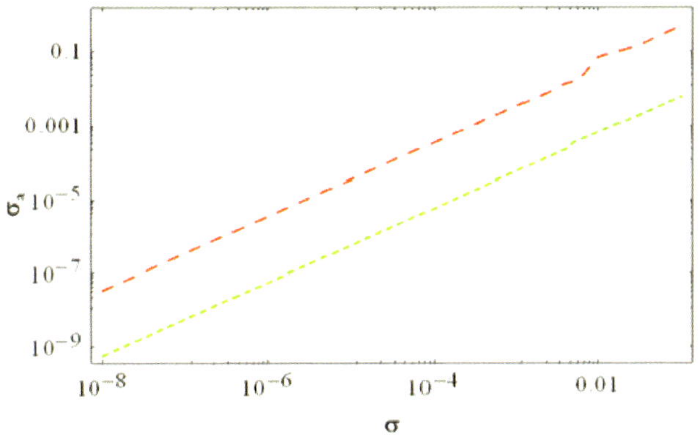

Fig. 10. Standard deviation of the amplitude error σ_α as a function of noise standard deviation σ for an input signal that consists of two complex sinusoids with amplitudes 1 and 0.01. The green, short dashed curve corresponds to the strong signal; red, long dashed, to the weak signal. Each curve is averaged over 20 realizations of the noise.

5.3 Signal composed of 2 sinusoids with closely spaced frequencies in noise

We have also investigated the ability of our algorithm to separate two closely spaced frequencies in the presence of noise. Fig. 11 shows σ_f and σ_a for the case with input frequencies {0.3389, 0.3390}, unity amplitude and a 16x1024 mixing matrix. Note that significant amplitude error occurs at $\sigma > 10^{-4}$ compared to the single frequency results. The frequencies are roughly correct but are not separated for $\sigma > 10^{-2}$.

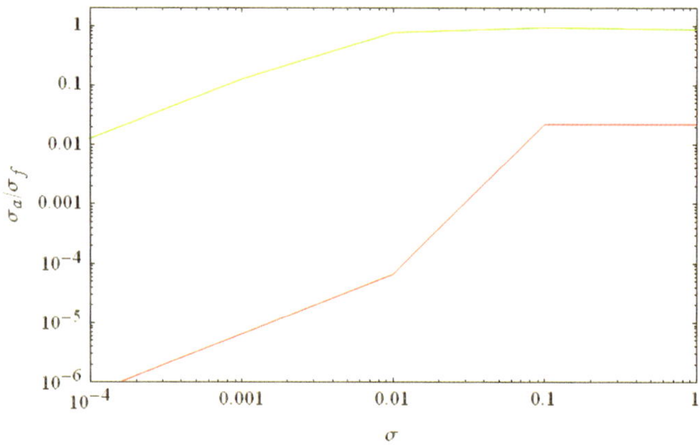

Fig. 11. Standard deviation in frequency σ_f (red-lower curve) and amplitude σ_a (green upper curve) for the case with input frequencies {0.3389, 0.3390}, unity amplitude and a 16x1024 mixing matrix.

6. Comparison with other recovery methods

In this section we compare our version of OMP with an NLS optimization step for the sinusoid frequency and amplitude at each iteration to two common methods for CS recovery: OMP with a linear least squares amplitude estimator at each iteration and convex optimization based on the ell-1 norm of the sparse target vector plus the ell-2 norm of the measurement constraint given by eq. (2). It should be noted that most of the cases presented in the previous sections cannot be solved with OMP/LS or penalized ell-1 norm methods so it is necessary to pick a special case to even perform the comparison. Consider a noise-free signal that consists of 5 unity amplitude sinusoids at 5 different frequencies. We assume N=1024 time samples and an M=30 x N=1024 complex measurement matrix made up of the sum of random reals plus i times different random reals, both sets of reals uniformly distributed between -1 and 1.

6.1 Baseline case OMP-NLS

We performed 100 different calculations with the frequencies chosen by a pseudo-random number generator. In order to control the number of significant figures, we took the frequencies from rational numbers uniformly distributed between 0 and 1 in steps of 10^{-6}. Table 2 shows the fraction of failed recoveries and the average standard deviation in the value of the recovered frequency as a function of the oversampling ratio.

6.2 OMP with Linear Least Squares

We performed the same 100 calcuations using conventional OMP in which the NLS step is replaced by LS as in Tropp and Gilbert (2007). Note that the number of failed recoveries is about the same as the baseline OMP-NLS but the frequency error is huge by comparison. This is the natural result of the frequency grid, which is the limit on the OMP resolution. Timing comparisons with our software show that OMP-NLS takes about 50% longer than conventional OMP. We have also windowed the OMP calculations in order to reduce „spectral leakage" and hopefully achieve better performance. Aside from the lowered failure fraction for N_f = 2, windowing OMP appears to have no statistically significant effect.

Method\ N_f	1	2	4	8
OMP with NLS	95	41	11	6
OMP	96	35	11	6
OMP with window	93	19	9	10

(a)

Method\ N_f	1	2	4	8
OMP with NLS	3.9 10^-15	3.9 10^-15	3.5 10^-15	3.7 10^-15
OMP	0.000150	0.000136	0.000085	0.000060
OMP with window	0.000168	0.000141	0.000084	0.000059

(b)

Table 2. Comparing OMP with NLS to OMP and OMP with windowing for 4 values of the overcomplete dictionary N_f = 1,2,4,8. (a) failure fraction, %. (b) rms error in recovered frequencies.

We have also compared windowed OMP to OMP/NLS in the presence of noise. Fig. 12 shows the frequency and amplitude errors, σ_f and σ_a, as a function of the noise standard deviation σ for OMP (blue) and OMP-NLS (red) for a signal composed of two sinusoids with N = 128, M = 20 and N_f = 4 averaged over 100 trials with randomly chosen input frequencies. Note that the OMP frequency error drops to an asymptote of about 6×10^{-4} and the OMP amplitude error to about 0.23 for $\sigma < 0.1$ while the OMP-NLS errors continue to drop linearly proportional to σ for $\sigma < 0.1$.

6.3 Convex optimization

We have performed the same calculations with a penalized ell-1 norm code (Loris, 2008). None of these calculations returns reliable estimates of frequencies off the grid. Windowing helps recover frequencies slightly off the grid but not arbitrary frequencies. Subdividing the frequency grid allows finer resolution in the recovery but only up to the fine frequency grid.

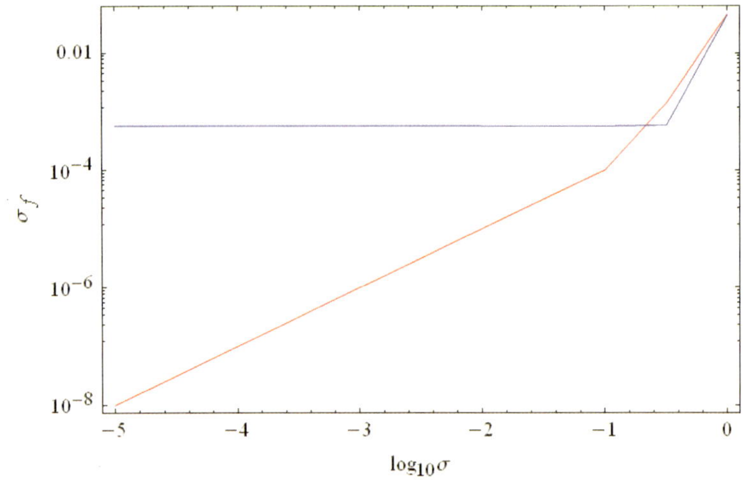

(a)

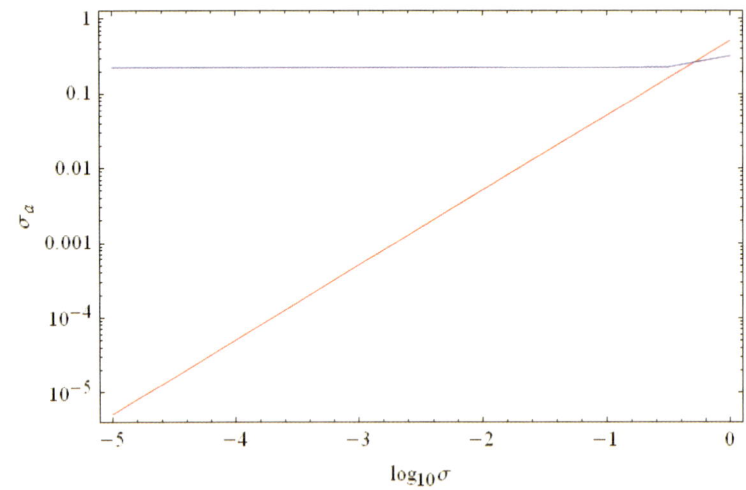

(b)

Fig. 12. Frequency and amplitude errors, σ_f and σ_a, as a function of the noise standard deviation σ for OMP (blue) and OMP-NLS (red) for a signal composed of a two sinusoids with $N = 128$, $M = 20$ and $N_f = 4$ averaged over 100 trials with randomly chosen input frequencies. (a) Frequency error. (b) Amplitude error.

The ell-1 norm code used in our studies (Loris, 2008) can be used with the frequency grid subdivided by 8 or more, but the results are not sparse for the test case described above. More frequencies are returned than in the input signal. Good approximations (consistent with the OMP estimates) can be obtained by precisely thresholding the recovered vector s in eq. (2), but the threshold is dependent on the oversampling ratio and on the random seed used to generate the frequencies.

7. Performance estimates

As discussed above, this study is based on experimental or empirical evaluation (i.e. numerical simulations) of a proposed technique for recovering compressively sensed signals. The weakness of such a study is that calculations alone do not provide performance guarantees while the strength of such a study is that calculations can evaluate practical cases that would be encountered in real applications. Regardless, it is necessary to know when and how an algorithm fails for it to be of much use, and we can use prior work on performance guarantees for CS, OMP and NLS to help us.

Consider first the noise-free case in which the number of sinusoids is known. Here the difference between success and failure is computationally obvious. If the recovery is successful, the residual after extraction of the known number of sinusoids collapses to near the machine precision. If it fails, the residual remains at about the level of the initial measurement vector y. In the presence of noise the situation is similar except the collapse is to the system noise level. If the number of sinusoids is unknown, then recovery proceeds until the system noise level is reached, but statistical testing must be used to determine if the residual at this threshold is due to noise or incorrectly recovered signals.

Practical use of the OMP/NLS algorithm requires at a minimum empirical knowledge of where the algorithm fails and ultimately, performance guarantees and complexity estimates (operation counts). Since this algorithm couples two well known algorithms, in principle we can rely on previous work. The problem can be divided into 3 parts. First, one has to assess the compressive sensing part of the problem. Does the mixing matrix $\boldsymbol{\Phi}$ satisfy the appropriate conditions? Is the value of $\boldsymbol{M}$ large enough to recover the $\boldsymbol{K}$ unknowns? Are the measurements really sparse in the chosen model or even is the model applicable to the signal of interest? Our empirical observations suggest that it is difficult for a random number generator to pick a bad mixing matrix. Observations also suggest that the requirement on $\boldsymbol{M}$ for recovery is on the same order as that derived for grid-based CS, $\boldsymbol{M} \sim \boldsymbol{K} \log(\boldsymbol{N}/\boldsymbol{K})$. Second, the sampling in the overcomplete dictionary must be fine enough that the first frequency found by the argmax of $\boldsymbol{G(f,r)}$ in (7) is near a true solution. If this is not the case due to insufficient granularity, multiple frequencies too close together, or high noise levels, the OMP cannot start. This issue is not restricted to our work but common to all matching pursuit algorithms. While we do not have performance guarantees here, we have noted empirically that lack of convergence is very easy to determine for a known number or sinusoids and known noise floor. Finally, the NLS must be able to converge. Here we can rely on the results given by (Stoica et al., 2000; Li et al., 2000; Chan and So, 2004; Christensen and Jensen, 2006) that the NLS achieves the Cramer Rao Bound. Empirically, we observe that the dictionary must be sufficiently overcomplete that the NLS is looking for a frequency solution in one local minimum.

8. Conclusion

The work reported in this chapter started with our work on compressive sensing for direction of arrival (DOA) detection with a phased array (Shaw and Valley, 2010). In that work, we realized that most work in compressive sensing concerned recovering signals on a sparse grid. In the DOA domain, that meant that targets had to be on a set of grid angles, which of course never happens in real problems. We found a recovery solution for a single target in that work by scanning the sparsifying DFT over an offset index that was a measure of the sine of the target angle but the solution was time consuming because the penalized ell-1 norm recovery algorithm had to be run multiple times until the best offset and best sparse solution was found and the procedure was not obviously extendable to multiple targets. This work led us to the concepts of orthogonal matching pursuit and removing one target (or sinusoid) at a time. But we still needed a reliable method to find arbitrary frequencies or angles not on a grid. The next realization was that nonlinear least squares could be substituted for the linear least squares used in most versions of OMP. This has proved to be an extremely reliable method and we have now performed 10's of thousands of calculations with this method. Since OMP is not restricted to finding sinusoids, it is natural to ask if OMP with NLS embedded in it works for other functions as well. We have not tried to prove this generally, but we have performed successful calculations using OMP-NLS with signals composed of multi-dimensional sinusoids such as would be obtained with 2D phased arrays (see also Li et al., 2001), signals composed of multiple sinusoids multiplied by chirps (i.e. sums of terms of the form $a_k exp(i\omega_k t+b_k t^2)$ and multiple Gaussian pulses within the same time window.

9. Acknowledgment

This work was supported under The Aerospace Corporation's Independent Research and Development Program.

10. References

Baraniuk, R. G.; (2007). Compressive sensing, *IEEE Signal Processing Mag.*, Vol.24, No.4, pp.118-120, 124.

Candes, E. J.; & Tao, T., (2006). Near optimal signal recovery from random projections: Universal encoding strategies? *IEEE Trans. Inform. Theory*, Vol.52, pp. 5406-5425.

Candes, E. J.; & Wakin, M. B., (2008). An introduction to compressive sampling, *IEEE Signal Processing Magazine*, Vol.21, pp. 21-30.

Candes, E. J.; Eldar, Y. C., Needell, D., & Randall, P., (2011). Compressed sensing with coherent and redundant dictionaries, submitted to *Applied and Computational Harmonic Analysis*.

Chan, K. W.; & So, H. C., (2004). Accurate frequency estimation for real harmonic sinusoids, *IEEE Signal Processing Lett.*, Vol.11, No.7, pp. 609-612.

Christensen, M. G.; & Jensen, S. H., (2006). On perceptual distortion minimization and nonlinear least-squares frequency estimation, *IEEE Trans. Audio, Speech, Language Processing*, Vol.14, No.1, pp. 99-109.

Davenport, M.; Wakin, M., and Baraniuk, R., (2006). Detection and estimation with compressive measurements, Rice ECE Department Technical Report, TREE 0610.

Davis, G.; Mallat, S., & Avellaneda, M., (1997). Greedy adaptive approximation, Const. Approx., Vol.13, pp. 57-98.

Donoho, D. L.; (2006). Compressed sensing, *IEEE Trans. Inform. Theory*, Vol.52, pp. 1289-1306, Sept. 2006.

Duarte, M. F.: & Baraniuk, R. G., (2010). Spectral Compressive Sensing, submitted to *IEEE Trans. Signal Processing.*

Hormati, A.; & M. Vetterli, (2007). Annihilating filter-based decoding in the compressed sensing framework, *Proc. SPIE*, Vol.6701, pp. 1-10.

Huang, S.; & J. Zhu, (2011). Recovery of sparse signals using OMP and its variants: convergence analysis based on RIP, Inverse Problems, Vol.27, 035003 (14pp).

Jacques, L.; & C. De Vleeschouwer, (2008). A Geometrical study of matching pursuit parameterization, IEEE Trans. Signal Proc., Vol.56, pp. 2835-2848.

Li, H.; Stoica, P., & Li, J., (2000). Computationally efficient parameter estimation for harmonic sinusoidal signals, *Signal Processing*, Vol.80, pp. 1937-1944.

Li, H.; Sun, W., Stoica, P. & Li, J., (2001). 2-D sinusoidal amplitude estimation with application to 2-D system identification, IEEE International Conf. On Acoustics, Speech, and Signal Proc., ICASSP, Vol.3, pp. 1921-1924.

Loris, I.; (2008). L1Packv2: A *Mathematica* package for minimizing an ell-1-penalized functional, *Computer Phys. Comm.*, Vol.79, pp. 895-902.

Mallat, S. G.; & Zhang, Z., (1993). Matching pursuits with time-frequency dictionaries, *IEEE Trans. Signal Processing*, Vol.41, No.12, pp. 3397-3415.

Needell, D.; & Tropp, J. A., (2009). CoSaMP: Iterative signal recovery from incomplete and inaccurate samples, Applied and Computational Harmonic Analysis, Vol.26, No.3, pp. 301-321.

Needell, D.; & Vershynin, R., (2009). Uniform uncertainty principle and signal recovery via regularized orthogonal matching pursuit, Found. Comput. Math., Vol.9, pp. 317-334.

Pati, Y.C.; Rezaifar, R., & Krishnaprasad, P.S., (1993). Orthogonal matching pursuit: Recursive function approximation with applications to wavelet decomposition, Proc. 27th Annu. Asilomar Conf. Signals. Systems, and Computers, Pacific Grove, CA. Vol.1, pp. 40-44.

Salzo, S.; & Villa, S., (2011). Convergence analysis of a proximal Gauss-Newton method, arXiv:1103.0414v1, pp. 1-29.

Shaw, T. J.; & Valley, G. C., (2010). Angle of arrival detection in sparse systems using compressed sensing, European Signal Processing Conference (EUSIPCO), Aalburg, Denmark, 23-27Aug. 2010, *EUSIPCO 2010 Digest*, pp. 1424-1428.

Stoica, P.; & Moses, R. L., (1997). *Introduction to Spectral Analysis*, Upper Saddle River, NJ: Prentice Hall, pp. 146-151.

Stoica, P.; Li, H., and Li, J. (2000). Amplitude estimation of sinusoidal signals: survey, new results, and an application, *IEEE Trans. Signal Processing*, Vol.48, No.2, pp. 338-352.

Tropp, J. A.; & Gilbert, A. C., (2007). Signal recovery from random measurements via orthogonal matching pursuit, *IEEE Trans. Information Theory*, Vol.53, No.12, pp. 4655-4666.

Vetterli, M.; Marziliano, P., & Blu, T., (2002). Sampling signals with finite rate of innovation, *IEEE Trans. Signal Processing*, Vol.50, No.6, pp. 1417-1428.

Part 3

DSP Filters

10

Complex Digital Filter Designs for Audio Processing in Doppler Ultrasound System

Baba Tatsuro
Toshiba Medical Systems Corporation
Japan

1. Introduction

A medical Doppler ultrasound system has a spectrum display that indicates the blood flow direction, whether the blood flows forward or away from a probe. It also has Doppler audio outputs. In particular, the latter is a special process peculiar to the Doppler ultrasound system and separates the blood flow direction and outputs from the left and right speakers. Owing to this function, the existence of a blood flow is quickly detectable. When changing conventional analog signal-processing into digital signal-processing, we researched many processing systems of Doppler audio. First, target performances, such as a response time and direction separation, were set up, and six kinds of digital signal-processing systems were examined. Further, we investigated some new anti-aliasing processing systems unique to Doppler ultrasound system. We compared three kinds of anti-aliasing processing systems. Consequently, we clarified that a complex IIR (infinite impulse response) filter system has an excellent response and a low calculation load.

2. Outline of Doppler ultrasound system and conventional analog signal-processing

Recently, the diagnostic ultrasound system has been popular in many diagnostic fields, such as cardiac, abdomen, and so on. In Section 2.1, an example of diagnostic image and its principle are introduced. In Section 2.2, the phase shift system that is an example of representation of conventional analog signal-processing is introduced.

2.1 Outline of Doppler ultrasound system

An example of diagnostic image of a carotid artery is shown in Fig. 1. The upper is a tomogram image and bottom is a spectrum Doppler image. This image expresses the time change of the flow velocity in the PWD (Pulse Wave Doppler) range gate set up in the central of a blood vessel in a tomogram. A horizontal axis and a vertical axis are the flow velocities corresponding to Doppler shift frequency and time, respectively.
Signal processing of the ultrasound echo signal is shown in Fig. 2. An ultrasonic wave is transmitted for every cycle of PRF (pulse repetition frequency: *fs*) in the transceiver processing part of Fig. 2(a), and a reflective echo is received. An ultrasonic beam is scanned in the transverse direction, and envelope detection of the received signal is carried out in the range direction. This scanning constitutes the tomogram image.

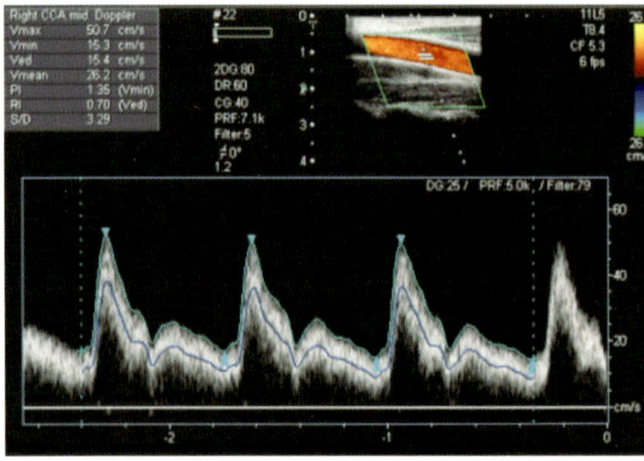

Fig. 1. Example of ultrasound diagnostic image of a carotid artery

Except for Doppler signal processing, as another method of blood-flow or tissue velocity detection, the cross-correlation method using the signal before quadrature-detection processing (*R (t)* in Fig. 2(a)) has been also reported. However, the base-band signal (*L (t)* in Fig. 2(a)) processing after quadrature-detection is the present mainstream, because of its narrow bandwidth and little processing load. All the direction separation systems examined this time are the IQ-signal processing after quadrature-detection. The received signal *R(t)* in a range gate is denoted by a formula (1). Here, a reflective echo signal is assumed to be the amplitude A_i, Doppler shift angle-frequency ω_i, and phase ϕ_i.

$$R(t) = \sum^{i} A_i \cdot \exp\left(j \cdot \left(\omega_p + \omega_i\right) \cdot t + j \cdot \phi_i\right) \tag{1}$$

The mixer output *M(t)* is denoted by a formula (2). Reference angle-frequency of a mixer is set to ω_p (same as probe Tx angle-frequency) here.

$$\begin{aligned} M(t) &= R(t) \cdot \exp\left(j \cdot \omega_p \cdot t\right) \\ &= \frac{1}{2}\sum^{i} A_i \cdot \exp\left(j \cdot \left(2 \cdot \omega_p + \omega_i\right) \cdot t + j \cdot \phi_i\right) + \frac{1}{2}\sum^{i} A_i \cdot \exp\left(j \cdot \omega_i \cdot t + j \cdot \phi_i\right) \end{aligned} \tag{2}$$

The LPF output *L(t)*, high frequency component is removed is denoted by a formula (3).

$$L(t) = \frac{1}{2}\sum^{i} A_i \cdot \exp\left(j \cdot \omega_i \cdot t + j \cdot \phi_i\right) \tag{3}$$

In Fig. 2(a) (R1), (R2), and (R3) show the position of the blood-vessel-wall upper part, the inside of a blood vessel, and the blood-vessel-wall lower part, respectively. Fig. 2(b) shows typical spectra of quadrature-detection output *L(t)*, when a range gate is set in each position. A vertical axis shows power and the horizontal axis shows frequency, respectively. Since the sampling is interlocked with PRF of transmission, the vertical axis has a frequency range of $\pm fs/2$. *L(t)* is mainly constituted from the low frequency component caused by the clatter (strong echo from tissue) and middle to high frequency component caused by weak blood-flow. Also inside of blood vessel, a blood vessel wall and a transmit-wavelength influence the blood-flow signal. Then, in order to prevent the saturation of the frequency analysis or the Doppler audio processing, a wall-filter is arranged in pre-processing of them. The wall-filter is HPF with high order cut-off property. The details of Spectrum Doppler signal processing are shown in Fig. 2(c). Range gate processing is the integration of *L(t)* in the range direction in the range gate. Wall-filter processing removes a clatter component. The complex IQ-signal *x(t)* after these processing is inputted into the spectrum Doppler display processing and the Doppler audio processing. The former displays the spectrum Doppler as a time-change image of a flow velocity. The latter separates the direction of Doppler signal, and outputs them as stereo sounds from a right-and-left speaker.

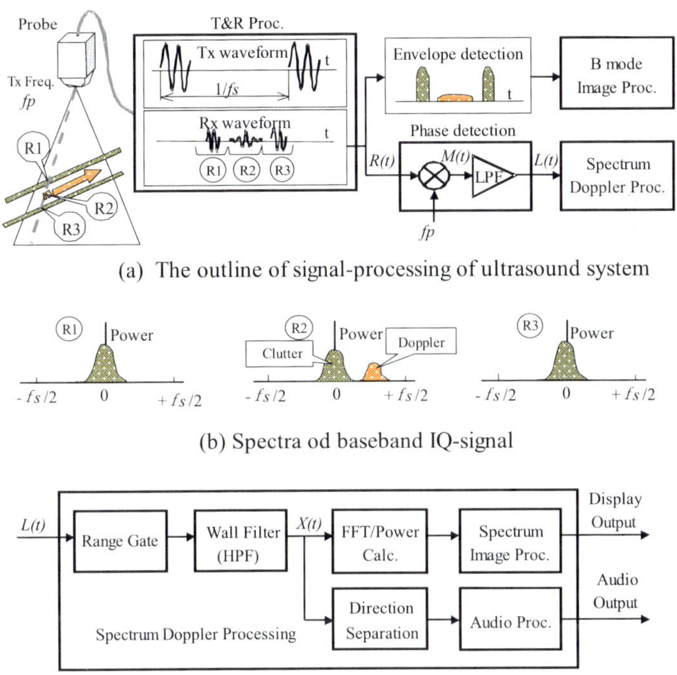

(a) The outline of signal-processing of ultrasound system

(b) Spectra od baseband IQ-signal

(c) Spectrum Doppler processing

Fig. 2. Doppler ultrasound signal-processing.

2.2 Conventional analog signal-processing

An analog phase-shift processing system that consists of all-pass filters has been used in the direction separation processing. The outline of it is shown in Fig. 3. This is a processing system that shifts the phase between the IQ-signals of 90 degree, and adds them or subtracts them. Since an all-pass filter has the characteristic that the phase reverses on cut-off frequency, this system shifts the phase in a target frequency range combining all-pass filter arrays. If it assumes that the input IQ-signal *x(t)* has a frequency component of ω_d .

$$x(t) = \exp(j \cdot \omega_d \cdot t) \tag{4}$$

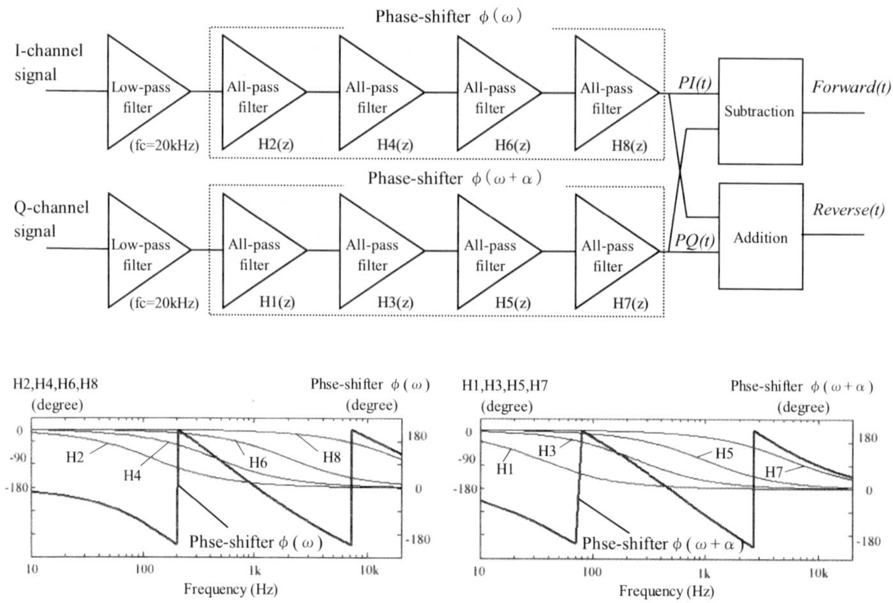

Fig. 3. Outline of analog direction separation system

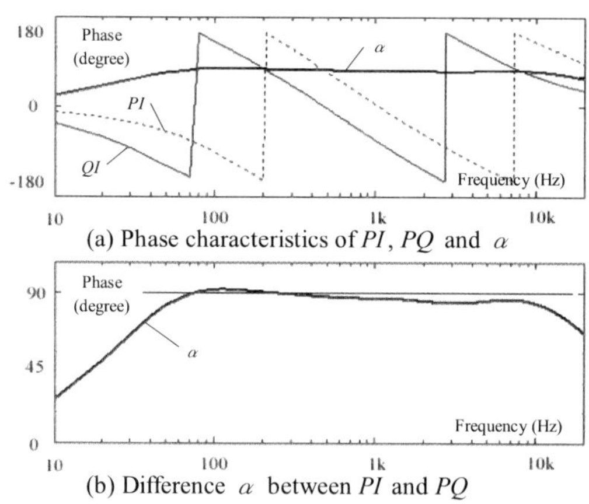

Fig. 4. Frequency characteristics of all-pass filters

In Fig. 4(b), the phase characteristics of I-channel and Q-channel are delayed as frequency becomes high. Here, the phase characteristics of I-channel and Q-channel are defined to be $\phi(\omega)+\alpha$ and $\phi(\omega)$, respectively. The output of I-channel and Q-channel are set to *PI(t)* and *PQ(t)*.

$$PI(t) = \mathrm{Re}\left[x(t)\cdot \exp(j\cdot(\phi(\omega)+\alpha))\right] = sign(\omega_d)\cdot \sin(\omega_d \cdot t + \phi(\omega)) \tag{5}$$

$$PQ(t) = \text{Im}\left[x(t) \cdot \exp(j \cdot (\phi(\omega)))\right] = -\sin(\omega_d \cdot t + \phi(\omega)) \tag{6}$$

Here, α is $-\pi / 2$ when Doppler frequency ω_d is positive, and α is $+\pi / 2$ when ω_d is negative. So $sign(\omega_d)$ means the polarity. The subtraction-output *Forward(t)* and the addition-output *Reverse(t)* are

$$Forward(t) = PI(t) + PQ(t) = (sign(\omega_d) + 1) \cdot \sin(\omega_d \cdot t + \phi(\omega)) \tag{7}$$

$$\text{Re}\,verse(t) = PI(t) - PQ(t) == (sign(\omega_d) - 1) \cdot \sin(\omega_d \cdot t + \phi(\omega)) \tag{8}$$

From the formulas (7) and (8), when ω_d is positive, only the *Forward(t)* serves as a non-zero output. And when ω_d is negative, only the *Reverse(t)* serves as a non-zero output. Thus, IQ-signals are separable into positive-component and negative-component. Comparison of direction separation performance is shown in Fig. 5. The frequency-characteristic in the velocity range 4kHz (-2kHz to +2kHz) that is well used in diagnosis of the cardiac or abdomen is shown. A solid line shows the positive-component (*Forward*) and a dashed line shows the negative-component (*Reverse*). The direction separation performance of the phase-shift system (conventional analog system) is shown in Fig. 5(a), and the direction separation performance of the complex IIR filter system (digital system referenced in section 3.2) is shown in Fig. 5(b).

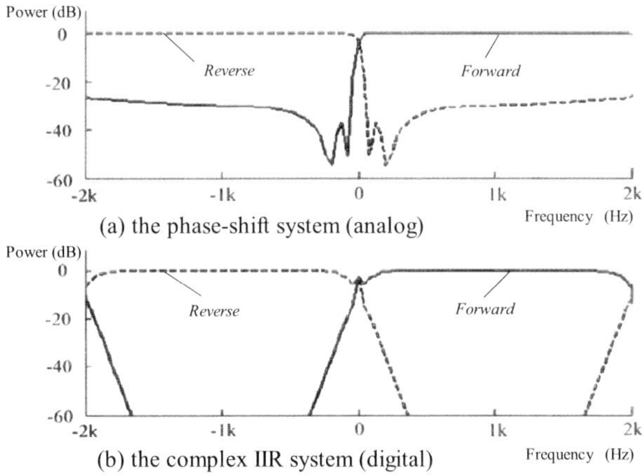

Fig. 5. Direction separation performance

In Fig. 5, a filter-order to which hardware size becomes same is set up. In the complex IIR filter system, sufficient separation performance (more than 30 dB) is got except for near a low frequency and near the Nyquist frequency. On the one hand a ringing has occurred by the phase shift system, there is little degradation near the Nyquist frequency. Although the direction separation performance near a low frequency and near the Nyquist frequency can improve if the filter-order is raised in the complex IIR filter system, the processing load becomes large. It is although the ringing will decrease if range of the phase-shift system is divided finely, processing load becomes large similarly.

3. Comparison of six kinds of Doppler audio processing

The digitization of Doppler ultrasound system had progressed in recent years, and the digital signal processing using DSP etc. can realize complex processing easily from the conventional analog-circuit. We made the target performances of the direction separating process of digital Doppler audio, and evaluated six kinds of digital-signal processing ideas that were pre-existing or were newly devised.

3.1 Design of a target specification

For the digitization, the target performance is investigated and taken up to Table 1.

item	target
1. time-delay	bellow 20ms (PRF 4kHz)
2. direction separation	above 30dB
3. frequency characterization	fs/128 to 63*fs/128 (both direction) flat as possible
4. frequency resolution	fs/100
5. calculation volume	light as possible

Table 1. Requirement specification of Doppler audio direction separation

Time-delay:

A user usually sets up the Doppler range gate on a tomogram, moves it, and performs blood flow diagnosis with the Doppler ultrasound system. In searching for a small blood vessel, the Doppler audio is effective, because its response is faster than that of the spectrum image. This is because a tomogram set with the Doppler audio delays the outputs of about 20 ms, compared with the spectrum image that has a typical delay of about 40 ms. The time delay of tomogram processing is a few cycle of one frame (13.3 - 16.7 ms). In the Doppler signal processing system, it has a total processing delay of 10 ms by quadrature-detection and HPF processing, except for the Doppler processing part. Therefore, to make the tomogram and audio agree, a time delay of 3.3 - 6.7 ms is required at the Doppler signal processing part. However, because the direction separation process, which is the main factor of the Doppler signal processing part delay, requires a number of series samplings for processing, a target time delay is theoretically difficult to achieve. Therefore, the target time delay was set to be 20 ms or smaller, so that the target delay time required for the direction separation process to store the Doppler audio is about one frame cycle at maximum in a tomogram.

Direction separation:

It has been reported that human's direction distinction requires a right-and-left signal difference of 15 to 20 dB or lager. In an actual Doppler ultrasound system, considering that the Doppler signal has a broad band, that the angle between the right-and-left speakers is small, and that blood flow velocity changes with time, a larger signal difference is required. The target performance of direction separation was set to be 30 dB or higher at observation frequency.

Frequency characteristic:

A signal processing frequency range is the range *fs* from negative-side Nyquist frequency to positive-side Nyquist frequency, where *fs* is input IQ-signal sampling frequency. We made the frequency characteristic flat in the region of $\pm fs/128$ to $\pm 63 \cdot fs/128$ range.

Frequency resolution:

Since spectrum image signal processing involves 256-point FFT, an acceptable frequency (velocity) resolution is obtained. However, when the frequency resolution of the Doppler audio is unacceptable, similar to that of a small-pitch Doppler image, we set the target resolution to be $fs/100$. The frequency range is determined from sample frequency. However, the frequency resolution is proportional to the reciprocal of observation time. For example, in FFT, it is equivalent to the main robe width of the sampling function determined from observation time width and the window function.

Calculation load:

Although operation load is dependent on the hardware-architecture, such as DSP, ASIC, and FPGA, lighter load is more advantageous to cost, size, and power consumption in common.

3.2 Six kinds of digital signal-processing ideas

Six kinds of digital signal-processing systems that were pre-existing or newly devised are examined. They are shown in Fig. 6.

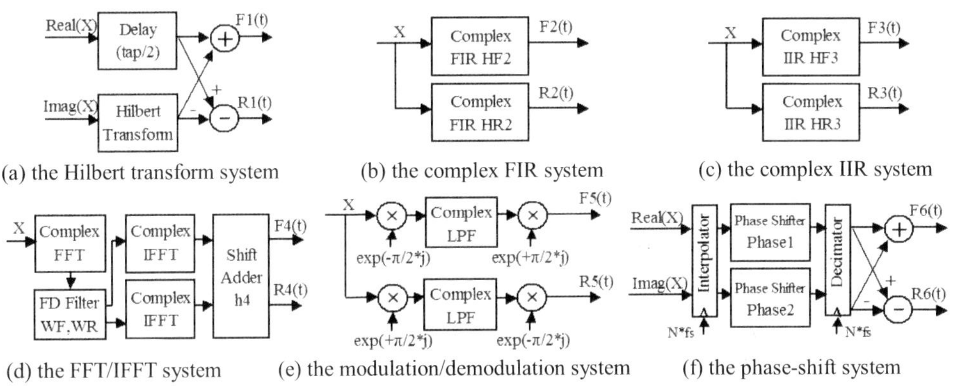

Fig. 6. Six kinds of digital signal-processing systems

Hilbert transform system:

The delay of (filter tap length)/2 is given to I-channel of IQ-signals. It and the Hilbert transform output of Q-signal are subtracted or added. The direction separated signals are calculates by formulas (9) and (10). Here a convolution is indicated ⊗. The tap number is set to 128 in the estimation of the calculation load shown in Table 3.

$$F1(n) = \mathrm{Re}\left(X(n - ntap / 2)\right) + \mathrm{Im}\left(X(n) \otimes h1(ntap)\right) \tag{9}$$

$$R1(n) = -\mathrm{Re}\left(X(n - ntap / 2)\right) + \mathrm{Im}\left(X(n) \otimes h1(ntap)\right) \tag{10}$$

The coefficient $h1$ of Hilbert transform is given by a formula (11).

$$h1(n) = \frac{2}{\pi} \cdot \frac{\sin^2(\pi \cdot n / 2)}{n} \quad (n \neq 0)$$
$$= 0 \quad (n = 0) \tag{11}$$

Complex FIR system:

There is a report of the Doppler audio separation processing using a complex FIR filter. However, since there is no description about a filter coefficient, we designed in a frequency domain and transformed into FIR coefficient in time domain using inverse Fourier transform. The output of complex FIR system is denoted by formulas (12) and (13). In the estimation of Table 3, the 128-tap coefficient sequence with the pass band of $\pm fs / 128$ to $\pm 63 \cdot fs / 128$ is used.

$$F2(n) = X(n) \otimes HF2(ntap) \tag{12}$$

$$R2(n) = X(n) \otimes HR2(ntap) \tag{13}$$

Complex IIR system:

Based on the shift theory of Fourier transform, frequency shift is applied to z operators. A real-LPF transfer function is changed into the positive-BPF and the negative-BPF. The complex IIR transfer functions become a formulas (14) and (15).

$$F3(z) = HF3(z) \cdot X(z) \tag{14}$$

$$R3(z) = HR3(z) \cdot X(z) \tag{15}$$

When the transfer function of real LPF is set to *RLPF (z)*, transfer functions of *HF3 (z')* and *FR3 (z'')* are calculated by transformed operators. In the estimation of Table 3, the filter with the 8th order Butterworth type is used.

$$HF3(z) = RLPF3(z') \quad \text{where} \quad z' = -j \cdot z \tag{16}$$

$$HR3(z) = RLPF3(z'') \quad \text{where} \quad z'' = j \cdot z \tag{17}$$

FFT/IFFT system:

The IQ-signal is separated by the positive-filter and negative-filter in a frequency domain. Next, the separated spectra are returned to waveforms in time domain by inverse-FFT. There is a report of this system aiming at the Doppler noise rejection. For the continuous output after inverse-FFT, a shift addition of the time waveform is carried out in time domain. The outputs of this system can be denoted by formulas (18) and (19). In estimation of Table 3, FFT/IFFT point number is set to 128, and used the frequency filter of $\pm fs / 128$ to $\pm 63 \cdot fs / 128$ for separation. Moreover, Hamming window (*h4*) is applied, and 32 time-series are shift-added.

$$F4(n) = \mathrm{Re}\left(IFFT\left(WF(\omega) \cdot FFT\left(X(n)\right)\right)\right) \cdot h4(n) \tag{18}$$

$$R4(n) = \mathrm{Re}\left(IFFT\left(WR(\omega)\cdot FFT\left(X(n)\right)\right)\right)\cdot h4(n) \tag{19}$$

Modulation/demodulation system:
IQ signal is modulated and frequency is shifted *+fs/4* and *–fs/4*. The positive-component (0 to *+fs/2*) and negative-component (*-fs/2* to *0*) are extracted by LPF. The *+fs/4* shift and the *–fs/4* shift are returned by demodulation. The direction separation outputs are calculated by formulas (20) and (21). The example of Table 3 is referred to the prior art. The 128-tap FIR low-pass filter, which has 63/128 cut-off, is used.

$$F5(n) = \left(\left(X(n)\cdot \exp\left(-\frac{\pi}{2}\cdot j\cdot n\right)\right)\otimes CLPF(ntap)\right)\cdot \exp\left(+\frac{\pi}{2}\cdot j\cdot n\right) \tag{20}$$

$$R5(n) = \left(\left(X(n)\cdot \exp\left(+\frac{\pi}{2}\cdot j\cdot n\right)\right)\otimes CLPF(ntap)\right)\cdot \exp\left(-\frac{\pi}{2}\cdot j\cdot n\right) \tag{21}$$

Phase-shift system:

There are two sets of phase-shifter with the transfer characteristic that makes relative phase difference of IQ-signal 90 degree. The addition-and-subtraction of these outputs is used. The direction separation outputs are calculated by formulas (22) and (23).

$$F6(z) = \mathrm{Re}\left(X(z)\right)\cdot Phase1(z) + \mathrm{Im}\left(X(z)\right)\cdot Phase2(z) \tag{22}$$

$$R6(z) = -\mathrm{Re}\left(X(z)\right)\cdot Phase1(z) + \mathrm{Im}\left(X(z)\right)\cdot Phase2(z) \tag{23}$$

The two sets of phase-shifter are the cascade connection of second-order all-pass filter arrays. They are denoted by formulas (24) and (25) as a *Phase1 (z)* and a *Phase2 (z)*. In the estimation of Table 3, the cascade connections of four steps of all-pass filters are used. Moreover, in order to improve the performance near the Nyquist frequency, an interpolator and a decimator are added before and after phase-shifter. Table 3 is calculated in *N*= 4, and the FIR filter of *2N* tap is used as an interpolator.

$$Phase1(z) = \prod_{k=1}^{n} \frac{z^{-1} - a_k}{1 - a_k \cdot z^{-1}} \tag{24}$$

$$Phase2(z) = \prod_{k=1}^{n} \frac{z^{-1} - b_k}{1 - b_k \cdot z^{-1}} \tag{25}$$

Above six kinds of signal-processing algorithms are confirmed by the simulation. The chirp-waveform that frequency and a direction are changed is used as an input. The result of a simulation is shown in Fig. 7. Fig. 7(a) is an input signal and the sign of frequency has inverted near 200ck (equivalent to the time shown in the Fig. 7 broken line). A solid line is I-signal and a dotted line is Q-signal. Figures 7(b) to (g) are output waveforms of each signal-processing system. A solid line is a positive-output (forward) of the Doppler audio, and a dotted line is a negative-output (reverse) of the Doppler audio. Amplitude of positive-output becomes large on the right-hand side of a broken line, and it becomes small on the left-hand side of the broken line. Amplitude of negative-output becomes small on the right-hand side of a broken line, and it becomes large on the left-hand side of the broken line. This result shows that each system works correctly. Moreover, it shows that the waveform and response time at the turning point of sign (near the DC) have a difference among the systems. As these causes, performance differences, such as the response characteristic and delay time, can be considered.

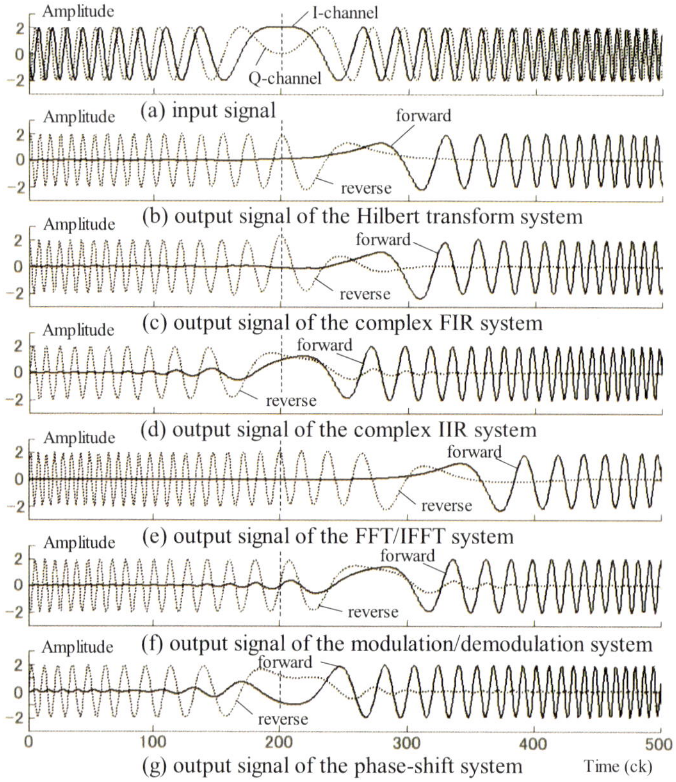

Fig. 7. Chirp wave responses

3.3 Comparison of time-delay and calculation load

The response is important for blood vessel detection, and the time-delay estimates it. The simulation result of the time-delay is shown in Fig. 8. They are response waveforms of the sinusoidal-waveform that changes discontinuously. Solid line and dotted line are I-signal and Q-signal in Fig. 8(a). Amplitude and frequency are changing near the 50ck. The solid lines of Fig. 8(b) and Fig. 8(c) are positive-output waveforms, and dotted lines are negative-output waveforms. The output waveform of the complex FIR system of Fig. 8 (b) changed from a turning point of the input shown with the dashed line after 64ck (time shown with the chain line among Fig. 8(b)), and is stable gradually. The output waveform of the complex IIR system of Fig. 8(c) is stable from the turning point after 8ck (time shown with the chain line among Fig. 8(c)).

The comparison of time-delay is shown in Table 2. Frequency resolution is adjusted by parameter of each system in accordance with the target performance of Table 1. Since the signal-processing inputs are sampled by *fs*, time-delay will become large if *fs* becomes low. Table 2 is calculated by *fs*=4kHz condition. Incidentally by *fs*=1kHz, time-delay increases 4 times. The time-delay caused by operation is assumed zero, and estimated only the delay caused by sampling simply. Moreover, the influence of the transient response of the complex IIR system and the phase-shift system is not taking into consideration here.

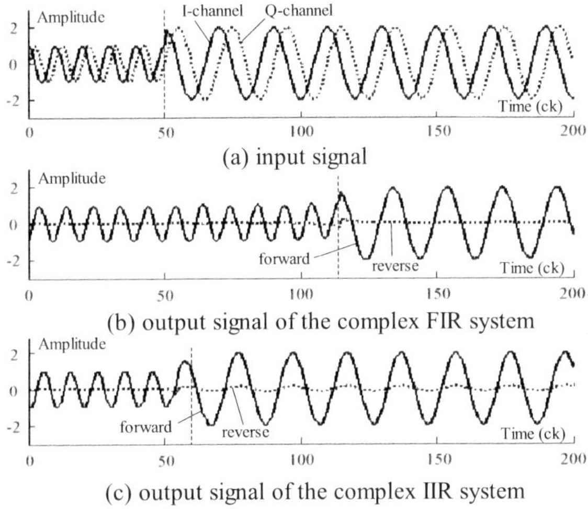

Fig. 8. Comparison of response between complex FIR method and complex IIR method

method	estimation	time-delay (ms)
Hilbert transform	*tap/fs*	32 (*tap*=128)
Complex FIR	*tap/fs*	32 (*tap*=128)
Complex IIR	*order/fs* (*1)	2 (*order*=8)
FFT/IFFT	*1.5*N/fs* (*2)	48 (*N*=128)
Moduration/Demoduration	*tap/fs*	32 (*tap*=128)
Phase shift	*max(2,order/N)/fs* (*1)	1 (*order*=4, *N*=1)

Estimated at *fs*=4kHz, (*1) not including transient response, (*2) IFFT shift addition pitch is *N/4*

Table 2. Comparison of time-delay

As calculation load depends on the hardware architecture, the multiplication and addition times per 1 second (floating point single precision) is used for this estimation. Moreover, the complex-multiplication is considered as 4 times, and complex-addition is considered as twice. The overhead of the processing which requires a lot of memory buffers is assumed to be 20%. The other overhead is assumed to be 10%. The calculation elements, estimation formula and calculation load for each signal-processing systems are shown in Table 3. Incidentally, at *fs*=52 kHz (maximum PRF in actual system), calculation load increases 13 times. The result of Table 2 and Table 3 shows that the complex IIR system and the phase-shift system are filling the target performance of time-delay. It turns out that calculation load is light in order of the phase-shift system, the complex IIR system, and the Hilbert transform system.

method	calculation component	estimation equation	load (MFLOPS)
Hilbert transform	R-add: (*tap*+1)**fs*, R-mul: *tap***fs* Ovh: 20%	*fs**(2**tap*+1)*1.2	1.26 (*tap*=128)
complex FIR	C-add: (*tap*-1)*2+*fs*, C-mul: *tap**2**fs* Ovh: 10%	*fs**(12**tap*-4)*1.1	6.74 (*tap*=128)
complex IIR	C-add: *order**4**fs*, C-mul: *order**4**fs* Ovh: 10%	*fs**(24**order*)*1.1	0.84 (*order*=8)
FFT/IFFT	C-add: *N***r**3, C-mul: (*N***r*/2)*3 Ovh: 20%, R-mul: *N**4	12**N***r**1.2*(*fs**4/*N*) (FFT shift addition, *N*/4 shift)	1.61 (*N*=128, *r*=7)
modulation/ demodulation	C-add: (*tap*-1)*2**fs* C-mul: (*tap*+2)*2**fs*, Ovh: 20%	*fs**(12**tap*-12)*1.2	7.32 (*tap*=128)
Phase-shift	R-add: [2**N**(2**N*+2**order*)+2]**fs* R-mul: 4**N**(*N*+*order*)**fs*, Ovh: 20%	*fs**[4**N**(*N*+*order*) +2*(*N*-1)]*1.2	0.64 (*order*=4,*N*=4)

R-add: real addition, R-mul: real multiplication, Ovh: over head, C-add: complex addition, C-mul: complex multiplication, Calculation load is estimated at *fs*=4kHz

Table 3. Comparison of calculation load

3.4 Comparison of a frequency characteristic and direction separation

Frequency characteristic and direction separation performance are largely dependent on the filter property that are related to time-delay and calculation load. If the number of filter taps of FIR and the filter order of IIR are reduced, time-delay and calculation load will decrease. But these become the trade-off of frequency resolution and frequency characteristic. The Hilbert transform system frequency characteristic when changing the number of taps is shown in Fig. 9. The frequency characteristic near the Nyquist and near the DC has deteriorated, when the number of taps is short. This is the same also about the taps of the complex FIR system, the modulation/demodulation system and the FFT point number of the FFT / IFFT system.

In order to compare the direction separation performance, the frequency characteristic simulation is performed. The frequency characteristics of positive-component (solid line: forward) and negative-component (dashed line: reverse) are shown in Fig. 10. The target performance of direction separation is filled except for the phase shift system. The stop-band property near the low frequency and near the Nyquist frequency is good in the Hilbert transform system, the complex FIR system, and the FFT/IFFT system. Exclude near the DC and near the Nyquist frequency, a sufficient separation performance (not less than 30 dB) and frequency characteristic are acquired by the complex IIR system and the modulation/demodulation system. The phase-shift system has generally insufficient separation performance. The separation performance is deteriorated especially near the Nyquist frequency.

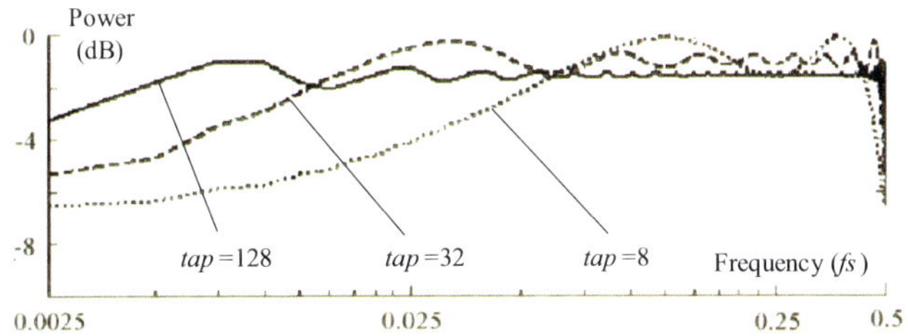

Fig. 9. Example of frequency response: the Hilbert transform system

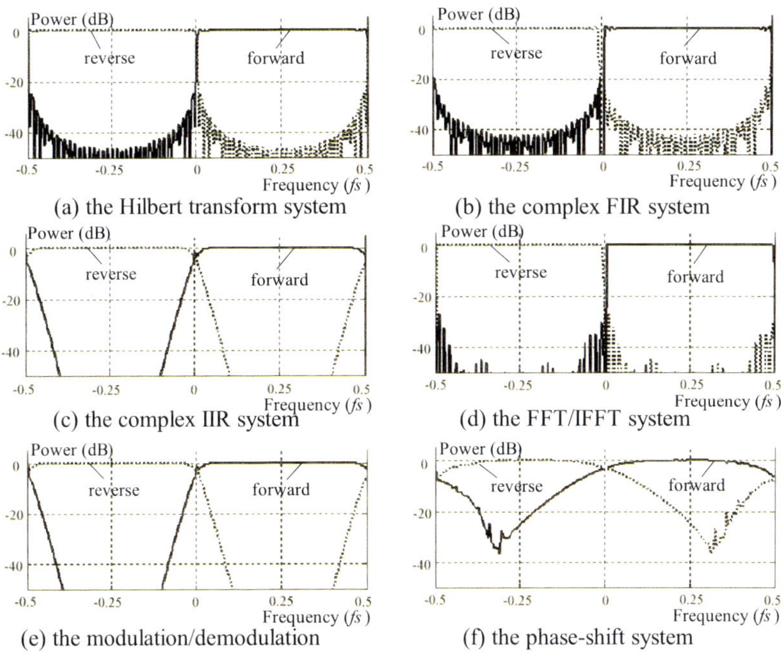

Fig. 10. Frequency characterization and direction separation performance

3.5 Conclusion

We made the target performances of the direction separating process of digital Doppler audio, and evaluated six kinds of digital-signal-processing ideas that were pre-existing or were newly devised. The performances of each processing were evaluated by comparing many responses such as chirp or step and so on. The results are following.

1. The complex IIR system and the phase-shift system are filling the target performance of response time.
2. The target performance of direction separation is filled except for the phase-shift system.

3. All the systems fill the frequency characteristic. However, the frequency characteristics near the DC and near the Nyquist region are dependent on the filter characteristics of each processing system.

4. Signal processing for Doppler audio anti-aliasing

The direction separation system of the foregoing section is developed further, and the Doppler audio technology exceeding the Nyquist frequency is examined. Some direction-separation systems for a Doppler audio that is interlocked with the baseline-shift of a spectrum image are investigated. First, section 4.1 explains a problem peculiar to the Doppler audio corresponding to the Doppler display processing. In section 4.2 we defined the target performance of anti-aliasing Doppler audio processing selected three kinds of signal-processing systems. In section 4.3 the various systems of the modulation/demodulation system, the FFT/IFFT system and the complex IIR Filter system are explained. Next, in section 4.4 the signal-processing algorithms are compared with the target performances. It was confirmed that the complex IIR band-pass filter system has an excellent response and a low calculation load. Finally, in section 4.5 using the blood-flow data collected from Doppler phantom, we performed functional and performance analyses by simulation shown in Fig. 22.

4.1 Anti-aliasing display and conventional problem

The Doppler ultrasound system extracts the blood flow component used in the quadrature-detection of the Doppler signal from the blood (mainly an erythrocyte), which moves inside a blood vessel, and removes a reflective signal from tissue, such as a blood vessel wall with a high-pass filter, and transforms the Doppler component into an image and sound. The Doppler ultrasound system is shown in Fig. 11. The signal obtained after HPF processing is divided into two lines. Spectrum image processing generates a Doppler signal as a spectrum time change image corresponding to blood velocity, and Doppler audio processing outputs direction separation signals as stereo sound from the right-and-left speakers.

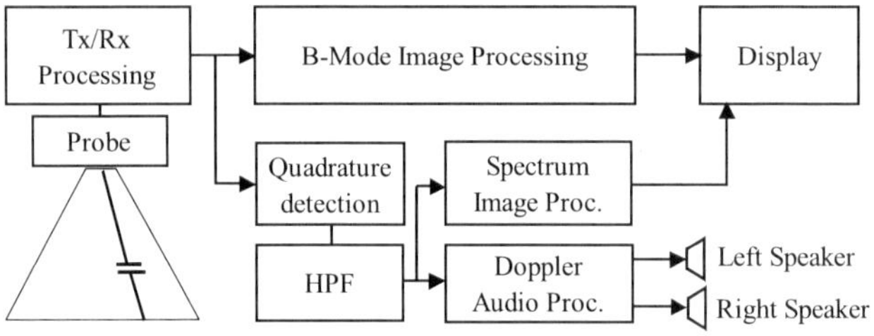

Fig. 11. Doppler ultrasound system.

Because the Doppler signal contains phase information, the signal includes both positive-side (forward) and negative-side (reverse) frequency components. If sampling frequency is set to be *fs*, the detection of a Doppler frequency component corresponding to the frequency range of -*fs*/2 to +*fs*/2 is possible. A spectrum image is shown in Fig. 12. The horizontal axis corresponds to time. The vertical axis corresponds to the velocity derived from Doppler shift frequency, and luminosity corresponds to the spectrum intensity of each time. Since a spectrum image is a power spectrum generated by complex FFT processing, it has the frequency range of *-fs/2* to *+fs/2* on the baseline (0Hz) shown in Fig. 12(a). At the time (A) in Fig. 12, the frequency of the spectrum exceeds *+fs/2* and aliasing is induced. The Doppler ultrasound system has an anti-aliasing display function (BLS: baseline-shift) that shifts a baseline to a negative side, as shown in Fig. 12(b), and expands a positive velocity range seemingly. Thus we can measure the peak velocity of blood flow easily. The power spectrum at the zero baseline-shift is shown in Fig. 13(a). The spectrum image at the *-0.25*fs* baseline-shift and the power spectrum corresponding to the time (A) in Fig. 12 are shown in Fig. 13(b). In the spectrum image, a baseline-shift is easily realized by changing the frequency read-out operation of the spectrum after FFT processing. However, since there is no baseline-shift function in the Doppler audio, a baseline-shift is not realized in spectrum imaging and Doppler audio processing. For example, although a negative-component is lost in the spectrum image shown in Fig. 13(b), since Doppler sound is still in the state shown in Fig. 13(a), it displays a negative-output and does not correspond to the Doppler image.

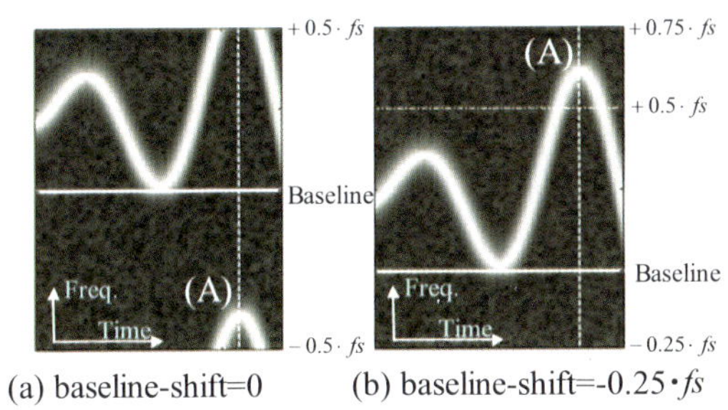

Fig. 12. Spectrum Doppler image

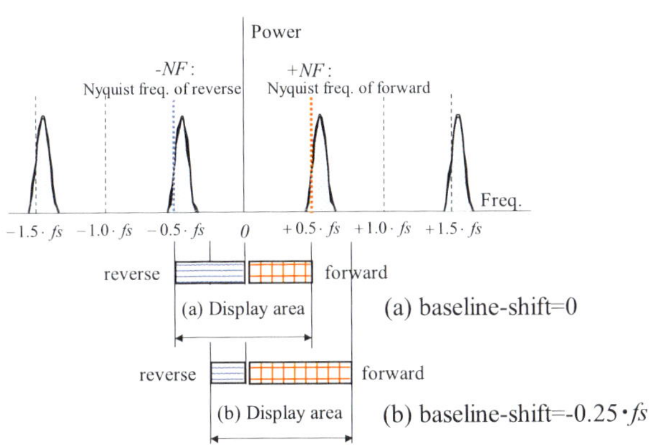

Fig. 13. Spectrum display area and baseline shift.

4.2 Anti-aliasing processing of Doppler audio and its target performance

To solve the problem of the spectrum image and Doppler audio not working together, we examined the signal processing system of the Doppler audio to determine the possible type of baseline-shift. On the other hand, since IQ-signals after quadrature-detection had little merit at a small operation load in narrow-band processing, we examined a realization method based on the IQ-signals. The Hilbert transform, complex FIR filter, phase-shift, complex IIR filter, FFT/IFFT and modulation/demodulation systems also indicated that the direction separation system of the Doppler audio does not allow a baseline-shift. Among these systems, the Hilbert transform and phase-shift systems enable direction separation by addition and subtraction between signals with a 180-degree phase-difference. Since an input IQ-signal has a 90-degree phase difference, these systems give a phase-difference of 90 degree between channels with a filter. Since the phase-difference of an IQ-signal stops being 90 degree when sampling frequency is doubled as a countermeasure, in the Hilbert transform and phase-shift systems, which make the phase-difference between channels a simple 90 degree, direction separation is difficult. Moreover, the complex FIR filter system involves the same pre-processing step as that in the complex IIR filter system, and anti-alias processing becomes possible. However, since the length of a FIR coefficient sequence doubles, the operation load increases. On the other hand, the FFT/IFFT system can reduce the operation load by diverting the FFT output of spectrum Doppler imaging processing. When the FFT output is diverted, the returning anti-alias processing can be performed only by inverse-FFT and shift-addition. The modulation/demodulation and the complex IIR filter systems mainly involve the multiplication of modulation/demodulation and IIR filter processing. Thus, their calculation processing is easy, and the increase in calculation load by anti-aliasing processing is small. As mentioned above, from the viewpoints of calculation load reduction and anti-alias processing feasibility, we chose and examined the following three systems: the modulation/demodulation, the FFT/IFFT, and the complex IIR systems. When evaluating these systems, we showed the same target performance required as that of the Doppler ultrasound system in Table 4. The items 1 to 4 (time-delay, direction separation, frequency characteristic, frequency resolution) are same as table 1.

item	target
1. time-delay	bellow 20ms (*fs*=4KHz)
2. direction separation	above 30dB
3. frequency characterization	-*fs*/128 to –127**fs*/128, *fs*/128 to 127**fs*/128 flat as possible
4. frequency resolution	*fs*/100
5. baseline-shift range	-*fs*/2 to +*fs*/2 (-0.5 to 0.5)

Table 4. Target specification of Doppler audio processing.

baseline-shift	-0.5	-0.25	0	0.25	0.5
FB: band-width of forward	4/8	3/8	2/8	1/8	0
FBC: center freq. of forward	4/16	3/16	2/16	1/16	0
RB: band-width of reverse	0	1/8	2/8	3/8	4/8
RBC: center freq. of reverse	0	-1/16	-2/16	-3/16	-4/16

Notes: Baseline shift, *FB*, *FBC*, *RB* and *RBC* are normalized by *fs*.

Table 5. Frequency shift and bandwidth table of baseline-shift

Baseline-shift range:

The baseline-shift range is considered to be *-0.5*fs* to *+0.5*fs* to enable range expansion on the positive and negative sides to twice the Nyquist frequency range. The ranges of both sides correspond to the baseline-shift shown in Table 5. *FB* and *RB* indicate the bandwidths on the positive (forward) and negative (reverse) sides, whereas *FBC* and *RBC*, the center frequencies on the same sides, respectively. These are normalized using *fs*. Although five stages were used from the baseline shift range of -0.5 to +0.5 in this example, a small setup is possible with the actual Doppler ultrasound system.

4.3 Three kinds of digital signal-processing ideas

4.3.1 The modulation/demodulation system

The block diagram of the modulation/demodulation system is shown in Fig. 14. The IQ-signal is modulated with two sets of quadrature modulators. Thereby, the frequency of the signal induces a +*FBC* shift on the positive-side and a -*RCB* shift on the negative-side. Next, Nyquist frequency is doubled by zero insertion, and applying band limitations on the positive and negative sides demodulates signals. The input signal (equivalent to (A) in Fig. 12) with the aliasing spectrum in Fig. 15(a) is modulated, and the spectra indicating the +*FBC*, and -*RCB* shifts of the frequency of the signal are shown in Figures 15(b) and 15(c), respectively. A positive-side component and a negative-side component are extracted by carrying out a baseline-shift and applying a band limitation using the bandwidths of $\pm FB$ and $\pm RB$ in the passage regions of *LPF1 (z)* and *LPF2 (z)*. The spectra of the *LPF1 (z)* and *LPF2 (z)* outputs are shown in Figures 15(d) and 15(e). Since sampling frequency has doubled after an LPF output, the direction separations on the positive and negative sides that shift the frequencies of -*FBC/2* and +*RCB/2* by demodulation, and are denoted by *BPF1 (z)* and *BPF2 (z)* in Fig. 15(f) are realizable. Although the spectrum in Fig. 15 (equivalent to the aliasing (A) in Fig. 12) is outputted to the negative side for the Nyquist frequency *fs/2*, it can extract the positive-side component beyond the Nyquist frequency in Fig. 15(f). The operation was changed and performed in the calculation example shown in Table 7. For response improvement, we did not use a FIR filter for LPF but the 8th IIR filter with an equivalent performance.

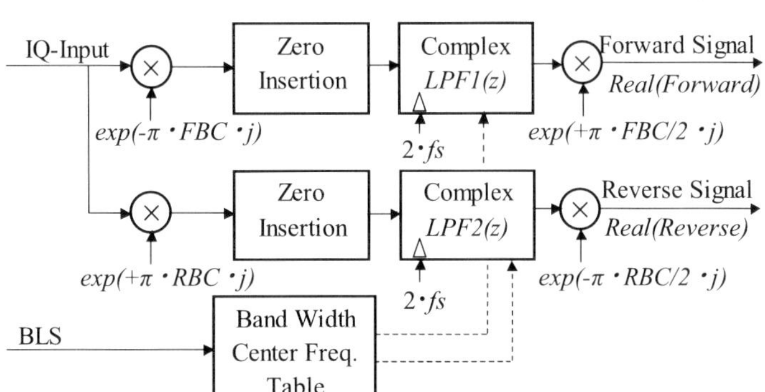

Fig. 14. Block diagram of the modulation/demodulation system

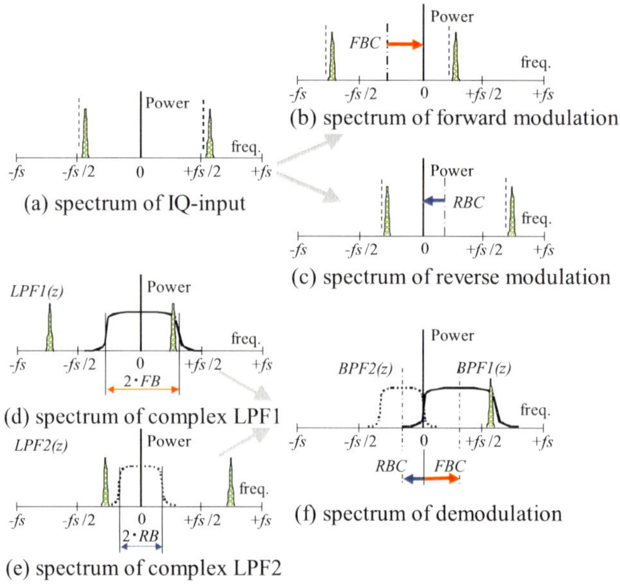

Fig. 15. Frequency design of the modulation/demodulation system

4.3.2 The FFT/IFFT system

The block diagram of the FFT/IFFT system is shown in Fig. 16. Two sets of filters corresponding to the baseline-shift separate the IQ-signal after FFT processing. These filters are realized by applying $WF(\omega)$ and $WR(\omega)$ with the characteristics of *FB*, *RB*, *FBC*, and *RBC* shown in Table 6. Next, the separated spectra are returned to the time domain signals by inverse-FFT. Since the frequency range expands on the basis of the baseline-shift, we perform twice-point inverse-FFT. Further shift in time waveform after inverse-FFT is carried out, and a continuous output is obtained. The power spectrum of the IQ-signal after FFT is shown in Fig. 17(a). When the baseline-shift is terminated, the spectrum in the figure (equivalent to the aliasing (A) in Fig. 12) is observed on the negative-side. However, by operating the read-out address of FFT, the positive display range is expanded and observed on the positive-side. Similarly, by carrying out inverse-FFT processing with $WF(\omega)$ and $WR(\omega)$ with a frequency twice that of sampling ($2 \cdot fs$), the frequency range of the Doppler

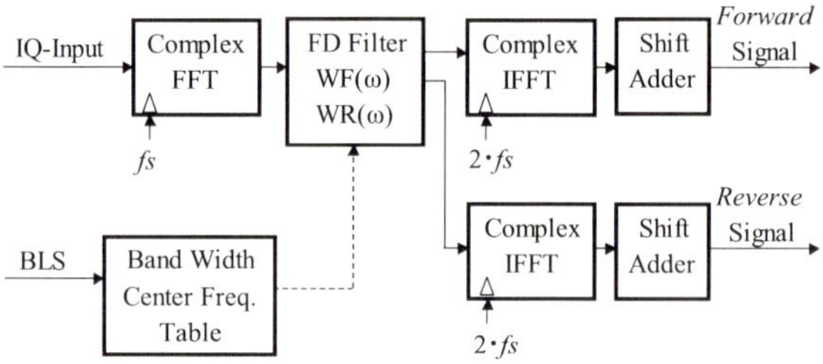

Fig. 16. Block diagram of the FFT/IFFT system

audio is expanded, and the positive-side component in Fig. 17 (b) and the negative-side component in Fig. 17 (c) are obtained. In the calculation example shown in Table 7, we perform 128-point FFT and 256-point inverse-FFT. Moreover, we perform the shift-addition of 32 time series data to which the Hamming window is applied after inverse-FFT.

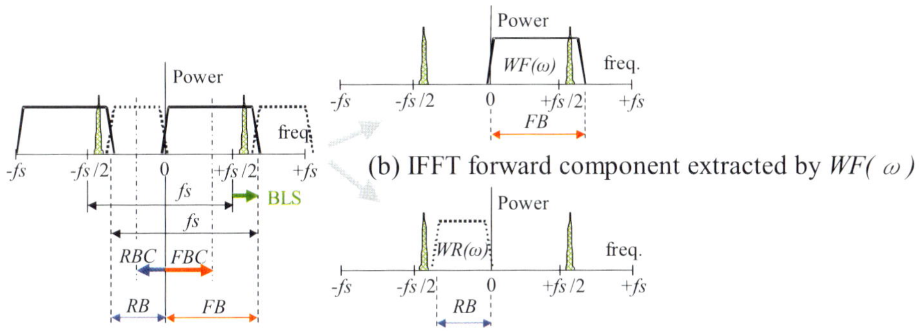

(a) spectrum of IQ-input (c) IFFT reverse component extracted by *WF(ω)*

Fig. 17. Frequency design of the FFT/IFFT system

4.3.3 The complex IIR filter system

The signal processing block diagram of the complex IIR filter system is shown in Fig. 18. Zero insertion is carried out with a pre-treatment, and Nyquist frequency is increased. Next, two complex band-pass filters separate both components directly. The frequency characteristics of the transfer functions *Hf (z)* and *Hr (z)* with the bandwidths of *FB* and *RB* (one side bandwidth) for LPF are shown in Figures 19(a) and 19(b). On the basis of the Fourier transform shift theory, the frequency shifts (*FBC* and *RBC*) are applied to *z* operators, and a transfer function of LPF changes to the positive-side and a negative-side band-pass filters. Operator z is transformed to $z' = z \cdot \exp(-j \cdot FBC)$ and $z'' = z \cdot \exp(-j \cdot RBC)$. The frequency characteristics of the complex band-pass filters *Hf (z')* and *Hr (z'')* enable the *+FBC* and *-RBC* frequency shifts are shown in Fig. 19(c). In the calculation example shown in Table 7, we use the 8th Butterworth filter by considering the response of direction separation.

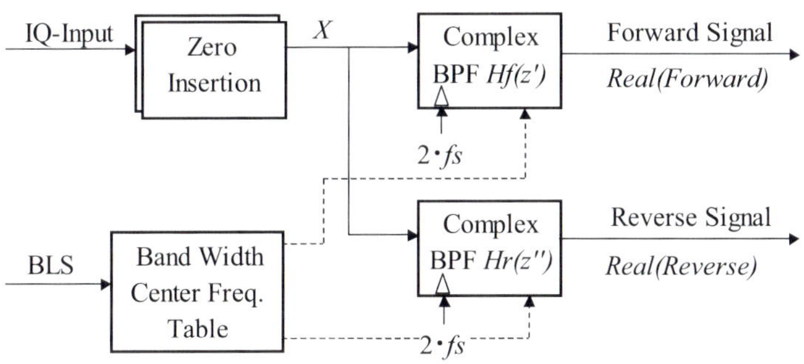

Fig. 18. Block diagram of the complex IIR filter system

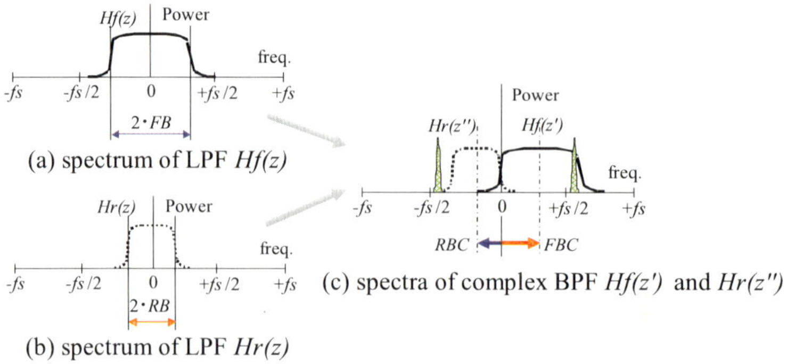

Fig. 19. Frequency design of the complex IIR filter system

4.4 Performances

To satisfy the target performances of frequency resolution and frequency characteristics shown in Table 6, we set up parameters for all the systems, such as the order of the filters and the FFT number. We use the 8th Butterworth filter with cut-off 0.495**FB***fs* and 0.495**RB***fs* for the LPFs of the modulation/demodulation and complex IIR filter systems. We perform 128-point FFT and 256-point inverse-FFT involved in the FFT/IFFT system, and we apply rectangular weight to $WF(\omega)$ and $WR(\omega)$.

system	estimation	delay (ms)
modulation/demodulation	*order*/*fs* (*1)	2 (*order*=8)
FFT/IFFT	0.75*N/*fs* (*2)	24 (*N*=128)
complex IIR	*order*/*fs* (*1)	2 (*order*=8)

Delay is estimated at *fs*=4 kHz. (*1) not including transient response
(*2) IFFT shift addition pitch is *N*/4.

Table 6. Time-delay of Doppler audio processing

System	calculation component	estimation equation	load (MFLOPS)
modulation/demodulation	C-add: *order**8**fs* C-mul: (*order**8+6)**fs* Ovh: 20%	*fs**(48**order*+24)*1.2	1.96 (*order*=8)
FFT/IFFT	C-add: *N***r1*+4**N***r2* C-mul: *N***r1*/2+2**N***r2* R-mul: 2*N**6 Ovh: 20%	(2*fs**4/*N*)**N**(12+6**r1* +12**r2*)*1.2	5.76 (*N*=128) (*r1*=7,*r2*=8)
complex IIR	C-add: *order**8**fs* C-mul: *order**8**fs*	*fs**(48**order*)*1.2	1.84 (*order*=8)

R-add: real-addition, Rmul: real-multiplication, Ovh: over head, C-add: complex-addition,
C-mul: complex-multiplication, IFFT shift addition pitch is *N*/4.
Calculation volume is estimated at *fs*=4 kHz.

Table 7. Calculation load of Doppler audio processing

First, the time-delays theoretically determined from the above-mentioned parameters and calculation loads are shown in Tables 6 and 7, respectively. Since the signal processing input is sampled using *fs*, delay time increases with a decrease in *fs*. Table 6 shows the time-delay calculation result for a typical *fs*=4 kHz diagnostic operation. Moreover, we simply estimate the time-delay from the calculation load itself considered to be zero by sampling, and the estimated values are not affected by the transient response. Since the operation load depends strongly on the hardware-architecture that performs signal processing, we evaluate the frequency of multiplication/addition for 1 s (single-accuracy floating point). The calculation element for every signal processing system, calculation-load estimated formula and operation load per second (*fs*=4 kHz) are shown in Table 7. The estimated results in Tables 6 and 7 show that the complex IIR filter system and the modulation/demodulation systems are fulfilling the time-delay performance goal. Regarding the calculation load, the complex IIR filter system is the smallest, the modulation/demodulation system is slightly larger, and the FFT/IFFT system is the largest, but still small compared with previously reported values. Next, we perform a simulation to check whether we can meet the frequency feature of the performance goal in Table 4. We sweep the frequency of the input IQ-signal and measure the powers of the positive-side and negative-side outputs.

We evaluate simultaneously the frequency features and direction separation performance at this time. The frequency features of the direction separation output according to the three signal processing systems are shown in Fig. 20. A solid line denotes the positive-side component, and a dashed line, the negative-side component. The horizontal axis indicates the frequency range from *-fs* to *+fs*. Moreover, the spectrum image display range corresponding to the frequency range is shown in the bottom rail. The output feature of the Doppler audio at the zero baseline-shift is shown in Figures 20(a), 20(c) and 20(e), and that of +0.4*fs baseline shift is shown in Figures 20(b), 20(d) and 20(f). From these results, we confirm that the frequency feature in each signal processing system of the Doppler audio corresponds to the baseline-shift of the spectrum image. Here, we consider that owing to the effect of the shift-addition in the Hamming window of the FFT/IFFT system, the component near DC in Figures 20(c) and 20(d) is missing. Since this missing part has a value lower than the typical setting value of cut-off frequency for the high-pass filter (equivalent to HPF in Fig. 11) of the preceding process, we do not encounter any problem. Moreover, we observe that the separation degrees of the positive-side component in Figures 20(b) and 20(f) are insufficient. We consider that the cut-off features (the 8th Butterworth filter is used in the simulation) of the modulation/demodulation and the complex IIR filter systems can be improved by making them steep. However, in the case of using an IIR filter, we should expand the internal bit length (dynamic range), because the increased load is expected to be affected by quantizing noise. For example, although Figures 20(e) and 20(f) are calculated using the single floating point (24-bit mantissa) in the simulation, by increasing cut-off frequency or filter order, mantissa bit length (accuracy) may be insufficient and the calculation load or hardware scale may increase. Although we use the Butterworth filter this time, we can choose the Chebysev filter and acquire a steep cut-off feature. On the other hand, the frequency feature and direction separation performance near cut-off frequency deteriorate with a ripple and rapid phase change.

From the above results, we observe that in choosing the response and calculation load, the complex IIR filter system is the most effective. On the other hand, the FFT/IFFT system is the most effective in choosing the frequency feature, although the response is poor. Since the response is more important than the frequency feature clinically, and the target performance in Table 4 is fulfilled mostly, we consider the complex IIR filter system to be the best device for the direction separation of the Doppler audio system.

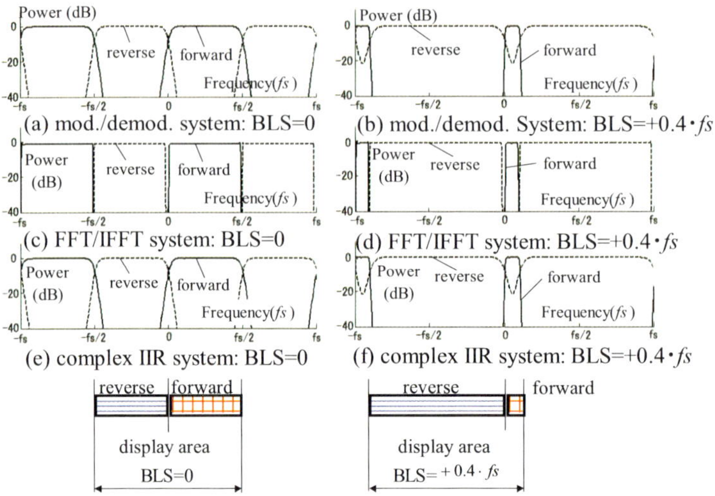

Fig. 20. Frequency characterization of Doppler audio output

4.5 Implementation of complex IIR filter system

4.5.1 Signal processing simulation

We examine the possibility of using the complex IIR filter system in signal processing simulation. The input signal is conceived to be for the actual venous blood model. The model consists of a noise component (white noise), a blood vessel wall component (clutter: low frequency high power), and a blood flow component. The powers and frequencies of these components are shown in Table 8. The input and output waveforms and power spectra of the processing blocks in the complex IIR filter system are shown in Fig. 21. The amplitude of the left-hand-side waveform is normalized by clutter amplitude to be 2. Moreover, 256-point FFT with a Hanning window is applied to the calculation of the right-hand-side power spectrum. Figures 21(a) and 21(c) show the input and output waveforms of zero insertion processing, respectively. A solid line denotes the I-component, and a dashed line, the Q-component. Figures 21(e) and 21(g) show the Doppler audio outputs of both directions at the zero baseline-shift. A solid line denotes the real component, and a dashed line, the imaginary component. Figures 21(i) and 21(k) show the Doppler audio outputs of both directions at the +0.4**fs* baseline-shift. A solid line denotes the real-component, and a dashed line, the imaginary-component. Figures 21(b), 21(d), 21(f), 21(h), 21(i) and 21(l) show power spectra corresponding to the waveforms in the time domain. The aliasing spectra of blood flow and clutter are observed in Fig. 21(d) for a zero insertion processing output. Moreover, the approximately –20 dB DC component is observed at the center of the spectra. This DC component, which is not removed using the Hanning window, does not affect the latter complex band-pass filter processing. From the positive-side output waveform at the zero baseline-shift shown in Fig. 21(e), we confirm that the blood flow component of +0.24**fs* frequency is separated on the positive-side. Moreover, in the power spectrum shown in Fig.

21(f) in addition to the blood flow component, we observe that the clutter component (-0.08*fs) remains on the negative-side under the effect of the filter element. In the negative-side output waveform at the zero baseline-shift in Fig. 21(g), the separation of the clutter component (-0.08*fs) is observed on the negative-side. Moreover, in the power spectrum in Fig. 21(h), a clutter component and a DC component are detected. When the baseline shift is +0.4*fs, the spectrum image and Doppler audio must generate a negative region larger than a positive region. The positive-side output waveform after the baseline shift in Fig. 21(i) shows the disappearance of the clutter component (+0.24*fs). Moreover, we confirm the absence of the blood flow component in the power spectrum shown in Fig. 21(j). We also confirm that a novel blood flow component (-0.76*fs), which is an alias component (+0.24*fs), is outputted into the negative-side output waveform after the baseline-shift in Fig. 21(k), except for the clutter component (-0.08*fs). Moreover, in the power spectrum in Fig. 21(l), we confirm that the blood flow and clutter components are separated on the negative-side.

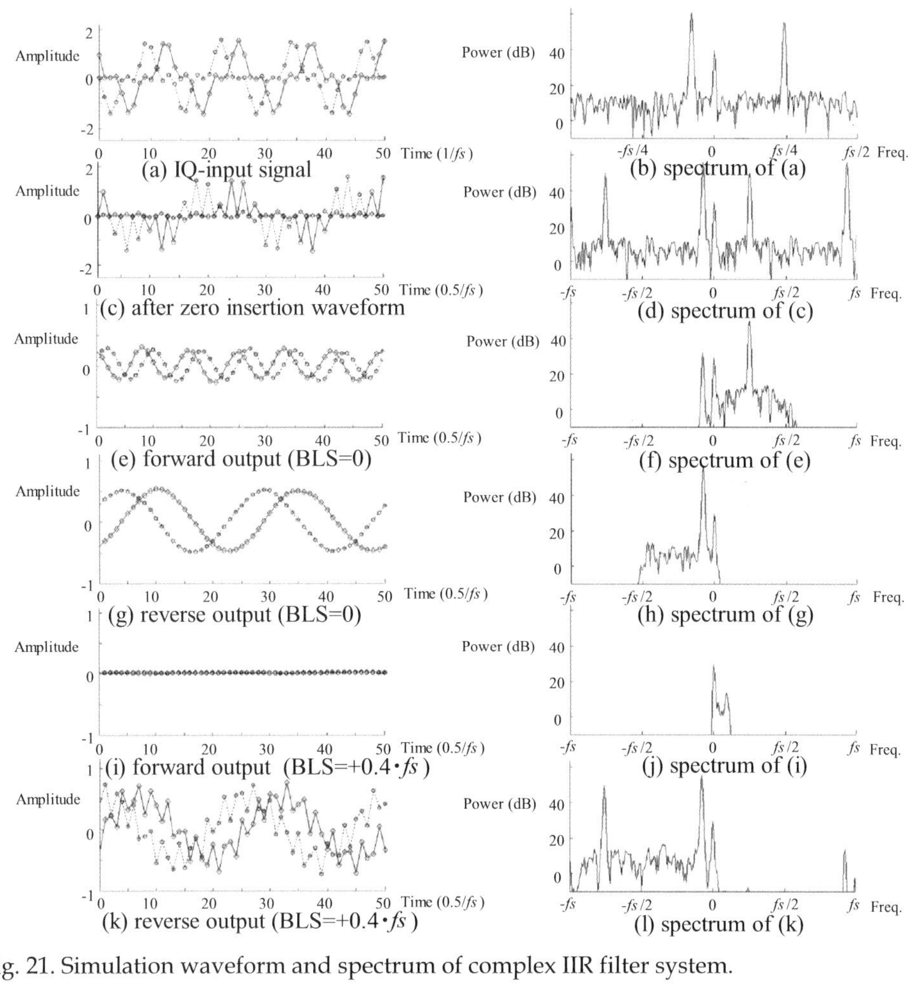

Fig. 21. Simulation waveform and spectrum of complex IIR filter system.

components	blood	noise	clutter
power	-6dB	-20dB	0dB
frequency	0.24*fs	(white noise)	-0.08*fs

Table 8. Components of simulation input model

4.5.2 Implementation

On the basis of the Doppler IQ-signal of the carotid artery collected with the actual Doppler ultrasound system, an example of anti-aliasing signal processing of the Doppler audio is shown in Fig. 22. We use a string phantom (Mark 4 Doppler Phantom: JJ&A Instrument Company) and the ultrasonic diagnosis equipment (SSA-770A: Toshiba Medical Systems Corporation) for generating and collecting the Doppler signal. We use PLT-604AT (6.0 MHz linear probe) at PRF=4 kHz equivalent to *fs*. We collect the IQ-data in PWD mode. Moreover, we set cut-off frequency at an HPF of 200 Hz for clutter removal. The output waveforms of both sides of the Doppler audio and spectrum image obtained from the IQ-data are shown in Fig. 22. In this figure, in the vicinity of 0.9 s, the baseline-shift is switched into -0.4*fs* from 0. At the zero baseline-shift, we observe aliasing in the spectrum image shown in Fig. 22(a) and a negative-side output in Fig. 22(c). However, we confirm that the positive-side display range of the spectrum image expands after a baseline-shift and is interlocked with the Doppler audio. Although it is not observed in Fig. 22, the characteristic of the band-pass filter changes immediately after a baseline-shift. We will continue to examine the transient response of the Doppler audio under this effect and to consider implementation technologies, such as muting.

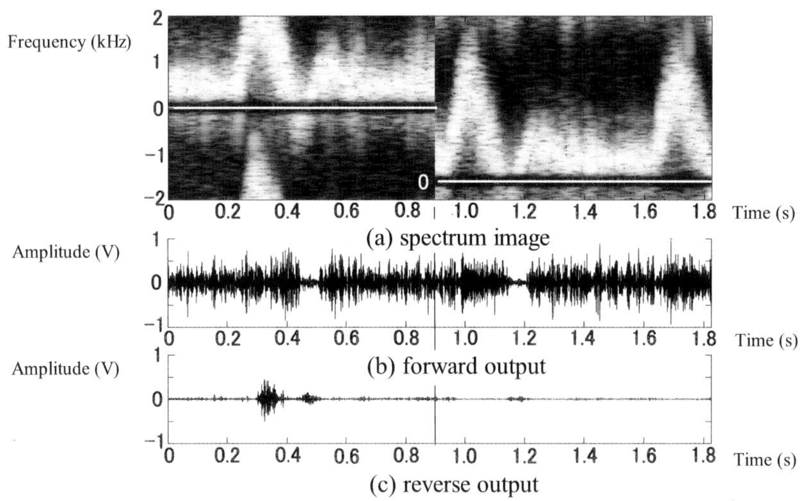

Fig. 22. Doppler spectrum display and audio output waveform

4.6 Conclusion

We developed the direction separation system of a Doppler audio interlocked with the anti-aliasing processing of a spectrum image using a complex IIR band-pass filter system.

First, we defined the target performance of Doppler audio processing and selected three signal-processing systems. We developed processing algorithms and compared their performances. Consequently, we confirmed that the complex IIR band-pass filter system has an excellent response and a low calculation load. Next, we performed functional and performance analyses by simulation with the data collected using a Doppler signal model and a phantom. Conventionally, although in the anti-aliasing process unique to a Doppler ultrasound system, the image and audio did not correspond, since it was applied only to a spectrum image, we could solve this problem by this signal processing.

5. References

Araki, T. (1985). Illustration: The Communication System Theory and Reality, Kogaku Tosho Co., Inc., Tokyo.

Baba, T., Miyajima, Y. & Toshiba Corp. (1998). Ultrasonic diagnosis equipment, *open patent official report of Japan,* Provisional Publication No. 10-99332.

Baba, T. & Toshiba Corp. (2002). Ultrasonic diagnostic equipment and the Doppler signal processing method, *open patent official report of Japan,* Provisional Publication of a Patent 2002-325767.

Baba, T. (2004). The investigation of the audio direction separation in the Doppler ultrasound system Part 1: The comparison of the digital signal processing algorithm, *Proceeding of Acoust. Soc. Jpn.,* Acoustic Imaging pp.29-33.

Baba, T. (2005). The investigation of the direction split technique of the Doppler ultrasound: Comparison of six kinds of Doppler audio processing, *J. Society of Signal Processing Applications and Technology of Japan,* Vol. 8, No. 2, pp.14-20.

Baba, T. (2006). Investigation of the audio direction separation in Doppler ultrasound system: Signal processing of Doppler audio for aliasing, *J. Acoust. Soc. of Jpn.,* Vol. 62, No. 3, 153-160.

Blauert, J. (1997). *Spatial hearing Revised edition,* The MIT Press, Cambridge, Massachusetts.

Bracewell R. N. (2000). *The Fourier Transform and Its Aplications,* McGraw-Hill Companies Inc., Boston.

Cappellini, V., Constantinides, A. G. & Emiliani, P. (1983). *DIGITAL FILTERS AND THEIR APPLICATIONS (3rd edition),* ACADEMIC PRESS INC. LTD., London

Jensen, J. A. (1996). *Estimation of blood velocities using ultrasound: A signal processing approach,* Cambridge University Press, New York

Jensen, J. A. (2001). A new estimator for vector velocity estimation, *IEEE transaction on UFFC,* Vol. 48, No. 4, pp.886-894.

Koo, J., Otterson S. D. & Siemens Medical Systems Inc. (1997). Method and system for Doppler ultrasound audio dealiasing, *United States Patent* US5676148.

Maeda, K., Sano, A., Takaie, H. & Hara, S. (2001). *Wavelet Transform and Its Application,* Asakura Publishing Co., Ltd., Tokyo.

Mo, L. Y. L. & General Electric Company. (2001). *Method and apparatus for dynamic noise reduction for Doppler audio output, United States Patent* US6251077.

Rabben, S. I. et al. (2002). Ultrasound-based vessel wall tracking: an auto-correlation technique with RF center frequency estimation, *Ultrasound in Med. & Biol.,* Vol. 28, No. 4, pp.507-517.

Takaie H. & Tsujii S. (1995). *Multirate Signal Processing* Shokodo Co., Ltd., Tokyo.

Zhang, Y., Wang, Y. & Wang, W. (2003). Denoising quadrature Doppler signals from bi-directional flow using the Wavelet frame, *IEEE Transactions on UFFC,* Vol.50, No.5, pp561-566.

11

Most Efficient Digital Filter Structures: The Potential of Halfband Filters in Digital Signal Processing

Heinz G. Göckler
Digital Signal Processing Group, Ruhr-Universität Bochum
Germany

1. Introduction

A digital halfband filter (HBF) is, in its basic form with real-valued coefficients, a lowpass filter with one passband and one stopband region of unity or zero desired transfer characteristic, respectively, where both specified bands have the same bandwidth. The zero-phase frequency response of a nonrecursive (FIR) halfband filter with its symmetric impulse response exhibits an odd symmetry about the quarter sample rate ($\Omega = \frac{\pi}{2}$) and half magnitude ($\frac{1}{2}$) point [Schüssler & Steffen (1998)], where $\Omega = 2\pi f / f_{\mathrm{n}}$ represents the normalised (radian) frequency and $f_{\mathrm{n}} = 1/T$ the sampling rate. The same symmetry holds true for the squared magnitude frequency response of minimum-phase (MP) recursive (IIR) halfband filters [Lutovac et al. (2001); Schüssler & Steffen (2001)]. As a result of this symmetry property, the implementation of a real HBF requires only a low computational load since, roughly, every other filter coefficient is identical to zero [Bellanger (1989); Mitra & Kaiser (1993); Schüssler & Steffen (2001)].

Due to their high efficiency, digital halfband filters are widely used as versatile building blocks in digital signal processing applications. They are, for instance, encountered in front ends of digital receivers and back ends of digital transmitters (software defined radio, modems, CATV-systems, etc. [Göckler & Groth (2004); Göckler & Grotz (1994); Göckler & Eyssele (1992); Renfors & Kupianen (1998)]), in decimators and interpolators for sample rate alteration by a factor of two [Ansari & Liu (1983); Bellanger (1989); Bellanger et al. (1974); Gazsi (1986); Valenzuela & Constantinides (1983)], in efficient multirate implementations of digital filters [Bellanger et al. (1974); Fliege (1993); Göckler & Groth (2004)] (cf. Fig. 1), where the input/output sampling rate f_{n} is decimated by I cascaded HBF stages by a factor of 2^I to $f_{\mathrm{d}} = 2^{-I} \cdot f_{\mathrm{n}}$ ($z_{\mathrm{d}} = z_{\mathrm{n}}^{2^I}$), in tree-structured filter banks for FDM de- and remultiplexing (e.g. in satellite communications) according to Fig. 2 and [Danesfahani et al. (1994); Göckler & Felbecker (2001); Göckler & Groth (2004); Göckler & Eyssele (1992)], etc. A frequency-shifted (complex) halfband filter (CHBF), generally known as Hilbert-Transformer (HT, cf. Fig. 3), is frequently used to derive an analytical bandpass signal from its real-valued counterpart [Kollar et al. (1990); Kumar et al. (1994); Lutovac et al. (2001); Meerkötter & Ochs (1998); Schüssler & Steffen (1998; 2001); Schüssler & Weith (1987)]. Finally, real IIR HBF or spectral factors of real FIR HBF, respectively, are used in perfectly reconstructing sub-band coder (cf. Fig. 4) and transmultiplexer filter banks [Fliege (1993); Göckler & Groth (2004);

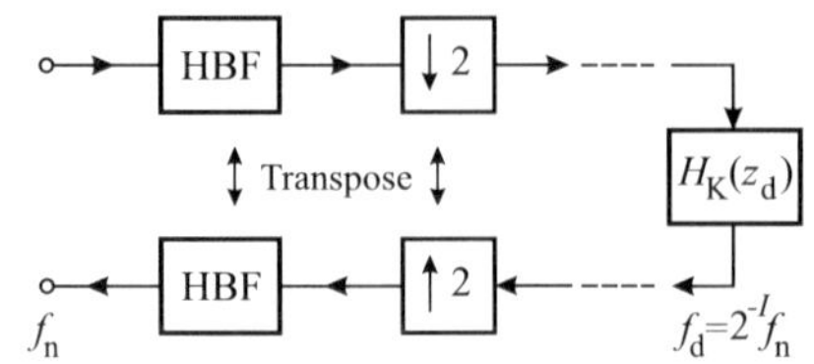

Fig. 1. Multirate filtering applying dyadic HBF decimators, a basic filter, and (transposed) HBF interpolators

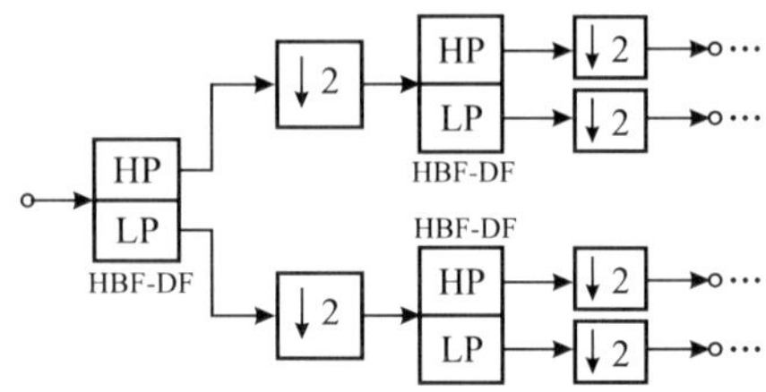

Fig. 2. FDM demultiplexer filter bank; LP/HP: lowpass/highpass directional filter block based on HBF

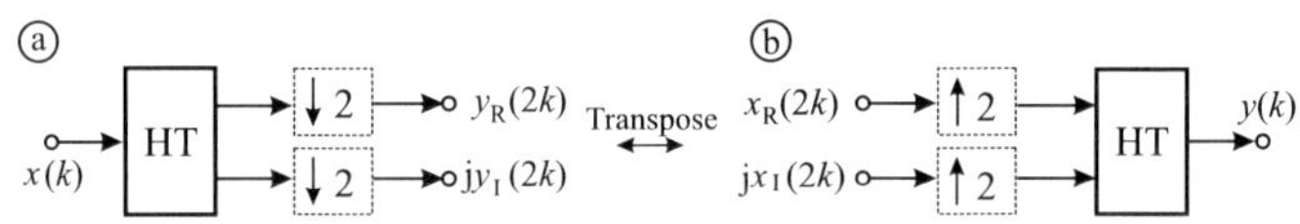

Fig. 3. Decimating Hilbert-Transformer (a) and its transpose for interpolation by two (b)

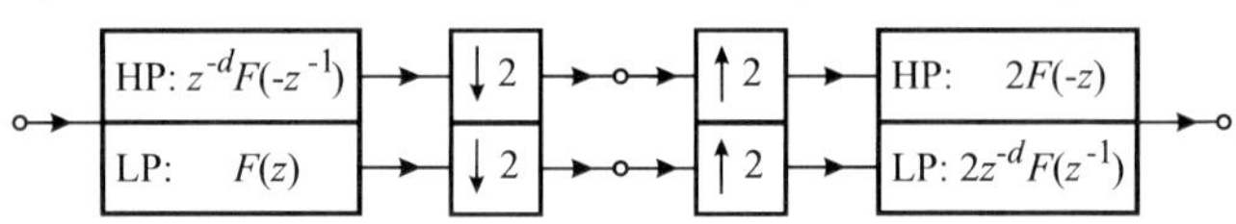

Fig. 4. Two-channel conjugated quadrature mirror filter sub-band coder (SBC) filter bank, where the filters $F(z)$ are spectral factors of a linear-phase FIR HBF

Mitra & Kaiser (1993); Vaidyanathan (1993)], which may apply the discrete wavelet transform [Damjanovic & Milic (2005); Damjanovic et al. (2005); Fliege (1993); Strang & Nguyen (1996)]. Digital linear-phase (LP) FIR and MP IIR HBF have thoroughly been investigated during the last three decades starting in 1974 [Bellanger et al. (1974)] and 1969 [Gold & Rader (1969)], respectively. An excellent survey of this evolution is presented in [Schüssler & Steffen (1998)]. However, the majority of these investigations deal with the properties and the design of HBF by applying allpass pairs [Regalia et al. (1988); Vaidyananthan et al. (1987)], also comprising IIR HBF with approximately linear-phase response [Schüssler & Steffen (1998; 2001); Schüssler & Weith (1987)]. Hence, only few publications on efficient structures e.g. [Bellanger (1989); Bellanger et al. (1974); Lutovac et al. (2001); Man & Kleine (1988); Milic (2009); Valenzuela & Constantinides (1983)], present elementary signal flow graphs (SFG) with minimum computational load. Moreover, only real-valued HBF and complex Hilbert-Transformers (HT) with a centre frequency of $f_c = f_n/4$ ($\Omega_c = \frac{\pi}{2}$) have been considered in the past.

The goal of Section 2 of this contribution is to show the existence of a family of real and complex HBF, where the latter are derived from the former ones by frequency translation, with their passbands (stopbands) centred at one point of an equidistant frequency grid

$$f_c = c \cdot \frac{f_n}{8}, \qquad c = 0, 1, 2, 3, 4, 5, 6, 7. \tag{1}$$

In addition, it is shown that the complex HBF defined by (1) require roughly the same amount of computation as their original real HBF prototype ($f_c = f_0 = 0$). Especially, we present the most efficient elementary SFG for sample rate alteration, their main application. The SFG will be given for LP FIR [Göckler (1996b)] as well as for MP IIR HBF for real- and complex-valued input and/or output signals, respectively. Detailed comparison of expenditure is included.

In Section 3 we combine *two* of those linear-phase FIR HBF investigated in Section 2 with different centre frequencies out of the set given by (2), to construct efficient SFG of directional filters (DF) for separation of one input signal into two output signals or for combination of two input signals to one output signal, respectively. These DF are generally referred to as two-channel frequency demultiplexer (FDMUX) or frequency multiplexer (FMUX) filter bank [Göckler & Eyssele (1992); Vaidyanathan & Nguyen (1987); Valenzuela & Constantinides (1983)].

In Section 4 of this chapter we consider the application of the two-channel DF as a building block of a multiple channel tree-structured FDMUX filter bank according to Fig. 2, typically applied for on-board processing in satellite communications [Danesfahani et al. (1994); Göckler & Felbecker (2001); Göckler & Groth (2004); Göckler & Eyssele (1992)]. In case of a great number of channels and/or challenging bandwidth requirements, implementation of the front-end DF is crucial, which must be operated at (extremely) high sampling rates. To cope with this issue, in Section 4 we present an approach to parallelise at least the front end of the FDMUX filter bank according to Fig. 2.

2. Single halfband filters[1]

In this Section 2 of this chapter we recall the properties of the well-known HBF with real coefficients (real HBF with centre frequencies $f_c \in \{f_0, f_4\} = \{0, f_n/2\}$ according to (1)), and investigate those of the complex HBF with their passbands (stopbands) centred at

$$f_c = c \cdot \frac{f_n}{8}, \qquad c = 1, 2, 3, 5, 6, 7 \tag{2}$$

that require roughly the same amount of computation as their real HBF prototype ($f_c = f_0 = 0$). In particular, we derive the most efficient elementary SFG for sample rate alteration. These will be given both for LP FIR [Göckler (1996b)] and MP IIR HBF for real- and complex-valued input and/or output signals, respectively. The expenditure of all eight versions of HBF according to (1) is determined and thoroughly compared with each other.

The organisation of Section 2 is as follows: First, we recall the properties of both classes of the afore-mentioned real HBF, the linear-phase (LP) FIR and the minimum-phase (MP) IIR approaches. The efficient multirate implementations presented are based on the polyphase decomposition of the filter transfer functions [Bellanger (1989); Göckler & Groth (2004); Mitra (1998); Vaidyanathan (1993)]. Next, we present the corresponding results on complex HBF (CHBF), the classical HT, by shifting a real HBF to a centre frequency according to (2) with $c \in \{2,6\}$. Finally, complex offset HBF (COHBF) are derived by applying frequency shifts according to (2) with $c \in \{1,3,5,7\}$, and their properties are investigated. Illustrative design examples and implementations thereof are given.

[1] Underlying original publication: Göckler & Damjanovic (2006b)

2.1 Real halfband filters (RHBF)

In this subsection we recall the essentials of LP FIR and MP IIR lowpass HBF with real-valued impulse responses $h(k) = h_k \longleftrightarrow H(z)$, where $H(z)$ represents the associated z-transform transfer function. From such a lowpass (prototype) HBF a corresponding real highpass HBF is readily derived by using the modulation property of the z-transform [Oppenheim & Schafer (1989)]

$$z_c^k h(k) \longleftrightarrow H(\frac{z}{z_c}) \tag{3}$$

by setting in accordance with (1)

$$z_c = z_4 = e^{j2\pi f_4/f_n} = e^{j\pi} = -1 \tag{4}$$

resulting in a frequency shift by $f_4 = f_n/2\,(\Omega_4 = \pi)$.

2.1.1 Linear-Phase (LP) FIR filters

Throughout this Section 2 we describe a real LP FIR (lowpass) filter by its non-causal impulse response with its centre of symmetry located at the time or sample index $k = 0$ according to

$$h_{-k} = h_k \quad \forall k \tag{5}$$

where the associated frequency response $H(e^{j\Omega}) \in \mathbb{R}$ is zero-phase [Mitra & Kaiser (1993); Oppenheim & Schafer (1989)].

Specification and properties

A real zero-phase (LP) lowpass HBF, also called Nyquist(2)filter [Mitra & Kaiser (1993)], is specified in the frequency domain as shown in Fig. 5, for instance, for an equiripple or constrained least squares design, respectively, allowing for a don't care transition band between passband and stopband [Mintzer (1982); Mitra & Kaiser (1993); Schüssler & Steffen (1998)]. Passband and stopband constraints $\delta_p = \delta_s = \delta$ are identical, and for the cut-off frequencies we have the relationship:

$$\Omega_p + \Omega_s = \pi. \tag{6}$$

As a result, the zero-phase desired function $D(e^{j\Omega}) \in \mathbb{R}$ as well as the frequency response $H(e^{j\Omega}) \in \mathbb{R}$ are centrosymmetric about $D(e^{j\pi/2}) = H(e^{j\pi/2}) = \frac{1}{2}$. From this frequency domain symmetry property immediately follows

$$H(e^{j\Omega}) + H(e^{j(\Omega-\pi)}) = 1, \tag{7}$$

indicating that this type of halfband filter is strictly complementary [Schüssler & Steffen (1998)].

According to (5), a real zero-phase FIR HBF has a symmetric impulse response of *odd* length $N = n + 1$ (denoted as type I filter in [Mitra & Kaiser (1993)]), where n represents the even filter order. In case of a minimal (canonic) monorate filter implementation, n is identical to the minimum number n_{mc} of delay elements required for realisation, where n_{mc} is known as the McMillan degree [Vaidyanathan (1993)]. Due to the odd symmetry of the HBF zero-phase frequency response about the transition region (don't care band according to Fig. 5), roughly every other coefficient of the impulse response is zero [Mintzer (1982); Schüssler & Steffen (1998)], resulting in the additional filter length constraint:

$$N = n + 1 = 4i - 1, \qquad i \in \mathbb{N}. \tag{8}$$

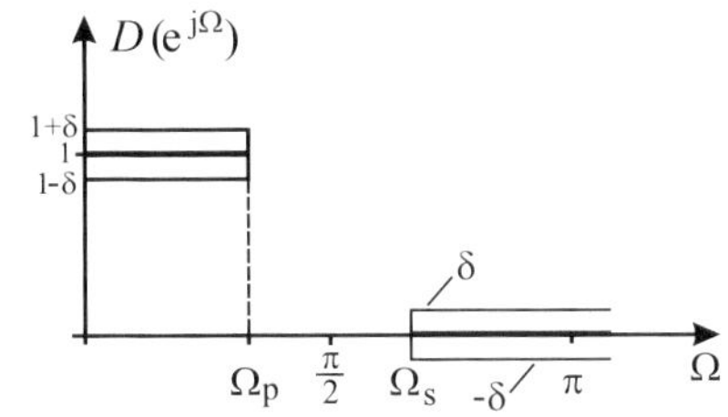

Fig. 5. Specification of a zero-phase FIR HBF; $\Omega_p + \Omega_s = \pi$

Hence, the non-causal impulse response of a real zero-phase FIR HBF is characterized by [Bellanger et al. (1974); Göckler & Groth (2004); Mintzer (1982); Schüssler & Steffen (1998)]:

$$h_k = h_{-k} = \begin{cases} \frac{1}{2} & k = 0 \\ 0 & k = 2l \quad l = 1, 2, \ldots, (n-2)/4 \\ h(k) & k = 2l - 1 \; l = 1, 2, \ldots, (n+2)/4 \end{cases} \tag{9}$$

giving rise to efficient implementations. Note that the name Nyquist(2)filter is justified by the zero coefficients of the impulse response (9). Moreover, if an HBF is used as an anti-imaging filter of an interpolator for upsampling by two, the coefficients (9) are scaled by the upsampling factor of two replacing the central coefficient with $h_0 = 1$ [Fliege (1993); Göckler & Groth (2004); Mitra (1998)]. As a result, independently of the application this coefficient does never contribute to the computational burden of the filter.

Design outline

Assuming an ideal lowpass desired function consistent with the specification of Fig. 5 with a cut-off frequency of $\Omega_t = (\Omega_p + \Omega_s)/2 = \pi/2$ and zero transition bandwidth, and minimizing the integral squared error, yields the coefficients [Göckler & Groth (2004); Parks & Burrus (1987)] in compliance with (9):

$$h_k = \frac{\Omega_t}{\pi} \frac{\sin(k\Omega_t)}{k\Omega_t} = \frac{1}{2} \frac{\sin(k\frac{\pi}{2})}{k\frac{\pi}{2}}, \quad |k| = 1, 2, \ldots, \frac{n}{2}. \tag{10}$$

This least squares design is optimal for multirate HBF in conjunction with spectrally white input signals since, e.g in case of decimation, the overall residual power aliased by downsampling onto the usable signal spectrum is minimum [Göckler & Groth (2004)]. To master the Gibbs' phenomenon connected with (10), a centrosymmetric smoothed desired function can be introduced in the transition region [Parks & Burrus (1987)]. Requiring, for instance, a transition band of width $\Delta\Omega = \Omega_s - \Omega_p > 0$ and using spline transition functions for $D(e^{j\Omega})$, the above coefficients (10) are modified as follows [Göckler & Groth (2004); Parks & Burrus (1987)]:

$$h_k = \frac{1}{2} \frac{\sin(k\frac{\pi}{2})}{k\frac{\pi}{2}} \left[\frac{\sin(k\frac{\Delta\Omega}{2\beta})}{k\frac{\Delta\Omega}{2\beta}} \right]^\beta, \; |k| = 1, 2, \ldots, \frac{n}{2}, \; \beta \in \mathbb{R}. \tag{11}$$

Least squares design can also be subjected to constraints that confine the maximum deviation from the desired function: The Constrained Least Squares (CLS) design [Evangelista (2001); Göckler & Groth (2004)]. This approach has also efficiently been applied to the design of high-order LP FIR filters with quantized coefficients [Evangelista (2002)].

Subsequently, all comparisons are based on equiripple designs obtained by minimization of the maximum deviation $\max\left|H(e^{j\Omega}) - D(e^{j\Omega})\right| \; \forall\Omega$ on the region of support according to [McClellan et al. (1973)]. To this end, we briefly recall the clever use of this minimax design procedure in order to obtain the exact values of the predefined (centre and zero) coefficients of (9), as proposed in [Vaidyanathan & Nguyen (1987)]: To design a two-band HBF of even order $n = N - 1 = 4m - 2$, as specified in Fig. 5, start with designing $i)$ a single-band zero-phase FIR filter $g(k) \longleftrightarrow G(z)$ of odd order $n/2 = 2m - 1$ for a passband cut-off frequency of $2\Omega_p$ which, as a type II filter [Mitra & Kaiser (1993)], has a centrosymmetric zero-phase frequency response about $G(e^{j\pi}) = 0$, $ii)$ upsample the impulse response $g(k)$ by two by inserting between any pair of coefficients an additional zero coefficient (without actually changing the sample rate), which yields an interim filter impulse response $h'(k) \longleftrightarrow H'(z^2)$ of the desired odd length N with a centrosymmetric frequency response about $H'(e^{j\pi/2}) = 0$ [Göckler & Groth (2004); Vaidyanathan (1993)], $iii)$ lift the passband (stopband) of $H'(e^{j\Omega})$ to 2 (0) by replacing the zero centre coefficient with $2h(0) = 1$, and $iv)$ scale the coefficients of the final impulse response $h(k) \longleftrightarrow H(z)$ with $\frac{1}{2}$.

Efficient implementations

Monorate FIR filters are commonly realized by using one of the direct forms [Mitra (1998)]. In our case of an LP HBF, minimum expenditure is obtained by exploiting coefficient symmetry, as it is well known [Mitra & Kaiser (1993); Oppenheim & Schafer (1989)]. The count of operations or hardware required, respectively, is included below in Table 1 (column MoR). Note that the "multiplication" by the central coefficient h_0 does not contribute to the overall expenditure.

The minimal implementation of an LP HBF decimator (interpolator) for twofold down(up)sampling is based on the decomposition of the HBF transfer function into two (type 1) polyphase components [Bellanger (1989); Göckler & Groth (2004); Vaidyanathan (1993)]:

$$H(z) = E_0(z^2) + z^{-1}E_1(z^2). \tag{12}$$

In the case of decimation, downsampling of the output signal (cf. upper branch of Fig. 1) is shifted from filter output to system input by exploiting the noble identities [Göckler & Groth (2004); Vaidyanathan (1993)], as shown in Fig. 6(a). As a result, all operations (including delay and its control) can be performed at the reduced (decimated) output sample rate $f_d = f_n/2$: $E_i(z^2) := E_i(z_d)$, $i = 0, 1$. In Fig. 6(b), the input demultiplexer of Fig. 6(a) is replaced with a commutator where, for consistency, the shimming delay $z_d^{-1/2} := z^{-1}$ must be introduced [Göckler & Groth (2004)].

As an example, in Fig. 7(a) an optimum, causal real LP FIR HBF decimator of $n = 10$th order and for twofold downsampling is recalled [Bellanger et al. (1974)]. Here, the odd-numbered coefficients of (9) are assigned to the zeroth polyphase component $E_0(z_d)$ of Fig. 6(b), whereas the only non-zero even-numbered coefficient h_0 belongs to $E_1(z_d)$.

For implementation we assume a digital signal processor as a hardware platform. Hence, the overall computational load of its arithmetic unit is given by the total number of operations $N_{Op} = N_M + N_A$, comprising multiplication (M) and addition (A), times the operational clock frequency f_{Op} [Göckler & Groth (2004)]. All contributions to the expenditure are listed in Table 1 as a function of the filter order n, where the McMillan degree includes the shimming delays. Obviously, both coefficient symmetry ($N_M < n/2$) and the minimum memory property ($n_{mc} < n$ [Bellanger (1989); Fliege (1993); Göckler & Groth (2004)]) are

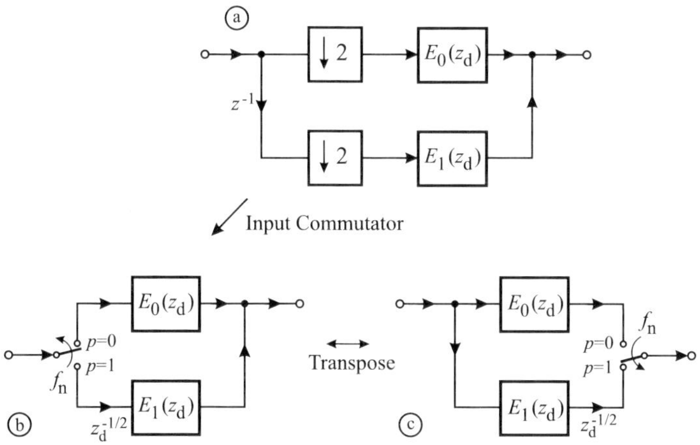

Fig. 6. Polyphase representation of a decimator (a,b) and an interpolator (c) for sample rate alteration by two; shimming delay: $z_d^{-1/2} := z^{-1}$

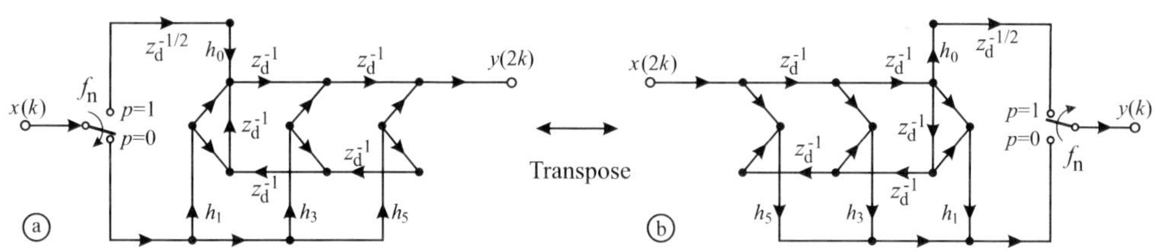

Fig. 7. Optimum SFG of LP FIR HBF decimator (a) and interpolator (b) of order $n = 10$

	MoR: $f_{Op} = f_n$	Dec: $f_{Op} = f_n/2$	Int: $f_{Op} = f_n/2$
n_{mc}	n	$n/2+1$	
N_M	$(n+2)/4$		
N_A	$n/2+1$		$n/2$
N_{Op}	$3n/4+3/2$		$3n/4+1/2$

Table 1. Expenditure of real linear-phase FIR HBF; n: order, n_{mc}: McMillan degree, $N_M(N_A)$: number of multipliers (adders), f_{Op}: operational clock frequency

concurrently exploited. (Note that this concurrent exploitation of coefficient symmetry *and* minimum memory property is not possible for Nyquist(M)filters with $M > 2$. As shown in [Göckler & Groth (2004)], for Nyquist(M)filters with $M > 2$ only *either* coefficient symmetry *or* the minimum memory property can be exploited.)

The application of the multirate transposition rules on the optimum decimator according to Fig. 7(a), as detailed in Section 3 and [Göckler & Groth (2004)], yields the optimum LP FIR HBF interpolator, as depicted in Fig. 6(c) and Fig. 7(b), respectively. Table 1 shows that the interpolator obtained by transposition requires less memory than that published in [Bellanger (1989); Bellanger et al. (1974)].

2.1.2 Minimum-Phase (MP) IIR filters

In contrast to FIR HBF, we describe an MP IIR HBF always by its transfer function $H(z)$ in the z-domain.

Specification and properties

The magnitude response of an MP IIR lowpass HBF is specified in the frequency domain by $\left|D(e^{j\Omega})\right|$, as shown in Fig. 8, again for a minimax or equiripple design. The constraints of the designed magnitude response $\left|H(e^{j\Omega})\right|$ are characterized by the passband and stopband deviations, δ_{p} and δ_{s}, according to [Lutovac et al. (2001); Schüssler & Steffen (1998)] related by

$$(1-\delta_{\mathrm{p}})^2 + \delta_{\mathrm{s}}^2 = 1. \tag{13}$$

The cut-off frequencies of the IIR HBF satisfy the symmetry condition (6), and the squared magnitude response $\left|H(e^{j\Omega})\right|^2$ is centrosymmetric about $\left|D(e^{j\pi/2})\right|^2 = \left|H(e^{j\pi/2})\right|^2 = \frac{1}{2}$. We consider real MP IIR lowpass HBF of odd order n. The family of the MP IIR HBF comprises Butterworth, Chebyshev, elliptic (Cauer-lowpass) and intermediate designs [Vaidyananthan et al. (1987); Zhang & Yoshikawa (1999)]. The MP IIR HBF is doubly-complementary [Mitra & Kaiser (1993); Regalia et al. (1988); Vaidyananthan et al. (1987)], and satisfies the power-complementarity

$$\left|H(e^{j\Omega})\right|^2 + \left|H(e^{j(\Omega-\pi)})\right|^2 = 1 \tag{14}$$

and the allpass-complementarity conditions

$$\left|H(e^{j\Omega}) + H(e^{j(\Omega-\pi)})\right| = 1. \tag{15}$$

$H(z)$ has a single pole at the origin of the z-plane, and $(n-1)/2$ complex-conjugated pole pairs on the imaginary axis within the unit circle, and all zeros on the unit circle [Schüssler & Steffen (2001)]. Hence, the odd order MP IIR HBF is suitably realized by a parallel connection of two allpass polyphase sections as expressed by

$$H(z) = \frac{1}{2}\left[A_0(z^2) + z^{-1}A_1(z^2)\right], \tag{16}$$

where the allpass polyphase components can be derived by alternating assignment of adjacent complex-conjugated pole pairs of the IIR HBF to the polyphase components. The polyphase components $A_l(z^2)$, $l = 0, 1$ consist of cascade connections of second order allpass sections:

$$H(z) = \frac{1}{2}\left(\underbrace{\prod_{i=0,2,\ldots}^{\frac{n-1}{2}-1} \frac{a_i + z^{-2}}{1 + a_i z^{-2}}}_{A_0(z^2)} + z^{-1} \underbrace{\prod_{i=1,3,\ldots}^{\frac{n-1}{2}-1} \frac{a_i + z^{-2}}{1 + a_i z^{-2}}}_{A_1(z^2)}\right), \tag{17}$$

where the coefficients a_i, $i = 0, 1, ..., (\frac{n-1}{2} - 1)$, with $a_i < a_{i+1}$, denote the squared moduli of the HBF complex-conjugated pole pairs in ascending order; the complete set of n poles is given by $\left\{0, \pm j\sqrt{a_0}, \pm j\sqrt{a_1}, ..., \pm j\sqrt{a_{\frac{n-1}{2}-1}}\right\}$ [Mitra (1998)].

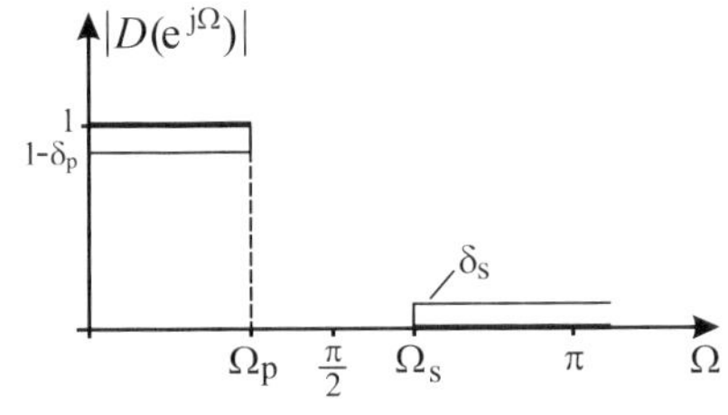

Fig. 8. Magnitude specification of minimum-phase IIR lowpass HBF; $(1-\delta_p)^2+\delta_s^2=1$, $\Omega_p+\Omega_s=\pi$

Design outline

In order to compare MP IIR and LP FIR HBF, we subsequently consider elliptic filter designs. Since an elliptic (minimax) HBF transfer function satisfies the conditions (6) and (13), the design result is uniquely determined by specifying the passband Ω_p (stopband Ω_s) cut-off frequency and one of the three remaining parameters: the odd filter order n, allowed minimal stopband attenuation $A_s = -20\log(\delta_s)$ or allowed maximum passband attenuation $A_p = -20\log(1-\delta_p)$.

There are two most common approaches to elliptic HBF design. The first group of methods is performed in the analogue frequency domain and is based on classical analogue filter design techniques: The desired magnitude response $\left|D(e^{j\Omega})\right|$ of the elliptic HBF transfer function $H(z)$ to be designed is mapped onto an analogue frequency domain by applying the bilinear transformation [Mitra (1998); Oppenheim & Schafer (1989)]. The magnitude response of the analogue elliptic filter is approximated by appropriate iterative procedures to satisfy the design requirements [Ansari (1985); Schüssler & Steffen (1998; 2001); Valenzuela & Constantinides (1983)]. Finally, the analogue filter transfer function is remapped to the z-domain by the bilinear transformation.

The other group of algorithms starts from an elliptic HBF transfer function, as given by (17). The filter coefficients a_i, $i=0,1,...,(\frac{n-1}{2}-1)$ are obtained by iterative nonlinear optimization techniques minimizing the peak stopband deviation. For a given transition bandwidth, the maximum deviation is minimized e.g. by the Remez exchange algorithm or by Gauss-Newton methods [Valenzuela & Constantinides (1983); Zhang & Yoshikawa (1999)].

For the particular class of elliptic HBF with *minimal Q-factor*, closed-form equations for calculating the exact values of stopband and passband attenuation are known allowing for straightforward designs, if the cut-off frequencies and the filter order are given [Lutovac et al. (2001)].

Efficient implementation

In case of a monorate filter implementation, the McMillan degree n_{mc} is equal to the filter order n. Having the same hardware prerequisites as in the previous subsection on FIR HBF, the computational load of hardware operations per output sample is given in Table 2 (column MoR). Note that multiplication by a factor of 0.5 does not contribute to the overall expenditure. In the general decimating structure, as shown in Fig. 9(a), decimation is performed by an input commutator in conjunction with a shimming delay according to Fig. 6(b). By the underlying exploitation of the noble identities [Göckler & Groth (2004); Vaidyanathan (1993)], the cascaded second order allpass sections of the transfer function (17) are transformed to first order allpass sections: $\frac{a_i+z^{-2}}{1+a_iz^{-2}} := \frac{a_i+z_d^{-1}}{1+a_iz_d^{-1}}$, $i=0,1,...,\frac{n-1}{2}-1$, as illustrated in Fig. 9(b).

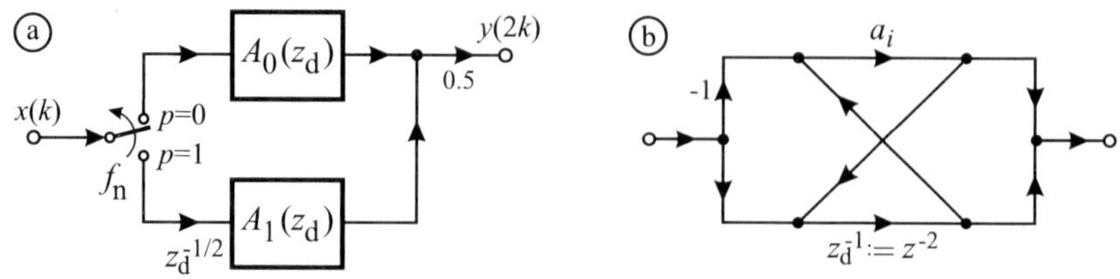

Fig. 9. Optimum minimum-phase IIR HBF decimator block structure (a) and SFG of the 1st (2nd) order allpass sections (b)

	MoR: $f_{\rm Op} = f_{\rm n}$	Dec: $f_{\rm Op} = f_{\rm n}/2$	Int: $f_{\rm Op} = f_{\rm n}/2$
$n_{\rm mc}$	n	$(n+1)/2$	
$N_{\rm M}$	$(n-1)/2$		
$N_{\rm A}$	$3(n-1)/2+1$		$3(n-1)/2$
$N_{\rm Op}$	$2n-1$		$2n-2$

Table 2. Expenditure of real minimum-phase IIR HBF; n: order, $n_{\rm mc}$: McMillan degree, N_M (N_A): number of multipliers (adders), $f_{\rm Op}$: operational clock frequency

Hence, the polyphase components $A_l(z^2) := A_l(z_{\rm d})$, $l = 0, 1$ of Fig. 9(a) operate at the reduced output sampling rate $f_{\rm d} = f_{\rm n}/2$, and the McMillan degree $n_{\rm mc}$ is almost halved. The optimum interpolating structure is readily derived from the decimator by applying the multirate transposition rules (cf. Section 3 and [Göckler & Groth (2004)]). Computational complexity is presented in Table 2, also indicating the respective operational rates $f_{\rm Op}$ for the $N_{\rm Op}$ arithmetical operations.
Elliptic filters also allow for multiplierless implementations with small quantization error, or implementations with a reduced number of shift-and-add operations in multipliers [Lutovac & Milic (1997; 2000); Milic (2009)].

2.1.3 Comparison of real FIR and IIR HBF

The comparison of the Tables 1 and 2 shows that $N_{\rm Op}^{FIR} < N_{\rm Op}^{IIR}$ for the same filter order n, where all operations are performed at the operational rate $f_{\rm Op}$, as given in these Tables. Since, however, the filter order $n_{\rm IIR} < n_{\rm FIR}$ or even $n_{\rm IIR} \ll n_{\rm FIR}$ for any type of approximation, the computational load of an MP IIR HBF is generally smaller than that of an LP FIR HBF, as it is well known [Lutovac et al. (2001); Schüssler & Steffen (1998)].
The relative computational advantage of equiripple minimax designs of monorate IIR halfband filters and polyphase decimators [Parks & Burrus (1987)], respectively, is depicted in Fig. 10 where, in extension to [Lutovac et al. (2001)], the expenditure $N_{\rm Op}$ is indicated as a parameter along with the filter order n. Note that the IIR and FIR curves of the lowest order filters differ by just one operation despite the LP property of the FIR HBF.
A specification of a design example is deduced from Fig. 10: $n_{\rm IIR} = 5$ and $n_{\rm FIR} = 14$, respectively, with a passband cut-off frequency of $f_{\rm p} = 0.1769 f_{\rm n}$ at the intersection point of the associated expenditure curves: Fig. 11. As a result, the stopband attenuations of both filters are the same (cf. Fig. 10). In addition, for both designs the typical pole-zero plots are shown [Schüssler & Steffen (1998; 2001)]. From the point of view of expenditure, the MP IIR HBF decimator ($N_{\rm Op} = 9$, $n_{\rm mc} = 3$) outperforms its LP FIR counterpart ($N_{\rm Op} = 12$, $n_{\rm mc} = 8$).

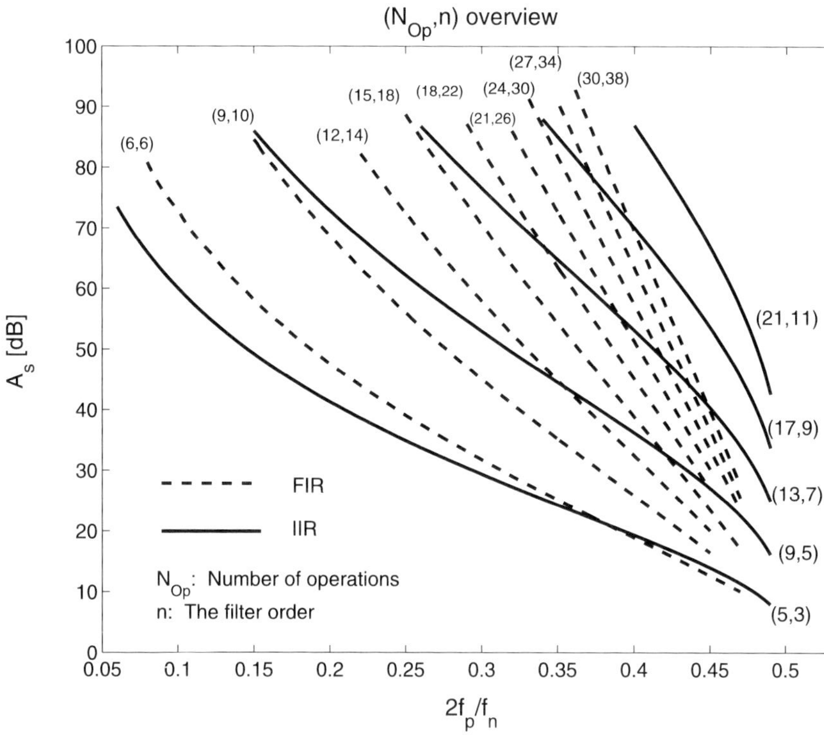

Fig. 10. Expenditure curves of real linear-phase FIR and minimum-phase IIR HBF decimators based on equiripple minimax designs [Parks & Burrus (1987)]

2.2 Complex Halfband Filters (CHBF)

A complex HBF, a classical Hilbert-Transformer [Lutovac et al. (2001); Mitra & Kaiser (1993); Schüssler & Steffen (1998; 2001); Schüssler & Weith (1987)], is readily derived from a real HBF according to Subsection 2.1 by applying the z-transform modulation theorem (3) by setting in compliance with (2)

$$z_c = z_{\pm 2} = z_{\mp 6} = e^{j2\pi f_{\pm 2}/f_n} = e^{\pm j\frac{\pi}{2}} = \pm j, \tag{18}$$

thus shifting the real prototype HBF to a passband centre frequency of $f_{\pm 2} = \pm f_n/4\,(\Omega_{\pm 2} = \pm\pi/2)$. For convenience, subsequently we restrict ourselves to the case $f_c = f_2$.

2.2.1 Linear-Phase (LP) FIR filters

In the FIR CHBF case the frequency shift operation (3) is immediately applied to the impulse response $h(k)$ in the time domain according to (3). As a result of the modulation of the impulse response (9) of any real LP HBF on a carrier of frequency f_2 according to (18), the complex-valued CHBF impulse response

$$\underline{h}_k = h(k)e^{jk\frac{\pi}{2}} \qquad -\frac{n}{2} \leq k \leq \frac{n}{2} \tag{19}$$

is obtained. (Underlining indicates complex quantities in time domain.) By directly equating (19) and relating the result to (9), we get:

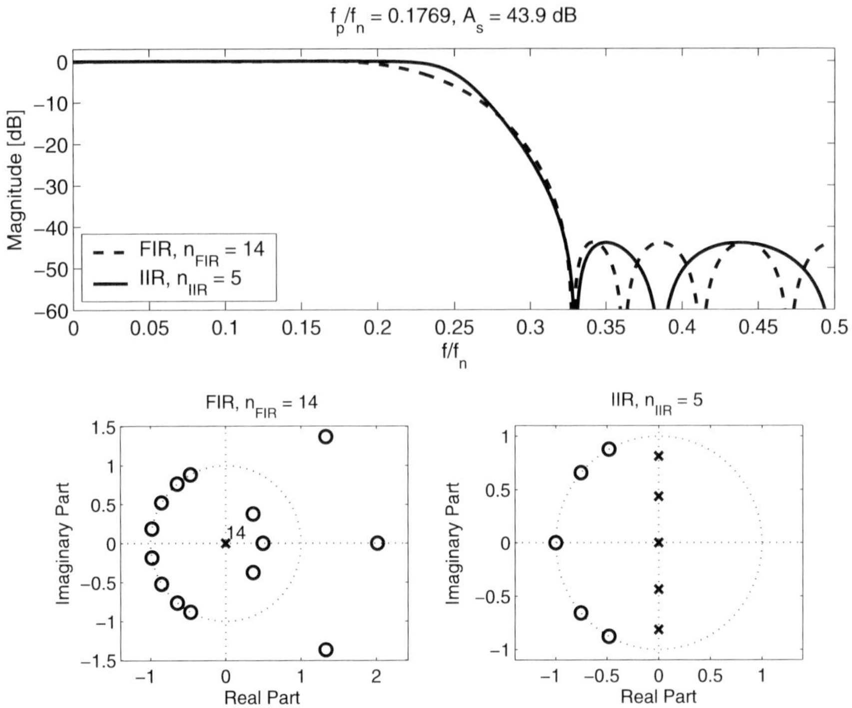

Fig. 11. RHBF design examples: Magnitude characteristics and pole-zero plots

$$\underline{h}_k = \begin{cases} \frac{1}{2} & k = 0 \\ 0 & k = 2l \quad l = 1, 2, \ldots, (n-2)/4 \\ j^k h(k) & k = 2l - 1 \; l = 1, 2, \ldots, (n+2)/4 \end{cases} \tag{20}$$

where, in contrast to (5), the imaginary part of the impulse response

$$\underline{h}_{-k} = -\underline{h}_k \quad \forall k > 0 \tag{21}$$

is skew-symmetric about zero, as it is expected from a Hilbert-Transformer. Note that the centre coefficient h_0 is still real, whilst all other coefficients are purely imaginary rather than generally complex-valued.

Specification and properties

All properties of the real HBF are basically retained except of those which are subjected to the frequency shift operation of (18). This applies to the filter specification depicted in Fig. 5 and, hence, (6) modifies to

$$\Omega_{\mathrm{p}} + \frac{\pi}{2} + \Omega_{\mathrm{s}} + \frac{\pi}{2} = \Omega_{\mathrm{p}+} + \Omega_{\mathrm{s}-} = 2\pi, \tag{22}$$

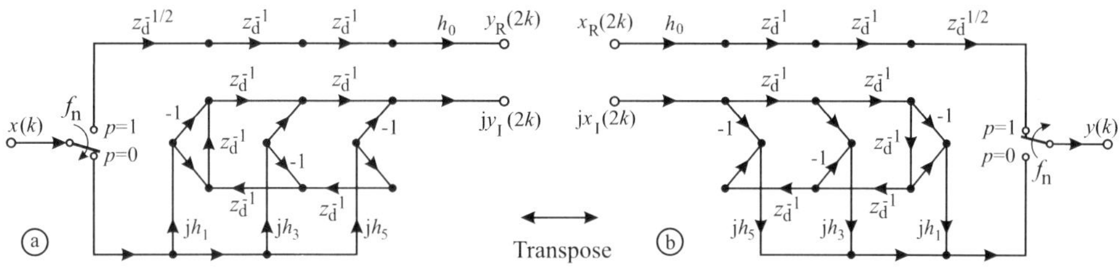

Fig. 12. Optimum SFG of decimating LP FIR HT (a) and its interpolating multirate transpose (b)

	Dec: $\mathbb{R} \mapsto \mathbb{C}$	Int: $\mathbb{C} \mapsto \mathbb{R}$	Dec: $\mathbb{C} \mapsto \mathbb{C}$	Int: $\mathbb{C} \mapsto \mathbb{C}$
n_{mc}	$3n/4+1/2$		$n+2$	
N_{M}	$(n+2)/4$		$(n+2)/2$	
N_{A}	$n/2$		$n+2$	n
N_{Op}	$3n/4+1/2$		$3n/2+3$	$3n/2+1$

Table 3. Expenditure of linear-phase FIR CHBF; n: order, n_{mc}: McMillan degree, $N_{\text{M}}(N_{\text{A}})$: number of multipliers (adders), operational clock frequency: $f_{\text{Op}} = f_{\text{n}}/2$

where $\Omega_{\text{p}+}$ represents the upper passband cut-off frequency and $\Omega_{\text{s}-}$ the associated stopband cut-off frequency. Obviously, strict complementarity (7) is retained as follows

$$\underline{H}(e^{j(\Omega \mp \frac{\pi}{2})}) + \underline{H}(e^{j(\Omega \pm \frac{\pi}{2})}) = 1, \tag{23}$$

where (3) is applied in the frequency domain.

Efficient implementations

The optimum implementation of an $n = 10$th order LP FIR CHBF for twofold downsampling is again based on the polyphase decomposition of (20) according to (12). Its SFG is depicted in Fig. 12(a) that exploits the odd symmetry of the HT part of the system. Note that all imaginary units are included deliberately. Hence, the optimal FIR CHBF interpolator according to Fig. 12(b), which is derived from the original decimator of Fig. 12(a) by applying the multirate transposition rules [Göckler & Groth (2004)], performs the dual operation with respect to the underlying decimator. Since, however, an LP FIR CHBF is strictly rather than power complementary (cf. (23)), the inverse functionality of the decimator is only approximated [Göckler & Groth (2004)].

In addition, Fig. 13 shows the optimum SFG of an LP FIR CHBF for decimation of a complex signal by a factor of two. In essence, it represents a doubling of the SFG of Fig. 12(a). Again, the dual interpolator is readily derived by transposition of multirate systems, as outlined in Section 3.

The expenditure of the half- ($\mathbb{R} \rightleftharpoons \mathbb{C}$) and the full-complex ($\mathbb{C} \mapsto \mathbb{C}$) CHBF decimators and their transposes is listed in Table 3. A comparison of Tables 1 and 3 shows that the overall numbers of operations $N_{\text{Op}}^{\text{CFIR}}$ of the half-complex CHBF sample rate converters (cf. Fig. 12) are almost the same as those of the real FIR HBF systems depicted in Fig. 7. Only the number of delays is, for obvious reasons, higher in the case of CHBF.

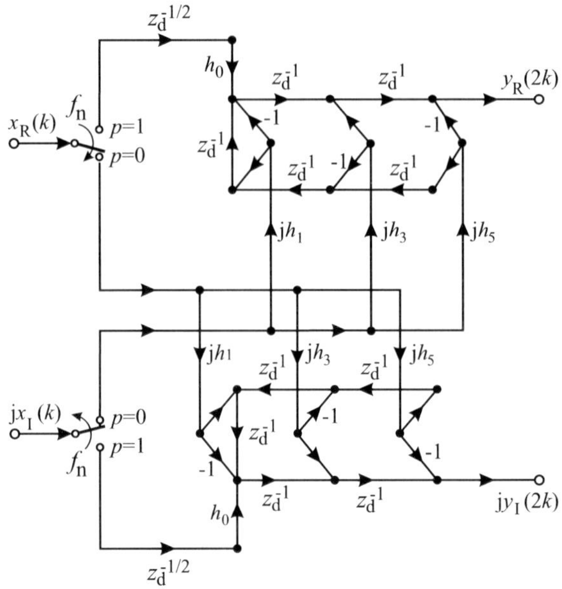

Fig. 13. Optimum SFG of decimating linear-phase FIR CHBF

2.2.2 Minimum-Phase (MP) IIR filters

In the IIR CHBF case the frequency shift operation (3) is again applied in the z-domain. Using (18), this is achieved by substituting the complex z-domain variable in the respective transfer functions $H(z)$ and all corresponding SFG according to:

$$z := \frac{z}{z_2} = ze^{-j\frac{\pi}{2}} = -jz. \tag{24}$$

Specification and properties

All properties of the real IIR HBF are basically retained except of those subjected to the frequency shift operation of (18). This applies to the filter specification depicted in Fig. 8 and, hence, (6) is replaced with (22). Obviously, power (14) and allpass (15) complementarity are retained as follows

$$|\underline{H}(e^{j(\Omega\mp\frac{\pi}{2})})|^2 + |\underline{H}(e^{j(\Omega\pm\frac{\pi}{2})})|^2 = 1, \tag{25}$$

$$\left|\underline{H}(e^{j(\Omega\mp\frac{\pi}{2})}) + \underline{H}(e^{j(\Omega\pm\frac{\pi}{2})})\right| = 1, \tag{26}$$

where (3) is applied in the frequency domain.

Efficient implementations

Introducing (24) into (16) performs a frequency-shift of the transfer function $H(z)$ by $f_2 = f_n/4\,(\Omega_2 = \pi/2)$:

$$\underline{H}(z) = \frac{1}{2}\left[A_0(-z^2) + jz^{-1}A_1(-z^2)\right]. \tag{27}$$

The optimum general block structure of a decimating MP IIR HT, being up-scaled by 2, is shown in Fig. 14(a) along with the SFG of the 1st (system theoretic 2nd) order allpass sections (b), where the noble identities [Göckler & Groth (2004); Vaidyanathan (1993)] are exploited. By

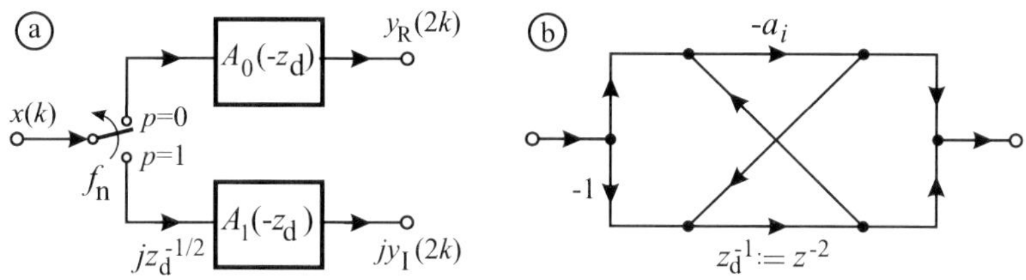

Fig. 14. Decimating allpass-based minimum-phase IIR HT: (a) optimum block structure (b) SFG of the 1st (2nd) order allpass sections

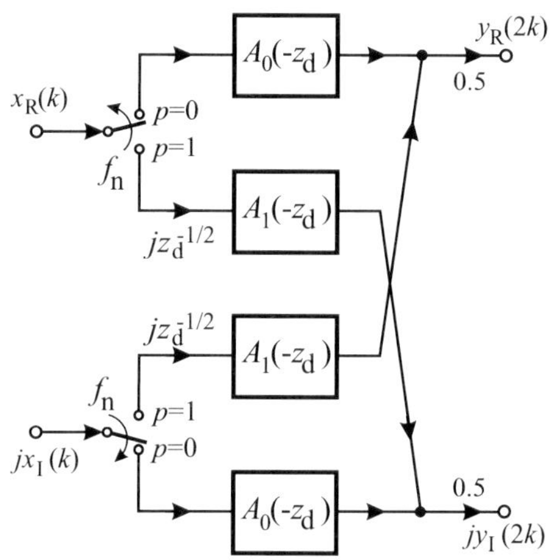

Fig. 15. Block structure of decimating minimum-phase IIR CHBF

	Dec: $\mathbb{R} \mapsto \mathbb{C}$	Int: $\mathbb{C} \mapsto \mathbb{R}$	Dec: $\mathbb{C} \mapsto \mathbb{C}$	Int: $\mathbb{C} \mapsto \mathbb{C}$
n_{mc}	$(n+1)/2$		$n+1$	
N_{M}	$(n-1)/2$		$n-1$	
N_{A}	$3(n-1)/2$		$3(n-1)+2$	$3(n-1)$
N_{Op}	$2n-2$		$4n-2$	$4n-4$

Table 4. Expenditure of minimum-phase IIR CHBF; n: order, n_{mc}: McMillan degree, $N_{\mathrm{M}}(N_{\mathrm{A}})$: number of multipliers (adders), operational clock frequency: $f_{\mathrm{Op}} = f_{\mathrm{n}}/2$

doubling this structure, as depicted in Fig. 15, the IIR CHBF for decimating a complex signal by two is obtained. Multirate transposition [Göckler & Groth (2004)] can again be applied to derive the corresponding dual structures for interpolation.
The expenditure of the half- ($\mathbb{R} \rightleftharpoons \mathbb{C}$) and the full-complex ($\mathbb{C} \mapsto \mathbb{C}$) CHBF decimators and their transposes is listed in Table 4. A comparison of Tables 2 and 4 shows that, basically, the half-complex IIR CHBF sample rate converters (cf. Fig. 14) require almost the same expenditure as the real IIR HBF systems depicted in Fig. 9.

2.2.3 Comparison of FIR and IIR CHBF

As it is obvious from the similarity of the corresponding expenditure tables of the previous subsections, the expenditure chart Fig. 10 can likewise be used for the comparison of CHBF decimators. Both for FIR and IIR CHBF, the number of operations has to be substituted: $N_{\text{Op}}^{CHBF} := N_{\text{Op}}^{HBF} - 1$.

2.3 Complex Offset Halfband Filters (COHBF)

A complex offset HBF, a Hilbert-Transformer with a frequency offset of $\Delta f = \pm f_{\text{n}}/8$ relative to an RHBF, is readily derived from a real HBF according to Subsection 2.1 by applying the zT modulation theorem (3) with $c \in \{1,3,5,7\}$, as introduced in (2):

$$z_c = e^{j2\pi f_c/f_{\text{n}}} = e^{jc\frac{\pi}{4}} = \cos(c\frac{\pi}{4}) + j\sin(c\frac{\pi}{4}) = \pm\frac{1 \pm j}{\sqrt{2}}. \tag{28}$$

As a result, the real prototype HBF is shifted to a passband centre frequency of $f_c \in \left\{\pm\frac{f_{\text{n}}}{8}, \pm\frac{3f_{\text{n}}}{8}\right\}$. In the sequel, we predominantly consider the case $f_c = f_1$ ($\Omega_1 = \pi/4$).

2.3.1 Linear-Phase (LP) FIR filters

Again, the frequency shift operation (3) is applied in the time domain. However, in order to get the smallest number of full-complex COHBF coefficients, we introduce an additional complex scaling factor of unity magnitude. As a result, the modulation of a carrier of frequency f_c according to (28) by the impulse response (9) of any real LP FIR HBF yields the complex-valued COHBF impulse response:

$$\underline{h}_k = e^{jc\frac{\pi}{4}} h(k) z_c^k = h(k) e^{j(k+1)c\frac{\pi}{4}} = h(k) j^{c(k+1)/2}, \tag{29}$$

where $-\frac{n}{2} \leq k \leq \frac{n}{2}$ and $c = 1,3,5,7$. By directly equating (39) for $c = 1$, and relating the result to (9), we get:

$$\underline{h}_k = \begin{cases} \frac{1}{2}\frac{1+j}{\sqrt{2}} & k = 0 \\ 0 & k = 2l \quad l = 1,2,\ldots,(n-2)/4 \\ j^{(k+1)/2} h(k) & k = 2l-1 \; l = 1,2,\ldots,(n+2)/4 \end{cases} \tag{30}$$

where, in contrast to (21), the impulse response exhibits the symmetry property:

$$\underline{h}_{-k} = -j^{ck}\underline{h}_k \quad \forall k > 0. \tag{31}$$

Note that the centre coefficient $\underline{h}_0$ is the only truly complex-valued coefficient where, fortunately, its real and imaginary parts are identical. All other coefficients are again either purely imaginary or real-valued. Hence, the symmetry of the impulse response can still be exploited, and the implementation of an LP FIR COHBF requires just one multiplication more than that of a real or complex HBF [Göckler (1996b)].

Specification and properties

All properties of the real HBF are basically retained except of those which are subjected to the frequency shift operation according to (28). This applies to the filter specification depicted in Fig. 5 and, hence, (6) modifies to

$$\Omega_{\text{p}} + c\frac{\pi}{4} + \Omega_{\text{s}} + c\frac{\pi}{4} = \Omega_{\text{p}+} + \Omega_{\text{s}-} = \pi + c\frac{\pi}{2}. \tag{32}$$

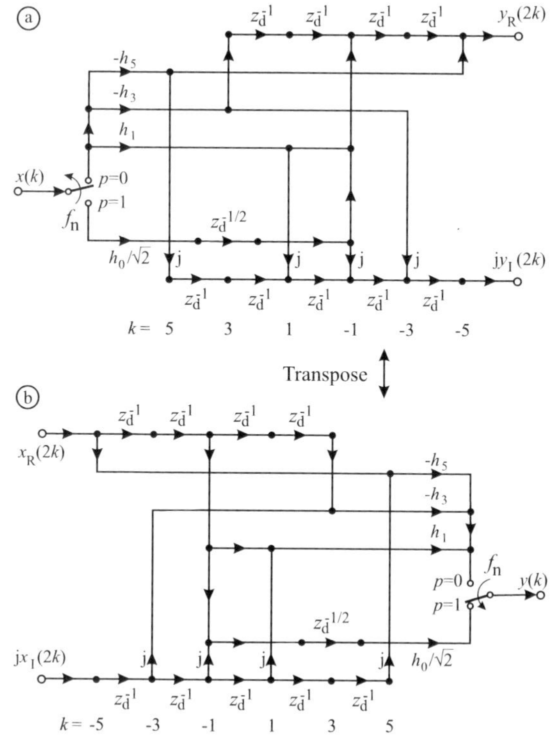

Fig. 16. Optimum SFG of decimating LP FIR COHBF (a) and its transpose for interpolation (b)

where Ω_{p+} represents the upper passband cut-off frequency and Ω_{s-} the associated stopband cut-off frequency. Obviously, strict complementarity (7) reads as follows

$$\underline{H}(e^{j(\Omega - c\frac{\pi}{4})}) + \underline{H}(e^{j(\Omega - \pi(1+c/4))}) = 1. \tag{33}$$

Efficient implementations

The optimum implementation of an n = 10th order LP FIR COHBF for twofold downsampling is again based on the polyphase decomposition of (40). Its SFG is depicted in Fig. 16(a) that exploits the coefficient symmetry as given by (41).

The optimum FIR COHBF interpolator according to Fig. 16(b) is readily derived from the original decimator of Fig. 16(a) by applying the multirate transposition rules, as discussed in Section 3. As a result, the overall expenditure is again retained (c.f. invariant property of transposition [Göckler & Groth (2004)]).

In addition, Fig. 17 shows the optimum SFG of an LP FIR COHBF for decimation of a complex signal by a factor of two. It represents essentially a doubling of the SFG of Fig. 16(a). The dual interpolator can be derived by transposition [Göckler & Groth (2004)].

The expenditure of the half- ($\mathbb{R} \rightleftharpoons \mathbb{C}$) and the full-complex ($\mathbb{C} \mapsto \mathbb{C}$) LP COHBF decimators and their transposes is listed in Table 5 in terms of the filter order n. A comparison of Tables 3 and 5 shows that the implementation of any type of COHBF requires just two or four extra

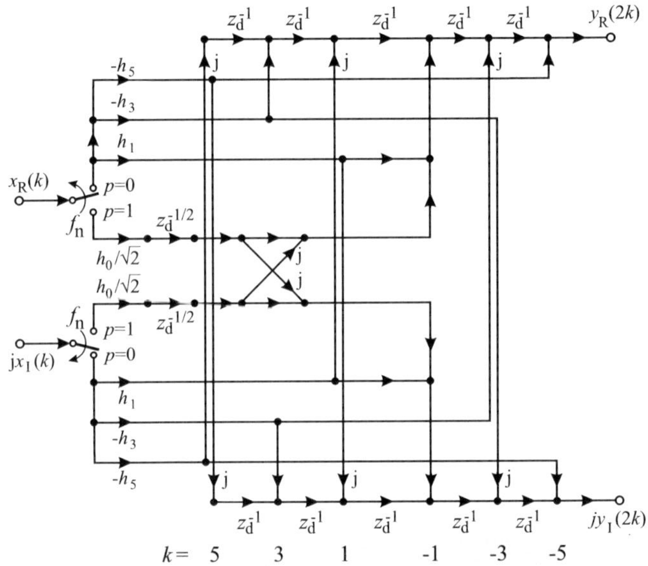

Fig. 17. Optimum SFG of linear-phase FIR COHBF decimating by two

	Dec: $\mathbb{R} \mapsto \mathbb{C}$	Int: $\mathbb{C} \mapsto \mathbb{R}$	Dec: $\mathbb{C} \mapsto \mathbb{C}$	Int: $\mathbb{C} \mapsto \mathbb{C}$
n_{mc}	n		$n+2$	
N_M	$(n+6)/4$		$(n+6)/2$	
N_A	$n/2+1$		$n+4$	$n+2$
N_{Op}	$3n/4+5/2$		$3n/2+7$	$3n/2+5$

Table 5. Expenditure of linear-phase FIR COHBF; n: order, n_{mc}: McMillan degree, $N_M(N_A)$: number of multipliers (adders), operational clock frequency: $f_{Op} = f_n/2$

operations over that of a classical HT (CHBF), respectively (cf. Figs. 12 and 13). This is due to the fact that, as a result of the transition from CHBF to COHBF, only the centre coefficient changes from trivially real ($h_0 = \frac{1}{2}$) to simple complex ($\underline{h}_0 = \frac{1+j}{2\sqrt{2}}$) calling for only one extra multiplication. The number n_{mc} of delays is, however, of the order of n, since a (nearly) full delay line is needed both for the real and imaginary parts of the respective signals. Note that the shimming delays are always included in the delay count. (The number of delays required for a monorate COHBF corresponding to Fig. 17 is $2n$.)

2.3.2 Minimum-Phase (MP) IIR filters

In the IIR COHBF case the frequency shift operation (3) is again applied in the z-domain. This is achieved by substituting the complex z-domain variable in the respective transfer functions $H(z)$ and all corresponding SFG according to:

$$z := \frac{z}{z_1} = ze^{-j\frac{\pi}{4}} = z\frac{1-j}{\sqrt{2}}. \tag{34}$$

	Dec: $\mathbb{R} \mapsto \mathbb{C}$	Int: $\mathbb{C} \mapsto \mathbb{R}$	Dec: $\mathbb{C} \mapsto \mathbb{C}$	Int: $\mathbb{C} \mapsto \mathbb{C}$
n_{mc}	n		$2n$	
N_{M}	n		$2n$	
N_{A}	$3(n-1)$		$6(n-1)+2$	$6(n-1)$
N_{Op}	$4n-3$		$8n-4$	$8n-6$

Table 6. Expenditure of minimum-phase IIR COHBF; n: order, n_{mc}: McMillan degree, $N_{\mathrm{M}}(N_{\mathrm{A}})$: number of multipliers (adders), operational clock frequency: $f_{\mathrm{Op}} = f_{\mathrm{n}}/2$

Specification and properties

All properties of the real IIR HBF are basically retained except of those subjected to the frequency shift operation of (28). This applies to the filter specification depicted in Fig. 8 and, hence, (6) is replaced with (32). Obviously, power (14) and allpass (15) complementarity are retained as follows

$$|\underline{H}(e^{j(\Omega - c\frac{\pi}{4})})|^2 + |\underline{H}(e^{j(\Omega - \pi(1+c/4))})|^2 = 1, \tag{35}$$

$$\left|\underline{H}(e^{j(\Omega - c\frac{\pi}{4})}) + \underline{H}(e^{j(\Omega - \pi(1+c/4))})\right| = 1, \tag{36}$$

where (3) is applied in the frequency domain.

Efficient implementations

Introducing (34) in (16), the transfer function is frequency-shifted by $f_1 = f_n/8$ $(\Omega = \pi/4)$:

$$\underline{H}(z) = \frac{1}{2}\left[\underline{A}_0(-jz^2) + \frac{1+j}{\sqrt{2}} z^{-1} \underline{A}_1(-jz^2)\right]. \tag{37}$$

The optimal structure of an $n = 5$th order MP IIR COHBF decimator for real input signals is shown in Fig. 18(a) along with the elementary SFG of the allpass sections Fig. 18(b). Doubling of the structure according to Fig. 19 allows for full-complex signal processing. Multirate transposition [Göckler & Groth (2004)] is again applied to derive the corresponding dual structure for interpolation.

The expenditure of the half- ($\mathbb{R} \rightleftharpoons \mathbb{C}$) and the full-complex ($\mathbb{C} \mapsto \mathbb{C}$) COHBF decimators and their transposes is listed in Table 6. A comparison of Tables 2 and 6 shows that the half-complex IIR COHBF sample rate converter (cf. Fig. 18(a)) requires almost twice, whereas the full-complex IIR COHBF (cf. Fig. 19) requires even four times the expenditure of that of the real IIR HBF system depicted in Fig. 9.

2.3.3 Comparison of FIR and IIR COHBF

LP FIR COHBF structures allow for implementations that utilize the coefficient symmetry property. Hence, the required expenditure is just slightly higher than that needed for CHBF. On the other hand, the expenditure of MP IIR COHBF is almost twice as high as that of the corresponding CHBF, since it is not possible to exploit memory and coefficient sharing. Almost the whole structure has to be doubled for a full-complex decimator (cf. Fig. 19).

2.4 Conclusion: Family of single real and complex halfband filters

We have recalled basic properties and design outlines of linear-phase FIR and minimum-phase IIR halfband filters, predominantly for the purpose of sample rate alteration by a factor of two, which have a passband centre frequency out of the specific set defined by (1). Our

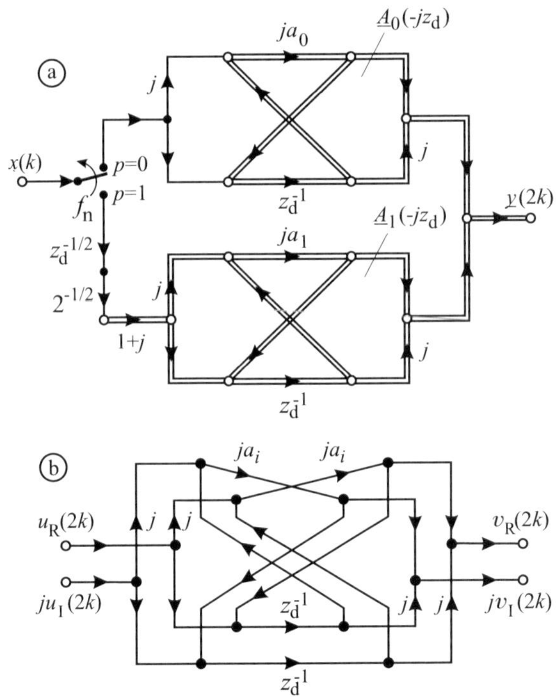

Fig. 18. Decimating allpass-based minimum-phase IIR COHBF, $n = 5$: (a) optimum SFG (b) the 1st (2nd) order allpass section, $i = 0, 1$

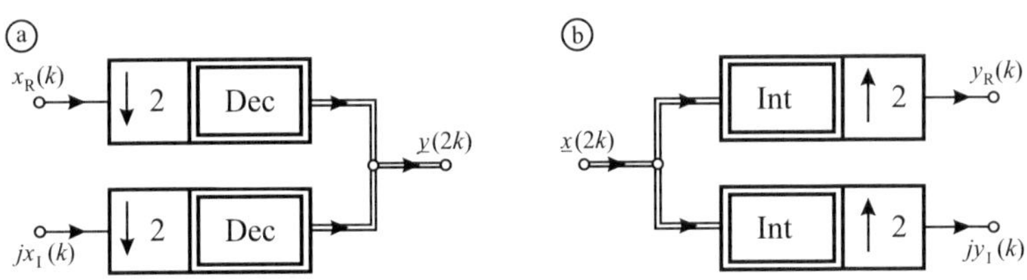

Fig. 19. Block structure of decimating (a) and interpolating (b) minimum-phase IIR COHBF

main emphasis has been put on the presentation of optimum implementations that call for minimum computational burden.

It has been confirmed that, for the even-numbered centre frequencies $c \in \{0, 2, 4, 6\}$, MP IIR HBF outperform their LP FIR counterparts the more the tighter the filter specifications. However, for phase sensitive applications (e.g. software radio employing quadrature amplitude modulation), the LP property of FIR HBF may justify the higher amount of computation to some extent.

In the case of the odd-numbered HBF centre frequencies of (2), $c \in \{1, 3, 5, 7\}$, there exist specification domains, where the computational loads of complex FIR HBF with frequency offset range below those of their IIR counterparts. This is confirmed by the two bottom rows of Table 7, where this table lists the expenditure of a twofold decimator based on the design examples given in Fig. 11 for all centre frequencies and all applications investigated in this

	LP FIR			MP IIR		
	N_{Op}	n_{mc}	Fig.	N_{Op}	n_{mc}	Fig.
HBF Decimator	12	8	7	9	3	9
CHBF: $\mathbb{R} \mapsto \mathbb{C}$	11	11	12(a)	8	3	14
CHBF: $\mathbb{C} \mapsto \mathbb{C}$	24	16	13	18	6	15
COHBF: $\mathbb{R} \mapsto \mathbb{C}$	13	14	16(a)	17	5	18
COHBF: $\mathbb{C} \mapsto \mathbb{C}$	28	16	17	36	10	19

Table 7. Expenditures of real and complex HBF decimators based on the design examples of Fig. 11; N_{Op}: number of operations, n_{mc}: McMillan degree; operational clock frequency: $f_{Op} = f_n/2$

contribution. This sectoral computational advantage of LP FIR COHBF is, despite $n_{IIR} < n_{FIR}$, due to the fact that these FIR filters still allow for memory sharing in conjunction with the exploitation of coefficient symmetry [Göckler (1996b)]. However, the amount of storage n_{mc} required for IIR HBF is always below that of their FIR counterparts.

3. Halfband filter pairs[2]

In this Section 3, we address a particular class of efficient directional filters (DF). These DF are composed of two real or complex HBF, respectively, of different centre frequencies out of the set given by (1). To this end, we conceptually introduce and investigate two-channel frequency demultiplexer filter banks (FDMUX) that extract from an incoming complex-valued frequency division multiplex (FDM) signal, being composed of up to *four* uniformly allocated independent user signals of identical bandwidth (cf. Fig. 20), two of its constituents by concurrently reducing the sample rate by *two* Göckler & Groth (2004). Moreover, the DF shall allow to select any pair of user signals out of the four constituents of the incoming FDM signal, where the individual centre frequencies are to be selectable with minimum switching effort. At first glance, there are two optional approaches: The selectable combination of two filter functions out of a pool of *i*) two RBF according to Subsection 2.1 and two CHBF (HT), as described in Subsection 2.2, where the centre frequencies of this filter quadruple are given by (1) with $c \in \{0, 2, 4, 6\}$, or *ii*) four COHBF, as described in Subsection 2.3, where the centre frequencies of this filter quadruple are given by (1) with $c \in \{1, 3, 5, 7\}$. Since centre frequency switching is more crucial in case one (switching between real and/or complex filters), we subsequently restrict our investigations to case two, where the FDM input spectrum must be allocated as shown in Fig. 20.

These DF with easily selectable centre frequencies are frequently used in receiver front-ends to meet routing requirements [Göckler (1996c)], in tree-structured FDMUX filter banks [Göckler & Felbecker (2001); Göckler & Groth (2004); Göckler & Eyssele (1992)], and, in modified form, for frequency re-allocation to avoid hard-wired frequency-shifting [Abdulazim & Göckler (2007); Eghbali et al. (2009)]. Efficient implementation is crucial, if these DF are operated at high sampling rates at system input or output port. To cope with this high rate challenge, we introduce a systematic approach to system parallelisation according to [Groth (2003)] in Section 4 .

In continuation of the investigations reported in Section 2, we combine two linear-phase (LP) FIR complex *offset* halfband filters (COHBF) with different centre frequencies, being characterized by (1) with $c \in \{1, 3, 5, 7\}$, to construct efficient directional filters for one input

[2] Underlying original publication: Göckler & Alfsmann (2010)

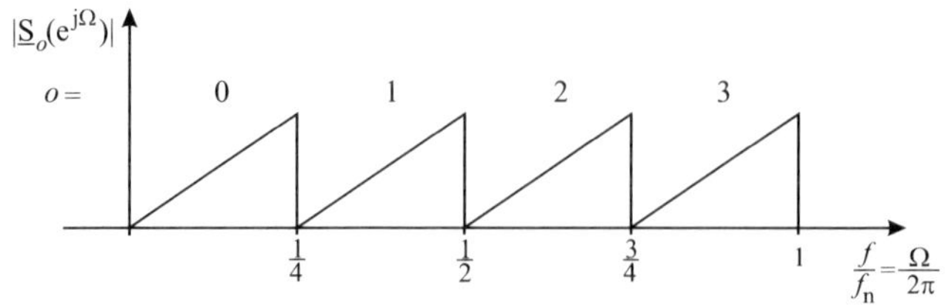

Fig. 20. FDM input spectrum for selection and separation by two-channel directional filter (DF)

and two output signals Göckler (1996a). For convenience, we map the original odd indices $c \in \{1,3,5,7\}$ of the COHBF centre frequencies to natural numbers as defined by

$$f_o = (2o+1) \cdot \frac{f_n}{8}, \quad o \in \{0,1,2,3\} \tag{38}$$

for subsequent use throughout Section 3.

Section 3 is organized as follows: In Subsection 3.1, we detail the statement of the problem, and recall the major properties of COHBF needed for our DF investigations. In the main Subsection 3.2, we present and compare two different approaches to implement the outlined LP DF for signal separation with selectable centre frequencies: *i*) A four-channel uniform complex-modulated FDMUX filter bank undercritically decimating by two, where the respective undesired two output signals are discarded, and *ii*) a synergetic connection of two COHBF that share common multipliers and exploit coefficient symmetry for minimum computation. In Subsection 3.3, we apply the transposition rules of [Göckler & Groth (2004)] to derive the dual DF for signal combination (FDM multiplexing). Finally, we draw some further conclusions in Subsection 3.4.

3.1 Statement of the DF problem

Given a uniform complex-valued FDM signal composed of up to four independent user signals $\underline{s}_o(kT_n) \longleftrightarrow \underline{S}_o(e^{j\Omega})$ centred at f_o, $o = \{0,1,2,3\}$, according to (38), as depicted in Fig. 20, the DF shall extract any freely selectable two out of the four user signals of the FDM input spectrum, and provide them at the two DF output ports separately and decimated by two: $\underline{s}_o(2kT_n) := \underline{s}_o(mT_d) \longleftrightarrow \underline{S}_o(e^{j\Omega^{(d)}})$; $T_d = 1/f_d = 2T_n$. Recall that complex-valued time-domain signals and spectrally transformed versions thereof are indicated by underlining.

Efficient signal separation and decimation is conceptually achieved by combining two COHBF with their differing passbands centred according to (38), where $o \in \{0,1,2,3\}$, along with twofold polyphase decomposition of the respective filter impulse responses [Göckler & Damjanovic (2006a); Göckler & Groth (2004)]. All COHBF are frequency-shifted versions of a real zero-phase (ZP) lowpass HBF prototype with symmetric impulse response $h(k) = h_k = h_{-k} \longleftrightarrow H_0(e^{j\Omega}) \in \mathbb{R}$ according to Subsection 2.1.1, as depicted in Fig. 21(a) as ZP HBF frequency response [Milic (2009); Mitra & Kaiser (1993)]. A frequency domain representation of a possible DF setting (choice of COHBF centre frequencies $o \in \{0,2\}$) is shown in Fig. 21(b), and Figs.21(c,d) present the output spectra at port I ($o = 0$) and port II ($o = 2$), respectively, related to the reduced sampling rate $f_d = f_n/2$.

A COHBF is derived from a real HBF (9) by applying the frequency shift operation in the time domain by modulating a complex carrier $z_o^k = e^{j2\pi k f_o/f_n} = e^{jk(2o+1)\frac{\pi}{4}}$ of a frequency prescribed

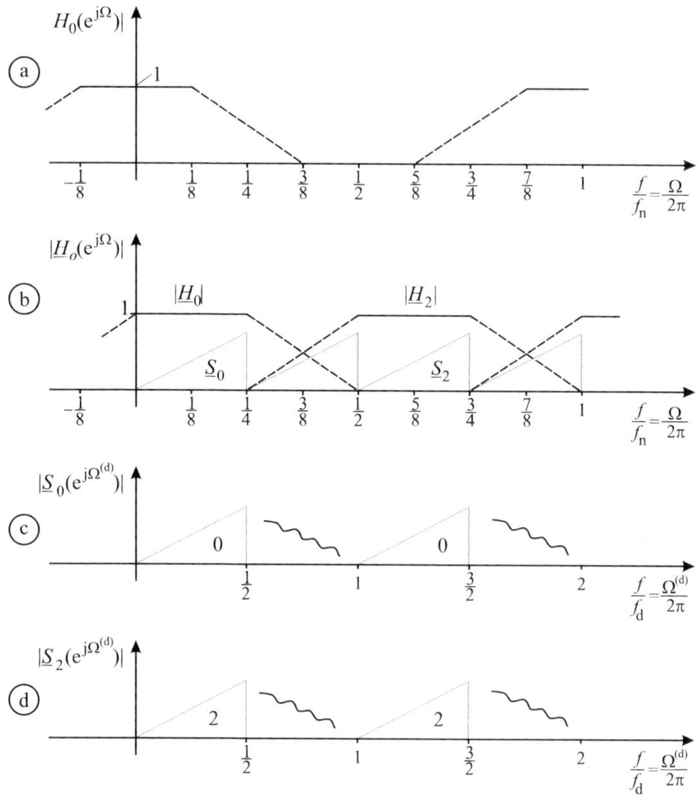

Fig. 21. DF operations: (a) Real HBF prototype centrosymmetric about $H_0(e^{j\pi/2}) = \frac{1}{2}$, (b) Two selected DF filter functions, (c,d) Spectra of decimated DF output signals

by (38), $o \in \{0,1,2,3\}$, with the RHBF impulse response $h(k)$ defined by (9). According to (39), highest efficiency is obtained by additionally introducing a suitable complex scaling factor of unity magnitude:

$$\underline{h}_{k,o,a} = e^{ja\frac{\pi}{4}} h(k) z_o^k = h(k) e^{j\frac{\pi}{4}[k(2o+1)+a]} = h(k) j^{k(o+\frac{1}{2})+\frac{a}{2}}, \tag{39}$$

where $-\frac{N-1}{2} \leq k \leq \frac{N-1}{2}$ and $o \in \{0,1,2,3\}$. By directly equating (39), and relating the result to (9) with a suitable choice of the constant $a = 2o+1$ compliant with (29), we get :

$$\underline{h}_{k,o} = \begin{cases} \frac{1}{2} j^{o+\frac{1}{2}} & k = 0 \\ 0 & k = 2l \quad l = 1,\ldots,(N-3)/4 \\ j^{(k+1)(o+\frac{1}{2})} h_k & k = 2l-1 \quad l = 1,\ldots,(N+1)/4 \end{cases} \tag{40}$$

with the symmetry property:

$$\underline{h}_{-k,o} = -j^{(2o+1)k} \underline{h}_{k,o} \quad \forall k > 0, \quad o \in \{0,1,2,3\}. \tag{41}$$

The respective COHBF centre coefficient

$$\underline{h}_{0,o} = \frac{1}{2}\{\cos[(2o+1)\frac{\pi}{4}] + j\sin[(2o+1)\frac{\pi}{4}]\}, \; o \in \{0,1,2,3\}, \tag{42}$$

is the only truly complex-valued coefficient, where its real and imaginary parts always possess identical moduli. All other coefficients are either purely imaginary or real-valued. Obviously, all frequency domain symmetry properties, including also those related to strict complementarity, are retained in the respective frequency-shifted versions, cf. Subsection 2.3.1 and [Göckler & Damjanovic (2006a)].

3.2 Linear-phase directional separation filter

We start with the presentation of the FDMUX approach [Göckler & Groth (2004); Göckler & Eyssele (1992)] followed by the investigation of a synergetic combination of two COHBF [Göckler (1996a;c); Göckler & Damjanovic (2006a)].

3.2.1 FDMUX approach

Using time-domain convolution, the $I = 4$ potentially required complex output signals, decimated by 2 and related to the channel indices $o \in \{0, 1, 2, 3\}$, are obtained as follows:

$$\underline{y}_o(mT_\text{d}) := \underline{y}_o(m) = \sum_{\kappa=0}^{N-1} \underline{x}(2m - \kappa)\underline{h}_o(\kappa - \tfrac{N-1}{2}), \tag{43}$$

where the complex impulse responses of channels o are introduced in causal (realizable) form. Replacing the complex impulse responses with the respective modulation forms (39), and setting the constant to $a = (2o + 1)(N - 1)/2$, we get:

$$\underline{y}_o(m) = \sum_{\kappa=0}^{N-1} \underline{x}(2m - \kappa)h(\kappa - \tfrac{N-1}{2})\text{e}^{\text{j}\frac{\pi}{4}\kappa(2o+1)}, \tag{44}$$

where $h[k - (N - 1)/2]$ represents the real HBF prototype (9) in causal form. Next, in order to introduce an I-component polyphase decomposition for efficient decimation, we split the convolution index κ into two indices:

$$\kappa = rI + p = 4r + p, \tag{45}$$

where $p = 0, 1, 2, I - 1 = 3$ and $r = 0, 1, \ldots, \lfloor (N - 1)/I \rfloor = \lfloor (N - 1)/4 \rfloor$. As a result, it follows from (44):

$$\underline{y}_o(m) = \sum_{p=0}^{3} \sum_{r=0}^{\lfloor \frac{N-1}{4} \rfloor} \underline{x}(2m - 4r - p)h(4r + p - \tfrac{N-1}{2}) \cdot \text{e}^{\text{j}\frac{\pi}{4}(4r+p)(2o+1)}. \tag{46}$$

Rearranging the exponent of the exponential term according to $\frac{\pi}{4}(4r + p)(2o + 1) = 2\pi ro + \pi r + p\frac{\pi}{4} + \frac{2\pi}{4}op$, (46) can compactly be rewritten as [Oppenheim & Schafer (1989)]:

$$\underline{y}_o(m) = \sum_{p=0}^{3} \underline{v}_p(m) \cdot \text{e}^{\text{j}\frac{2\pi}{4}op} = 4 \cdot \text{IDFT}_4\{\underline{v}_p(m)\}, \tag{47}$$

where the quantity

$$\underline{v}_p(m) = \sum_{r=0}^{\lfloor \frac{N-1}{4} \rfloor} \underline{x}(2m - 4r - p)h(4r + p - \tfrac{N-1}{2})(-1)^r \text{e}^{\text{j}p\frac{\pi}{4}} \tag{48}$$

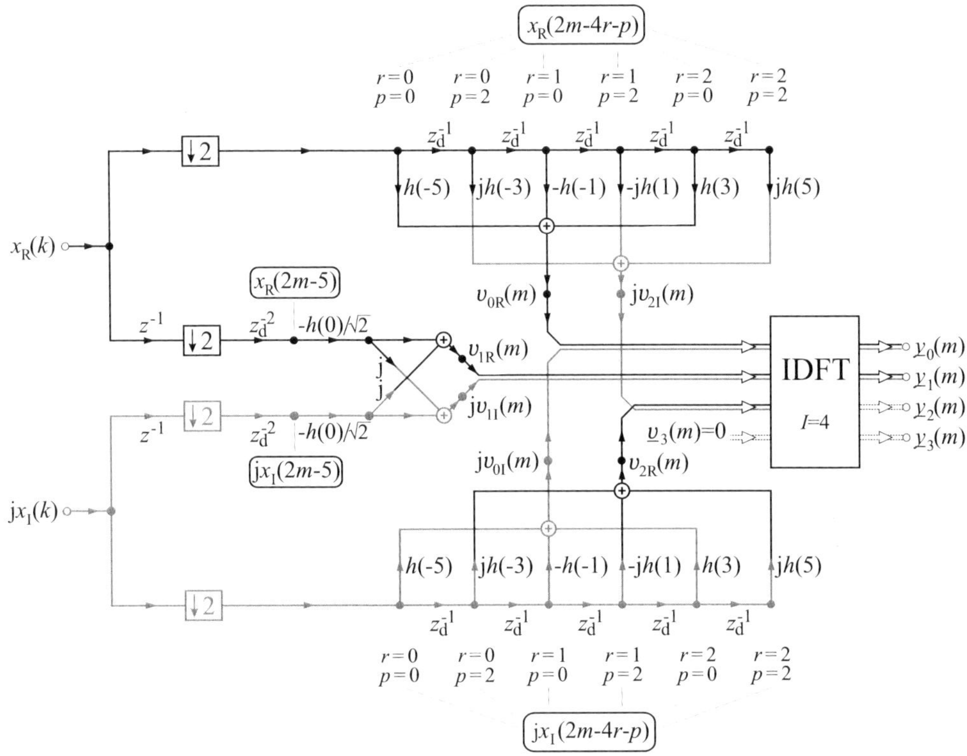

Fig. 22. SFG of directional filter with allowing for 2-out-of-4 channel selection: FDMUX approach; $N = 11$

encompasses all complex signal processing to be performed by the modified causal HBF prototype.
An illustrative example with an underlying HBF prototype filter of length $N = n + 1 = 11$ is shown in Fig. 22 [Göckler & Groth (2004)]. Due to polyphase decomposition (45) and (46), sample rate reduction can be performed in front of any signal processing (shimming delays: z^{-1}). Always two polyphase components of the real and the imaginary parts of the complex input signal share a delay chain in the direct form implementation of the modified causal HBF, where all coefficients are either real- or imaginary-valued except for the centre coefficient $\underline{h}_0 = \frac{1}{2}e^{j\frac{\pi}{4}}$. As a result, only $N + 3$ real multiplications must be performed to calculate a set of complex output samples at the two (i.e. all) DF output ports. Furthermore, for the FDMUX DF implementation a total of $(3N - 5)/2$ delays are needed (not counting shimming delays). The calculation of $\underline{v}_p(m)$, $p = 0, 1, 2, 3$, is readily understood from the signal flow graph (SFG) Fig. 22, where for any filter length N *always one* of these quantities vanishes as a result of the zero coefficients of (9). Hence, the $I = 4$ point IDFT, depicted in Fig. 23(a,b) in detailed form, requires only 4 real additions to provide a complex output sample at any of the output ports $o \in \{0, 1, 2, 3\}$; Fig. 23(b). Channel selection, for instance as shown in Fig. 21, is simply achieved by selection of the respective two output ports of the SFG of Figs.22 and 23(a), respectively. Moreover, the remaining two unused output ports may be deactivated by disconnection from power supply.

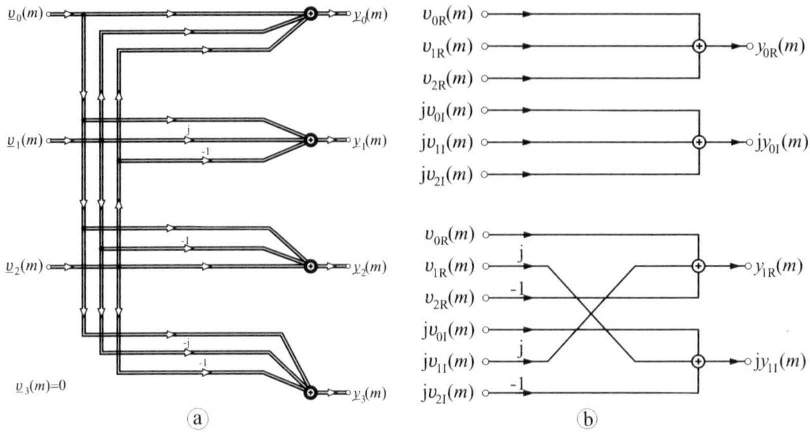

Fig. 23. $I = 4$ point IDFT of FDMUX approach; $N = 11$: (a) general (b) pruned for channels $o = 0, 1$

k	-5	-3	-1	0	1	3	5
$\frac{h_{k,o}}{h_k}$	-1	$-\frac{\mathrm{j}}{(-1)^o}$	1	$\frac{1+\mathrm{j}}{\sqrt{2}}\mathrm{j}^o$	$\frac{\mathrm{j}}{(-1)^o}$	-1	$-\frac{\mathrm{j}}{(-1)^o}$
type	$\mathbb{R}$	$\mathbb{I}$	$\mathbb{R}$	$\mathbb{C}$	$\mathbb{I}$	$\mathbb{R}$	$\mathbb{I}$

Table 8. Properties of COHBF coefficients in dependence of channel index $o \in \{0, 1, 2, 3\}$; $\mathbb{I}$: $\mathbb{C}$ with $Re\{\bullet\} = 0$

3.2.2 COHBF approach

For this novel approach, we combine two decimating COHBF of different centre frequencies f_o, $o \in \{0, 1, 2, 3\}$, according to (38) in a synergetic manner to construct a DF for signal separation that requires *minimum* computation. To this end, we first study the commonalities of the impulse responses (40) of the four transfer functions $\underline{H}_o(z)$, $o \in \{0, 1, 2, 3\}$ (underlying constant in (39) subsequently: $a = 2o + 1$). These impulse responses are presented in Table 8 as a function of the channel number $o \in \{0, 1, 2, 3\}$ for the non-zero coefficients of (40), related to the respective real RHBF coefficients.

Except for the centre coefficient exhibiting identical real and imaginary parts, one half of the coefficients is real ($\mathbb{R}$) and *independent of the desired centre frequency* represented by the channel indices $o \in \{0, 1, 2, 3\}$. Hence, these coefficients are common to all four transfer functions. The other half of the coefficients is purely imaginary ($\mathbb{I}$: i.e., their real parts are zero) and dependent of the selected centre frequency. However, this dependency on the channel number is identical for all these coefficients and just requires a simple sign operation. Finally, the repetitive pattern of the coefficients, as a result of coefficient symmetry (41), is reflected in Table 8.

A COHBF implementation of a demultiplexing DF aiming at *minimum computational load* must exploit the inherent coefficient symmetry (41), cf. Table 8. To this end, we consider the COHBF as depicted in Fig. 17 of Subsection 2.3.1, applying input commutators for sample rate reduction. In contrast to the FDMUX approach of Fig. 22, the SFG of Fig. 17 is based on the transposed FIR direct form Bellanger (1989); Mitra (1998), where the incoming signal samples are concurrently multiplied by the complete set of all coefficients, and the delay chains are directly connected to the output ports. When combining two of these COHBF

SFG, the coefficient multipliers can obviously be shared with all transfer functions $\underline{H}_o(z)$, $o \in \{0,1,2,3\}$; however, the respective outbound delay chains must essentially be duplicated. Merging all of the above considerations, a signal separating DF requiring minimum computation that, in addition, allows for simple channel selection or switching, respectively, is readily developed as follows:

1. Multiply the incoming decimated polyphase signal samples concurrently and consecutively by the complete set of all real coefficients (9) to allow for the exploitation of coefficient symmetry (41) in compliance with Table 8.
2. Form a real and imaginary (R/I) sub-sequence of DF output signals being *independent* of the selected channel transfer functions, i.e. $o_{\mathrm{I}}, o_{\mathrm{II}} \in \{0,1,2,3\}$, by using all $\mathbb{R}$-set coefficients of Table 8.
3. Form an R and I sub-sequence of DF output signals being likewise *independent* of the selected channels $o_{\mathrm{I}}, o_{\mathrm{II}}$ by using all $\mathbb{I}$-set coefficients of Table 8 multiplied by $(-1)^o$ to eliminate channel dependency.
4. Form R/I sub-sequences of DF output signals being *dependent* of the selected channels $o_{\mathrm{I}}, o_{\mathrm{II}}$ that are derived from centre coefficients $\underline{h}_{0,o}$.
5. Combine all of the above R/I sub-sequences considering the sign rules of Table 8 to select the desired DF transfer functions $\underline{H}_{o_i}(z)$, $o_i \in \{0,1,2,3\}$, $i \in \{\mathrm{I}, \mathrm{II}\}$.

Based on the outlined DF implementation strategy, an illustrative example is presented in Fig. 24 with an underlying RHBF of length $N = 11$. The front end for polyphase decomposition and sample rate reduction by 2 is identical to that of the FDMUX approach of Fig. 22. Contrary to the former approach, the delay chains for the odd-numbered coefficients are outbound *and* duplicated (rather than interlaced) to allow for simple channel selection. As a result, channel selection is performed by combining the respective sub-sequences that have passed the $\mathbb{R}$-set coefficients (cf. Table 8) with those having passed the corresponding $\mathbb{I}$-set coefficients, where the latter sub-sequences are pre-multiplied by $b_i = (-1)^{o_i}$; $o_i \in \{0,1,2,3\}$, $i \in \{\mathrm{I}, \mathrm{II}\}$.
Multipliers and delays for the centre coefficient $\underline{h}_{0,o_i}$ signal processing are implemented similarly to Fig. 22 without need for duplication of delays. However, the post-delay inner lattice must be realized for each transfer function individually; its channel dependency follows from Table 8 and (40):

$$\underline{h}_{0,o_i} = \frac{h_0}{\sqrt{2}}(1+\mathrm{j})\mathrm{j}^{o_i} = \frac{h_0}{\sqrt{2}}\left[(-1)^{\lceil o_i/2 \rceil} + \mathrm{j}(-1)^{\lfloor o_i/2 \rfloor}\right], \tag{49}$$

where $o_i \in \{0,1,2,3\}$, $i \in \{\mathrm{I}, \mathrm{II}\}$ and $h_0 = 1/2$ according to (9). Rearranging (49) yields with obvious abbreviations:

$$\underline{h}_{0,o_i} = \frac{h_0}{\sqrt{2}}\left[(-1)^{o_i} + \mathrm{j}\right](-1)^{\lfloor o_i/2 \rfloor} = \frac{h_0}{\sqrt{2}}\left[b_i + \mathrm{j}\right] d_i. \tag{50}$$

It is easily recognized that the inner lattices of Fig. 24 implement the operations within the brackets of (50) with their results displayed at the respective inner nodes A, B, C, D. In compliance with (50), these inner node sequences must be multiplied by the respective signs $d_i = (-1)^{\lfloor o_i/2 \rfloor}$; $o_i \in \{0,1,2,3\}$, $i \in \{\mathrm{I}, \mathrm{II}\}$, prior to their combination with the above R/I sub-sequences.
To calculate a set of complex output samples at the two DF output ports, obviously the minimum number of $(N+5)/2$ real multiplications must be carried out. Furthermore, for

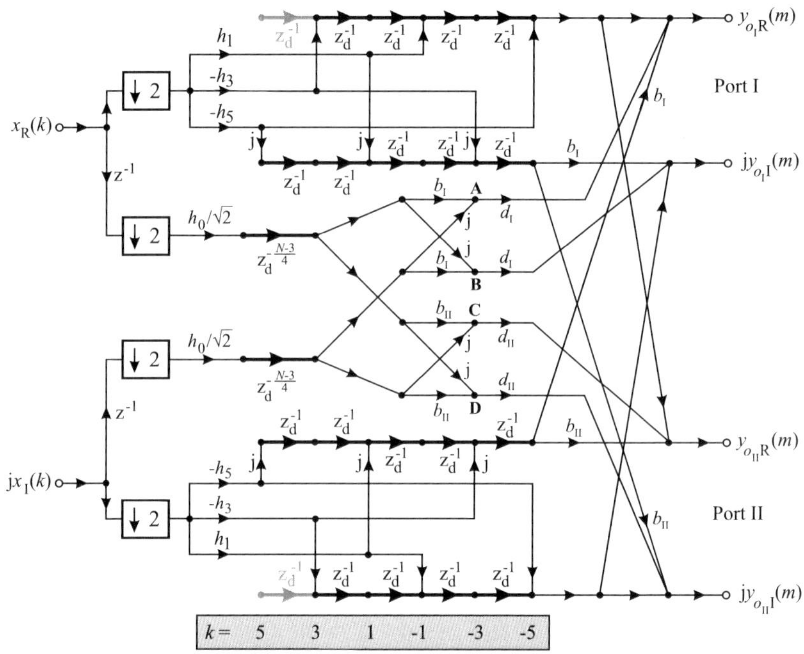

Fig. 24. COHBF approach to *demultiplexing* DF implementation with selectable transfer functions; $N = 11$, $b_i = (-1)^{o_i}$, $d_i = (-1)^{\lfloor o_i/2 \rfloor}$; $o_i \in \{0, 1, 2, 3\}$, $i \in \{\mathrm{I}, \mathrm{II}\}$

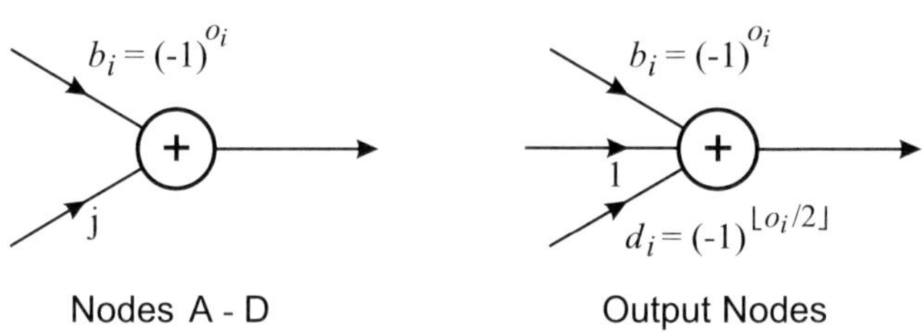

Fig. 25. DF separator: Sign-setting for selection of desired channel transfer functions

the COHBF approach to DF implementation a total of $(5N - 11)/2$ delays are needed (not counting shimming delays, z^{-1}, and the two superfluous delays at the input nodes of the outer delay chains, indicated in grey).

Finally, we want to show and emphasise the simplicity of the channel selection procedure. There is a total of 8 summation points, the inner 4 lattice output nodes A, B, C, and D, and the 4 system output port nodes, where the signs of some input sequences of the output port nodes must be set compliant to the desired channel transfer functions: $o_i \in \{0, 1, 2, 3\}$, $i \in \{\mathrm{I}, \mathrm{II}\}$. The sign selection is most easily performed as shown in Fig. 25.

A concise survey of the required expenditure of the two approaches to the implementation of a demultiplexing DF is given in Table 9, not counting sign manipulations for channel selection. Obviously, the COHBF approach requires the minimum number of multiplications

APPROACH	multiplications/sample	delays
FDMUX	$N+3$	$(3N-5)/2$
FDMUX ex.: $N=11$	14	14
COHBF	$(N+5)/2$	$(5N-11)/2$
COHBF ex.: $N=11$	8	22

Table 9. Comparison of expenditure of FDMUX and COHBF DF approaches

at the expense of a higher count of delay elements. Finally, it should be noticed that the DF group delay is independent of its (FDMUX or COHBF) implementation.

3.3 Linear-phase directional combination filter

Using transposition techniques, we subsequently derive DF being complementary (dual) to those presented in Subsection 3.2: They combine two complex-valued signals of identical sampling rate f_d that are likewise oversampled by at least 2 to an FDM signal, where different oversampling factors allow for different bandwidths.

An example can be deduced from Fig. 21 by considering the signals $\underline{s}_o(mT_\mathrm{d}) \longleftrightarrow \underline{S}_o(\mathrm{e}^{\mathrm{j}\Omega^{(\mathrm{d})}})$, $o = 0,2$, of Figs.21(c,d) as input signals. The multiplexing DF increases the sampling rates of both signals to $f_\mathrm{n} = 2f_\mathrm{d}$, and provides the filtering operations shown in Fig. 21(b), $\underline{h}_o(kT_\mathrm{n}) \longleftrightarrow \underline{H}_o(\mathrm{e}^{\mathrm{j}\Omega})$, $c = 0,2$, to form the FDM output spectrum being exclusively composed of $\underline{S}_o(\mathrm{e}^{\mathrm{j}\Omega})$, $o = 0,2$.

3.3.1 Transposition of complex multirate systems

The goal of transposition is to derive a system that is complementary or dual to the original one: The various filter transfer functions must be retained, demultiplexing and decimating operations must be replaced with the dual operations of multiplexing and interpolation, respectively [Göckler & Groth (2004)].

The types of systems we want to transpose, Figs.22 and 24, represent complex-valued 4×2 multiple-input multiple-output (MIMO) multirate systems. Obviously, these systems are composed of *complex monorate* sub-systems (complex filtering of polyphase components) and *real multirate* sub-systems (down- and upsampler), cf. [Göckler & Groth (2004)].

While the transposition of real MIMO monorate systems is well-known and unique [Göckler & Groth (2004); Mitra (1998)], in the context of *complex* MIMO monorate systems the *Invariant* (ITr) and the *Hermitian* (HTr) transposition must be distinguished, where the former retains the original transfer functions, $\underline{H}_o^{\mathrm{T}}(z) = \underline{H}_o(z)\,\forall o$, as desired in our application. As detailed in [Göckler & Groth (2004)], the ITr is performed by applying the transposition rules known for real MIMO monorate systems *provided that* all imaginary units "j", both of the complex input and output signals *and* of the complex coefficients, are conceptually considered and treated as multipliers within the SFG[3] (denoted as truly complex implementation), as to be seen from Figs.22 and 24.

The transposition of an M-downsampler, representing a real single-input single-output (SISO) multirate system, uniquely leads to the corresponding M-upsampler, the complementary (dual) multirate system, and vice versa [Göckler & Groth (2004)].

---

[3] The imaginary units of the input signals and the coefficients *must not* be eliminated by simple multiplication and consideration of the correct signs in subsequent adders; this approach would transform the original complex MIMO SFG to a corresponding real SFG, where the direct transposition of the latter would perform the HTr [Göckler & Groth (2004)].

Connecting all of the above considerations, the ITr transposition of a complex-valued MIMO multirate system is performed as follows [Göckler & Groth (2004)]:

- The system SFG to be transposed must be given as truly complex implementation.
- Reverse *all* arrows of the given SFG, both the arrows representing signal flows and those symbolic arrows of down- and upsamplers or rotating switches (commutators), respectively.

As a result of transposition [Göckler & Groth (2004)]

- all input (output) nodes become output (input) nodes, a 4×2 MIMO system is transformed to a 2×4 MIMO system,
- the number of delays and multipliers is retained,
- the overall number of branching and summation nodes is retained, and
- the overall number of down- and upsamplers is retained.

Obviously, the original *optimality* (minimality) is *transposition invariant*.

3.3.2 Transposition of the SFG of the COHBF approach to DF

As an example, we transpose the SFG of the COHBF approach to the implementation of a separating DF, as depicted in Fig. 24. The application of the transposition rules of the preceding Subsection 3.3.1 to the SFG of Fig. 24 results in the COHBF approach to a multiplexing DF shown in Fig. 26. The invariant properties are easily confirmed by comparing the original and the transposed SFG. Hence, the numbers of delays and multipliers required by both DF systems being mutually dual are identical. As expected, the numbers of adders required are different, since the *overall* number of branching *and* summation nodes is retained only.

Moreover, it should be noted that also the simplicity of the channel selection procedure is retained. To this end, we have shifted the channel-dependent sign-setting operators $d_i = (-1)^{\lfloor o_i/2 \rfloor}$, $o_i \in \{0, 1, 2, 3\}$, $i \in \{\mathrm{I}, \mathrm{II}\}$, to more suitable positions in front of the summation nodes G and H. Again, there is a total of 8 summation points, where the signs of the respective input sequences must be adjusted: The 4 inner lattice output nodes A, B, C, and D, the 2 input summation nodes E and F immediately fed by the imaginary parts of the input sequences, and the 2 inner post-lattice summing nodes G and H. At all these summation nodes, the signs of some or all input sequences must be set in compliance with the desired channel transfer functions: $\underline{H}_o(z)$, $o_i \in \{0, 1, 2, 3\}$, $i \in \{\mathrm{I}, \mathrm{II}\}$, cf. Fig. 26. The sign selection is again most easily performed, as shown in Fig. 27.

3.4 Conclusion: Halfband filter pair combined to directional filter

In this Section 3, we have derived and analyzed two different approaches to linear-phase directional filters that separate from a complex-valued FDM input signal two complex user signals, where the FDM signal may be composed of up to four independent user signals: The FDMUX approach (Subsection 3.2.1) needs the least number of delays, whereas the synergetic COHBF approach (Subsection 3.2.2) requires minimum computation. Signal extraction is always combined with decimation by two.

While the four frequency slots of the user signals to be processed (corresponding to the four potential DF transfer functions $\underline{H}_o(z)$, $o_i \in \{0, 1, 2, 3\}$, $i \in \{\mathrm{I}, \mathrm{II}\}$, centred according to (38); cf. Fig. 21) are equally wide and uniformly allocated, as indicated in Fig. 28, the individual

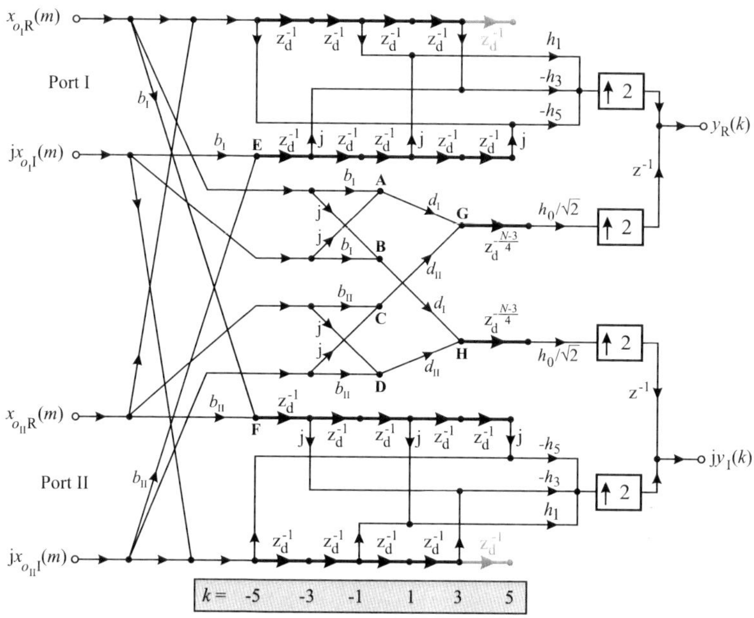

Fig. 26. COHBF approach to *multiplexing* DF implementation with selectable transfer functions derived by transposition from corresponding separating DF; $N = 11$, $b_i = (-1)^{o_i}$, $d_i = (-1)^{\lfloor o_i/2 \rfloor}$; $o_i \in \{0, 1, 2, 3\}$, $i \in \{\mathrm{I}, \mathrm{II}\}$

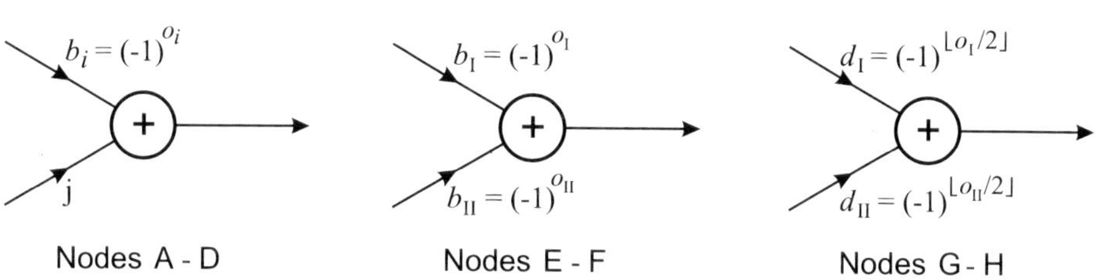

Fig. 27. DF combiner: Sign-setting for selection of desired channel transfer functions

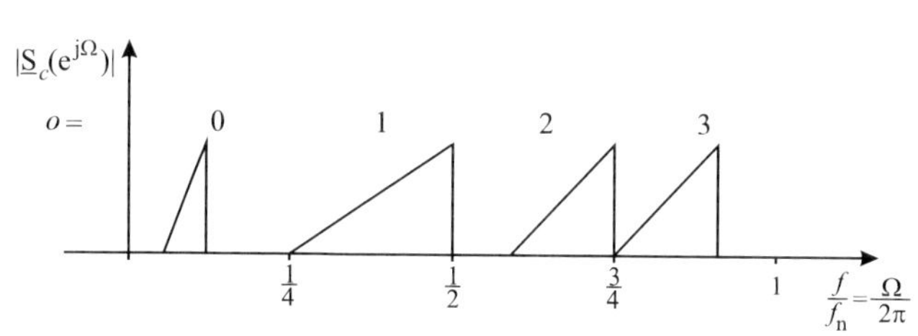

Fig. 28. Generally permissible FDM input spectrum to separation DF

user signals may possess different bandwidths. However, each user signal must completely be contained in one of the four frequency slots, as exemplified in Fig. 28.
Furthermore, by applying the transposition rules of [Göckler & Groth (2004)], the corresponding complementary (dual) combining directional filters have been derived, where the multiplication rates and the delay counts of the original structures are always retained. Obviously, transposing a system allows for the derivation of an optimum dual system by applying the simple transposition rules, provided that the original system is optimal. Thus, a tedious re-derivation and optimization of the complementary system is circumvented. Nevertheless, it should be noted that by transposition always just one particular structure is obtained, rather than a variety of structures [Göckler & Groth (2004)].
Finally, to give an idea of the required filter lengths required, we recall the design result reported in [Göckler & Eyssele (1992)] where, as depicted in the above Fig. 21(a,b), the passband, stopband and transition bands were assumed equally wide: With an HBF prototype filter length of $N = 11$ and 10 bit coefficients, a stopband attenuation of > 50dB was achieved.

4. Parallelisation of tree-structured filter banks composed of directional filters [4]

In the subsequent Section 4 of this chapter we consider the combination of multiple two-channel DF investigated in Section 3 to construct tree-structured filter banks. To this end, we cascade separating DF in a hierarchical manner to demultiplex (split) a frequency division multiplex (FDM) signal into its constituting user signals: this type of filter bank (FB) is denoted by FDMUX FB; Fig. 2. Its transposed counterpart (cf. Subsection 3.3.1), the FMUX FB, is a cascade connection of combining DF considered in Subsection 3.3 to form an FDM signal of independent user signals. Finally, we call an FDMUX FB followed by an FMUX FB an FDFMUX FB, which may contain a switching unit for channel routing between the two FB. Subsequently, we consider an application of FDFMUX FB for on-board processing in satellite communications. If the number of channels and/or the bandwidth requirements are high, efficient implementation of the high-end DF is crucial, if they are operated at (extremely) high sampling rates. To cope with this issue, we propose to parallelise the at least the front-end (back-end) of the FDMUX (FMUX) filter bank. For this outlined application, we give the following introduction and motivation.
Digital signal processing on-board communication satellites (OBP) is an active field of research where, in conjunction with frequency division multiplex (FDMA) systems, presently two trends and challenges are observed, respectively: *i*) The need of an ever-increasing number of user channels makes it necessary to digitally process, i.e. to demultiplex, cross-connect and remultiplex, ultra-wideband FDM signals requiring high-end sampling rates that range considerably beyond 1GHz [Arbesser-Rastburg et al. (2002); Maufroid et al. (2004; 2003); Rio-Herrero & Maufroid (2003); Wittig (2000)], and *ii*) the desire of flexibility of channel bandwidth-to-user assignment calling for simply reconfigurable OBP systems [Abdulazim & Göckler (2005); Göckler & Felbecker (2001); Johansson & Löwenborg (2005); Kopmann et al. (2003)]. Yet, overall power consumption must be minimum demanding highly efficient FB for FDM demultiplexing (FDMUX) and remultiplexing (FMUX).
Two baseline approaches to most efficient uniform digital FB, as required for OBP, are known: *a*) The complex-modulated (DFT) polyphase (PP) FB applying single-step sample rate alteration [Vaidyanathan (1993)], and *b*) the multistage tree-structured FB as depicted in Fig. 2, where its directional filters (DF) are either based on the DFT PP method

[4] Underlying original publication: Göckler et al. (2006)

[Göckler & Groth (2004); Göckler & Eyssele (1992)] according to Subsection 3.2.1, or on the COHBF approach investigated in Subsection 3.2.2. For both approaches it has been shown that bandwidth-to-user assignment is feasible within reasonable constraints [Abdulazim et al. (2007); Johansson & Löwenborg (2005); Kopmann et al. (2003)]: A minimum user channel bandwidth, denoted by slot bandwidth b, can stepwise be extended by any integer number of additional slots up to a desired maximum overall bandwidth that shall be assigned to a single user.

However, as to challenge *i*), the above two FB approaches fundamentally differ from each other: In a DFT PP FDMUX (*a*) the overall sample rate reduction is performed in compliance with the number of user channels in a single step: all arithmetic operations are carried out at the (lowest) output sampling rate [Vaidyanathan (1993)]. In contrast, in the multistage FDMUX (*b*) the sampling rate is reduced stepwise, in each stage by a factor of two [Göckler & Eyssele (1992)]. As a result, the polyphase approach (*a*) inherently represents a completely parallelised structure, immediately usable for extremely high front-end sampling frequencies, whereas the high-end stages of the tree-structured FDMUX (*b*) cannot be implemented with standard space-proved CMOS technology. Hence, the tree structure, FDMUX as well as FMUX, calls for a parallelisation of the high rate stages.

As motivated, this contribution deals with the parallelisation of multistage multirate systems. To this end, we recall a general systematic procedure for multirate system parallelisation [Groth (2003)], which is deployed in detail in Subsection 4.1. For proper understanding, in Subsection 4.2 this procedure is applied to the high rate front-end stages of the FDMUX part of the recently proposed tree-structured SBC-FDFMUX FB [Abdulazim & Göckler (2005); Abdulazim et al. (2007)], which uniformly demultiplexes an FDM signal always down to slot level (of bandwidth b) and that, after on-board switching, recombines these independent slot signals to an FDM signal (FMUX) with different channel allocation – *FDFMUX functionality*. If a single user occupies a multiple slot channel, the corresponding parts of FDMUX and FMUX are matched for (nearly) perfect reconstruction of this wideband channel signal – *SBC functionality* [Vaidyanathan (1993)]. Finally, some conclusions are drawn.

4.1 Sample-by-sample approach to parallelisation

In this subsection, we introduce the novel sample-by-sample processing (SBSP) approach to parallelisation of digital multirate systems, as proposed by [Groth (2003)] where, without any additional delay, all incoming signal samples are directly fed into assigned units for immediate signal processing. Hence, in contrast to the widely used block processing (BP) approach, SBSP does not increase latency.

In order to systematically parallelise a (multirate) system, we distinguish four procedural steps [Groth (2003)]:

1. *Partition the original system* in (elementary SISO or MIMO) subsystems $E(z)$ with single or multiple input and/or output ports, respectively, still operating at the original high clock frequency $f_n = 1/T$ that are simply amenable to parallelisation. To enumerate some of these: Delay, multiplier, down- and up-sampler, summation and branching, but also suitable compound subsystems such as SISO filters and FFT transform blocks.

2. *Parallelise each subsystem* $E(z)$ in an SBSP manner according to the desired individual degree of parallelisation P, where $P \in \mathbb{N}$. To this end, each subsystem is cascaded with a P-fold SBSP serial-to-parallel (SP) commutator for signal decomposition (demultiplexing) followed by a consistently connected P-fold parallel-to-serial (PS) commutator for recomposition (remultiplexing) of the original signal, as depicted in Fig. 29(a). Here, obviously $P =$

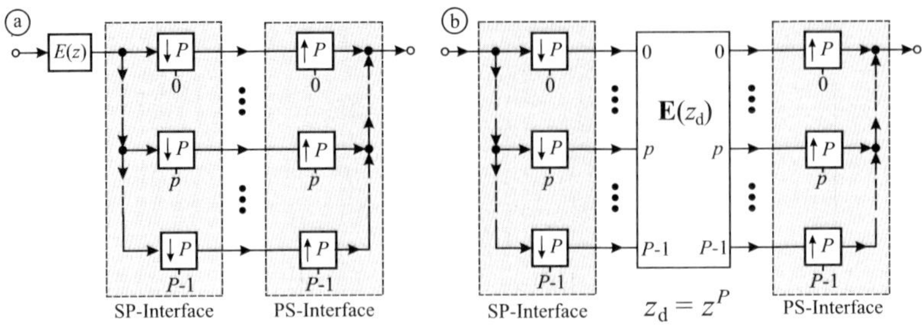

Fig. 29. P-Parallelisation of SISO subsystem $E(z)$ to $P \times P$ MIMO system $\mathbf{E}(z_d)$

$P_{SP} = P_{PS}$, and $p \in [0, P-1]$ denotes the relative time offsets of connected pairs of down- and up-samplers, respectively. Evidently, the P output signals of the SP interface comprise all polyphase components of its input signal in a time-interleaved (SBSP) manner at a P-fold lower sampling rate $f_d = f_n/P$ [Göckler & Groth (2004); Vaidyanathan (1993)]. Since the subsequent PS interface is inverse to the preceding SP interface [Göckler & Groth (2004)], the SP-PS commutator cascade has unity transfer with zero delay in contrast to the $(P-1)$-fold delay of the BP Delay-Chain Perfect-Reconstruction system [Göckler & Groth (2004); Vaidyanathan (1993)], as anticipated (cf. also Fig. 30).

After this preparation, P-fold parallelisation is readily achieved by shifting the (SISO) subsystem $E(z)$ between the SP and PS interfaces by exploiting the noble identities [Göckler & Groth (2004); Vaidyanathan (1993)] and some novel generalized SBSP multirate identities [Groth (2003); Groth & Göckler (2001)]. Thus, as shown in Fig. 29(b), the two interfaces are interconnected by an equivalent $P \times P$ MIMO system $\mathbf{E}(z_d)$, which represents the P-fold parallelisation of $E(z)$, where all operations of which are performed at the P-fold reduced operational clock frequency f_d.

3. *Reconnect all parallelised subsystems* exactly in the same manner as in the original system. This is always given, since parallelisation does not change the original numbers of input and output ports of SISO or MIMO subsystems, respectively.

4. *Eliminate all interfractional cascade connections of PS-SP interfaces* using the obvious multirate identity depicted in Fig. 30. Note that this elimination process requires identical up- and down-sampling factors, $P_{PS}^{out,a} = P_{SP}^{in,b}$, of each PS-SP interface cascade restricting free choice of P for subsystem parallelisation. As a result of parallelisation, all input signals of the original (possibly MIMO) system are decomposed into P time-interleaved polyphase components by a SP demultiplexer for subsequent parallel processing at a P-fold lower rate, and all system output ports are provided with a PS commutator to interleave all low rate subsignals to form the high speed output signals.

For illustration, we present the parallelisation of a unit delay $z^{-1} := z_d^{-1/P}$, and of an M-fold down-sampler with zero time offset [Groth (2003)], as shown in Fig. 31. The unit delay (a) is realized by P parallel time-interleaved shimming delays to be implemented by suitable system control:

$$\mathbf{E}_{P\times P}(z_d) = z_d^{-1/P} \begin{pmatrix} \mathbf{0} & 1 \\ \mathbf{I}_{(P-1)\times(P-1)} & \mathbf{0} \end{pmatrix},$$

where permutation is introduced for straightforward elimination of interfractional PS-SP cascades according to Fig. 30 ($\mathbf{I}$: Identity matrix). In case of down-sampling Fig. 31(b), to increase efficiency, the P parallel down-samplers of the diagonal MIMO system $\mathbf{E}(z_d)$ are merged with the P down-samplers of the SP interface. Hence, by using suitable multirate

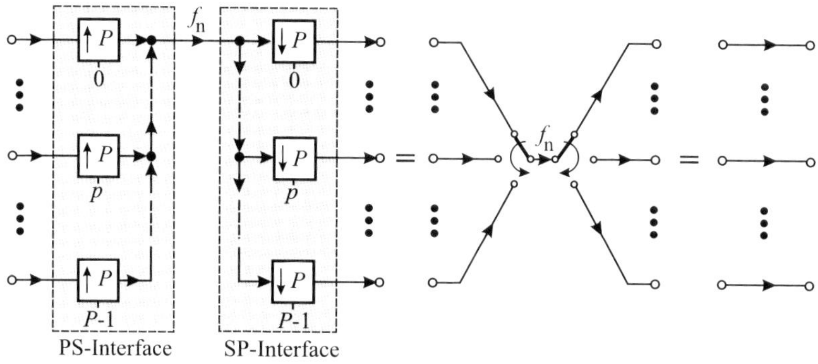

Fig. 30. Identity for elimination of P-fold interfractional PS-SP cascades

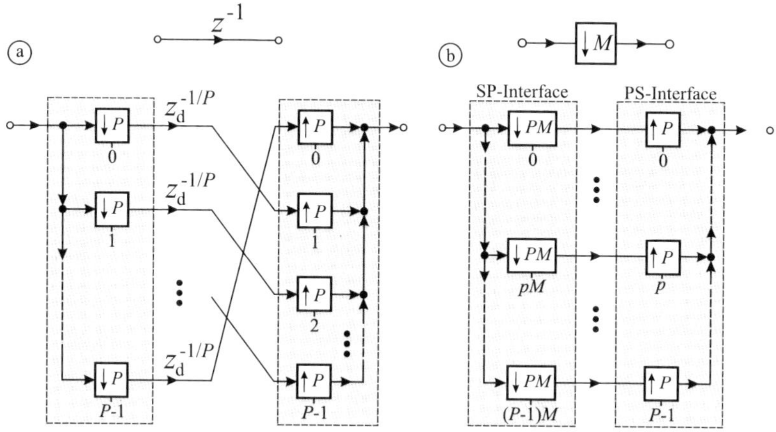

Fig. 31. Parallelisation of unit delay (a) and M-fold down-sampler (b) with zero time offset ($p = 0$)

identities [Groth (2003)], the contiguous PM-fold down-samplers of the SP demultiplexer have a relative time offset of M.

4.2 Parallelisation of SBC-FDFMUX filter bank

Subsequently, we deploy the parallelisation of the high rate FDMUX front-end section of the versatile tree-structured SBC-FDFMUX FB for flexible channel and bandwidth allocation [Abdulazim & Göckler (2005); Abdulazim et al. (2007)]. The first three hierarchically cascaded stages of the FDMUX are shown in Fig. 32 in block diagram form applying BP. In each stage, $\nu = 1, 2, 3$, the respective input spectrum is split into two subbands of equal bandwidth in conjunction with decimation by two. For convenience of presentation, all DF have identical coefficients and, in contrast to Section 3, are assumed as critically sampling 2-channel DFT PP FB with zero frequency offset (cf. [Abdulazim et al. (2007)]). The branch filter transfer functions $H_\lambda(z_\nu)$, $\lambda = 0, 1$, represent the two PP components of the prototype filter [Göckler & Groth (2004); Vaidyanathan (1993)] where, by setting $z_\nu := e^{j\Omega^{(\nu)}}$ with $\Omega^{(\nu)} = 2\pi f / f_\nu$ and $\nu = 1, 2, 3$, the respective frequency responses $H_\lambda(e^{j\Omega^{(\nu)}})$ are obtained, which are related to the operational sampling rate f_ν of stage ν. The respective DF lowpass

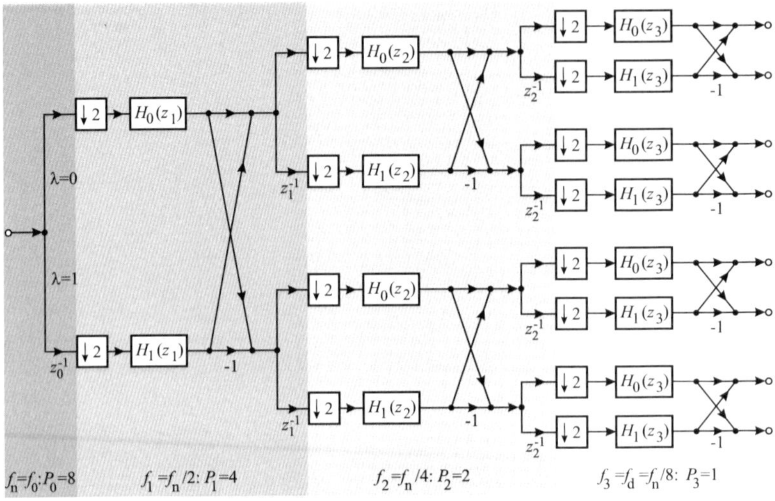

Fig. 32. FDMUX front end of SBC-FDFMUX filter bank according to Abdulazim et al. (2007)); $z_\nu := e^{j\Omega^{(\nu)}}$, $\Omega^{(\nu)} = 2\pi f / f_\nu$, $\nu = 0, 1, 2, 3$, $f_3 = f_d = f_n/8$

and highpass filter transfer functions of stage ν, related to the original sampling rate $2f_\nu$, are generated by the two branch filter transfer functions $H_\lambda(z_\nu)$, $\lambda = 0, 1$, in combination with the simple "butterfly" across the output ports of each DF: Summation produces the lowpass, subtraction the complementary highpass filter transfer function Bellanger (1989); Kammeyer & Kroschel (2002); Mitra (1998); Schüssler (2008); Vaidyanathan (1993).

Assuming, for instance, a high-end input sampling frequency of $f_n = f_0 = 2.4$GHz [Kopmann et al. (2003); Maufroid et al. (2003)], the operational clock rate of the third stage is $f_3 = f_n/2^3 = 300$MHz, which is deemed feasible using present-day CMOS technology. Hence, front-end parallelisation has to reduce operational clock of all subsystems preceding the third stage down to $f_d = f_3 = 300$MHz. This is achieved by 8-fold parallelisation of input branching and blocking (delay z_0^{-1}), 4-fold parallelisation of the first stage of the FDMUX tree (comprising input decimation by two, the PP branch filters $H_\lambda(z_1)$, $\lambda = 0, 1$, and butterfly), and of the input branching and blocking (delay z_1^{-1}) of the second stage and, finally, corresponding 2-fold parallelisation of the two parallel 2-channel FDMUX FB of the second stage of the tree, as indicated in Fig. 32.

The result of parallelisation, as required above, is shown in Fig. 33, where all interfractional interfaces have been removed by straightforward application of identity of Fig. 30. Subsequently, parallelisation of elementary subsystems is explained in detail:

1. *Down-Sampling by* $M = 2$: In compliance with Fig. 31(b), each 2-fold down-sampler is replaced with P_ν units in parallel for $2P_\nu$-fold down-sampling with even time offset $2p$, where $p = 0, 1, 2, 3$ applies to the first tree stage ($P_1 = 4$), and $p = 0, 1$ to the second stage ($P_2 = 2$). The result of 4-fold parallelisation of the front end input down-sampler of the upper branch ($\nu = 1$, $\lambda = 0$) is readily visible in Fig. 33 preceding filter MIMO block $\mathbf{H}_0^1(z_d)$: In fact, it represents an 8-to-4 parallelisation, where all odd PP components are removed according to Fig. 31(b) Groth (2003).

2. *Cascade of unit blocking delay and 2-fold down-sampler*: For proper explanation, we first focus on the input section of the first tree stage, lower branch ($\nu = \lambda = 1$) in front of filter block

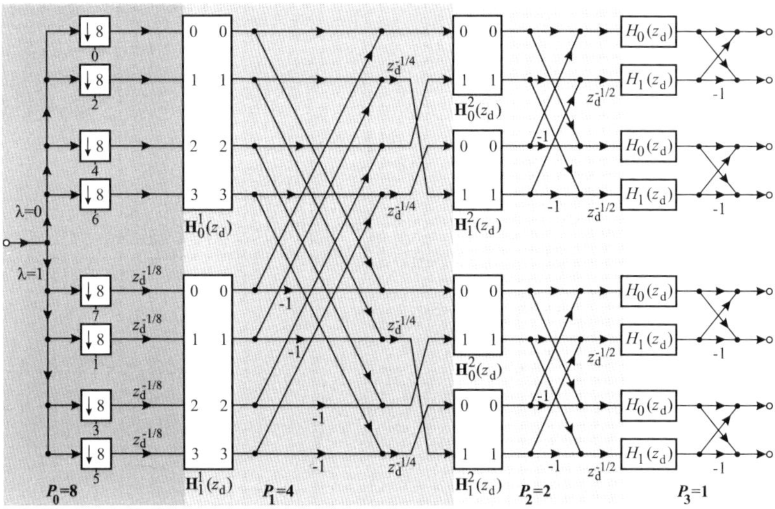

Fig. 33. Complete parallelisation of FDMUX front-end of SBC-FDFMUX filter bank (Fig. 32); $z_d := e^{j\Omega^{(d)}}$, $\Omega^{(d)} = 2\pi f/f_d$, $f_d = f_n/8$

$H_1(z_1)$. To this end, as required by Fig. 32, the unit delay z_0^{-1} is parallelised by $P_0 = 8$, as shown in Fig. 31(a), while the subsequent down-sampler applies $P_1 = 4$, as described above w.r.t. Fig. 31(b). Immediate cascading of parallelised unit delay ($P_0 = 8$) and down-sampling ($P_1 = 4$, $M = 2$) (as induced by Fig. 31) shows that only those four PP components of the parallelised delay with *even* time offset ($p = 0, 2, 4, 6$) are transferred via the 4-branch SP-input interface of down-sampling ($2P_1 = 8$) to its PS-output interface with naturally ordered time offsets $p = 0, 1, 2, 3$ w.r.t. $P_1 = 4$. Hence, only those retained 4 out of 8 PP components of odd time index $p = 7, 1, 3, 5$, being provided by the unit delay's SP-input interface and delayed by $z_0^{-1} = z_d^{-1/8}$, are transferred (mapped) to the $P_1 = 4$ up-samplers with timing offset $p = 0, 1, 2, 3$ of the 4-branch PS-output interface of the down-sampler. Fig. 33 shows the correspondingly rearranged signal flow graph representation of stage 1 input section ($\nu = \lambda = 1$).

As a result, the upper branch of stage 1, $H_0(z_1) \rightarrow \mathbf{H}_0^1(z_d)$, is fed by the even-indexed PP components of the high rate FDMUX input signal, whereas the lower branch $H_1(z_1) \rightarrow \mathbf{H}_1^1(z_d)$ is provided with the delayed versions of the PP components of odd index, as depicted in Fig. 33. Hence, as in the original system Fig. 32, the input sequence is completely fed into the parallelised system.

This procedure is repeated with the input branching and blocking sections of the subsequent stages $\nu = 2, 3$: The PP branch filters $H_0(z_\nu) \rightarrow \mathbf{H}_0^\nu(z_d)$ parallelised by P_ν, where $P_2 = 2$ and $P_3 = 1$ ($P_1 = 4$), are provided with the even-numbered PP components of the respective input signals with timing offsets in natural order. Contrary, the set of PP components of odd index is always delayed by $z_d^{-1/P_{\nu-1}}$ and fed into filter blocks $H_1(z_\nu) \rightarrow \mathbf{H}_1^\nu(z_d)$ in crossed manner (cf. input section $\lambda = 1$).

3. *P_ν-fold Parallelisation of PP branch filters* $H_\lambda(z_\nu) \rightarrow \mathbf{H}_\lambda^\nu(z_d)$, $\lambda = 0, 1$; $\nu = 1, 2$, is achieved by systematic application of the procedure condensed in Fig. 29 (for details cf. Göckler & Groth (2004); Groth (2003)). To this end, $H_\lambda(z_\nu)$ is decomposed in P_ν PP components of correspondingly reduced order, which are arranged to a MIMO system by

exploiting a multitude of multirate identities Groth (2003); Groth & Göckler (2001). The resulting $P_\nu \times P_\nu$ MIMO filter transfer matrix $\mathbf{H}_\lambda^\nu(z_\mathrm{d})$ contains each PP component of $H_\lambda(z_\nu)$ P_ν times: Thus, the amount of hardware is increased P_ν times whereas, as desired for feasibility, the operational clock rate is concurrently reduced by P_ν. Hence, the overall expenditure, i.e. the number of operations times the respective operational clock rate Göckler & Groth (2004), is not changed.
4. *Parallelisation of butterflies* combining the output signals of associated PP filter blocks is straightforward: For each (time-interleaved) PP component of the respective signals a butterfly has to be foreseen, as shown in Fig. 33.

4.3 Conclusion: Parallelisation of multirate systems

In this Section 4, a general and systematic procedure for parallelisation of multirate systems, for instance as investigated in Sections 2 and 3, has been presented . Its application to the high rate decimating FDMUX front end of the tree-structured SBC-FDFMUX FB Abdulazim & Göckler (2005); Abdulazim et al. (2007) has been deployed in detail. The stage ν degree of parallelisation P_ν, $\nu = 0, 1, 2, 3$, is diminished proportionally to the operational clock frequency f_ν of stage ν and is, thus, adapted to the actual sampling rate. As a result, after suitable decomposition of the high rate front end input signal by an input commutator in $P_0 = P_{\mathrm{max}}$ polyphase components (as depicted for $P_{\mathrm{max}} = 8$ in Fig. 33), all subsequent processing units are likewise operated at the same operational clock rate $f_\mathrm{d} = f_\mathrm{n}/P_0 = f_0/P_0$. Since inherent parallelism of the original tree-structured FDMUX (Fig. 32) has attained $P_{\mathrm{max}} = 8$ in the third stage, and the output signals of this stage represent the desired eight demultiplexed FDM subsignals, interleaving PS-output commutators are no longer required, as to be seen in Fig. 33. Finally, it should be noted that parallelisation does not change overall expenditure; yet, by multiplying stage ν hardware by P_ν, the operational clock rates are reduced by a factor of P_ν to a feasible order of magnitude, as desired.
Applying the rules of multirate transposition (cf. Subsection 3.3.1 or Göckler & Groth (2004)) to the parallelised FDMUX front end, the high rate interpolating back end of the tree-structured SBC-FDFMUX FB is obtained likewise and exhibits the same properties as to expenditure and feasibility Groth (2003). Hence, the versatile and efficient tree-structured filter bank (FDMUX, FMUX, SBC, wavelet, or any combination thereof) can be used in any (ultra) wide-band application without any restriction.

5. Summary and conclusion

In Section 2 we have introduced and investigated a special class of real and complex FIR and IIR halfband bandpass filters with the particular set of centre frequencies defined by (1). As a result of the constraint (1), almost all filter coefficients are either real-valued or purely imaginary-valued, as opposed to fully complex-valued coefficients. Hence, this class of halfband filters requires only a small amount of computation.
In Section 3, two different options to combine two of the above FIR halfband filters with different centre frequencies to form a directional filter (DF) have been investigated. As a result, one of these DF approaches is optimum w.r.t. to computation (most efficient), whereas the other requires the least number of delay elements (minimum McMillan degree). The relation between separating DF and DF that combine two independent signals to an FDM signal via multirate transposition rules has extensively been shown.
Finally, in Section 4, the above FIR directional filters (DF) have been combined to tree-structured multiplexing and demultiplexing filter banks. While this procedure is

straightforward, the operating clock rates within the front- or back-ends may be too high for implementation. To this end, we have introduced and described to some extent the systematic graphically induced procedure to parallelise multirate systems according to [Groth (2003)]. It has been applied to a three-stage demultiplexing tree-structured filter bank in such a manner that all operations throughout the overall system are performed at the operational output clock. As a result, parallelisation makes the system feasible but retains the computational load.

6. References

Abdulazim, M. N. & Göckler, H. G. (2007). Tree-structured MIMO FIR filter banks for flexible frequency reallocation, *Proc. of the 5th Int. Symposium on Image and Signal Processing and Analysis (ISPA 2007)*, Istanbul, Turkey, pp. 69–74.

Abdulazim, M. N. & Göckler, H. G. (2005). Efficient digital on-board de- and remultiplexing of FDM signals allowing for flexible bandwidth allocation, *Proc. Int. Comm. Satellite Systems Conf.*, Rome, Italy.

Abdulazim, M. N., Kurbiel, T. & Göckler, H. G. (2007). Modified DFT SBC-FDFMUX filter bank systems for flexible frequency reallocation, *Proc. EUSIPCO'07*, Poznan, Poland, pp. 60–64.

Ansari, R. (1985). Elliptic filter design for a class of generalized halfband filters, *IEEE Trans. Acoust., Speech, Sign. Proc.* ASSP-33(4): 1146–1150.

Ansari, R. & Liu, B. (1983). Efficient sampling rate alternation using recursive IIR digital filters, *IEEE Trans. Acoustics, Speech, and Signal Processing* ASSP-31(6): 1366–1373.

Arbesser-Rastburg, B., Bellini, R., Coromina, F., Gaudenzi, R. D., del Rio, O., Hollreiser, M., Rinaldo, R., Rinous, P. & Roederer, A. (2002). R&D directions for next generation broadband multimedia systems: An ESA perspective, *Proc. Int. Comm. Satellite Systems Conf.*, Montreal, Canada.

Bellanger, M. (1989). *Digital Processing of Signals - Theory and Practice*, 2nd edn, John Wiley & Sons, New York.

Bellanger, M. G., Daguet, J. L. & Lepagnol, G. P. (1974). Interpolation, extrapolation, and reduction of computation speed in digital filters, *IEEE Trans. Acoust., Speech, and Sign. Process.* ASSP-22(4): 231–235.

Damjanovic, S. & Milic, L. (2005). Examples of orthonormal wavelet transform implemented with IIR filter pairs, *Proc. SMMSP*, ICSP Series No.30, Riga, Latvia, pp. 19–27.

Damjanovic, S., Milic, L. & Saramäki, T. (2005). Frequency transformations in two-band wavelet IIR filter banks, *Proc. EUROCON*, Belgrade, Serbia and Montenegro, pp. 87–90.

Danesfahani, G. R., Jeans, T. G. & Evans, B. G. (1994). Low-delay distortion recursive (IIR) transmultiplexer, *Electron. Lett.* 30(7): 542–543.

Eghbali, A., Johansson, H., Löwenborg, P. & Göckler, H. G. (2009). Dynamic frequency-band reallocation and allocation: From satellite-based communication systems to cognitive radios, *Journal of Signal Processing Systems* (10.1007/s11265-009-0348-1, Springer NY).

Evangelista, G. (2001). *Zum Entwurf digitaler Systeme zur asynchronen Abtastratenumsetzung*, PhD thesis, Ruhr-Universität Bochum, Bochum, Germany.

Evangelista, G. (2002). Design of optimum high-order finite-wordlength digital FIR filters with linear phase, *EURASIP Signal Processing* 82(2): 187–194.

Fliege, N. (1993). *Multiraten-Signalverarbeitung: Theorie und Anwendungen*, B. G. Teubner, Stuttgart.

Gazsi, L. (1986). Quasi-bireciprocal and multirate wave digital lattice filters, *Frequenz* 40(11/12): 289–296.

Göckler, H. G. (1996a). Digitale Filterweiche. German patent P 19 627 784.

Göckler, H. G. (1996b). Nichtrekursives Halb-Band-Filter. German patent P 19 627 787.

Göckler, H. G. (1996c). Umschaltbare Frequenzweiche. German patent P 19 627 788.

Göckler, H. G. & Alfsmann, D. (2010). Efficient linear-phase directional filters with selectable centre frequencies, *Proc. 1st Int. Conf. Green Circuits and Systems (ICGCS 2010)*, Shanghai, China, pp. 293–298.

Göckler, H. G. & Damjanovic, S. (2006a). Efficient implementation of real and complex linear-phase FIR and minimum-phase IIR halfband filters for sample rate alteration, *Frequenz* 60(9/10): 176–185.

Göckler, H. G. & Damjanovic, S. (2006b). A family of efficient complex halfband filters, *Proc. 4th Karlsruhe Workshop on Software Radios*, Karlsruhe, Germany, pp. 79–88.

Göckler, H. G. & Felbecker, B. (2001). Digital on-board FDM-demultiplexing without restrictions on channel allocation and bandwidth, *Proc. 7th Int. Workshop on Dig. Sign. Proc. Techn. for Space Communications*, Sesimbra, Portugal.

Göckler, H. G. & Groth, A. (2004). *Multiratensysteme: Abtastratenumsetzung und digitale Filterbänke*, J. Schlembach Fachverlag, Wilburgstetten, Germany, ISBN 3-935340-29-X (Chinese Edition: ISBN 978-7-121-08464-5).

Göckler, H. G., Groth, A. & Abdulazim, M. N. (2006). Parallelisation of digital signal processing in uniform and reconfigurable filter banks for satellite communications, *Proc. IEEE Asia Pacific Conf. Circuits and Systems (APCCAS 2006)*, Singapore, pp. 1061–1064.

Göckler, H. G. & Grotz, K. (1994). DIAMANT: All digital frequency division multiplexing for 10 Gbit/s fibre-optic CATV distribution system, *Proc. EUSIPCO'94*, Edinburgh, UK, pp. 999–1002.

Göckler, H. G. & Eyssele, H. (1992). Study of on-board digital FDM-demultiplexing for mobile SCPC satellite communications (Part I & II), *Europ. Trans. Telecommunic.* ETT-3: 7–30.

Gold, B. & Rader, C. M. (1969). *Digital Processing of Signals*, McGraw-Hill, New York.

Groth, A. (2003). *Effiziente Parallelisierung digitaler Systeme mittels äquivalenter Signalflussgraph-Transformationen*, PhD thesis, Ruhr-Universität Bochum, Bochum, Germany.

Groth, A. & Göckler, H. G. (2001). Signal-flow-graph identities for structural transformations in multirate systems, *Proc. Europ. Conf. Circuit Theory Design*, Vol. II, Espoo, Finland, pp. 305–308.

Johansson, H. & Löwenborg, P. (2005). Flexible frequency-band reallocation networks based on variable oversampled complex-modulated filter banks, *Proc. IEEE Int. Conf. on Acoustics, Speech, and Signal Processing*, Philadelphia, USA.

Kammeyer, K. D. & Kroschel, K. (2002). *Digitale Signalverarbeitung*, Teubner, Stuttgart.

Kollar, I., Pintelon, R. & Schoukens, J. (1990). Optimal FIR and IIR Hilbert Transformer design via LS and minimax fitting, *IEEE Trans. Instrumentation and Measurement* 39(6): 847–852.

Kopmann, H., Göckler, H. G. & Abdulazim, M. N. (2003). Analogue-to-digital conversion and flexible FDM demultiplexing algorithms for digital on-board processing of ultra-wideband FDM signals, *Proc. 8th Int. Workshop on Signal Processing for Space Commun.*, Catania, Italy, pp. 277–292.

Kumar, B., Roy, S. C. D. & Sabharwal, S. (1994). Interrelations between the coefficients of FIR digital differentiators and other FIR filters and a versatile multifunction configuration, *EURASIP Signal Processing* 39(1/2): 247–262.

Lutovac, M. D. & Milic, L. D. (1997). Design of computationally efficient elliptic IIR filters with a reduced number of shift-and-add opperations in multipliers, *IEEE Trans. Sign. Process.* 45(10): 2422–2430.

Lutovac, M. D. & Milic, L. D. (2000). Approximate linear phase multiplierless IIR halfband filter, *IEEE Trans. Sign. Process. Lett.* 7(3): 52–53.

Lutovac, M. D., Tosic, D. V. & Evans, B. L. (2001). *Filter Design for Signal Processing Using MATLAB and Mathematica*, Prentice Hall, NJ.

Man, E. D. & Kleine, U. (1988). Linear phase decimation and interpolation filters for high-speed application, *Electron. Lett.* 24(12): 757–759.

Maufroid, X., Coromina, F., Folio, B., Hughes, R., Couchman, A., Stirland, S. & Joly, F. (2004). Next generation of transparent processors for broadband satellite access networks, *Proc. Int. Comm. Satellite Systems Conf.*, Monterey, USA.

Maufroid, X., Coromina, F., Folio, B.-M., Göckler, H. G., Kopmann, H. & Abdulazim, M. N. (2003). High throughput bent-pipe processor for future broadband satellite access networks, *Proc. 8th Int. Workshop on Signal Processing for Space Commun.*, Catania, Italy, pp. 259–275.

McClellan, H. J., Parks, T. W. & Rabiner, L. R. (1973). A computer program for designing optimum FIR linear phase digital filters, *IEEE Trans. Audio and Electroacoustics* AU(21): 506–526.

Meerkötter, K. & Ochs, K. (1998). A new digital equalizer based on complex signal processing, *in* Z. Ghassemlooy & R. Saatchi (eds), *Proc. CSDSP98*, Vol. 1, pp. 113–116.

Milic, L. (2009). *Multirate Filtering for Digital Signal Processing*, Information Science Reference, Hershey, NY, ISBN 978-1-60566-178-0.

Mintzer, F. (1982). On half-band, third-band, and Nth-band FIR-filters and their design, *IEEE Trans. Acoustics, Speech, and Signal Processing* ASSP-30(5): 734–738.

Mitra, S. K. (1998). *Digital Signal Processing: A Computer Based Approach*, McGraw-Hill, New York.

Mitra, S. K. & Kaiser, J. F. (eds) (1993). *Handbook for Digital Signal Processing*, John Wiley & Sons, New York.

Oppenheim, A. V. & Schafer, R. W. (1989). *Discrete-Time Signal Processing*, Signal Processing Series, Prentice Hall, NJ.

Parks, T. W. & Burrus, C. S. (1987). *Digital Filter Design*, John Wiley & Sons, New York.

Regalia, P. A., Mitra, S. K. & Vaidyanathan, P. P. (1988). The digital all-pass filter: A versatile signal processing building block, *Proc. of the IEEE* 76(1): 19–37.

Renfors, M. & Kupianen, T. (1998). Versatile building blocks for multirate processing of bandpass signals, *Proc. EUSPICO '98*, Rhodos, Greece, pp. 273–276.

Rio-Herrero, O. & Maufroid, X. (2003). A new ultra-fast burst switched processor architecture for meshed satellite networks, *Proc. 8th Int. Workshop on Signal Processing for Space Commun.*, Catania, Italy.

Schüssler, H. W. (2008). *Digitale Signalverarbeitung 1: Analyse diskreter Signale und Systeme*, 5th edn, Springer, Heidelberg.

Schüssler, H. W. & Steffen, P. (1998). Halfband filters and Hilbert Transformers, *Circuits Systems Signal Processing* 17(2): 137–164.

Schüssler, H. W. & Steffen, P. (2001). Recursive halfband-filters, *AEÜ* 55(6): 377–388.

Schüssler, H. W. & Weith, J. (1987). On the design of recursive Hilbert-transformers, *Proc. ICASSP 87, Dallas, TX*, pp. 876–879.

Strang, G. & Nguyen, T. (1996). *Wavelets and Filter Banks*, Wellesly-Cambridge Press, Wellesley, MA.

Vaidyananthan, P. P., Regalia, P. A. & Mitra, S. K. (1987). Design of doubly-complementary IIR digital filters using a single complex allpass filter, with multirate applications, *IEEE Trans. Circuits and Systems* CAS-34(4): 378–389.

Vaidyanathan, P. P. (1993). *Multirate Systems and Filter Banks*, Englewood Cliffs, NJ: Prentice Hall.

Vaidyanathan, P. P. & Nguyen, T. Q. (1987). A trick for the design of FIR half-band filters, *IEEE Trans. Circuits and Systems* CAS-34: 297–300.

Valenzuela, R. A. & Constantinides, A. G. (1983). Digital signal processing schemes for efficient interpolation and decimation, *IEE Proc.* 130(6): 225–235.

Wittig, M. (2000). Satellite on-board processing for multimedia applications, *IEEE Commun. Mag.* 38(6): 134–140.

Zhang, X. & Yoshikawa, T. (1999). Design of orthonormal IIR wavelet filter banks using allpass filters, *EURASIP Signal Processing* 78(1): 91–100.

12

Min-Max Design of FIR Digital Filters by Semidefinite Programming

Masaaki Nagahara
Kyoto University
Japan

1. Introduction

Robustness is a fundamental issue in signal processing; unmodeled dynamics and unexpected noise in systems and signals are inevitable in designing systems and signals. Against such uncertainties, *min-max optimization*, or *worst case optimization* is a powerful tool. In this light, we propose an efficient design method of FIR (finite impulse response) digital filters for approximating and inverting given digital filters. The design is formulated by *min-max optimization* in the frequency domain. More precisely, we design an FIR filter which minimizes the maximum gain of the frequency response of an error system.

This design has a direct relation with H^∞ *optimization* (Francis, 1987). Since the space H^∞ is not a Hilbert space, the familiar projection method in conventional signal processing cannot be applied. However, many studies have been made on the H^∞ optimization, and nowadays the optimal solution to the H^∞ problem is deeply analysed and can be easily obtained by numerical computation. Moreover, as an extension of H^∞ optimization, a min-max optimization on a *finite* frequency interval has been proposed recently (Iwasaki & Hara, 2005). In both optimization, the *Kalman-Yakubovich-Popov (KYP) lemma* (Anderson, 1967; Rantzer, 1996; Tuqan & Vaidyanathan, 1998) and the *generalized KYP lemma* (Iwasaki & Hara, 2005) give an easy and fast way of numerical computation; *semidefinite programming* (Boyd & Vandenberghe, 2004). Semidefinite programming can be efficiently solved by numerical optimization softwares.

In this chapter, we consider two fundamental problems of signal processing: FIR approximation of IIR (infinite impulse response) filters and inverse FIR filtering of FIR/IIR filters. Each problems are formulated in two types of optimization: H^∞ optimization and finite-frequency min-max one. These problems are reduced to semidefinite programming in a similar way. For this, we introduce state-space representation. Semidefinite programming is obtained by the generalized KYP lemma. We will give MATLAB codes for the proposed design, and will show design examples.

2. Preliminaries

In this chapter, we frequently use notations in control systems. For readers who are not familiar to these, we here recall basic notations and facts of control systems used throughout the chapter. We also show MATLAB codes for better understanding.

Let us begin with a linear system $\mathcal{G}$ represented in the following *state-space equation* or *state-space representation* (Rugh, 1996):

$$\mathcal{G} : \begin{cases} x[k+1] = Ax[k] + Bu[k], \\ \quad y[k] = Cx[k] + Du[k], \quad k = 0, 1, 2, \ldots. \end{cases} \tag{1}$$

The nonnegative number k denotes the time index. The vector $x[k] \in \mathbb{R}^n$ is called the state vector, $u[k] \in \mathbb{R}$ is the input and $y[k] \in \mathbb{R}$ is the output of the system $\mathcal{G}$. The matrices $A \in \mathbb{R}^{n \times n}$, $B \in \mathbb{R}^{n \times 1}$, $C \in \mathbb{R}^{1 \times n}$, and $D \in \mathbb{R}$ are assumed to be static, that is, independent of the time index k. Then the *transfer function* $G(z)$ of the system $\mathcal{G}$ is defined by

$$G(z) := C(zI - A)^{-1}B + D, \quad z \in \mathbb{C}.$$

The transfer function $G(z)$ is a rational function of z of the form

$$G(z) = \frac{b_n z^n + b_{n-1}z^{n-1} + \cdots + b_1 z + b_0}{z^n + a_{n-1}z^{n-1} + \cdots + a_1 z + a_0}.$$

Note that $G(z)$ is the Z-transform of the impulse response $\{g[k]\}_{k=0}^{\infty}$ of the system $\mathcal{G}$ with the initial state $x[0] = 0$, that is,

$$G(z) = \sum_{k=0}^{\infty} g[k]z^{-k} = D + \sum_{k=1}^{\infty} CA^{k-1}Bz^{-k}.$$

To convert a state-space equation to its transfer function, one can use the above equations or the MATLAB command `tf`. On the other hand, to convert a transfer function to a state-space equation, one can use realization theory which provides a method to derive the state space matrices from a given transfer function (Rugh, 1996). An easy way to obtain the matrices is to use MATLAB or Scilab with the command `ss`.

Example 1. *We here show an example of MATLAB commands. First, we define state-space matrices:*

```
>A=[0,1;-1,-2]; B=[0;1]; C=[1,1]; D=0;
>G=ss(A,B,C,D,1);
```

This defines a state-space (ss) representation of $\mathcal{G}$ with the state-space matrices

$$A = \begin{bmatrix} 0 & 1 \\ -1 & -2 \end{bmatrix}, \; B = \begin{bmatrix} 0 \\ 1 \end{bmatrix}, \; C = \begin{bmatrix} 1 & 1 \end{bmatrix}, \; D = 0.$$

The last argument `1` *of* `ss` *sets the sampling time to be 1.*

To obtain the transfer function $G(z) = C(zI - A)^{-1}B + D$, we can use the command `tf`

```
>> tf(G)

Transfer function:
    z + 1
-------------
z^2 + 2 z + 1

Sampling time (seconds): 1
```

On the other hand, suppose that we have a transfer function at first:

```
>> z=tf('z',1);
>> Gz=(z^2+2*z+1)/(z^2+0.5*z+1);
```

The first command defines the variable z of Z-transform with sampling time 1, and the second command defines the following transfer function:

$$G(z) = \frac{z^2 + 2z + 1}{z^2 + 0.5z + 1}.$$

To convert this to state-space matrices A, B, C, and D, use the command `ss` as follows:

```
>> ss(Gz)

a =
          x1     x2
   x1   -0.5     -1
   x2      1      0

b =
        u1
   x1    1
   x2    0

c =
         x1    x2
   y1   1.5     0

d =
        u1
   y1    1

Sampling time (seconds): 1
Discrete-time model.
```

These outputs shows that the state-space matrices are given by

$$A = \begin{bmatrix} -0.5 & -1 \\ 1 & 0 \end{bmatrix},\ B = \begin{bmatrix} 1 \\ 0 \end{bmatrix},\ C = \begin{bmatrix} 1.5 & 0 \end{bmatrix},\ D = 1,$$

with sampling time 1. □

Note that the state-space representation in **Example 1** is *minimal* in that the state-space model describes the same input/output behavior with the minimum number of states. Such a system is called *minimal realization* (Rugh, 1996).

We then introduce a useful notation, called *packed notation* (Vidyasagar, 1988), describing the transfer function from state-space matrices as

$$G(z) = C(zI - A)^{-1}B + D =: \left[\begin{array}{c|c} A & B \\ \hline C & D \end{array}\right](z).$$

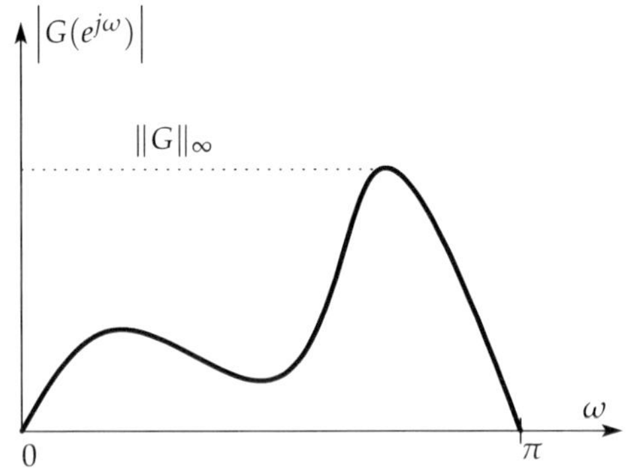

Fig. 1. The H^∞ norm $\|G\|_\infty$ is the maximum gain of the frequency response $G(e^{j\omega})$.

By the packed notation, the following formulae are often used in this chapter:

$$\left[\begin{array}{c|c} A_1 & B_1 \\ \hline C_1 & D_1 \end{array}\right] \times \left[\begin{array}{c|c} A_2 & B_2 \\ \hline C_2 & D_2 \end{array}\right] = \left[\begin{array}{cc|c} A_2 & 0 & B_2 \\ B_1C_2 & A_1 & B_1D_2 \\ \hline D_1C_2 & C_1 & D_1D_2 \end{array}\right],$$

$$\left[\begin{array}{c|c} A_1 & B_1 \\ \hline C_1 & D_1 \end{array}\right] \pm \left[\begin{array}{c|c} A_2 & B_2 \\ \hline C_2 & D_2 \end{array}\right] = \left[\begin{array}{cc|c} A_1 & 0 & B_1 \\ 0 & A_2 & \pm B_2 \\ \hline C_1 & C_2 & D_1 \pm D_2 \end{array}\right].$$

Next, we define *stability* of linear systems. The state-space system $\mathcal{G}$ in (1) is said to be *stable* if the eigenvalues $\lambda_1, \ldots, \lambda_n$ of the matrix A lie in the open unit circle $\mathbb{D} = \{z \in \mathbb{C} : |z| < 1\}$. Assume that the transfer function $G(z)$ is irreducible. Then $\mathcal{G}$ is stable if and only if the poles of the transfer function $G(z)$ lie in $\mathbb{D}$. To compute the eigenvalues of A in MATLAB, use the command `eig(A)`, and for the poles of $G(z)$ use `pole(Gz)`.
The H^∞ *norm* of a stable transfer function $G(z)$ is defined by

$$\|G\|_\infty := \max_{\omega \in [0,\pi]} \left|G(e^{j\omega})\right|.$$

This is the maximum gain of the frequency response $G(e^{j\omega})$ of $\mathcal{G}$ as shown in Fig. 1. The MATLAB code to compute the H^∞ norm of a transfer function is given as follows:

```
>> z=tf('z',1);
>> Gz=(z-1)/(z^2-0.5*z);
>> norm(Gz,inf)

ans =

    1.3333
```

This result shows that for the stable transfer function

$$G(z) = \frac{z-1}{z^2 - 0.5z},$$

the H^∞ norm is given by $\|G\|_\infty \approx 1.3333$.

H^∞ optimization is thus minimization of the maximum value of a transfer function. This leads to robustness against uncertainty in the frequency domain. Moreover, it is known that the H^∞ norm of a transfer function $G(z)$ is equivalent to the *ℓ^2-induced norm* of $\mathcal{G}$, that is,

$$\|G\|_\infty = \|\mathcal{G}\| := \sup_{\substack{u\in\ell^2 \\ u\neq 0}} \frac{\|\mathcal{G}u\|_2}{\|u\|_2},$$

where $\|u\|_2$ is the ℓ^2 norm of u:

$$\|u\|_2 := \left(\sum_{n=0}^{\infty} |u[k]|^2\right)^{1/2}.$$

The H^∞ optimization is minimization of the system gain when the worst case input is applied. This fact implies that the H^∞ optimization leads to robustness against uncertainty in input signals.

3. H^∞ Design problems of FIR digital filters

In this section, we consider two fundamental problems in signal processing: filter approximation and inverse filtering. The problems are formulated as H^∞ optimization by using the H^∞ norm defined in the previous section.

3.1 FIR approximation of IIR filters

The first problem we consider is *approximation*. In signal processing, there are a number of design methods for IIR (infinite impulse response) filters, e.g., Butterworth, Chebyshev, Elliptic, and so on (Oppenheim & Schafer, 2009). In general, to achieve a given characteristic, IIR filters require fewer memory elements, i.e., z^{-1}, than FIR (finite impulse response) filters. However, IIR filters may have a problem of instability since they have feedbacks in their circuits, and hence, we prefer an FIR filter to an IIR one in implementation. For this reason, we employ FIR approximation of a given IIR filter. This problem has been widely studied (Oppenheim & Schafer, 2009). Many of them are formulated by H^2 optimization; they aim at minimizing the average error between a given IIR filter and the FIR filter to be designed. This optimal filter works well *averagely*, but in the worst case, the filter may lead a large error. To guarantee the worst case performance, H^∞ optimization is applied to this problem (Yamamoto et al., 2003). The problem is formulated as follows:

Problem 1 (FIR approximation of IIR filters). *Given an IIR filter $P(z)$, find an FIR (finite impulse response) filter $Q(z)$ which minimizes*

$$\|(P-Q)W\|_\infty = \max_{\omega\in[0,\pi]} \left|\left(P(e^{j\omega}) - Q(e^{j\omega})\right) W(e^{j\omega})\right|,$$

where W is a given stable weighting function.

The procedure to solve this problem is shown in Section 4.

3.2 Inverse filtering

Inverse filtering, or *deconvolution* is another fundamental issue in signal processing. This problem arises for example in direct-filter design in spline interpolation (Nagahara & Yamamoto, 2011).

Suppose a filter $P(z)$ is given. Symbolically, the inverse filter of $P(z)$ is $P(z)^{-1}$. However, real design is not that easy.

Example 2. *Suppose $P(z)$ is given by*

$$P(z) = \frac{z+0.5}{z-0.5}.$$

Then, the inverse $Q(z) := P(z)^{-1}$ becomes

$$Q(z) = P(z)^{-1} = \frac{z-0.5}{z+0.5},$$

which is stable and causal. Then suppose

$$P(z) = \frac{z-2}{z-0.5},$$

then the inverse is

$$Q(z) = P(z)^{-1} = \frac{z-0.5}{z-2}.$$

This has the pole at $|z| > 1$, and hence the inverse filter is unstable. On the other hand, suppose

$$P(z) = \frac{1}{z-0.5},$$

then the inverse is

$$Q(z) = P(z)^{-1} = z - 0.5,$$

which is noncausal.

By these examples, the inverse filter $P(z)^{-1}$ may unstable or noncausal. Unstable or noncausal filters are difficult to implement in real digital device, and hence we adopt approximation technique; we design an FIR digital filter $Q(z)$ such that $Q(z)P(z) \approx 1$. Since FIR filters are always stable and causal, this is a realistic way to design an inverse filter. Our problem is now formulated as follows:

Problem 2 (Inverse filtering). *Given a filter $P(z)$ which is necessarily not bi-stable or bi-causal (i.e., $P(z)^{-1}$ can be unstable or noncausal), find an FIR filter $Q(z)$ which minimizes*

$$\|(QP-1)W\|_\infty = \max_{\omega\in[0,\pi]} \left|\left(Q(e^{j\omega})P(e^{j\omega}) - 1\right) W(e^{j\omega})\right|,$$

where W is a given stable weighting function.

The procedure to solve this problem is shown in Section 4.

4. KYP lemma for H^∞ design problems

In this section, we show that the H^∞ design problems given in the previous section are efficiently solved via *semidefinite programming* (Boyd & Vandenberghe, 2004). For this purpose, we first formulate the problems in state-space representation reviewed in Section 2. Then we bring in *Kalman-Yakubovich-Popov* (KYP) lemma (Anderson, 1967; Rantzer, 1996; Tuqan & Vaidyanathan, 1998) to reduce the problems into semidefinite programming.

4.1 State-space representation

The transfer functions $(P(z) - Q(z))\,W(z)$ and $(Q(z)P(z) - 1)\,W(z)$ in **Problems 1** and **2**, respectively, can be described in a form of

$$T(z) = T_1(z) + Q(z)T_2(z), \tag{2}$$

where

$$T_1(z) = P(z)W(z), \quad T_2(z) = -W(z),$$

for **Problem 1** and

$$T_1(z) = -W(z), \quad T_2(z) = P(z)W(z),$$

for **Problem 2**. Therefore, our problems are described by the following min-max optimization:

$$\min_{Q(z)\in\mathcal{F}_N} \|T_1 + QT_2\|_\infty = \min_{Q(z)\in\mathcal{F}_N} \max_{\omega\in[0,\pi]} \left|T_1(e^{j\omega}) + Q(e^{j\omega})T_2(e^{j\omega})\right|, \tag{3}$$

where $\mathcal{F}_N$ is the set of N-th order FIR filters, that is,

$$\mathcal{F}_N := \left\{ Q(z) : Q(z) = \sum_{i=0}^{N} \alpha_i z^{-i}, \alpha_i \in \mathbb{R} \right\}.$$

To reduce the problem of minimizing (3) to semidefinite programming, we use state-space representations for $T_1(z)$ and $T_2(z)$ in (2). Let $\{A_i, B_i, C_i, D_i\}$ $(i = 1, 2)$ are state-space matrices of $T_i(z)$ in (2), that is,

$$T_i(z) = C_i(zI - A_i)^{-1}B_i + D_i =: \left[\begin{array}{c|c} A_i & B_i \\ \hline C_i & D_i \end{array}\right](z), \quad i = 1, 2.$$

Also, a state-space representation of an FIR filter $Q(z)$ is given by

$$Q(z) = \sum_{n=0}^{N} \alpha_n z^{-n} = \left[\begin{array}{ccccc|c} 0 & 1 & 0 & \dots & 0 & 0 \\ 0 & 0 & 1 & \ddots & \vdots & \vdots \\ 0 & 0 & \ddots & \ddots & 0 & \vdots \\ \vdots & \vdots & \ddots & 0 & 1 & 0 \\ 0 & 0 & \dots & 0 & 0 & 1 \\ \hline \alpha_N & \alpha_{N-1} & \dots & \alpha_2 & \alpha_1 & \alpha_0 \end{array}\right](z) =: \left[\begin{array}{c|c} A_q & B_q \\ \hline \alpha_{N:1} & \alpha_0 \end{array}\right](z), \tag{4}$$

where $\alpha_{N:1} := \begin{bmatrix} \alpha_N & \alpha_{N-1} & \dots & \alpha_1 \end{bmatrix}$.

By using these state-space matrices, we obtain a state-space representation of $T(z)$ in (2) as

$$T(z) = \left[\begin{array}{ccc|c} A_1 & 0 & 0 & B_1 \\ 0 & A_2 & 0 & B_2 \\ 0 & B_q C_2 & A_q & B_q D_2 \\ \hline C_1 & \alpha_0 C_2 & \alpha_{N:1} & D_1 + \alpha_0 D_2 \end{array}\right](z) =: \left[\begin{array}{c|c} A & B \\ \hline C(\alpha_{N:0}) & D(\alpha_0) \end{array}\right](z). \tag{5}$$

Note that the FIR parameters $\alpha_0, \alpha_1, \ldots, \alpha_N$ depend affinely on C and D, and are independent of A and B. This property is a key to describe our problem into semidefinite programming.

4.2 Semidefinite programming by KYP lemma

The optimization in (3) can be equivalently described by the following minimization problem:

$$\begin{aligned} &\text{minimize } \gamma \text{ subject to } Q(z) \in \mathcal{F}_N \text{ and} \\ &\max_{\omega \in [0,\pi]} \left| T_1(e^{j\omega}) + Q(e^{j\omega}) T_2(e^{j\omega}) \right| \leq \gamma. \end{aligned} \tag{6}$$

To describe this optimization in semidefinite programming, we adopt the following lemma (Anderson, 1967; Rantzer, 1996; Tuqan & Vaidyanathan, 1998):

Lemma 1 (KYP lemma). *Suppose*

$$T(z) = \left[\begin{array}{c|c} A & B \\ \hline C & D \end{array}\right](z)$$

is stable, and the state-space representation $\{A, B, C, D\}$ of $T(z)$ is minimal[1]. *Let $\gamma > 0$. Then the following are equivalent conditions:*

1. $\|T\|_\infty \leq \gamma$.
2. *There exists a positive definite matrix X such that*

$$\begin{bmatrix} A^\top X A - X & A^\top X B & C^\top \\ B^\top X A & B^\top X B - \gamma^2 & D \\ C & D & -1 \end{bmatrix} \leq 0.$$

By using this lemma, we obtain the following theorem:

Theorem 1. *The inequality (6) holds if and only if there exists $X > 0$ such that*

$$\begin{bmatrix} A^\top X A - X & A^\top X B & C(\alpha_{N:0})^\top \\ B^\top X A & B^\top X B - \gamma^2 & D(\alpha_0) \\ C(\alpha_{N:0}) & D(\alpha_0) & -1 \end{bmatrix} \leq 0, \tag{7}$$

where A, B, $C(\alpha_{N:0})$, and $D(\alpha_0)$ are given in (5).

By this, the optimal FIR parameters $\alpha_0, \alpha_1, \ldots, \alpha_N$ can be obtained as follows. Let x be the vector consisting of all variables in $\alpha_{N:0}$, X, and γ^2 in (7). The matrix in (7) is *affine* with respect to these variables, and hence, can be rewritten in the form

$$M(x) = M_0 + \sum_{i=1}^{L} M_i x_i,$$

[1] For minimality of state-space representation, see Section 2 or Chapter 26 in (Rugh, 1996).

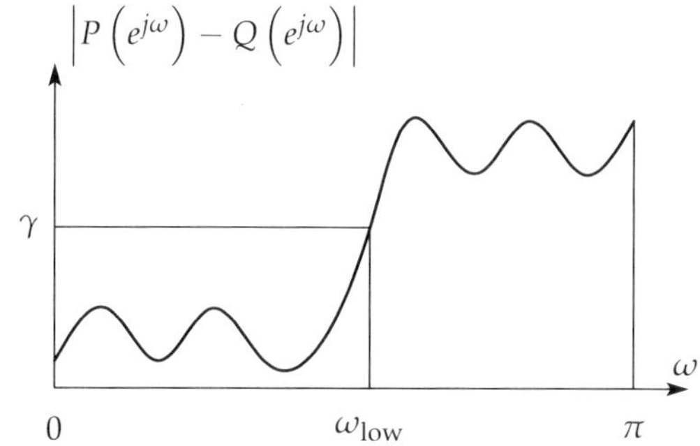

Fig. 2. Finite frequency approximation (**Problem 3**): the gain of the error $P(e^{j\omega}) - Q(e^{j\omega})$ is minimized over the finite frequency range $\Omega_{\text{low}} = [0, \omega_{\text{low}}]$.

where M_i is a symmetric matrix and x_i is the i-th entry of x. Let $v \in \{0,1\}^L$ be a vector such that $v^\top x = \gamma^2$. Our problem is then described by semidefinite programming as follows:

$$\text{minimize } v^\top x \text{ subject to } M(x) \leq 0.$$

By this, we can effectively approach the optimal parameters $\alpha_0, \alpha_1, \ldots, \alpha_N$ by numerical optimization softwares. For MATLAB codes of the semidefinite programming above, see Section 7.

5. Finite frequency design of FIR digital filters

By the H^∞ design discussed in the previous section, we can guarantee the maximum gain of the frequency response of $T = (P - Q)W$ (approximation) or $T = (QP - 1)W$ (inversion) over the *whole frequency range* $[0, \pi]$. Some applications, however, do not need minimize the gain over the whole range $[0, \pi]$, but a finite frequency range $\Omega \subset [0, \pi]$. Design of noise shaping $\Delta\Sigma$ modulators is one example of such requirement (Nagahara & Yamamoto, 2009). In this section, we consider such optimization, called *finite frequency optimization*. We first consider the approximation problem over a finite frequency range.

Problem 3 (Finite frequency approximation). *Given a filter $P(z)$ and a finite frequency range $\Omega \subset [0, \pi]$, find an FIR filter $Q(z)$ which minimizes*

$$V_\Omega(P - Q) := \max_{\omega \in \Omega} \left|P(e^{j\omega}) - Q(e^{j\omega})\right|.$$

Figure 2 illustrates the above problem for a finite frequency range $\Omega = \Omega_{\text{low}} = [0, \omega_{\text{low}}]$, where $\omega_{\text{low}} \in (0, \pi]$. We seek an FIR filter which minimizes the gain of the error $P(e^{j\omega}) - Q(e^{j\omega})$ over the finite frequency range Ω, and do not care about the other range $[0, \pi] \setminus \Omega$. We can also formulate the inversion problem over a finite frequency range.

Problem 4 (Finite frequency inversion). *Given a filter $P(z)$ and a finite frequency range $\Omega \subset [0, \pi]$, find an FIR filter $Q(z)$ which minimizes*

$$V_\Omega(QP - 1) := \max_{\omega \in \Omega} \left|Q(e^{j\omega})P(e^{j\omega}) - 1\right|.$$

These problems are also fundamental in digital signal processing. We will show in the next section that these problems can be also described in semidefinite programming via generalized KYP lemma.

6. Generalized KYP lemma for finite frequency design problems

In this section, we reduce the problems given in the previous section to semidefinite programming. As in the H^∞ optimization, we first formulate the problems in state-space representation, and then derive semidefinite programming via *generalized KYP lemma* (Iwasaki & Hara, 2005).

6.1 State-space representation

As in the H^∞ optimization in Section 4, we employ state-space representation. Let $T(z) = P(z) - Q(z)$ for the approximation problem or $T(z) = P(z)Q(z) - 1$ for the inversion problem. Then $T(z)$ can be described by $T(z) = T_1(z) + Q(z)T_2(z)$ as in (2). Then our problems are described by the following min-max optimization:

$$\min_{Q(z)\in\mathcal{F}_N} V_\Omega(T_1 + QT_2) = \min_{Q(z)\in\mathcal{F}_N} \max_{\omega\in\Omega} \left| T_1(e^{j\omega}) + Q(e^{j\omega})T_2(e^{j\omega}) \right|. \tag{8}$$

Let $\{A_i, B_i, C_i, D_i\}$, $i = 1, 2$ be state-space matrices of $T_i(z)$. By using the same technique as in Section 4, we can obtain a state-space representation of $T(z)$ as

$$T(z) = \left[\begin{array}{c|c} A & B \\ \hline C(\alpha_{N:0}) & D(\alpha_0) \end{array}\right](z), \tag{9}$$

where $\alpha_{N:0} = [\alpha_N, \ldots, \alpha_0]$ is the coefficient vector of the FIR filter to be designed as defined in (4).

6.2 Semidefinite programming by generalized KYP lemma

The optimization in (8) can be equivalently described by the following problem:

$$\begin{aligned} &\text{minimize } \gamma \text{ subject to } Q(z) \in \mathcal{F}_N \text{ and} \\ &\max_{\omega\in\Omega} \left| T_1(e^{j\omega}) + Q(e^{j\omega})T_2(e^{j\omega}) \right| \leq \gamma \end{aligned} \tag{10}$$

To describe this optimization in semidefinite programming, we adopt the following lemma (Iwasaki & Hara, 2005):

Lemma 2 (Generalized KYP Lemma). *Suppose*

$$T(z) = \left[\begin{array}{c|c} A & B \\ \hline C & D \end{array}\right](z)$$

is stable, and the state-space representation $\{A, B, C, D\}$ of $T(z)$ is minimal. Let Ω be a closed interval $[\omega_1, \omega_2] \subset [0, \pi]$. Let $\gamma > 0$. Then the following are equivalent conditions:

1. $V_\Omega(T) = \max_{\omega\in[\omega_1,\omega_2]} \left| T(e^{j\omega}) \right| \leq \gamma.$
2. *There exist symmetric matrices $Y > 0$ and X such that*

$$\begin{bmatrix} M_1(X,Y) & M_2(X,Y) & C^\top \\ M_2(X,Y)^\top & M_3(X,\gamma^2) & D \\ C & D & -1 \end{bmatrix} \leq 0,$$

where

$$\begin{aligned} M_1(X,Y) &= A^\top XA + YAe^{-j\omega_0} + A^\top Ye^{j\omega_0} - X - 2Y\cos r, \\ M_2(X,Y) &= A^\top XB + YBe^{-j\omega_0}, \quad \overline{M}_2(X,Y) = A^\top XB + YBe^{j\omega_0}, \\ M_3(X,\gamma^2) &= B^\top XB - \gamma^2, \quad \omega_0 = \frac{\omega_1+\omega_2}{2}, \quad r = \frac{\omega_2-\omega_1}{2}. \end{aligned} \tag{11}$$

By using this lemma, we obtain the following theorem:

Theorem 2. *The inequality (10) holds if and only if there exist symmetric matrices $Y > 0$ and X such that*

$$\begin{bmatrix} M_1(X,Y) & M_2(X,Y) & C(\alpha_{N:0})^\top \\ \overline{M}_2(X,Y)^\top & M_3(X,\gamma^2) & D(\alpha_0) \\ C(\alpha_{N:0}) & D(\alpha_0) & -1 \end{bmatrix} \leq 0,$$

where M_1, M_2, and M_3 are given in (11), A, B, $C(\alpha_{N:0})$, and $D(\alpha_0)$ are given in (9).

By this theorem, we can obtain the coefficients $\alpha_0, \ldots, \alpha_N$ of the optimal FIR filter by semidefinite programming as mentioned in Section 4. MATLAB codes for the semidefinite programming are shown in Section 7.

7. MATLAB codes for semidefinite programming

In this section, we give MATLAB codes for the semidefinite programming derived in previous sections. Note that the MATLAB codes for solving **Problems 1** to **4** are also available at the following web site:

`http://www-ics.acs.i.kyoto-u.ac.jp/~nagahara/fir/`

Note also that to execute the codes in this section, Control System Toolbox (Mathworks, 2010), YALMIP (Löfberg, 2004), and SeDuMi (Sturm, 2001) are needed. YALMIP and SeDuMi are free softwares for solving optimization problems including semidefinite programming which is treated in this chapter.

7.1 FIR approximation of IIR filters by H^∞ norm

```
function [q,gmin] = approxFIRhinf(P,W,N);
% [q,gmin]=approxFIRhinf(P,W) computes the
% H-infinity optimal approximated FIR filter Q(z) which minimizes
%   J(Q) = ||(P-Q)W||,
% the maximum frequency gain of (P-Q)W.
% This design uses SDP via the KYP lemma.
%
% Inputs:
%   P: Target stable linear system in SS object
%   W: Weighting stable linear system in SS object
%   N: Order of the FIR filter to be designed
%
% Outputs:
%   q: The optimal FIR filter coefficients
%   gmin: The optimal value
%
```

```
%% Initialization
T1 = P*W;
T2 = -W;
[A1,B1,C1,D1]=ssdata(T1);
[A2,B2,C2,D2]=ssdata(T2);
n1 = size(A1,1);
n2 = size(A2,1);

%% FIR filter to be designed
Aq = circshift(eye(N),-1);
Aq(N,1) = 0;
Bq = [zeros(N-1,1);1];

%% Semidefinite Programming
A = [A1, zeros(n1,n2), zeros(n1,N);
     zeros(n2,n1), A2, zeros(n2,N);
     zeros(N,n1),Bq*C2, Aq];
B = [B1;B2;Bq*D2];

NN = size(A,1);

X = sdpvar(NN,NN,'symmetric');
alpha_N1 = sdpvar(1,N);
alpha_0 = sdpvar(1,1);
gamma = sdpvar(1,1);

M1 = A'*X*A-X;
M2 = A'*X*B;
M3 = B'*X*B-gamma;

C = [C1, alpha_0*C2, alpha_N1];
D = D1 + alpha_0*D2;

M = [M1, M2, C'; M2', M3, D; C, D, -gamma];

F = set(M < 0) + set(X > 0) + set(gamma > 0);

solvesdp(F,gamma);

%% Optimal FIR filter coefficients
q = fliplr([double(alpha_N1),double(alpha_0)]);
gmin = double(gamma);
```

7.2 Inverse FIR filtering by H^∞ norm

```
function [q,gmin] = inverseFIRhinf(P,W,N,n);
% [q,gmin]=inverseFIRhinf(P,W,N,n) computes the
```

```
% H-infinity optimal (delayed) inverse FIR filter Q(z) which minimizes
%   J(Q) = ||(QP-z^(-n))W||,
% the maximum frequency gain of (QP-z^(-n))W.
% This design uses SDP via the KYP lemma.
%
% Inputs:
%   P: Target stable linear system in SS object
%   W: Weighting stable linear system in SS object
%   N: Order of the FIR filter to be designed
%   n: Delay (this can be omitted; default value=0);
%
% Outputs:
%   q: The optimal FIR filter coefficients
%   gmin: The optimal value
%

if nargin==3
    n=0
end

%% Initialization
z = tf('z');
T1 = -z^(-n)*W;
T2 = P*W;
[A1,B1,C1,D1]=ssdata(T1);
[A2,B2,C2,D2]=ssdata(T2);
n1 = size(A1,1);
n2 = size(A2,1);

%% FIR filter to be designed
Aq = circshift(eye(N),-1);
Aq(N,1) = 0;
Bq = [zeros(N-1,1);1];

%% Semidefinite Programming
A = [A1, zeros(n1,n2), zeros(n1,N);
     zeros(n2,n1), A2, zeros(n2,N);
     zeros(N,n1),Bq*C2, Aq];
B = [B1;B2;Bq*D2];

NN = size(A,1);

X = sdpvar(NN,NN,'symmetric');
alpha_N1 = sdpvar(1,N);
alpha_0 = sdpvar(1,1);
gamma = sdpvar(1,1);
```

```
M1 = A'*X*A-X;
M2 = A'*X*B;
M3 = B'*X*B-gamma;

C = [C1, alpha_0*C2, alpha_N1];
D = D1 + alpha_0*D2;

M = [M1, M2, C'; M2', M3, D; C, D, -gamma];

F = set(M < 0) + set(X > 0) + set(gamma > 0);

solvesdp(F,gamma);

%% Optimal FIR filter coefficients
q = fliplr([double(alpha_N1),double(alpha_0)]);
gmin = double(gamma);
```

7.3 FIR approximation of IIR filters by finite-frequency min-max

```
function [q,gmin] = approxFIRff(P,Omega,N);
% [q,gmin]=approxFIRff(P,Omega,N) computes the
% Finite-frequency optimal approximated FIR filter Q(z)
  which minimizes
%   J(Q) = max{|P(exp(jw))-Q(exp(jw))|, w in Omega}1.
% the maximum frequency gain of P-Q in a frequency band Omega.
% This design uses SDP via the generalized KYP lemma.
%
% Inputs:
%   P: Target stable linear system in SS object
%   Omega: Frequency band in 1x2 vector [w1,w2]
%   N: Order of the FIR filter to be designed
%
% Outputs:
%   q: The optimal FIR filter coefficients
%   gmin: The optimal value
%

%% Initialization
[A1,B1,C1,D1]=ssdata(P);
n1 = size(A1,1);

%% FIR filter to be designed
Aq = circshift(eye(N),-1);
Aq(N,1) = 0;
Bq = [zeros(N-1,1);1];

%% Semidefinite Programming
A = blkdiag(A1,Aq);
```

```
B = [B1;-Bq];

NN = size(A,1);

omega0 = (Omega(1)+Omega(2))/2;
omegab = (Omega(2)-Omega(1))/2;

P = sdpvar(NN,NN,'symmetric');
Q = sdpvar(NN,NN,'symmetric');
alpha_N1 = sdpvar(1,N);
alpha_0 = sdpvar(1,1);
g = sdpvar(1,1);

C = [C1, alpha_N1];
D = D1 - alpha_0;

M1r = A'*P*A+Q*A*cos(omega0)+A'*Q*cos(omega0)-P-2*Q*cos(omegab);
M2r = A'*P*B + Q*B*cos(omega0);
M3r = B'*P*B-g;
M1i = A'*Q*sin(omega0)-Q*A*sin(omega0);
M21i = -Q*B*sin(omega0);
M22i = B'*Q*sin(omega0);
Mr = [M1r,M2r,C';M2r',M3r,D;C,D,-1];
Mi = [M1i, M21i, zeros(NN,1);M22i, 0, 0; zeros(1,NN),0,0];
M = [Mr, Mi; -Mi, Mr];

F = set(M < 0) + set(Q > 0) + set(g > 0);

solvesdp(F,g);

%% Optimal FIR filter coefficients
q = fliplr([double(alpha_N1),double(alpha_0)]);
gmin = double(g);
```

7.4 Inverse FIR filtering by finite-frequency min-max

```
function [q,gmin] = inverseFIRff(P,Omega,N,n);
% [q,gmin]=inverseFIRff(P,Omega,N,n) computes the
% Finite-frequency optimal (delayed) inverse FIR filter Q(z) which minimiz
%   J(Q) = max{|Q(exp(jw)P(exp(jw))-exp(-jwn)|, w in Omega}.
% the maximum frequency gain of QP-z^(-n) in a frequency band Omega.
% This design uses SDP via the generalized KYP lemma.
%
% Inputs:
%   P: Target stable linear system in SS object
%   Omega: Frequency band in 1x2 vector [w1,w2]
%   N: Order of the FIR filter to be designed
%   n: Delay (this can be omitted; default value=0);
```

```
%
% Outputs:
%   q: The optimal FIR filter coefficients
%   gmin: The optimal value
%

if nargin==3
    n=0
end

%% Initialization
z = tf('z');
T1 = -z^(-n);
T2 = P;
[A1,B1,C1,D1]=ssdata(T1);
[A2,B2,C2,D2]=ssdata(T2);
n1 = size(A1,1);
n2 = size(A2,1);

%% FIR filter to be designed
Aq = circshift(eye(N),-1);
Aq(N,1) = 0;
Bq = [zeros(N-1,1);1];

%% Semidefinite Programming
A = [A1, zeros(n1,n2), zeros(n1,N);
     zeros(n2,n1), A2, zeros(n2,N);
     zeros(N,n1),Bq*C2, Aq];
B = [B1;B2;Bq*D2];

NN = size(A,1);

omega0 = (Omega(1)+Omega(2))/2;
omegab = (Omega(2)-Omega(1))/2;

P = sdpvar(NN,NN,'symmetric');
Q = sdpvar(NN,NN,'symmetric');
alpha_N1 = sdpvar(1,N);
alpha_0 = sdpvar(1,1);
g = sdpvar(1,1);

C = [C1, alpha_0*C2, alpha_N1];
D = D1 + alpha_0*D2;

M1r = A'*P*A+Q*A*cos(omega0)+A'*Q*cos(omega0)-P-2*Q*cos(omegab);
M2r = A'*P*B + Q*B*cos(omega0);
M3r = B'*P*B-g;
```

```
M1i = A'*Q*sin(omega0)-Q*A*sin(omega0);
M21i = -Q*B*sin(omega0);
M22i = B'*Q*sin(omega0);
Mr = [M1r,M2r,C';M2r',M3r,D;C,D,-1];
Mi = [M1i, M21i, zeros(NN,1);M22i, 0, 0; zeros(1,NN),0,0];
M = [Mr, Mi; -Mi, Mr];

F = set(M < 0) + set(Q > 0) + set(g > 0);

solvesdp(F,g);

%% Optimal FIR filter coefficients
q = fliplr([double(alpha_N1),double(alpha_0)]);
gmin = double(g);
```

8. Examples

By the MATLAB codes given in the previous section, we design FIR filters for **Problems 1** and **3**. Let the FIR filter order $N = 8$. The target filter is the second order lowpass Butterworth filter with cutoff frequency $\pi/2$. This can be computed by `butter(2,1/2)` in MATLAB. The weighting transfer function in **Problem 1** is chosen by an 8th-order lowpass Chebyshev filter, computed by `cheby1(8,1/2,1/2)` in MATLAB. The frequency band for **Problem 3** is $\Omega = [0, \pi/2]$. Figure 3 shows the gain of the error $E(z) := P(z) - Q(z)$. We can see that the H^∞ optimal filter (the solution of **Problem 1**), say $Q_1(z)$, shows the lower H^∞ norm than the finite-frequency min-max design (the solution of **Problem 3**), say $Q_2(z)$. On the other hand, in the frequency band $[0, \pi/2]$, $Q_1(z)$ shows the larger error than $Q_2(z)$.

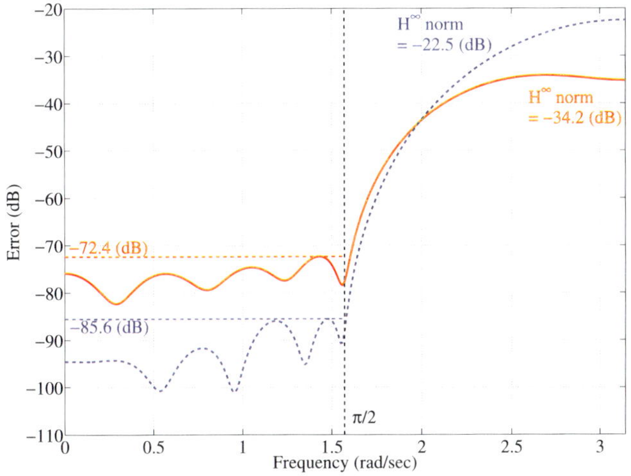

Fig. 3. The gain of the error $E(z) = P(z) - Q(z)$ for H^∞ optimization (solid) and finite-frequency min-max optimization (dash)

9. Conclusion

In this chapter, we consider four problems, FIR approximation and inverse FIR filtering of FIR/IIR filters by H^∞ and finite-frequency min-max, which are fundamental in signal processing. By using the KYP and generalized KYP lemmas, the problems are all solvable via semidefinite programming. We show MATLAB codes for the programming, and show examples of designing FIR filters.

10. References

Anderson, B. D. O. (1967). A system theory criterion for positive real matrices, *Siam Journal on Control and Optimization* 5: 171–182.

Boyd, S. & Vandenberghe, L. (2004). *Convex Optimization*, Cambridge University Press.

Francis, B. A. (1987). *A Course in H_∞ Control Theory*, Springer.

Iwasaki, T. & Hara, S. (2005). Generalized KYP lemma: unified frequency domain inequalities with design applications, *IEEE Trans. Autom. Control* 50: 41–59.

Löfberg, J. (2004). Yalmip : A toolbox for modeling and optimization in MATLAB, *Proc. IEEE International Symposium on Computer Aided Control Systems Design* pp. 284–289. URL: *http://users.isy.liu.se/johanl/yalmip/*

Mathworks (2010). *Control System Toolbox Users Guide.* URL: *http://www.mathworks.com/products/control/*

Nagahara, M. & Yamamoto, Y. (2009). Optimal noise shaping in $\Delta\Sigma$ modulators via generalized KYP lemma, *Proc. of IEEE ICASSP* III: 3381–3384.

Nagahara, M. & Yamamoto, Y. (2011). H^∞ optimal approximation for causal spline interpolation, *Signal Processing* 91(2): 176–184.

Oppenheim, A. V. & Schafer, R. W. (2009). *Discrete-Time Signal Processing*, 3rd edn, Prentice Hall.

Rantzer, A. (1996). On the Kalman–Yakubovich–Popov lemma, *Systems & Control Letters* 28(1): 7–10.

Rugh, W. J. (1996). *Linear Systems Theory*, Prentice Hall.

Sturm, J. F. (2001). *Using SeDuMi 1.02, a MATLAB toolbox for optimization over symmetric cones.* URL: *http://sedumi.ie.lehigh.edu/*

Tuqan, J. & Vaidyanathan, P. P. (1998). The role of the discrete-time Kalman-Yakubovitch-Popov lemma in designing statistically optimum FIR orthonormal filter banks, *Proc. of ISCAS* 5: 122–125.

Vidyasagar, M. (1988). A state-space interpretation of simultaneous stabilization, *IEEE Trans. Autom. Control* 33(5): 506–508.

Yamamoto, Y., Anderson, B. D. O., Nagahara, M. & Koyanagi, Y. (2003). Optimizing FIR approximation for discrete-time IIR filters, *IEEE Signal Process. Lett.* 10(9).

13

Applications of Interval-Based Simulations to the Analysis and Design of Digital LTI Systems

Juan A. López[1], Enrique Sedano[1], Luis Esteban[2], Gabriel Caffarena[3], Angel Fernández-Herrero[1] and Carlos Carreras[1]
[1]*Departamento de Ingeniería Electrónica, Universidad Politécnica de Madrid,*
[2]*Laboratorio Nacional de Fusión, Centro de Investigaciones Energéticas Medioambientales y Tecnológicas (CIEMAT),*
[3]*Departamento de Ingeniería de Sistemas de Información y de Telecomunicación, Universidad CEU-San Pablo,*
Spain

1. Introduction

As the complexity of digital systems increases, the existing simulation-based quantization approaches soon become unaffordable due to the exceedingly long simulation times. Thus, it is necessary to develop optimized strategies aimed at significantly reducing the computation times required by the algorithms to find a valid solution (Clark et al., 2005; Hill, 2006). In this sense, interval-based computations are particularly well-suited to reduce the number of simulations required to quantize a digital system, since they are capable of evaluating a large number of numerical samples in a single interval-based simulation (Caffarena et al., 2009, 2010; López, 2004; López et al., 2007, 2008).

This chapter presents a review of the most common interval-based computation techniques, as well as some experiments that show their application to the analysis and design of digital Linear Time Invariant (LTI) systems. One of the main features of these computations is that they are capable of significantly reducing the number of simulations needed to characterize a digital system, at the expense of some additional complexity in the processing of each operation. On the other hand, one of the most important problems associated to these computations is interval oversizing (i.e., the computed bounds of the intervals are wider than required), so new descriptions and methods are continuously being proposed. In this sense, each description has its own features and drawbacks, making it suitable for a different type of processing.

The structure is as follows: Section 2 presents a general review of the main interval-based computation methods that have been proposed in the literature to perform fast evaluation of system descriptions. For each technique, the representation of the different types of computing elements is given, as well as the main advantages and disadvantages of each approach. Section 3 presents three groups of interval-based experiments: (i) a comparison of the results provided by two different interval-based approaches to show the main problem of interval-based computations; (ii) an analysis of the application of interval-based computations to measure and compare the sensitivity of the signals in the frequency domain; and (iii) an analysis of the application of interval-based techniques to the Monte-Carlo method. Finally, Section 4 concludes this work.

2. General overview of interval-based computations

2.1 Interval arithmetic

Since its formalization in 1962 by R. Moore (Moore, 1962), Interval Arithmetic (IA) has been widely used to bound uncertainties in complex systems (Moore, 1966). The main advantage of traditional IA is that it is able to obtain the range of all the possible results of a given function. On the other hand, it suffers from three different types of problems (Neumaier, 2002): the dependency problem, the cancellation problem, and the wrapping effect.

The dependency problem expresses that IA computations overestimate the output range of a given function whenever it depends on one or more of its variables through two or more different paths. The cancellation problem occurs when the width of the intervals is not canceled in the inverse functions. In particular, this situation occurs in the subtraction operations (i.e., given the non-empty interval $I_1 - I_1 \neq 0$), what can be seen as a particular case of the dependency problem, but its effect is clearly identified. The wrapping effect occurs because the intervals are not able to accurately represent regions of space whose boundaries are not parallel to the coordinate axes.

These overestimations are propagated in the computations and make the results inaccurate, and even useless in some cases. For this reason, the Overestimation Factor (*OF*) (Makino & Berz, 2003; Neumaier, 2002) has been defined as

$$\text{OF} = (\text{Estimated Range} - \text{Exact Range}) \,/\, (\text{Exact Range}), \tag{1}$$

to quantify the accuracy of the results. Another interesting definition used to evaluate the performance of these methods is the Approximation Order (Makino & Berz, 2003; Neumaier, 2002), defined as the minimum order of the monomial $C\varepsilon^{S}$ (where C is constant, and $\varepsilon \in [0,1]$) that contains the difference between the bounds of the interval function and the target function in the range of interest.

2.2 Extensions of interval arithmetic

The different extensions of IA try to improve the accuracy of the computed results at the expense of more complex representations. A classification of the main variants of IA is given in Figure 1.

According to the representation of the uncertainties, the extensions of IA can be classified in three different types: Extended IA (EIA), Parameterized IA and Centered Forms (CFs). In a further division, these methods are further classified as follows. In the first group, Directed Intervals (DIs) and Modal Intervals (MIs); in the second group, Generalized IA (GIA); and in the third group, Mean Value Forms (MVFs), slopes, Taylor Models (TMs) and Affine Arithmetic (AA). A brief description of each formulation is given below.

DIs (Kreinovich, 2004) include the direction or sign of each interval to avoid the cancellation problem in the subtraction operations ($I_1^+ - I_1^+ = 0$), which is the most important source of overestimation (Kaucher, 1980; Ortolf, Bonn, 1969).

- Interval Arithmetic (IA)
  - Extended IA (EIA)
    - Directed Intervals (DIs)
    - Modal Intervals (MIs)
  - Parameterized IA
    - Generalized IA (GIA)
  - Centered Forms (CFs)
    - Mean Value Forms (MVFs)
    - Slopes
    - Taylor Models (TMs)
    - Affine Arithmetic (AA)

Fig. 1. Classification of interval-based computations methods.

In MIs (Gardenes, 1985; Gardenes & Trepat, 1980; SIGLA/X, 1999a, 1999b), each element is composed of one interval and a parameter called "modality" that indicates if the equation of the MIs holds for a single value of the interval or for all its values. These two descriptions are used to generate equations that bound the target function. If both descriptions exist and are equal, the result is exact. Among the publications on MIs, the underlying theoretical formulation and the justifications are given in (SIGLA/X, 1999a) and the applications, particularly for control systems, are given in (Armengol, et al., DX-2001; SIGLA/X, 1999b; Vehí, 1998)

GIA (Hansen, 1975; Tupper, 1996) is based on limiting the regions of the represented domain using intervals with parameterizable endpoints, such as $[1 - 2x, 3 + 4x]$ with $x \in [0,1]$. The authors define different types of parameterized intervals (constant, linear, quadratic, linear, multi-dimensional, functional and symbolic), but their analysis has focused on evaluating whether the target function is increasing or decreasing, concave or convex, in the region of interest using constant, linear and polynomial parameters. In the experiments, they have obtained the areas where the existence of the function is impossible, but they conclude that this type of analysis is too complex for parameterizations greater than the linear case.

In the different representations, CFs are based on representing a function as a Taylor Series expansion with one or more intervals that incorporate the uncertainties. Therefore, all these techniques are composed of one independent value (the central point of the function) and a set of summands that incorporate the intervals in the representation.

MVFs (Alefeld, 1984; Coconut_Group, 2002; Moore, 1966; Neumaier, 1990; Schichl & Neumaier, 2002) are based on developing an expression of a first-order Taylor Series that bounds the region of interest. The general expression is as follows:

$$f(x) = f(x_0) + f'(x)(x - x_0) \quad \in \quad f_{MVF}(I_x) = f(x_0) + f'(I_x)(I_x - x_0) \tag{2}$$

where x is the point or region where $f(x)$ must be evaluated, x_0 is the central point of the Taylor Series, and I_x is the interval that bounds the uncertainty range. The computation of the derivative is not complex when the function is polynomial, as it is usually the case in function approximation methods. Since the approximation error is quadratic, this method does not provide good results when the input intervals are large. However, if the input intervals are small, it provides better results than traditional IA.

The slopes (Moore, 1966; Neumaier, 1990; Schichl & Neumaier, 2002) also use a first-order Taylor Series expansion, but they apply the Newton's method to recursively compute the values of the derivatives. Its general expression is as follows:

$$f(x) = f(x_0) + f'(x)(x - x_0) \quad \in \quad f_S(I_S, I_x) = f(x_0) + I_S(I_x - x_0) \qquad (3)$$

where I_S is determined according to the expression (Garloff, 1999):

$$I_S = \begin{cases} \dfrac{f(x) - f(x_0)}{x - x_0} & if\ x \neq x_0 \\ x_0 & if\ x = x_0 \end{cases} \qquad (4)$$

It is worth mentioning that slopes typically provide better estimates than MVFs by a factor of 2, and that the results can be further improved by combining their computation with IA (Schichl & Neumaier, 2002)

TMs (Berz, 1997, 1999; Makino & Berz, 1999) combine a N-order Taylor Series expansion with an interval that incorporates the uncertainty in the function under analysis. Its mathematical expression is as follows:

$$f_{TM}(x, I_n) = a_n x^n + a_{n-1} x^{n-1} + \ldots + a_1 x + a_0 + I_n \qquad (5)$$

where a_i is the i-th coefficient of the interpolation polynomial of order n, and I_n is the uncertainty interval for this polynomial. The approximation error has now order N+1, rather than quadratic as in previous cases. In addition, TMs improve the representation of the domain regions, which reduces the wrapping effect. The applications of TMs have been largely studied thanks to the development of the tool COSY INFINITY (Berz, 1991, 1999; Berz, et al., 1996; Berz & Makino, 1998, 2004; Hoefkens, 2001; Hoefkens, et al., 2001, 2003; Makino, 1998, 1999). The main features of this tool include the resolution of Ordinary Differential Equations (ODEs), higher order ODEs and systems, multivariable integration, and techniques for relieving the wrapping effect, the dimensionality course, and the cluster effect (Hoefkens, 2001; Makino & Berz, 2003; Neumaier, 2002). Another relevant contributor in the development of the TMs is the GlobSol project (Corliss, 2004; GlobSol_Group, 2004; Kearfott, 2004; Schulte, 2004; Walster, 2004), focused on the application of interval computations to different applications, including systems modeling, computer graphics, gene prediction, missile design tips, portfolio management, foreign exchange market, parameter optimization in medical measures, software development of Taylor operators, interval support for the GNU Fortran compiler, improved methods of automatic differentiation, resolution of chemical models, etc. (GlobSol_Group, 2004).

There are discussions about the capabilities of TMs to solve the different theoretical and applied problems. In this sense, it is worth mentioning that "the TMs only reduce the problem of bounding a factorable function to bounding the range of a polynomial in a small box centered at 0. However, they are good or bad depending on how they are applied to solve each problem." (Neumaier, 2002). This statement is also applicable to the other uncertainty computation methods.

In AA (Comba & Stolfi, 1993; Figuereido & Stolfi, 2002; Stolfi & Figuereido, 1997), each element or affine form consists of a central value plus a set of noise terms (NTs). Each NT is composed of one uncertainty source identifier, called Noise Symbol (NS), and a constant coefficient associated to it. The mathematical expression is:

$$f_{AA}(\varepsilon_i) = x' = x_c + x_0\,\varepsilon_0 + x_1\,\varepsilon_1 + \varepsilon_2\,x_2 + \ldots + \varepsilon_n\,x_n \tag{6}$$

where x' represents the affine form, x_c is the central point, and each ε_i and x_i are the NS and its associated coefficient. In AA the operations are classified in two types: affine and non-affine operations. Affine operations (addition and constant multiplication) are computed without error, but non-affine operations need to include additional NTs to provide the bounds of the results. The main advantage of AA is that it keeps track of the different noise symbols and cancels all the first-order uncertainties, so it is capable of providing accurate results in linear sequences of operations. In nonlinear systems, AA obtains quadratic convergence, but the increment of the number of NTs in the nonlinear operations makes the computations less accurate and more time-consuming. A detailed analysis of the implementation of AA and a description of the most relevant computation algorithms is given in (Stolfi & Figuereido, 1997).

Among other applications, AA has been successfully used to evaluate the tolerance of circuit components (Femia & Spagnuolo, 2000), the sizing of analog circuits (Lemke, et al., Nov. 2002), the evolution of deformable models (Goldenstein, et al., 2001), the evaluation of polynomials (Shou, et al., 2002), and the analysis of the Round-Off Noise (RON) in Digital Signal Processing (DSP) systems (Fang, 2003; López, 2004; López et al., 2007, 2008), etc.

Modified AA (MAA) (Shou, et al., 2003) has been proposed to accurately compute the evolution of the uncertainties in nonlinear descriptions. Its mathematical expression is as follows:

$$f_{MAA}(e_i^{\,k}) = x' = x_c + x_0 e_0 + x_1 e_1 + x_2 e_0^{\,2} + x_3 e_0 e_1 + x_4 e_1^{\,2} + \ldots + x_n \prod_{i,k} \varepsilon_i^k \tag{7}$$

It is easy to see that MAA is an extension of AA that includes the polynomial NTs in the description. Thus, it is capable of computing the evolution of higher-order uncertainties that appear in polynomial descriptions (of a given smooth system), but the number of terms of the representation grows exponentially with the number of uncertainties and the order of the polynomial description. Thus, in this case it is particularly important to keep the number of NTs of the representation under a reasonable limit.

Obviously, the higher order NTs are not required when computing the evolution of the uncertainties in LTI systems, so MAA is less convenient than AA in this case.

3. Interval-based analysis of DSP systems

This Section examines the variations of the properties of the signals that occur in the evaluation of the DSP systems when Monte-Carlo Simulations (MCS) are performed using Extensions of IA (EIA) instead of the traditional numerical simulations. The simulations based on IA and EIA can handle the uncertainties and nonlinearities associated, for example, to the quantization operations of fixed-point digital filters, and other types of systems in the general case.

The most relevant advantages of using EIA to evaluate DSP systems can be summarized in the following points:

1. It is capable of managing the uncertainties associated with the quantization of coefficients, signals, complex computations and nonlinearities.
2. It avoids the cancellation problem of IA.
3. It provides faster results than the traditional numerical simulations.

The intuitive reason that determines the benefits of EIA is simple. Since EIA is capable of processing large sets of data in a single interval-based simulation, the results are obtained faster than in the separate computation of the numerical samples. Although the use of intervals imposes a limitation of connectivity on the computation of the results, both the speed and the accuracy are improved with respect to the numerical processing of the same number of samples.

Section 3.1 discusses the cancellation problem in the analysis of digital filter structures using IA, and justifies the selection of AA for such analysis, indicating the cases in which it can be used, and under what types of restrictions. Section 3.2 examines how the Fourier Transform is affected when uncertainties are included in one or all of the samples. Section 3.3 evaluates the changes that occur in the parameters of the random signals (mean, variance and Probability Density Function (PDF)) when a specific width is introduced in the samples, and how these changes affect the computed estimates using the Monte-Carlo method. Finally, Section 3.4 provides a brief discussion to highlight the capabilities of interval-based simulations.

3.1 Analysis of digital filter structures using IA and AA

The main problem that arises when performing interval-based analyses of DSP systems using IA is that the addition and subtraction operations always increase the interval widths. If there are variables that depend on other variables through two or more different paths, such as in $z(k) = x(k) - x(k)$, the ranges provided by IA are oversized. This problem, called the cancellation problem, is particularly severe when there are feedback loops in the realizations, a characteristic which is common in most DSP systems.

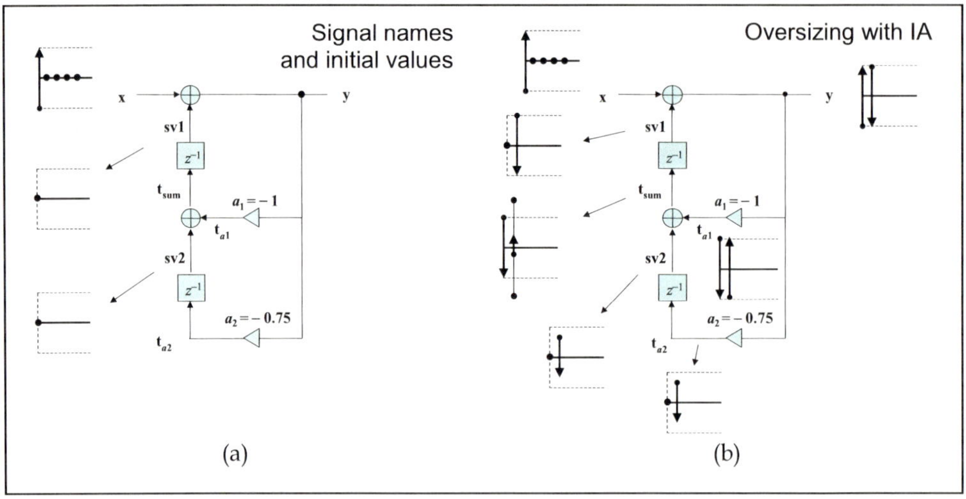

Fig. 2. Interval oversizing due to the cancellation effect of IA: (a) Signal names and initial (interval) values. (b) Computed intervals until the oversizing in the variable t_{sum} is detected. In each small figure, the abscissa axis represents the sampled time, and the ordinate axis represents the interval values. A dot in a given position represents the interval [0,0].

Figure 2.a shows a second-order Infinite Impulse Response (IIR) filter realized in direct form, whose transfer function is

$$H(z) = \frac{1}{1 + a_1 z^{-1} + a_2 z^{-2}} = \frac{1}{1 + z^{-1} + 0.75 z^{-2}} . \tag{8}$$

It is initially assumed that the filter is implemented using infinite precision, which implies that the quantization effects are negligible and that all signals are generated as linear combinations of the input and the state variables. This assumption allows: (i) to perform a separate analysis of the mean and the width of the intervals; and (ii) to generalize the results obtained in the simulation of a normalized interval to larger or smaller ones.

Figure 2.b shows the oversizing that occurs in the IA simulation. The input is set to the normalized interval [-1, 1], and the state variables are initially set to zero. Here, the representations are based on oriented intervals to keep track of the position of the samples in each interval, and to detect the overestimations. The initial values and the evolution of the intervals are:

$$x = [-1, 1] \Rightarrow y = [-1, 1] \Rightarrow \begin{cases} t_{a1} = [1, -1] \Rightarrow t_{sum} = [1, -1] \Rightarrow sv_1 = [1, -1] \\ sv_2 = [0.75, -0.75] \end{cases} \tag{9}$$

and in the next sampled time the values are:

$$\left. \begin{array}{l} sv_1 = [1, -1] \Rightarrow y = [1, -1] \Rightarrow t_{a1} = [-1, 1] \\ sv_2 = [0.75, -0.75] \end{array} \right\} \Rightarrow t_{sum} = [-1.75, 1.75] \tag{10}$$

instead of t_{sum} = [-0.25, 0.25], which is the correct value. Figure 2.b also shows that this oversizing occurs because signal t_{sum} depends on the input signal through two different paths.

Since AA includes a separate signed identifier per uncertainty source, it avoids such overestimations and provides the smallest intervals. In this case, the initial values and the evolution of the affine forms are:

$$x = 2\varepsilon \Rightarrow y = 2\varepsilon \Rightarrow \begin{cases} t_{a1} = 2\varepsilon \Rightarrow t_{sum} = 2\varepsilon \Rightarrow sv_1 = 2\varepsilon \\ sv_2 = -1.5\varepsilon \end{cases} \tag{11}$$

and in the next sampled time

$$\left. \begin{array}{l} sv_1 = 2\varepsilon \Rightarrow y = 2\varepsilon \Rightarrow t_{a1} = 2\varepsilon \\ sv_2 = -1.5\varepsilon \end{array} \right\} \Rightarrow t_{sum} = 0.5\varepsilon \tag{12}$$

which corresponds to the most accurate interval [-0.25, 0.25].

This simple example confirms the selection of AA instead of IA, particularly in structures with feedback loops. Although the cancellation effect is not necessarily present in all the structures, it commonly appears in most DSP realizations. For this reason, it is highly recommended to use this arithmetic when performing interval-based analysis of DSP systems.

When there are multiple simultaneous uncertainty sources, it is necessary to use an oriented identifier for each source, in addition to the average value of the signals, which are the elements offered by AA to perform the computations. Moreover, the objective of AA is to accurately determine the results of the linear operations (additions, subtractions, constant multiplications and delays), and the purpose of the filters is to perform a given linear transformation of the input signal. Consequently, the features offered by AA match perfectly with the requirements of the interval-based simulations of the unquantised digital filter structures.

When the quantization operations are included in this type of analysis, the affine forms must be adjusted to include all the possible values of the results. Since AA keeps track of the effects of the uncertainty sources (the noise terms can be seen as the first-order relationship between each uncertainty source and the signals), the affine forms are easily modified to simulate the effects of the quantization operations in the structures containing feedback loops.

In summary, one of the most important problems of IA to perform accurate interval-based simulations of the DSP realizations is the cancellation problem. The use of AA, in combination with the modification of the affine forms in the quantization operations, solves this problem and allows performing accurate analysis of the linear structures, even when they contain feedback loops.

3.2 Computation of the fourier transform of deterministic interval-based signals

The analysis of deterministic signals in DSP systems is of great importance, since most systems use or modify their properties in the frequency domain to send the information. In this sense, the decomposition of the signals using the Fourier transform as finite or infinite sums of sinusoids allows to evaluate these properties. Conversely, it is also widely known that a sufficient condition to characterize the linear systems is to determine the variations of the properties of the sinusoids of the different frequencies.

The following experiment shows the variations of the properties of deterministic signals when intervals of a given width are included in one or all of their samples. These widths represent the possible uncertainties in these signals and their effect on their associated signals in the transformed domain.

First, we evaluate the effects of including uncertainties of the same width in all the samples of the sequence. The steps required to perform this example are as follows:

1. Generate the Fast Fourier Transform (FFT) program file, specifying the number of stages.
2. Generate the sampled sinusoidal signals to be used as inputs.
3. Include the uncertainty specifications in the input signals.
4. Compute the Fourier Transform (run the interval-based simulation).
5. Repeat the steps 1-4 modifying the widths of the intervals of step 3.
6. Repeat the previous steps modifying the periods of the sinusoids of step 2.

Steps 1 to 4 generate the FFT of the interval-based sinusoidal signals. Step 5 has been included to investigate the effects of incorporating uncertainties of a given width to all input samples of the FFT. By superposition, this should be equal to the numerical FFT of the mean values of the original signal, plus another FFT in which all the input intervals are centered in zero and they all have the same width. Finally, step 6 allows us to investigate the variations of the computed results according to the periods of the sinusoids.

Figure 3 shows two examples of cosine signals with equal-width intervals in all the samples and their respective computed FFTs. Figure 3.a corresponds to a cosine signal of amplitude 1, length 1024, period 32, and width 1/8 in all the samples, and Figure 3.c shows another cosine signal of the same amplitude and width, length 256 and period 8. Figures 3.b and 3.d show the computed FFTs for each case, where each black line represents a data interval.

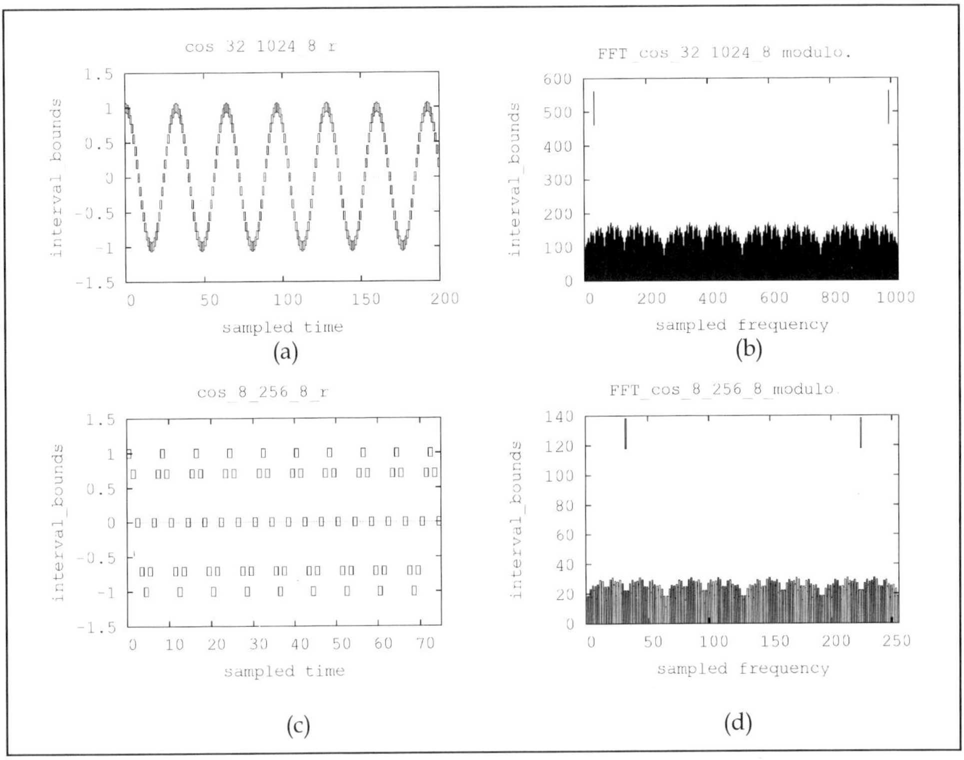

Fig. 3. Examples of FFTs of deterministic interval signals: (a) First 200 samples of a cosine signal of length 1024, period 32, and interval widths 1/8 in all the samples. (b) FFT of the previous signal. (c) First 75 samples of a cosine signal of length 256, period 8, and interval widths 1/8 in all the samples. (d) FFT of the previous signal.

As expected, these figures clearly show that the output intervals in the transformed domain have the form of the numerical transform, plus a given level of uncertainty in all the samples. In addition, Figures 3.b and 3.d also provide: (i) the values of the deviations in the transformed domain in each sample with respect to the numerical case, and (ii) the maximum levels of uncertainty associated with the uncertainties of the inputs.

The second part of this experiment evaluates how each uncertainty separately affects to the FFT samples. As mentioned above, by performing a separate analysis of how each uncertainty affects to the input samples, we are characterizing the quantization effects of the FFT. In this case, step 3 is replaced by the following statement:

3. Include one uncertainty in the specified sample of the input signals.

which is performed by generating a delta interval in the specified position, and adding it to the input signal.

Figure 4.a shows a cosine signal of length 1024 and period 32, in which only an interval of width 1/5 in the sample 27 has been included, and Figure 4.b shows the computed FFT of the previous interval trace. In this case, two small intervals appear in the sampled frequencies 32 and 968, as well as in the values near 0 in the other frequencies. Unlike the results shown in Figure 3, the uncertainties associated with the input interval are very small in this case.

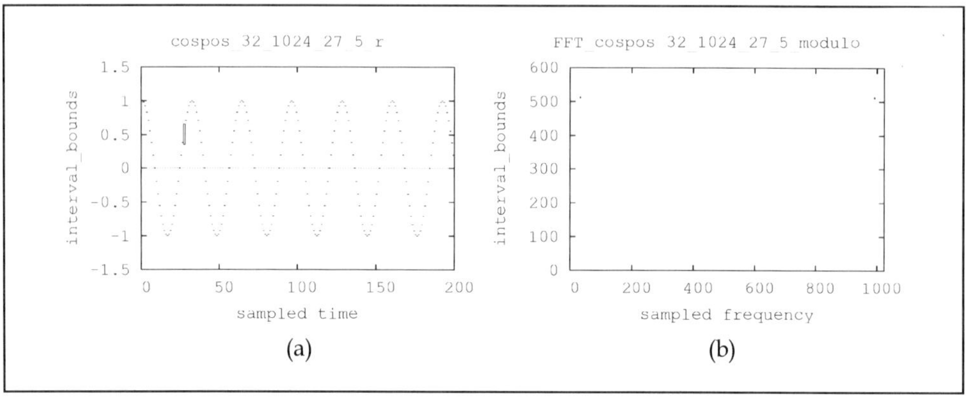

Fig. 4. Example of an FFT of a deterministic signal with a single interval: (a) First 200 samples of a cosine signal of length 1024, period 32 and interval width 1/5 in the sample 27. (b) FFT of the previous signal, with two small uncertainties in the sampled frequencies 32 and 968.

Figure 5 shows the details of the ripples generated by the uncertainties according to their positions in each trace. In the first case (Figure 5.a), the interval has been included in sample 16, which is a factor of the number of FFT points. In this case, there is no ripple. In the other three cases (Figures 5.b-d), the interval has been included in three different positions (17, 20 and 27, respectively), and there is a small ripple in the transformed domain, different in each case. Since the FFTs are linear systems, the large ripples that appear in the Figures 3.b and 3.d are the sum of all the possible equal-width ripples in the frequency domain.

In summary, the inclusion of intervals in sinusoidal signals and the computation of the FFTs show the maximum and minimum deviations in the frequency domain due to the different uncertainties. It has been found that the uncertainties do not affect to all the frequencies of the FFT in the same way, and that their effects depend on their positions in the trace. Although the intervals represent the maximum values of the uncertainties and the noise is commonly associated to the second-order statistics, the variations in the computed interval widths implies that the noise generated by the FFT is not white, but follows a deterministic pattern.

3.3 Analysis of the statistical parameters of random signals using interval-based simulations

The following experiments show the variations of the statistical parameters of random signals (mean, variance and PDF) when random sequences are generated using the Monte-Carlo method, using intervals of a specified width instead of the traditional numerical simulations.

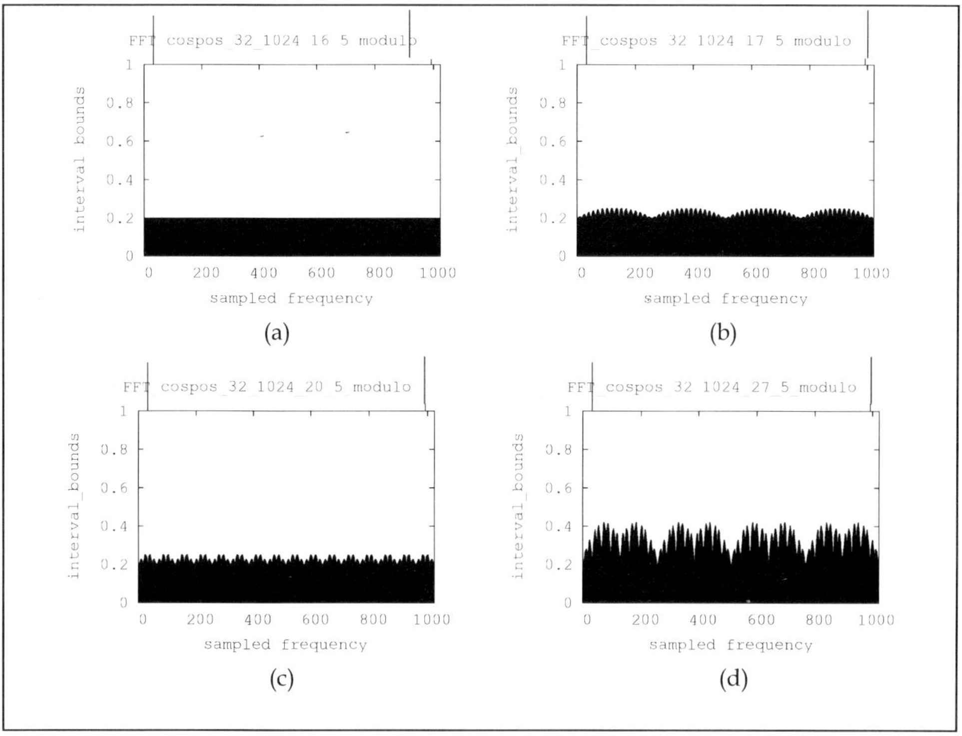

Fig. 5. Details of the ripples that occur in the transformed domain due to the presence of uncertainty intervals in the deterministic signals: (a) in a position which is a factor of the number of FFT points (16). (b) - (d) in other non-factor positions (17, 20 and 27, respectively). The vertical lines above the figures indicate the positions of the deltas, whose heights exceed the representable values in the graph.

The first part of this section analyzes the changes in the PDFs. To do this, data sequences following a particular PDF are generated, and they are later reconstructed and compared with the original results. The steps used to perform the experiments are as follows:

1. Generate the traces of the random samples following the specified PDF, and assign the width of the intervals.
2. Obtain the histogram of the trace, group the samples and plot the computed PDF.
3. Repeat steps 1 and 2 to reduce the variance of the parameters (*M* times).
4. Average the histograms obtained in step 3.
5. Repeat the previous steps assigning other interval widths.

Step 1 generates the sequences of samples that follow the specified PDF, and in step 2 the PDFs are recomputed from these samples. In this experiment, three types of PDFs have been used: (i) a uniform PDF in [-1, 1], a normalized normal PDF (mean 0 and variance 1), and a bimodal PDF composed of two normal PDFs, with means -3 and 3 and variance 1. Steps 3 and 4 have been included to reduce the variance of the results. Finally, step 5 allows selecting other interval widths.

Figure 6 presents the results of the three histograms using the Monte-Carlo method with: (i) numerical samples, (ii) intervals whose width is set to 1/8 of the variance, and (iii) intervals whose width is set to the variance of the distribution. All the histograms have been computed using 20 averages of 5000 data items each. It can be seen that the areas near the edges on the uniform distribution are modified, but the remaining parts of the distribution are also computed taking into account a larger number of points. It is also noticeable that the new PDFs are smoother than the ones computed using the numerical traces, which can be explained from the Central Limit Theorem.

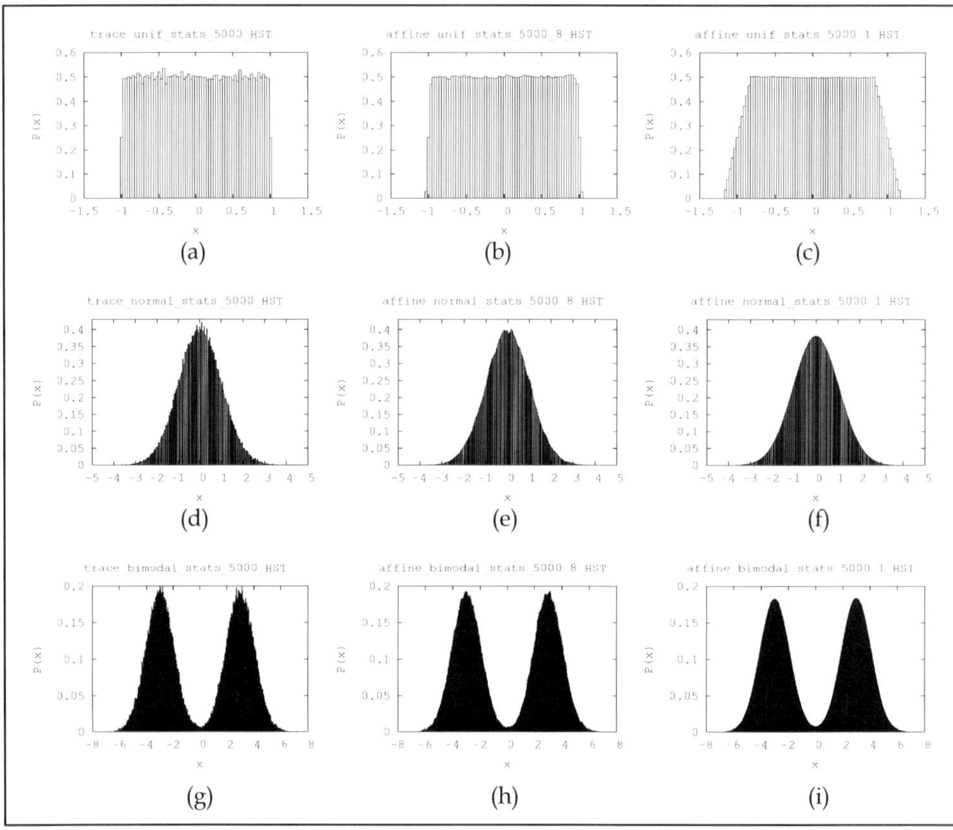

Fig. 6. Distributions generated using traces of numbers, traces of intervals whose widths are set to 1/8 of the variance, and traces of intervals whose widths are set to the variance of the distribution. These traces are applied using the Monte Carlo Method to: (a) - (c) a uniform distribution in [-1, 1]; (d) - (f) a normal distribution with mean 0 and variance 1; (g) - (i) a bimodal distribution with modes 3 and -3 and variance 1.

Figure 7 details the central part and the tails of a normal distribution generated using traces of 100000 numbers and 5000 intervals. It can be observed that the transitions of the histograms are much smoother in the distribution generated using intervals. Although there are slight deviations from the theoretical values, these deviations (approximately 5% in the central part and 15% in the tails) are comparable to the deviations obtained by the numerical trace using 100000 numbers.

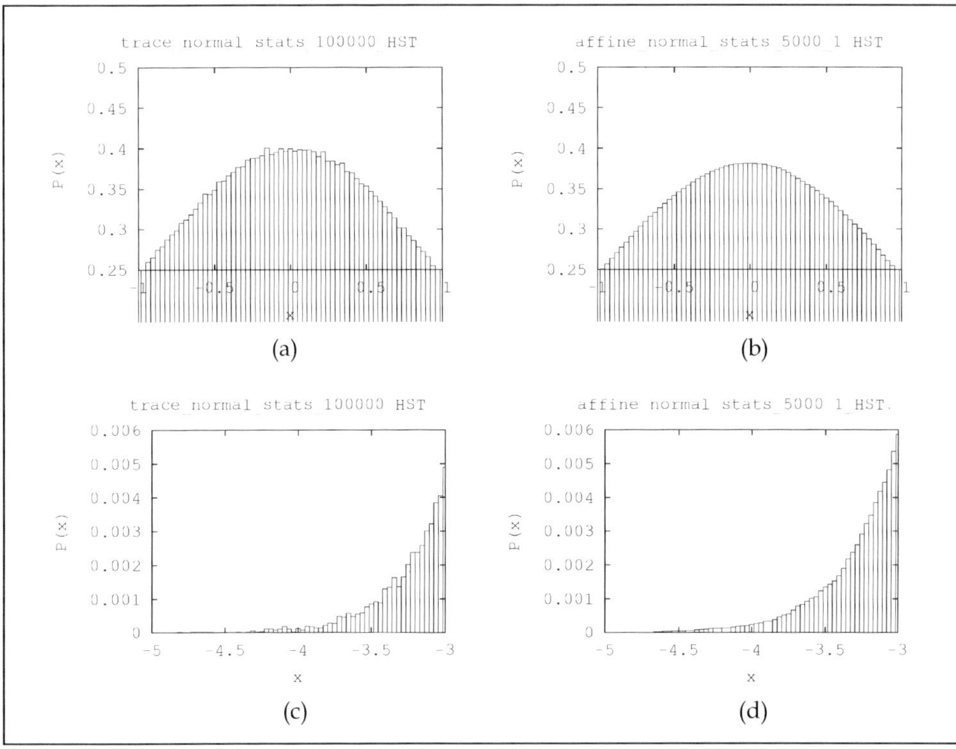

Fig. 7. Details of the normal distribution generated with numerical and interval traces: (a) and (b) Central part of the distribution; (c) and (d) Tail of the distribution.

Therefore, this experiment has shown that signals with normal distributions maintain their shape and statistical parameters in the interval-based simulations, but they require fewer computations to obtain similar degrees of accuracy.

The second part of this section evaluates the variations of the statistical estimators when interval samples of a specific width are used to compute the mean and variance of the random signals in the simulations. Now, the sequence of steps is as follows:

1. Generate the traces of the random samples following the specified PDF, and assign the width of the intervals.
2. Compute the mean and the variance of the trace.
3. Repeat steps 1 and 2 to reduce the variance of the parameters (M times).
4. Group the means and variances of the computed traces, and obtain the estimation and the variations of the statistical parameters.
5. Repeat the previous steps assigning other interval widths.

These steps allow the computation of the means and variances of the estimators, instead of averaging the computed histograms. Step 2 computes the mean and variance of the signals specified in step 1, and step 4 averages the results of the mean and variance of the estimators (in this experiment M is high, to ensure the reliability of estimator statistics).

Figure 8 shows the evolution of the estimators of the mean and the variance as a function of the lengths of the traces (500, 1000 and 5000 samples) and the widths of the intervals

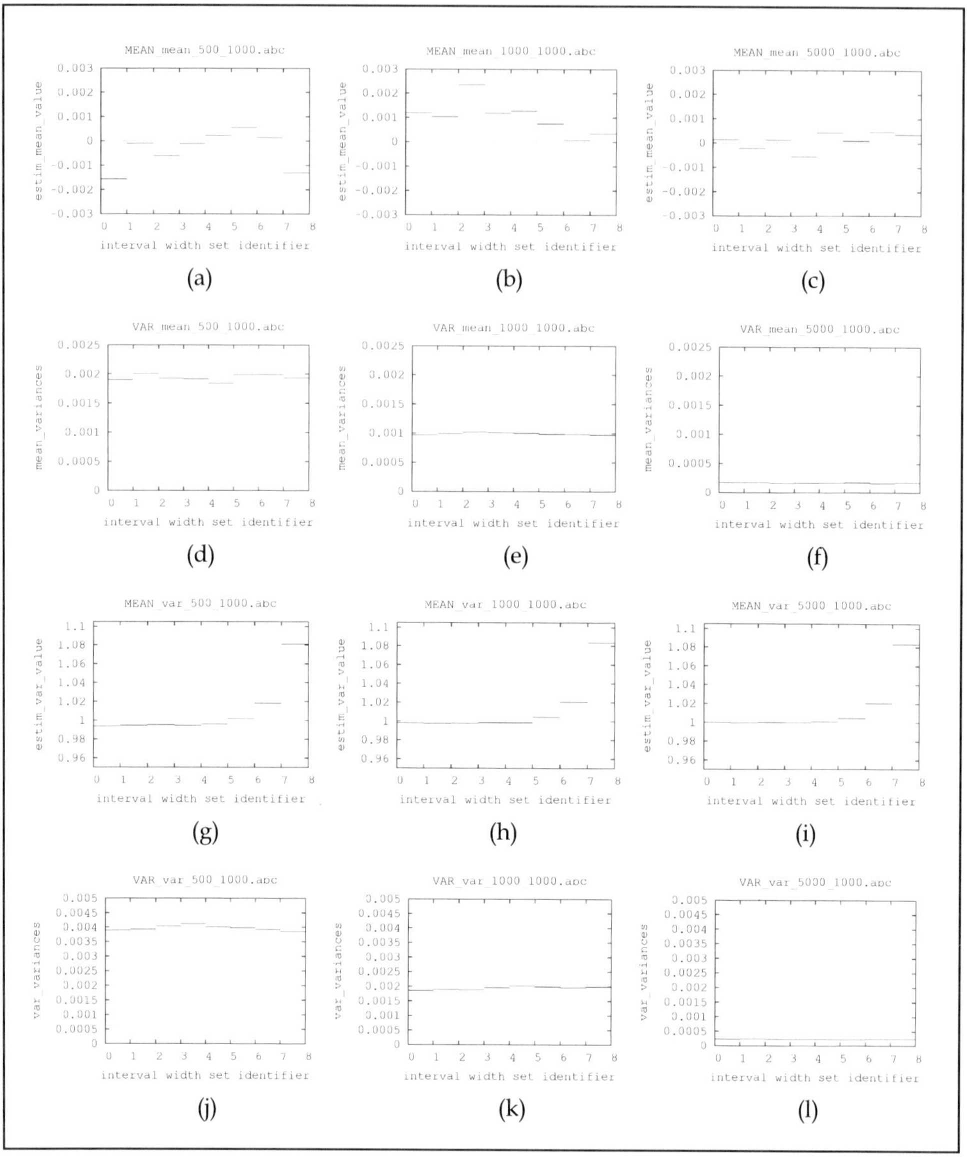

Fig. 8. Analysis of the values provided by the mean and variance interval-based estimators depending on the lengths of the traces: (a) - (c) average of the mean estimator, (d) - (f) variance of the mean, (g) - (i) mean of the variance of the estimator, (j) - (l) variance of the variance. In the four cases, the first column represents the average of 1000 simulations using traces of 500 samples; the second column, of 1000 samples; and the third column, of 5000 samples. The values of the abscissa (1 to 8) respectively represent the interval widths: 0, 1/64, 1/32, 1/16, 1/8, 1/4, 1/2 and 1.

(between 0 and 1). Figures 8.a-c show the averaged mean values computed by the estimator for the previous three lengths. It can be observed that the interval-based estimators tend to obtain slightly better results than the ones of the numerical simulation, although they are roughly of the same order of magnitude. Figures 8.d-f show the variances of these computations. In this case, all the results are approximately equal, and the values decrease (i.e. they become more precise) with longer simulations. Figures 8.g-i show the mean of the variance of the interval-based simulations estimator. It can be observed that when the intervals have small widths, the ideal values are obtained, but when the interval widths are comparable to the variance of the distribution (approximately from 1/4 of its value) the computed values increase significantly the variance of the estimator. Figures 8.j-l show the evolution of the variance estimator. The results are approximately equal in all cases, and decrease with the longer simulations.

Therefore, interval-based simulations tend to reduce the edges of the PDFs and to equalize the other parts of the distribution according to the interval widths. If no additional operation is performed, the edges of the PDFs may change significantly, particularly in uniform distributions. However, since these effects are known, they can possibly be compensated. When using normal signals, the mean and variance of the MC method are similar to the ones obtained in numerical simulations, but the mean of the variance tends to grow for widths above 1/8 of the variance. However, since the improvement in the computed accuracy is small, it does not seem to compensate the increased complexity of the process.

3.4 Discussion on interval-based simulations

Section 3.1 has revealed the importance of using EIA in the interval-based simulation of DSP systems, particularly when they contain feedback loops. It has also shown that traditional IA provides overestimated results due to the cancellation problem. Although the analysis has been performed through a simple example, it can be shown that this problem occurs in most IIR realizations of order equal or greater than two. If there are no dependencies, IA provides the same results than AA, but AA is recommended to be used in the general case. In interval-based simulations of quantized systems, the affine forms must be modified to include all the possible values of the quantization operations without increasing the number of noise terms. The proposed approach solves the overestimation problem, and allows performing accurate analysis of linear systems with feedback loops.

Another important conclusion is that, since the propagation of uncertainties in AA is accurate for linear computations, the features of AA perfectly match with the requirements of the interval-based simulations of digital filters and transforms.

Section 3.2 has evaluated the effects of including one or more uncertainties in a deterministic signal. In addition to determining the maximum and minimum bounds of the variations of the signals in the frequency domain, the analyses have shown the position of the largest uncertainties. Since these amplitudes are not equal, the noise at the output of the FFT does not seem to be white. Moreover, its effect seems to be dependent on the position of the uncertainties in the time domain. The analyses based on interval computations have detected this effect, but they must be combined with statistical techniques to verify the results. A more precise understanding of these effects would help to recover weak signals in environments with low signal-to-noise ratios.

In Section 3.3 the effects of using intervals or extended intervals of a given width in the Monte-Carlo method instead of the traditional numerical simulations has been analyzed. In the first part, the results show that this type of processing softens the edges and the peaks of the PDFs, although these effects can be reduced by selecting smaller intervals or by preprocessing the probability function. In particular, normal distributions are better defined (due to the Central Limit Theorem) and, if the widths of the intervals are significantly smaller than the variance of the distribution, the differences with respect to the theoretical PDFs are smaller than with numerical simulations using the same number of samples. In the second part, the evolution of the mean and the variance of the mean and variance estimators has been studied for a normal PDF using the Monte-Carlo method for different interval widths. These estimators behave similarly than their numerical counterparts (slightly better in most cases), but the mean of the variance increases when the interval widths are greater than 1/8 of the variance of the distribution. Moreover, the increased complexity associated to the interval-based computations does not seem to compensate the small improvement of the accuracy of the statistical estimators in the general case.

In summary, interval-based simulations are preferred when the PDFs are being evaluated, but these improvements are not significant when only the statistical parameters are computed. If the distributions contain edges (for example in the uniform or histogram-based distributions), a pre-processing or post-processing stage can be included to cancel the smoothing performed by the interval sets. Otherwise (such in normally distributed signals), this step can be avoided.

4. Conclusions and future work

This chapter has presented a detailed review of the interval-based simulation techniques and their application to the analysis and design of DSP systems. First, the main extensions of the traditional IA have been explained, and AA has been selected as the most suitable arithmetic for the simulation of linear systems. MAA has also been introduced for the analysis of nonlinear systems, but in this case it is particularly important to keep the number of noise terms of the affine forms under a reasonable limit.

Second, three groups of experiments have been performed. In the first group, a simple IIR filter has been simulated using IA and AA to detail the causes of the oversizing of the IA-based simulations, and to determine why AA is particularly well suited to solve this problem. In the second group, different deterministic traces have been simulated using intervals of different widths in some or all the samples. This experiment has revealed the most sensitive frequencies to the small variations of the signals. In the third group, the effect of including intervals in the computation of the statistical parameters using the Monte-Carlo method has been studied. Thanks to these experiments, it has been shown that interval-based simulations can reduce the number of samples of the simulations, but the edges of the distributions are softened by this type of processing.

Finally, it is important to remark that interval-based simulations can significantly reduce the computation times in the analysis of DSP systems. Due to their features, they are particularly well suited to perform rapid system modeling, verification of the system stability, and fast and accurate determination of finite wordlength effects.

5. Acknowledgment

This work has been partially supported by the Ministerio de Ciencia e Innovación of Spain under project TEC2009-14219-C03-02, and the E.T.S.I. Telecomunicación of the Universidad Politécnica de Madrid under the FastCFD project.

6. References

Alefeld, G. (1984), The Centered Form and the Mean Value Form - A Necessary Condition that They Yield the Range, *Computing*, 33, 165-169.

Armengol, J.; Vehí, J.; Travé-Massuyès, L. & Sainz, M. A. (DX-2001), Application of Multiple Sliding Time Windows to Fault Detection Based on Interval Models, *12th International Workshop on Principles of Diagnosis.*

Berz, M. (1991), *Forward Algorithms for High Orders and Many Variables.*

Berz, M. (1997), *COSY INFINITY Version 8 Reference Manual.*

Berz, M. (1999), *Modern Map Methods in Particle Beam Phisics*, Academic Press, San Diego.

Berz, M.; Bischof, C.; Griewank, A. & Corliss, G. (1996), *Computational Differentiation: Techniques, Applications and Tools.*

Berz, M. & Makino, K. (1998), "Verified Integration of ODEs and Flows Using Differential Algebraic Methods on High-Order Taylor Models", *Reliable Computing*, 4, 4, 361-369.

Berz, M. & Makino, K. (2004), *Taylor Model Research: Results and Reprints.*

Caffarena, G.; López, J.A.; Leyva, G.; Carreras C.; Nieto-Taladriz, O., (2009), Architectural Synthesis of Fixed-Point DSP Datapaths using FPGAs, *International Journal of Reconfigurable Computing*, vol. 2009, 14 pages.

Caffarena, G.; López, J.A.; Leyva, G.; Carreras C.; Nieto-Taladriz, O., (2010), SQNR Estimation of Fixed-Point DSP Algorithms, *EURASIP Journal on Advances in Signal Processing*, vol. 2010, article 21, 12 pages.

Clark, M.; Mulligan, M.; Jackson, D.; & Linebarger, D. (2005), Accelerating Fixed-Point Design for MB-OFDM UWB Systems. *CommsDesign.* Online available at: *http://www.commsdesign.com/showArticle.jhtml?articleID=57703818.*

Coconut_Group (2002), *COCONUT, COntinuous COnstraints - UpdatiNg the Technology - IST Project funded by the European Union.*

Comba, J. L. D. & Stolfi, J. (1993), *Affine Arithmetic and Its Applications to Computer Graphics*, 9-18.

Corliss, G. F. (2004), *G.F. Corliss Homepage*, http://www.eng.mu.edu/corlissg/

Fang, C. F.; Chen, T. & Rutenbar, R. A. (2003), "Floating-point error analysis based on affine arithmetic", *Proc. IEEE Int. Conf. on Acoustics, Speech, and Signal Processing (ICASSP '03)*, 2, 561-564.

Femia, N. & Spagnuolo, G. (2000), "True Worst-Case Circuit Tolerance Analysis Using Genetic Algorithms and Affine Arithmetic", *IEEE Trans. Circuits and Systems I: Fundamental Theory and Applications*, 47, 9, 1285-1296.

Figuereido, L. H. d. & Stolfi, J. (2002), "Affine Arithmetic: Concepts and Applications", *10th GAMM - IMACS International Symposium on Scientific Computing, Computer Arithmetic, and Validated Numerics, SCAN 2002.*

Gardenes, E. (1985), "Modal Intervals: Reasons and Ground Semantics", *Lecture Notes in Computer Science*, 212, 27-35.

Gardenes, E. & Trepat, A. (1980), "Fundamentals of SIGLA, an Interval Computing System over the Completed Set of Intervals", *Computing*, 24, 161-179.

Garloff (1999), *Introduction to Interval Computations.*

GlobSol_Group (2004), *GlobSol Homepage*, http://caneos.mcmaster.ca/solvers/GLOB:GLOBSOL/

Goldenstein, S.; Vogler, C. & Metaxas, D. (2001), *Affine Arithmetic Based Estimation of Cue Distributions in Deformable Model Tracking.*

Hansen, E. R. (1975), *A Generalized Interval Arithmetic*, 7-18.

Hill, T. (2006), *Acceldsp synthesis tool floating-point to fixed-point conversion of matlab algorithms targeting fpgas.* White paper, Xilinx.

Hoefkens, J. (2001), *Verified Methods for Differential Algebraic Equations.*
Hoefkens, J.; Berz, M. & Makino, K. (2001), *Verified High-Order Integration of DAEs and Higher-Order ODEs*, 281-292.
Hoefkens, J.; Berz, M. & Makino, K. (2003), "Controlling the Wrapping Effect in the Solution of ODEs of Asteriods", *Reliable Computing*, 9, 1, 21-41.
Kaucher, E. (1980), "Interval Analysis in the Extended Interval Space IR", *Computing Suppl.*, 2, 33-49.
Kearfott, R. B. (2004), *R.B. Kearfott Homepage.* http://interval.louisiana.edu/kearfott.html
Kreinovich, V. (2004), *The Interval Computations Homepage.* http://www.cs.utep.edu/interval-comp/
Lemke, A.; Hedrich, L. & Barke, E. (Nov. 2002), *Analog Circuit Sizing Based on Formal Methods Using Affine Arithmetic.*
López, J.A. (2004), *Evaluación de los Efectos de Cuantificación en las Estructuras de Filtros Digitales Utilizando Técnicas de Cuantificación Basadas en Extensiones de Intervalos*, Ph.D. Thesis, Univ. Politécnica de Madrid.
López, J.A.; Carreras, C. & Nieto-Taladriz O. (2007), Improved Interval-Based Characterization of Fixed-Point LTI Systems With Feedback Loops, *IEEE Trans. Computer-Aided Design of Integrated Circuits and Systems*, vol. 26, pp. 1923-1933.
López, J.A.; Caffarena, G.; Carreras, C. & Nieto-Taladriz O. (2008), Fast and accurate computation of the roundoff noise of linear time-invariant systems, *IET Circuits, Devices & Systems*, vol. 2, pp. 393-408.
Makino, K. (1998), *Rigurous Analysis of Nonlinear Motion in Particle Accelerators*,
Makino, K. (1999), "Efficient Control of the Dependency Problem Based on Taylor Model Methods", *Reliable Computing*, 5, 1, 3-12.
Makino, K. & Berz, M. (1999), "COSY INFINITY Version 8", *Nuclear Instruments and Methods*, A427, 338-343.
Makino, K. & Berz, M. (2003), "Taylor Models and Other Validated Functional Inclusion Methods", *Int. J. of Pure and Applied Mathematics*, 4, 4, 379-456.
Moore, R. E. (1966), *Interval analysis*, Prentice-Hall.
Moore, R. E. (1962), *Interval Arithmetic and Automatic Error Analysis in Digital Computing*,
Neumaier, A. (1990), *Interval Methods for Systems of Equations.*
Neumaier, A. (2002), "Taylor Forms, Use and Limits", *Reliable Computing*, 9, 43-79.
Ortolf, J. H. (Bonn, 1969), "Eine Verallgemeinerung der Intervallarithmetik", *Geselschaft fuer Mathematik und Datenverarbeitung*, 11, 1-71.
Schichl, H. & Neumaier, A. (2002), *Interval Analysis – Basics.*
Schulte, M. J. (2004), *M.J. Schulte Homepage.* http://www.engr.wisc.edu/ece/faculty/ schulte_michael.html
Shou, H.; Lin, H.; Martin, R. & Wang, G. (2003), "Modified Affine Arithmetic Is More Accurate than Centered Interval Arithmetic or Affine Arithmetic", *Mathematics of Surfaces, 10th IMA International Conference, Proceedings*, 2768 / 2003, 355 - 365.
Shou, H.; Martin, R.; Voiculescu, I.; Bowyer, A. & Wang, G. (2002), "Affine Arithmetic in Matrix Form for Polynomial Evaluation and Algebraic Curve Drawing", *Progress in Natural Science*, 12, 1, 77-81.
SIGLA/X (1999a), *Ground Construction of Modal Intervals.*
SIGLA/X (1999b), *Applications of Interval Analysis to Systems and Control*, 127-227.
Stolfi, J. & Figuereido, L. H. d. (1997), *Self-Validated Numerical Methods and Applications.*
Tupper, J. A. (1996), *Graphing Equations with Generalized Interval Arithmetic.*
Vehí, J. (1998), *Anàlisi i Disseny de Controladors Robustos Mitjançant Interval Modals*,
Walster, G. W. (2004), *G.W. Walster Papers, http://www.mscs.mu.edu/globsol/walster-papers.html.*

Part 4

DSP Algorithms and Discrete Transforms

14

An Emotional Talking Head for a Humoristic Chatbot

Agnese Augello[1], Orazio Gambino[1], Vincenzo Cannella[1], Roberto Pirrone[1], Salvatore Gaglio[1] and Giovanni Pilato[2]

[1]*DICGIM - University of Palermo, Palermo*
[2]*ICAR - Italian National Research Council, Palermo*
Italy

1. Introduction

The interest about enhancing the interface usability of applications and entertainment platforms has increased in last years. The research in human-computer interaction on conversational agents, named also chatbots, and natural language dialogue systems equipped with audio-video interfaces has grown as well. One of the most pursued goals is to enhance the realness of interaction of such systems. For this reason they are provided with catchy interfaces using humanlike avatars capable to adapt their behavior according to the conversation content. This kind of agents can vocally interact with users by using Automatic Speech Recognition (ASR) and Text To Speech (TTS) systems; besides they can change their "emotions" according to the sentences entered by the user. In this framework, the visual aspect of interaction plays also a key role in human-computer interaction, leading to systems capable to perform speech synchronization with an animated face model. These kind of systems are called Talking Heads.

Several implementations of talking heads are reported in literature. Facial movements are simulated by rational free form deformation in the 3D talking head developed in Kalra et al. (2006). A Cyberware scanner is used to acquire surface of a human face in Lee et al. (1995). Next the surface is converted to a triangle mesh thanks to image analysis techniques oriented to find reflectance local minima and maxima.

In Waters et al. (1994) the DECface system is presented. In this work, the animation of a wireframe face model is synchronized with an audio stream provided by a TTS system. An input ASCII text is converted into a phonetic transcription and a speech synthesizer generates an audio stream. The audio server receives a query to determine the phoneme currently running and the shape of the mouth is computed by the trajectory of the main vertexes. In this way, the audio samples are synchronized with the graphics. A nonlinear function controls the translation of the polygonal vertices in such a way to simulate the mouth movements. Synchronization is achieved by calculating the deformation length of the mouth, based on the duration of an audio samples group.

BEAT (Behavior Expression Animation Toolkit) an intelligent agent with human characteristics controlled by an input text is presented in Cassell et al. (2001). A talking head for the Web with a client-server architecture is described in Ostermann et al. (2000). The client application comprises the browser, the TTS engine, and the animation renderer. A coarticulation model determines the synchronization between the mouth movements and the synthesized voice. The 3D head is created with a Virtual Reality Modeling Language (VRML) model.

LUCIA Tisato et al. (2005) is a MPEG-4 talking head based on the INTERFACE Cosi et al. (2003) platform. Like the previous work, LUCIA consists in a VRML model of a female head. It speaks Italian thanks to the FESTIVAL Speech Synthesis System Cosi et al. (2001). The animation engine consists in a modified Cohen-Massaro coarticulation model. A 3D MPEG-4 model representing a human head is used to accomplish an intelligent agent called SAMIR (Scenographic Agents Mimic Intelligent Reasoning) Abbattista et al. (2004). SAMIR is used as a support system to web users. In Liu et al. (2008) a talking head is used to create a man-car-entertainment interaction system. The facial animation is based on a mouth gesture database.

One of the most important features in conversations between human beings is the capability to generate and understand humor: "Humor is part of everyday social interaction between humans" Dirk (2003). Since having a conversation means having a kind of social interaction, conversational agents should be capable to understand and generate also humor. This leads to the concept of *computational humor*, which deals with automatic generation and recognition of humor.

Verbally expressed humor has been analyzed in literature, concerning in particular very short expressions (jokes) Ritchie (1998): a one-liner is a short sentence with comic effects, simple syntax, intentional use of rhetoric devices (e.g., alliteration, rhyme), and frequent use of creative language constructions Stock & Strapparava (2003). Since during a conversation the user says short sentences, one-liners, jokes or gags can be good candidates for the generation of humorous sentences. As a consequence, literature techniques about computational humor regarding one-liners can be customized for the design of a humorous conversational agent.

In recent years the interest in creating humorous conversational agents has grown. As an example in Sjobergh & Araki (2009) an humorous Japanese chat-bot is presented, implementing different humor modules, such as a database of jokes and conversation-based jokes generation and recognition modules. Other works Rzepka et al. (2009) focus on the detection of emotions in user utterances and puns generation.

In this chapter we illustrate a humorous conversational agent, called *EHeBby*, equipped with a realistic talking head. The conversational agent is capable to generate humorous expressions, proposing to the user riddles, telling jokes, ironically answering to the user. Besides, the chatbot is capable to detect, during the conversation with the user, the presence of humorous expressions, listening and judging jokes and react changing the visual expression of the talking head, according to the perceived level of humor. The chatbot reacts accordingly to the user jokes, adapting the expression of its talking head. Our talking head offers a realistic presentation layer to mix emotions and speech capabilities during the conversation with the user. It shows a smiling expression if it considers the user's sentence "funny", indifferent if it does not perceive any humor in the joke, or angry if it considers the joke in poor taste. In the following paragraphs we illustrate both the talking head features and the humorous agent brain.

2. EHeBby architecture

The system is composed by two main components, as shown in figure 1, a reasoner module and a Talking Head (TH) module. The reasoner processes the user question by means of the A.L.I.C.E. (Artificial Linguistic Internet Computer Entity)engine ALICE (2011), which has been extended in order to manage humoristic and emotional features in conversation. In particular the reasoner is composed by a humoristic area, divided in turn in a humoristic recognition area and in a humoristic evocation area, and an emotional area. The first area allows the chatbot to search for the presence of humoristic features in the user sentences, and to produce an appropriate answer. Therefore, the emotional area allows the chatbot to elaborate information related to the produced answer and a correspondent humor level in order to produce the correct information needed for the talking head animation. In particular prosody and emotional information, necessary to animate the chatbot and express emotions during the speech process, are communicated to the Talking Head component. The TH system relies on a web application where a servlet selects the basis facial meshes to be animated, and integrates with the reasoner to process emotion information, expresses using ad hoc AIML (Artificial Intelligence Markup Language) tags, and to obtain the prosody that are needed to control animation. On the client side, all these data are used to actually animate the head. The presented animation procedure allows for considerable computational savings, so both plain web, and mobile client have been implemented.

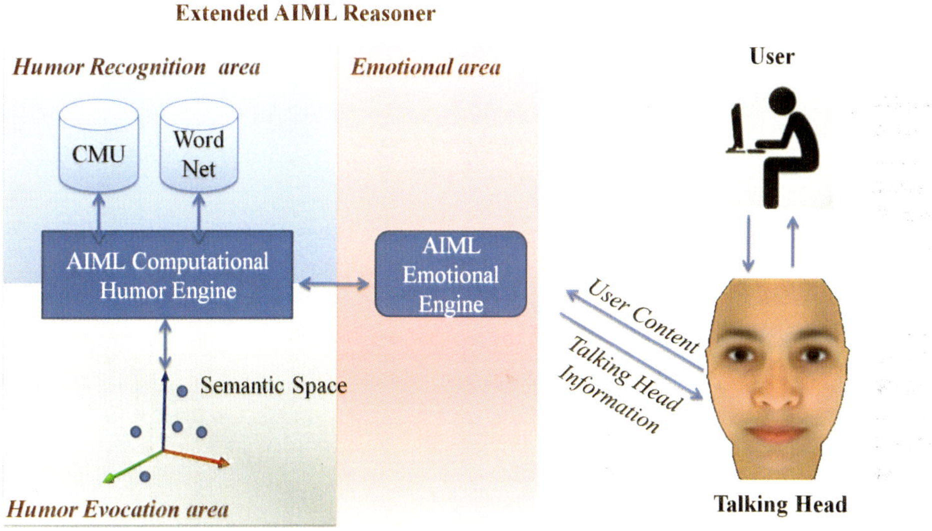

Fig. 1. EHeBby Architecture

3. EHeBby reasoner

The chatbot brain has been implemented using an extended version of the ALICE ALICE (2011) architecture, one of the most widespread conversational agent technologies.
The ALICE dialogue engine is based on a pattern matching algorithm which looks for a match between the user's sentences and the information stored in the chatbot knowledge base. Alice knowledge base is structured with an XML-like language called AIML (Artificial Intelligence Mark-up Language). Standard AIML tags make possible for the chatbot understanding user questions, to properly give him an answer, save and get values of variables, or store the context of conversation. The basic item of knowledge in ALICE is the *category*, which represents a question-answer module, composed a *pattern* section representing a possible user question, and a *template* section which identifies the associated chatbot answer. The AIML

reasoner has been extended defining *ad hoc* tags for computational humor and emotional purposes.
The chabot implements different features, by means of specific reasoning areas, shown in figure 1. The areas called *Humor Recognition Area* and *Humor Evocation Area*, deal with the recognition and generation of humor during the conversation with the user. A set of AIML files, representing the chatbot KB are processed during the conversation. Humor recognition and generation features are triggered when the presence of specific AIML tags is detected. The humorous tags are then processed by a *Computational Humor Engine*, which in turn queries other knowledge repositories, to analyze or generate humor during the conversation. In particular the *AIML Computational Humor Engine* exploits both WordNet MultiWordNet (2010) and the a pronouncing dictionary of the Carnegie Mellon University (CMU) CMU (2010) in order to recognize humorous features in the conversation, and a semantic space in oder to retrieve humorous sentences related to the user utterances. The area called *Emotional Area* deals with the association of chabot emotional reaction to the user sentences. In particular it allows for a binding of a conversation humor level with a set of *ad hoc* created emotional tags, which are processed by the *AIML Emotional Engine* in order to send the necessary information to the Talking Head. In particular in the proposed model we have considered only three possible humor levels, and three correspondent emotional expressions.

3.1 AIML KB

The AIML knowledge base of our humorous conversational agent is composed of four kinds of AIML categories:

1. the standard set of ALICE categories, which are suited to manage a general conversation with the user;
2. a set of categories suited to generate humorous sentences by means of jokes. The generation of humor is obtained writing specific funny sentences in the template of the category.
3. a set of categories suited to retrieve humorous or funny sentences through the comparison between the user input and the sentences mapped in a semantic space belonging to the evocative area. The chatbot answers with the sentence which is semantically closer to the user input.
4. a set of categories suited to to recognize an humorous intent in the user sentences. This feature is obtained connecting the chatbot knowledge base to other resources, like the WordNet lexical dictionary MultiWordNet (2010) and the CMU pronouncing dictionary CMU (2010).
5. a set of categories suited to generate emotional expressions in the talking head.

3.2 Humour recognition area

The humour recognition consists in the identification, inside the user sentences, of particular humorous texts features. According to Mihalcea and Strapparava Mihalcea et al. (2006) we focus on three main humorous features: alliteration, antinomy and adult slang. Special tags inserted in the AIML categories allows the chatbot to execute modules aimed to detect the humorous features.

3.2.1 Alliteration recognition module

The phonetic effect induced by the alliteration, the rhetoric figure consisting in the repetition of a letter, a syllable or a phonetic sound in consecutive words, captures the attention of

people listening it, often producing a funny effect Mihalcea et al. (2006). This module removes punctuation marks and stopwords (i.e. word that do not carry any meaning) from the sentence, and then analyzes its phonetic transcription, obtained by using the CMU dictionary CMU (2010). This technique is aimed at discovering possible repetitions of the beginning phonemes in subsequent words. In particular the module searches the presence of at least three words have in common the first one, the first two or the first three phonemes.
As an example the module consider the following humorous sentences:

```
Veni, Vidi, Visa: I came, I saw, I did a little shopping
Infants don't enjoy infancy like adults do adultery
```

detecting in the first sentence three words having the first phoneme in common, and in the second sentence two pairs of words having the first three phonemes in common. The words infancy and infants have the same following initial phonemes *ih1 n f ah0 n* while the words adultery and adults begin with the following phonemes *ah0 d ah1 l t*.

3.2.2 Antinomy recognition module

This module detects the presence of antinomies in a sentence has been developed exploiting the lexical dictionary WordNet. In particular the module searches into a sentence for:

- a direct antinomy relation among nouns, verbs, adverbs and adjectives;
- an extended antinomy relation, which is an antinomy relation between a word and a synonym of its antonym. The relation is restricted to the adjectives;
- an indirect antinomy relation, which is an antinomy relation between a word and an antonym of its synonym. The relation is restricted to the adjectives.

These humorous sentences contain antinomy relation:

```
A clean desk is a sign of a cluttered desk drawer
Artificial intelligence usually beats real stupidity
```

3.2.3 Adult slang recognition module

This module analyzes the presence of adult slang searching in a set of pre-classified words. As an example the following sentences are reported:

```
The sex was so good that even the neighbors had a cigarette
Artificial Insemination: procreation without recreation
```

3.3 Humor evocation area

This area allows the chatbot to evocate funny sentences that are not directly coded as AIML categories, but that are encoded as vectors in a semantic space, created by means of Latent Semantic Analysis (LSA) Dumais & Landauer (1997). In fact, if none of the features characterizing a humorous phrase is recognized in the sentence through the humor recognition area, the user question is mapped in a semantic space. The humor evocation area then computes the semantic similarity between what is said by the user and the sentences encoded in the semantic space; subsequently it tries to answer to the user with a funny expression which is conceptually close to the user input. This procedure allows to go beyond the rigid pattern-matching rules, generating the funniest answers which best semantically fit the user query.

3.3.1 Semantic space creation

A semantic representation of funny sentences has been obtained mapping them in a semantic space. The semantic space has been built according to a Latent Semantic Analysis (LSA) based approach described in Agostaro (2005)Agostaro (2006). According to this approach, we have created a semantic space applying the truncated singular value decomposition (TSVD) on a $m \times n$ co-occurrences matrix obtained analyzing a specific texts corpus, composed of humorous texts, where each *(i, j)-th* entry of the matrix represents square root of the number of times the *i-th* word appears in the *j-th* document.
After the decomposition we obtain a representation of words and documents in the reduced semantic space. Moreover we can automatically encode in the space new items, such as sentences inserted into AIML categories, humorous sentences and user utterances. In fact, a vectorial representation can be obtained evaluating the sum of the vectors associated to words composing each sentence.
To evaluate the similarity between two vectors v_i and v_j belonging to this space according to Agostaro et al. we use the following similarity measure Agostaro (2006):

$$sim\left(v_i, v_j\right) = \begin{cases} \cos^2\left(v_i, v_j\right) & \text{if } \cos\left(v_i, v_j\right) \geq 0 \\ 0 & otherwise \end{cases} \tag{1}$$

The closer this value is to 1, the higher is the similarity grade. The geometric similarity measure between two items establishes a semantic relation between them. In particular given a vector **s**, associated to a user sentence s, the set CR(s) of vectors sub-symbolically conceptually related to the sentence s is given by the q vectors of the space whose similarity measure with respect to **s** is higher than an experimentally fixed threshold T.

$$CR(s) = v_i | sim(s, v_i) > T \quad with \quad i = 1 \ldots q \tag{2}$$

To each of these vectors will correspond a funny sentence used to build the space. Specific AIML tags called *relatedSentence* and *randomRelatedSentence* allow the chatbot to query the semantic space to retrieve respectively the semantically closer riddle to the user query or one of the most conceptually related riddles. Tha chatbot can also improve its own AIML KB mapping in the evocative area new items like jokes, riddles and so on introduced by the user during the dialogue.

3.4 Emotional area

This area is suited to the generation of emotional expressions in the Talking Head. Many possible models of emotions have been proposed in literature. We can distinguish three different categories of models. The first one includes models describing emotions through collections of different dimensions (intensity, arousal, valence, unpredictability, potency, ...). The second one includes models based on the hypothesis that a human being is able to express only a limited set of primary emotions. All the range of the human emotions should be the result of the combination of the primary ones. The last category includes mixed models, according to which an emotion is generated by a mixture of basic emotions parametrized by a set of dimensions. One of the earlier model of the second category is the model of Plutchik Ekman (1999). He listed the following primary emotions: acceptance, anger, anticipation, disgust, joy, fear, sadness, surprise. Thee emotions can be combined to produce secondary emotions, and in their turn those can be combined to produce ternary emotions. Each emotion can be characterized by an intensity level. After this pioneering model, many other similar

models have been developed. An interesting overview can be found in Ortony (1997). Among the models cited in Ortony (1997), the model by Ekman have been chosen as basis for our work. According to Ekman's model, there are six primary emotions: anger, disgust, fear, joy, sadness, surprise. We have developed a reduced version of this model, including only three of the listed basic emotions: anger, joy, sadness. We selected them as basis to express humor.
At this moment our agent is able to express one of these three emotions at a time, with a variable intensity level. The emotional state of the agent is represented by a couple of values: the felt emotion, and its corresponding intensity. The state is established on the basis of the humor level detected in the conversation. As just said, there are only three possible values for the humor level. These levels have to correspond to a specific emotion in the chatbot, with an intensity level. The correspondence should to be defined according to a collection of psychological criteria. At this moment, the talking head has a predefined behavior for its humorist attitude useful to express these humor levels. Each level is expressed with a specific emotion at a certain intensity level. This emotional patterns represent a default behavior for the agent. The programmer can create a personal version of emotional behavior defining different correspondences between humor levels and emotional intensities. Moreover, he can also program specialized behaviors for single steps of the conversation or single witticisms, as exceptions to the default one.
The established emotional state has to be expressed by prosody and facial expressions. Both of them are generated by the *Emotional Area*. This task is launched by *ad hoc* AIML tags.

4. EHeBby talking head

Our talking head is conceived to be a multi-platform system that is able to speak several languages, so that various implementations have been realized. In what follows the different components of our model are presented: model generation, animation technique, coarticulation, and emotion management.

4.1 Face model generation

The FaceGen Modeler FaceGen (2010) has been used to generate graphic models of the 3D head. FaceGen is a special tool for the creation of 3D human heads and characters as polygon meshes. The facial expressions are controlled by means of numerical parameters. Once the head is created, it can be exported as a Wavefront Technologies .obj file containing the information about vertexes, normals and textures of the facial mesh. The .obj is compliant with the most popular high level graphics libraries such as Java3D and OpenGL. A set of faces with different poses is generated to represent a "viseme", which is related to a phoneme or a groups of phonemes. A phoneme is the elementary speech sound, that is the smallest phonetic unit in a language. Indeed, the spoken language can be thought as a sequence of phonemes. The term "viseme" appeared in literature for the first time in Fischer (1968) and it is equivalent to the phoneme for the face gesture. The viseme is the facial pose obtained by articulatory movements during the phoneme emission. Emotional expressions can be generated by FaceGen also. In our work we have implemented just 4 out of the Ekman basic emotions Ekman & Friesen (1969): joy, surprise, anger, sadness. The intensity of each emotion can be controlled by a parameter or mixed to each other, so that a variety of facial expressions can be obtained. Such "emotional visemes" will be used during the animation task. Some optimizations can be performed to decrease amount of memory necessary to store such a set of visemes. Just the head geometry can be loaded from the .obj file. Lights and virtual camera parameters are set within the programming code. A part of the head mesh can be loaded as a background mesh and after the 3 sub-meshes referred to face, tongue and teeth are loaded.

Indeed, these 3 parts of the head are really involved in the animation. The amount of vertexes can be reduced with a post-processing task with a related decrease of quality, which is not severe if this process involves the back and top sides of the head. Moreover, for each polygon mesh a texture should be loaded, but all the meshes can use the same image file as texture to save memory. A basic viseme can provide both the image texture and the texture coordinates to allow the correct position of the common texture for the other ones.

4.2 Animation

The facial movements are performed by morphing . Morphing starts from a sequence of geometry objects called "keyframes". Each keyframe's vertex translates from its position to occupy the one of the corresponding vertex in the subsequent keyframe. For this reason we have to generate a set of visemes instead of modifying a single head geometric model. Such an approach is less efficient than an animation engine able to modify the shape according to facial parameters (tongue position, labial protrusion and so on) but it simplifies strongly the programming level: First, the whole mesh is considered in the morphing process, and efficient morphing engines are largely present in many computer graphics libraries. Various parameters have to be set to control each morphing step between two keyframes, i.e. the translation time. In our animation scheme, the keyframes are the visemes related to the phrase to be pronounced but they cannot be inserted in the sequence without considering the facial coarticulation to obtain realistic facial movements. The coarticulation is the natural facial muscles modification to generate a succession of fundamental facial movements during phonation. The Löfqvist gestural model described in Löfqvist (1990) controls the audio-visual synthesis; such a model defines the "dominant visemes", which influence both the preceding and subsequent ones. Each keyframe must be blended dynamically with the adjacent ones. The next section is devoted to this task, showing a mathematical model for the coarticulation.

4.2.1 Cohen-Massaro model

The Cohen-Massaro model Cohen & Massaro (1993) computes the weights to control the keyframe animation. Such weights determine the vertexes positions of an intermediate mesh between two keyframes. It is based on the coarticulation, which is the influence of the adjacent speech sounds to the actual one during the phonation. Such a phenomenon can be also considered for the interpolation of a frame taking into account the adjacent ones in such a way that the facial movement appear more natural. Indeed, the Cohen-Massaro model moves from the work by Löfqvist, where a speech segment shows the strongest influence on the organs of articulation of the face than the adjacent segments. Dominance is the name given to such an influence and can be mathematically defined as a time dependent function. In particular, an exponential function is adopted as the dominance function. The dominance function proposed in our approach is simplified with respect to the original one. Indeed, it is symmetric. The profile of a dominance function for given speech segment s and facial parameter p is expressed by the following equation:

$$D_{sp} = \alpha \cdot exp(-\theta \left|\tau\right|^{c}) \tag{3}$$

where α is the peak for $\tau = 0$, θ and c control the function slope and τ is the time variable referred to the mid point of the speech segment duration. In our implementation we set $c = 1$ to reduce the number of parameters to be tuned. The dominance function reachs its maximum value (α) in the mid point of speech segment duration, where $\tau = 0$. In the present approach, we assume that the time interval of each viseme is the same of the duration of the respective phoneme. The coarticulation can be thought as composed by two sub-phenomenons: the pre- and post- articulation. The former consists in the influence of the present viseme on the facial parameters to be used for interpolating the preceding keyframe towards the present one (τ<0). The latter regards the dominance of the next viseme on the parameters used morph the present keyframe towards the next one (τ>0). Our implementation doesn't make use of an animation engine to control the facial parameters (labial opening, labial protrusion and so on) but the interpolation process acts on the translation of all the vertexes in the mesh. The prosodic sequence S of time intervals $[t_{i-1}, t_i[$ associated to each phoneme can be expressed as follows:

$$S = \{f_1 \in [0, t_1[; f_2 \in [t_1, t_2[; \ldots ; f_n \in [t_{n-1}, t_n[\} \tag{4}$$

A viseme is defined "active" when t falls into the corresponding time interval. The preceding and the following visemes are defined as "adjacent visemes". Due to the negative exponential nature of the dominance function, just the adjacent visemes are considered for computing weights. For each time instant, 3 weights must be computed on the basis of the respective dominance functions of 3 visemes at a time. The weights are computed as follows:

$$w_i(t) = D_i(t) = \alpha_i \cdot \exp(-\theta_i \cdot |t - \tau_i|) \tag{5}$$

where τ_i the mid point of the i-th time interval. The w_i must be normalized:

$$w_i^{'}(t) = \frac{w_i(t)}{\sum_{j=-1}^{+1} w_{i-j}(t)} \tag{6}$$

so that for each time instant the coordinates of the interpolating viseme vertexes $v_{int}^{(l)}(t) \in \{V_{int}(t)\}$ will be computed as follows:

$$v_{int}^{(l)}(t) = \sum_{k=i-1}^{i+1} w_i^{'}(t) \cdot v_k^{(l)}(t) \tag{7}$$

where the index l indicates corresponding vertexes in all the involved keyframes.
Our implementation simplifies also this computation. It is sufficient to determine the result of the coarticulation just for the keyframes, because the interpolation is obtained using directly the morphing engine with a linear control function. Once the dominance functions are determined, each coarticulated keyframe is computed and its duration is the same as in the corresponding phoneme.

4.2.2 Diphthongs and dominant visemes

A sequence of two adjacent vowels is called diphthong. The word "euro" contains one diphthong. The vowels in a diphthong must be visually distinct as two separate entities. The visemes belonging to the vowels in a diphthong mustn't influence each other. Otherwise, both the vowel visemes wouldn't be distinguishable due to their fusion. In order to avoid this problem, the slope of the dominance function belonging to each vocal viseme in a diphthong must be very steep (see Fig.2). On the contrary, the sequence vowel-consonant requires a different profile of the dominant function. Indeed, the consonant is heavily influenced by the preceding vowel: a vowel must be dominant with respect to the adjacent consonants, but not with other vowels. As shown in Fig.3, the dominance of a vowel with respect to a consonant is accomplished with a less steep curve than the consonant one.

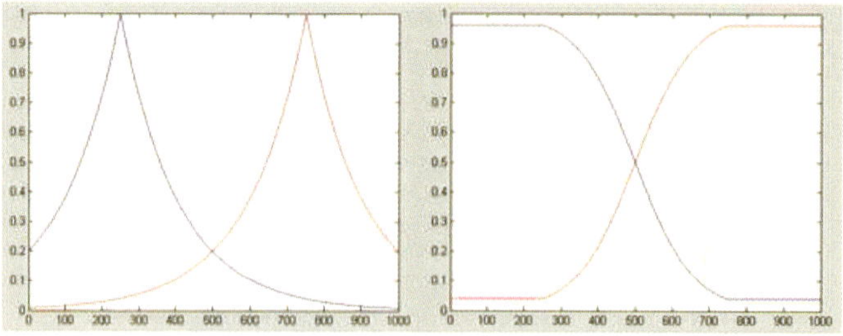

(a) (b)

Fig. 2. The dominance function for the diphtong case (a) and the weights diagram (b) for the diphthong case

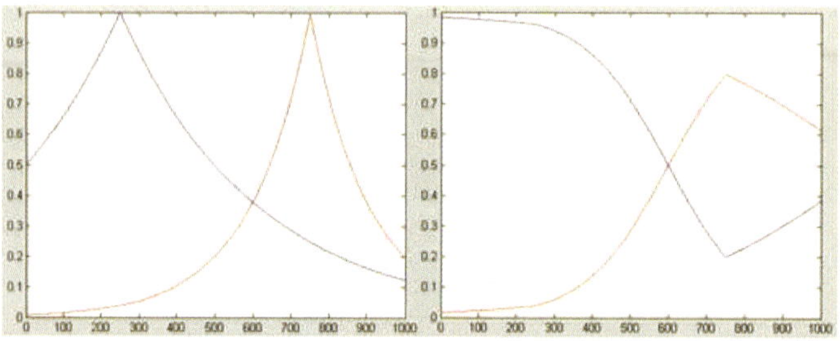

(a) (b)

Fig. 3. The same of Fig.2 for the vowel-consonant case.

4.3 The emotional talking head

Emotions can be considered as particular visemes, called emotional visemes. They must be "mixed" with the phonetic visemes to express an emotion during the facial animation. Such a process can be performed in two different ways. FaceGen can generate also facial modification to express an emotion, so a phonetic viseme can be modified using FaceGen to include an emotion. As result, different sets of modified phonetic visemes can be produced. Each of them are different both as type and intensity of a given emotion. Such a solution is very accurate but it requires an adequate amount of memory and time to create a large emotional/phonetic visemes database. The second approach considers a single emotional viseme whose mesh vertexes coordinate are blended with a viseme to produce a new keyframe. Even though such a solution is less accurate than the previous one, it is less expensive on the computational side, and allows to include and mix "on the fly" emotional and phonetic visemes at run-time.

4.4 Audio streaming synchronization

Prosody contains all the information about the intonation and duration to be assigned to each phoneme in a sentence. In our talking head model, the prosody is provided by Espeak espeak (2010), a multilanguage and multiplatform tool that is able to convert the text into a .pho prosody file. The Talking Head is intrinsically synchronized with the audio streaming because the facial movements are driven by the .pho file, which determines the phoneme (viseme) and its duration. Espeak provides a variety of options to produce the prosody for the language and speech synthesizer to use. As an example it can generate a prosody control for the couple Italian/Mbrola, which is a speech synthesizer based on concatenation of diphones. It takes as input a list of phonemes, together with prosodic information (duration and intonation), and produces an audio file .wav which is played during the facial animation.

5. Some example of interaction

5.1 Example of humorous sentences generation

The following is an example of an humorous dialogue

```
User: What do you think about robots?
EHeBby: Robots will be able to buy happiness,
         but in condensed chip form!!
```

obtained writing an *ad hoc* AIML category:

```
<category>
  <pattern>WHAT DO YOU THINK ABOUT ROBOTS</pattern>
  <template>Robots will be able to buy happiness,
        but in condensed chip form!!
  </template>
< /category>
```

The *pattern* delimits what the user can say. Every time the *pattern* is matched, the corresponding *template* is activated.

5.2 Example of humor recognition

The recognition of humorous sentences is obtained using specific tag inserted into the template, as shown in the following categories:

```
<category>
  <pattern>CAN I TELL YOU A JOKE</pattern>
  <template>Yes you can</template>
< /category>

<category>
  <pattern>*</pattern>
    <that>YES YOU CAN</that>
  <template>
    <srai> <humorlevel> <star/> </humorlevel> <srai>
  </template>
< /category>
```

The second category is activated if the previous answer of the chatbot was "Yes you can" (according to the *that* tag behavior), and the *humorlevel* tag evaluates the level of humor of the sentence matched with the * wildcard (i.e. what the user said). The humor level can assume three different values, *low*, *medium* and *high*. Depending on the humor level value, the category will recursively call, by means of the *srai* tag, another category, which will explicit an emotional tag, including the information needed to the talking head expression, and a *prosody* tag to produce the prosody file. In particular we have extended the AIML language to include three emotional tags *joy*, *anger* and *sadness*. Each of them also includes a mandatory *intensity* attribute. The value assigned to the attribute is a measure of how much that emotion combines to produce the overall emotional state of the chatbot. The called tag link the proper

.m3g emotional file to the talking head, with the corresponding intensity level. In particular a low humor level will be associated to the *anger* tag, a medium level to a *sadness* tag and finally a high level to a *joy* tag. In a default behavior the emotional intensity for each of these emotions is fixed to a specific value. The following example shows the category related to a high humor level.

```
<category>
  <pattern>HIGH *</pattern>
  <template> <think>
   <prosody>
<star/>
</prosody>
    </think>
<joy intensity="080" />
  </template>
</category>
```

This is a dialogue example that shows the chatbot reaction to a funny joke, while in figure 4 is shown the corresponding TH expression:

```
User: Doctor gave a man six months to live. The man couldn't pay
      his bill, so he gave him another six months.
EHeBby: Ehehe! You are very funny!! Good.. your joke is fantastic!!
```

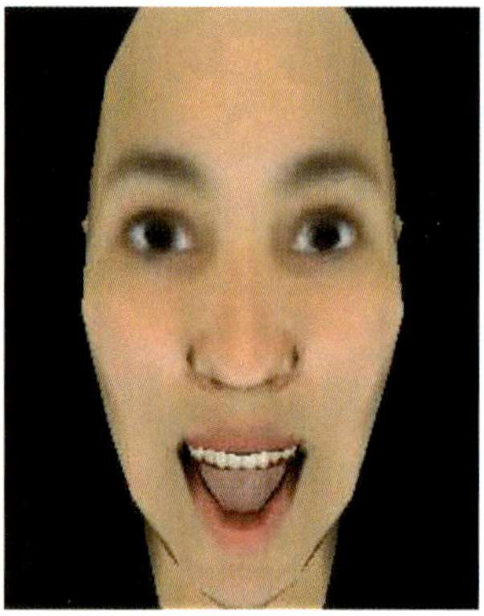

Fig. 4. TH reaction to a funny joke

This other example shows the chatbot reaction to a non-humorous user sentence (see figure 5:

```
User: The book is on the table.
EHeBby: I don't enjoy myself.. !! Your joke is stupid!!
```

The last example shows the chatbot reaction to a rude joke, while in figure 6 is shown a disgusted talking head expression:

```
User: 100,000 sperm and you were the fastest?
EHeBby: What a disgusting joke!!
```

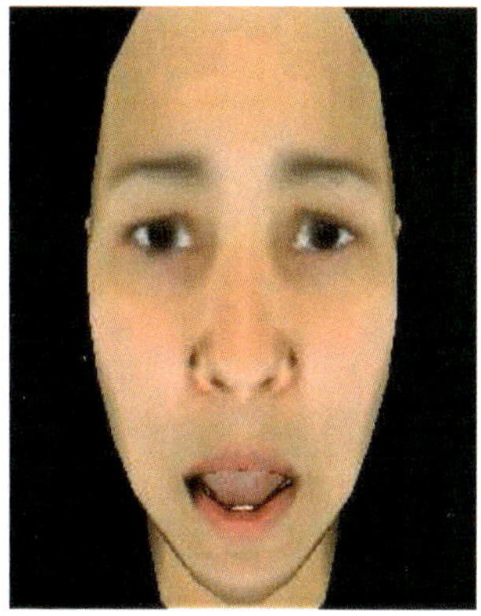

Fig. 5. TH reaction to a non-humorous sentence

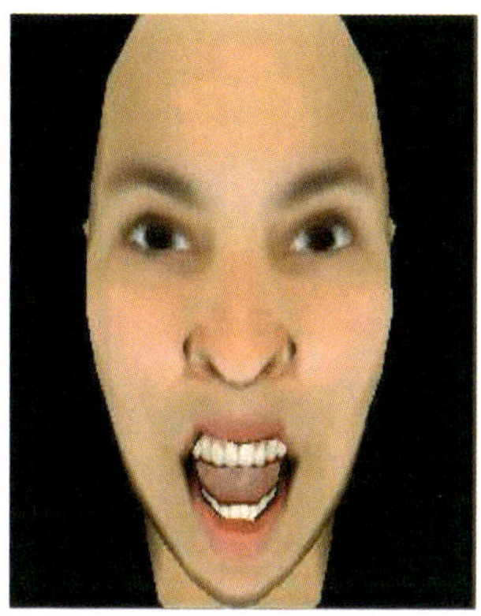

Fig. 6. TH reaction to a rude joke

5.3 AIML categories for the evocation of humorous sentences

The semantic space of the humorous chatbot evocative area has been created by using 3835 funny sentences downloaded from internet one-liners repositories onliners (2011) as well as the standard categories stored in the AIML knowledge base. Their retrieval is obtained through the definition and the use of specifically created AIML tags. As an example, the *relatedSentence* tag in the category template triggers the comparison between the user inputs with the sentences mapped in the semantic space belonging to the evocative area by choosing the sentence semantically closer to the user question. The *randomRelatedSentence* is defined in order to make possible for the chatbot to answer with one of the sentences which is semantically related to the user input.
Below we report the following example of category:

```
<category>
  <pattern>What do you think about *</pattern>
  <template>
   <randomRelatedSentence> <star/></randomRelatedSentence>
```

```
  </template>
< /category>
```

If the user question is:

```
User: What do you think about marriage?
```

The chat-bot will answer randomically in one of the following ways:

```
EHeBby: Marriage is an expensive way of getting your laundry done
        for free.
EHeBby: Benjamin Franklin said "'Keep your eyes wide open before
        marriage, half shut afterwards"'.
EHeBby: Love may be blind but marriage is a real eye-opener.
```

While the standard Alice chat-bot answers:

```
Alice: The state of being married; relation between husband and wife;
       married life.
```

The following dialogue shows a comparison between the answers given with the traditional pattern matching by the chat-bot Alice and the answers given by the evocative chat-bot using the <relatedSentence> tag.

```
User: You are ignorant.
Alice: We can learn new things.
EHeBby: Everybody is ignorant, only on different subjects
User: what is the difference between woman and men?
Alice: I think you already know the answer. I will search for it.
EHeBby: Men always want to be a woman's first love. Women have a more
        subtle instinct; what they like to be is a man's last romance.
User: See you tomorrow.
Alice: Goodbye.
EHeBby: Tomorrow is a mistery...
User: Do you enjoy working?
Alice: I enjoy working 24 hours every day.
EHeBby: I only work to enjoy when I am not working
User: My computer has the windows operating system.
Alice: Where did he get it?
EHeBby: What do Windows and a handgun have in common? Both are
        harmless while not loaded.
User:Are you a pessimist or an optimist?
Alice: Am I a pessimist or an optimist? Yes.
EHeBby: A pessimist is one who makes difficulties of his
        opportunities and an optimist is one who makes
        opportunities of his difficulties. (Harry Truman)
```

5.4 AIML categories for targeting

The humorous chatbot is able to update its own sub-symbolic knowledge through a targeting process, which maps new acquired riddles in the semantic space. Targeting is obtained by means of the ad-hoc created AIML tag *addRiddle*, as shown in the following chunk of AIML code:

```
<category>
  <pattern>Listen this joke *</pattern>
  <template> <humorlevel><star/> </humorlevel>
    <think>
    <addRiddle> <star/> </addRiddle>
    < /think>
  < /template>
< /category>
```

The *addRiddle* tag has the effect to code a sentence introduced by the user as a vector in the semantic space by means of the folding-in procedure. The entire procedure is hidden to the user by the *think* tag. In this manner, the user will see only the chat-bot reaction to his joke.

6. Results

In order to validate the humor recognition, two appropriate data set have been created. The former, called *DataSet1*, is composed of 100 humorous phrases extracted by ad-hoc selected web sites, characterized by the presence of a percentual of humoristic features, as shown in table 1.

Features	Percentual in DataSet1
Alliteration	45%
Direct Antinomy	10%
Extended Antinomy	6%
Indirect Antinomy	5%
Adult Slang	6%
No Feature	28%

Table 1. Humoristic Features Distribution in DataSet1

The latter, called *DataSet2*, is composed of 200 phrases, equally distributed between humorous and non-humorous sentences, where the non-humorous examples are always extracted from Internet and chosen among titles of newspapers, proverbs and sets definitions. The implemented algorithms well performed in humor recongition, as shown in table 2.

DataSet	Correct Classification	No Recognition
DataSet1	66%	34%
DataSet2	73%	27%

Table 2. Humor Recognition in the analyzed DataSets

The results can be considered satisfactory. Moreover if with the humor recognition area cannot identify the humorous level of sentences written by the user, the dialogue can continue in a funny way. In fact, the chatbot exploits the semantic space, which allows retrieving amusing sentences so that the dialogue can continue in a funny manner.
The following results show a set of funny answers extracted by the chatbot exploiting the semantic space, and their semantic closeness to the user entry, compared to the answers given by standard Alice chatbot: Example 1:

```
User: What do you think about chocolate?
Alice: Cocoa flavoring used in a variety of food products.
EHeBby:
```

```
0.94 A quarter-ounce of chocolate = four pounds of fat.
    (Slick's Second Universe Law)
0.82 Equal amounts of dark chocolate and white chocolate,
     is a balanced diet.

User: What can you say about children?
Alice: I will think about it and get back to you.
EHeBby:
0.68 Praise your children openly, reprove them secretly.
    (W. Cecil)
0.68 When speaking to your children: say what you mean,
     mean what you say,mbut don't be mean when you say it.

User: What do you think of life?
Alice: Life is the opposite of death.
EHeBby:
0.987 Life is an echo.
0.986 Life is a roller coaster.

User: Have you ever seen Santa Claus?
Alice: No I haven't.
EHeBby:
0.983 The Bermuda Triangle got tired of warm weather.
      It moved to Finland. Now Santa Claus is missing.
0.746 There are 4 Santa Claus stages:1) When you believe in Santa.
      2) When you donŠt believe in Santa.3) When you are Santa.
      4) When you look like Santa.ve not seen it. What's it like?
```

7. Conclusion

A complete framework for an emotional talking head able to manage humor while conversing with the user has been presented along with its implementation. The whole architecture relies on a suitable AIML-based chatbot, and an animation engine for the talking head.
The chatbot reasoner module is based on an extended AIML architecture where both humor, and emotions can be dealt with using suitable tags. A computational humor engine is able both to detect and to generate humorous sentences. Humor detection relies on the presence of alliteration, antinomy, or adult slang in the user's utterances, which are searched for using suitable thesauri like CMU and WordNet. Generation of humor makes use of a LSA based semantic space where humorous sentences have been placed along with the conversation topics.
The system can also select the control parameters for the animation engine, regarding the mesh deformation due to the emotive state to be expressed, the prosody for controlling speech generation, and the coarticulation model that is used to morph a set of key visemes related to phonemes.
The whole system has been tested on the humor recognition task with satisfactory results. However, our system is currently under development and much work has to be done in order to improve the whole architecture. Humor recognition algorithms can be enhanced, in order to capture different grades of humor, and to fully exploit the different levels of intensity in Talking Head emotional expressions.
The emotion database has to be completed al least with all the six Ekman basic emotions. Moreover, the most recent emotion models Ekman (1999) use more than six basis emotional

states so we plan to investigate these models using compositions of our current emotion database visemes. Finally, also web technology is going along with emotions management, and new standards like the W3C EmotionML emotionML (2011) are going to be released. In consideration of this, we plan to modify our AIML extensions towards these standards in order to enable interoperability with other emotion-oriented web systems.

8. References

Abbattista F, Catucci G, Semeraro G, Zambetta F. SAMIR: A Smart 3D Assistant on the Web. Psychnology Journal, 2(1):43-60, 2004.

F. Agostaro, A. Augello, G. Pilato, G. Vassallo, S. Gaglio. A Conversational Agent Based on a Conceptual Interpretation of a Data Driven Semantic Space. Lecture Notes in Artificial Intelligence, Springer-Verlag GmbH, vol. 3673/2005, pp 381-392, ISSN: 0302-9743.

Francesco Agostaro. Metriche per l'Analisi della Semantica Latente finalizzata ai Modelli del Linguaggio. PhD thesis, Università degli Studi di Palermo. Dipartimento di Ingegneria Informatica, 2006. Supervisor: Prof. S. Gaglio.

Alice website: www.alicebot.org

Cassell J, Vilhjálmsson H H, Bickmore T. BEAT: the Behavior Expression Animation Toolkit. s.l. : Proceedings of the 28th annual conference on Computer graphics and interactive techniques (2001), pp. 477-486. doi:10.1145/383259.383315

CMU Dictionary: (2010) http://www.speech.cs.cmu.edu/cgi-bin/cmudict

Cohen, M. M., and Massaro, D. W. (1993) Modeling coarticulation in synthetic visual speech. In N. M. Thalmann and D. Thalmann (Eds.) Models and Techniques in Computer Animation. pp 139-156. Springer-Verlag.

Cosi P., Tesser F., Gretter R., Avesani C. (2001). FESTIVAL Speaks Italian. In Proceedings Eurospeech 2001, Aalborg, Denmark, September 3-7 2001 (pp. 509-512)

Cosi P., Fusaro A., Tisato G. (2003). LUCIA a New Italian Talking-Head Based on a Modified Cohen-Massaro's Labial Coarticulation Model. In Proceedings of Eurospeech 2003, Geneva, Switzerland, September 1-4, 2003 (pp. 2269-2272).

Heylen Dirk. (2003)Talking Head Says Cheese! Humor as an impetus for Embodied Conversational Agent Research. CHI-2003 WorkShop: Humor Modeling In the Interface.

Dumais Susan T. Thomas K. Landauer (1997). A Solution to Plato's Problem: The Latent Semantic Analysis Theory of the Acquisition, Induction, and Representation of Knowledge. Psychological Review, 104(2)

Pawel Dybala, Michal Ptaszynski, Jacek Maciejewski, Mizuki Takahashi, Rafal Rzepka and Kenji Araki. Multiagent system for joke generation: Humor and emotions combined in humanagent conversation. Journal of Ambient Intelligence and Smart Environments 2 (2010) 31-48. DOI 10.3233/AIS-2010-0053. IOS Press

Ekman, P., and Friesen, W. V (1969). The repertoire of nonverbal behavior: Categories, origins, usage, and coding. Semiotica, 1, 49Ð98.

Ekman, P., Basic Emotions, in Dalgleish, T., Power, M., Handbook of Cognition and Emotion, Sussex, UK: John Wiley and Sons, (1999)

http://www.w3.org/TR/2011/WD-emotionml-20110407/

espeak.sourceforge.net/download.html

Singular Inversions Inc., (2010) FaceGen Modeller: www.facegen.com/modeller.htm

G, Fisher C. Confusions among visually perceived consonants. Journal of Speech and Hearing Research, 11(4):796-804.

Kalra P, Mangili A., Magnetat-Thalmann N, Thalmann D. Simulation of facial muscle actions based on rational free form deformations. SCA '06 Proceedings of the 2006 ACM SIGGRAPH Eurographics symposium on Computer animation ISBN:3-905673-34-7

Löfqvist, A. (1990) Speech as audible gestures. In W.J. Hardcastle and A. Marchal (Eds.) Speech Production and Speech Modeling. Dordrecht: Kluwer Academic Publishers, 289-322.

Lee Y, Terzopoulos D, Waters K. Realistic modeling for facial animation. Proc. ACM SIGGRAPH'95 Conference, Los Angeles, CA, August, 1995, in Computer Graphics Proceedings, Annual Conference Series, 1995, 55-62.

Liu K, Ostermann J. Realistic Talking Head for Human-Car-Entertainment Services. IMA 2008 Informationssysteme für mobile Anwendungen, GZVB e.V. (Hrsg.), pp. 108-118, Braunschweig, Germany

Mihalcea R. and C.Strapparava. (2006) Learning to laugh (automatically): Computational Models for Humor Recognition. Computer Intelligence, Volume 22, 2006

MultiWordNet (2010): http://multiwordnet.itc.it/english/home.php

http://www.oneliners-and-proverbs.com/ and http://www.bdwebguide.com/jokes/1linejokes-1.htm.

Ortony, A. and Turner, T. J. (1990) What's basic about basic emotions? In Psychological Review, Vol. 97, pp. 315–331, ISSN 0033-295X

Ostermann J, Millen D. Talking heads and synthetic speech: an architecture for supporting electronic commerce.. ICME 2000. 2000 IEEE International Conference on Multimedia and Expo, 2000. 71 - 74 vol.1 ISBN: 0-7803-6536-4

Ritchie G. (1998). Prospects for Computational Humour. Pp. 283-291 in Proceedings of 7th IEEE International Workshop on Robot and Huma Communication (ROMAN-98), Takamatsu, Japan, October 1998.

Rafal Rzepka, Wenhan Shi, Michal Ptaszynski, Pawel Dybala, Shinsuke Higuchi, and Kenji Araki. 2009. Serious processing for frivolous purpose: a chatbot using web-mining supported affect analysis and pun generation. In Proceedings of the 14th international conference on Intelligent user interfaces (IUI '09). ACM, New York, NY, USA, 487-488. DOI=10.1145/1502650.1502728 http://doi.acm.org/10.1145/1502650.1502728

Jonas Sjobergh and Kenji Araki. A Very Modular Humor Enabled Chat-Bot for Japanese. Pacling 2009

Stock O. and C.Strapparava. (2003). Getting serious about the development of computational humor. In proceedings of the 8th International Joint Conference on Artificial Intelligence (IJCAI-03) pp. 59-64, Acapulco, Mexico,2003.

Tisato G, Cosi P, Drioli C, Tesser F. INTERFACE: a New Tool for Building Emotive/Expressive Talking Heads. INTERFACE: a New Tool for Building Emotive/Expressive Talking Heads. In CD Proceedings INTERSPEECH 2005 Lisbon, Portugal, 2005 (pp. 781-784).

Waters K, Levergood T M. An automatic lip-synchronization algorithm for synthetic faces. s.l. : MULTIMEDIA '94 Proceedings of the second ACM international conference on Multimedia ISBN:0-89791-686-7

15

Digital Camera Identification Based on Original Images

Dmitry Rublev, Vladimir Fedorov and Oleg Makarevich
Technological Institute of Southern Federal University
Russia

1. Introduction

The development of instruments for copyright protection and pirated copies detection requires new methods of intellectual property protection. Specific of execution of that analysis considerably depends on media type – whether it material or energy and recording device (analog or digital).

1.1 Identification task

To analyse the capability of identification due to direct dependence of ID-procedure on media nature it is feasible to select two groups of media: material (physical bodies) and energetical (physical fields: electric currents, sound fields, etc). The common property of any field is wavelike pattern so it can be named wave media type. Electric current both the power carrier and an media. On the material media the information if fixed by changing physical properties according to character alphabet. Information transfer by material media is done by transfer of changed matter. Information fixation by wavelike media is done by environmental changes. Information transfer by wavelike media is performed by energy transfer. According to abovementioned, the analog recording device identification (for example microcassette dictophone) is done by traces leaved on by material media by on-off impulses, transients, inactivity noises, noise of clear media (magnetic tape), high-frequency current for magnetic bias, speed parameters of deck. For analog cameras that type of parameters includes frame margin, film-feeding and optical system specific features. For printers this type of parameters includes features of methods and alogorithms of rasterizing amd printing methods implementation.
Devices which uses energy media are also identifiable, for example radio transmitting devices are identified by transients of modulated signal.

1.2 Digital recording identification features

Easy bit-to-bit copying process of digital information and inapplicability of traditional "original vs copy" division both with non availability of automated procedures of digital sourcing had led to wide distribution of counterfeit production. Identification based on format features, metadata fields, etc is unreliable because of its removal and forgery simplicity. Use of digital watermarks for content protection is not always possible due to computational complexity of embedding procedure.

Widening of digital audio and videorecording devices distribution and an abrupt increase of storage density had led to situation where the most frequently identification case is identifiable records that are external to identifiable device, leading to complete absence of the primary physical state of the primary "source" and file system properties. Than the rest identification way are the identification based on file format features and identification based on features of recording path and postprocessing.
Copyright protection task operates with the same features, but the signal can be presented after multiple format conversions, which preserve consumption quality but changes the physical representation of original signal, so the identifiable and applicable features are ones containing in digital *content* rather than *format* representation.
Currently questions of identification of analog audio, still images and video recording devices are well researched and are based on traces which the recording device leaves on the carrier in process of writing at change of its physical properties. It is widely used at any carrying out of the expertizes which example is, in particular, phototechnical examination. Phototechnical expert appraisal represents judicial expertize on research of facsimiles of various property and the content, photos (including numeral), paper pictures (photo), for definite purposes of criminal, civil or arbitration legal proceedings. Each picture contains information about the circumstances concerning procedure of manufacturing. Phototechnical expert appraisal is produced with a view of identification of objects by their images photos, photographic materials and laboratory accessories on traces on negatives and positives, ascertainment of type and mark of "unknown" photofilms, detections on photos traces of tampering , ascertainment of other circumstances linked to photographing and handling of photographic materials (photos, photographic paper).
Thus, phototechnical expert appraisal tasks are subdivided on:

- Identification – associated with identification of specific object (a picture, a negative, a film);
- Classification - associated with specific object (a photo) belonging to certain group according to existing classification;
- Diagnostic - associated with determination of object properties (a picture, the facsimile), a method of detection its manufacturer, original form recovery.

1.2.1 Practical tasks of identification

The immediate practical task of identification of records can be put in various variants. In practice of ascertainment and protection of copyrights, and also detections of a source of media object the most often situations are when record on the initial carrier is exposed to identification - ascertainment or a refutation of the fact of an origin of record from the presented device is required, or the record copied on other carrier (possibly with automatic format conversion, compression of dynamic range or other variants of postprocessing) is exposed to identification. In the latter case initial record obtaining, as a rule is complicated, and frequently impossible. It is required to determine a record ownership to the device presented by means of another records set certainly acquired with it.

1.2.2 Digital watermarking as a technique for digital media data identification

The most known decision for maintenance of such protection, in particular the rights to the media information presented in a digital form, is application of digital watermarks (DW). Robust DW represent some information which is built in readout of a signal marked by them. DW, as a rule, contain some authentic code, the information on the proprietor or the operating information for reproduction and copying tools. Unlike usual watermarks, DW can be not only visible, but also (as a rule) invisible because by the nature DW are distortions of a signal, which is intended for perception by the person in the first place, and, hence, for preservation of consumer qualities of protected audiovisual production should be as less as possible. Invisible DW are analyzed by the special decoder which renders the decision on their presence, and if necessary, extracts the hidden message. The most suitable objects of protection by means of DW are static images, files of audio- and the video data[1-3]. DW applications are not limited to information security applications. The basis areas of DW technology can be united in four groups:

- Copy protection;
- Hidden labeling of documents;
- Proof of authenticity of the information;
- Hidden communication channels.

Definition of the received information authenticity, plays a special role in a modern information exchange. Usually the digital signature is used for authentication. However it is not quite appropriate for authentication of multimedia information. The message with attached digital signature should be stored and transferred absolutely precisely, «bit-to-bit», while multimedia information can slightly be changed both at storage (at the expense of compression and due to insufficient correcting ability of a code), and at transfer (influence of single or package errors in a communication channel). Thus its quality remains admissible for the user, but the digital signature will not work, so the addressee cannot distinguish true, though and a little changed message from the completely false one. Besides, the multimedia data can be transformed from one format to another, thus traditional means of definition of integrity also will not work.

It is possible to tell that DW are capable to protect the content of digital audio/video, instead of its digital representation in the form of sequence of bits. An essential lack of the digital signature is also that it is easy to completely remove it from the message and attach the new signature. Signature removal will allow the infringer to refuse authorship or to mislead the lawful addressee concerning authorship of the message. Modern systems of DW are projected so that to minimise possibility of similar infringements without simultaneous essential deterioration of record. DW should be robust or fragile (depending on application) to deliberate and casual influences. If DW is used for authenticity acknowledgement, inadmissible change of the container should lead to DW destruction (fragile DW). If DW contains an identification code, a firm logo, etc. it should remain at the maximum distortions of the container, of course, not leading to essential distortions of an initial signal. Thus, at use DW the basic problem are the attacks, which aim is infringement of their integrity. It is possible to distinguish the following attacks: the attacks directed on DW removal, the geometrical attacks directed on distortion of the container, cryptographic attacks, attacks against the used embedding method and DW checking procedure [4-6]. Researching new methods of embedding DW, robust against malicious attacks is base problem in researching new methods of protection of the multimedia information presented in a digital form.

Along with clear advantages of a digital watermarks embedding, its application demands inclusion of the additional block of embedding in structure of each recording device. For already existing modern mobile digital recording devices it leads to at least updating of the microprogram and it can be impossible if computing resources of the device are limited.

Besides it, embedding worsens consumer characteristics of received record that is not always tolerable, and, at special importance of originality of digital record, can be inadmissible.

Other way of authenticity ascertainment is identification on the basis of recording path features, which are presented in a digital record.

1.3 Digital images creation in photo cameras

The image on a photosensitive matrix of a photocamera is formed after light passage through a lens and the blurring filter (LF-filter), further postprocessing of digital signal received from a matrix [21]. At the analysis of the given circuit it is possible to select the following main sections of a recording path in digital photographic cameras which can be used for identification on a basis of features induced in resultant images [22]. The lens and bayonet joint form identifiable signs (low-frequency defects of the image, vignetting). Usage of the given signs for the automated and automatic identification is inconvenient in view of complexity of their extraction from context and built-in compensating circuits and algorithms in a majority of the modern cameras.

LF-filter ("blurring filter") is applied to lower moire formed due to space sampling by a photomatrix of image components with frequencies near and above Nyquist frequency. The filter forms average and high-frequency stable signs (the shade of the settled dust, filter spot defects). In view of it placement and, in most cases, impossibility of replacement, the features imported by it, are similar to the signs imported by the matrix. The photosensitive matrix unit with ADC forms stable signs in broad band of frequencies (additive and multiplicative noise of a matrix, defects of sensor elements - pointwise, cluster, column, line). In the majority of digital photocameras for color image forming the Bayer's [7] method is used, thus there is only one photosensitive sensor before which the lattice color filter (color filter array - CFA) is placed. Bayer's grid uses layout of filters of three primary colors allocated shown on a picture 1.3, where *R*, *G* and *B* accordingly filters of red, green and blue colors. The number of pixels with filters of green color is twice more than number of pixels for red and blue components, that reflects spectral sensitivity features of a human eye. Along with base Bayer pattern there is a set of other variants of a Bayer's matrix, created for the purpose of increasing sensitivity and color rendition accuracy, generally reached at the expense of space resolution of chromaticity.

Algorithms of interpolation form average and high-frequency features (correlative dependences of adjacent pixels, context-dependent interpolation heuristics).

The non-linear processing including noise reduction, color correction, levels correction (brightness, saturation, contrast). Forms low-frequency (gamma correction) and high-frequency (increase of contour sharpness), equalizing.

Compression stage features at the given stage are features of a used format (JPEG or other) such as specific quantization matrixes, a set and placement of the metadada fields.

In the most general case for the analysis of the image received from the real camera, the only accessible image is image in one of storage formats with lossy compression. On occasion (cameras of the upper consumer segment, semiprofessional and professional) also the RAW-version of the image subjected to correction of matrix defects, or compressed by lossless compression methods (TIFF) the image which has transited all steps of processing, except compression with quality loss can be the accessible.

Thus it is possible to formulate the requirements necessary for practically applicable systems of image identification:

1. A basis of camera-by-image identification is the analysis of features leaved to area of pixels of the given image.
2. As an input image format for creation of image identifying the camera, and subsequent identification of belonging the arbitrary checked image to the camera, the most suitable is the raster format without any compression. In view of that similar formats of representation are last formats at logical level for the majority of visual information output devices. It is possible to convert any format of digital images without quality loss.

Thus, for digital photocameras it is possible to select two classes of features which could be used as a basis for identification:

1. Hardware features are reflections of deviations of characteristics of a sensor control steady in time and the subsequent units of handling, including ADC, as separate device in the received digital image. Generally sensor control signs allow to identify a specific copy of the device. In particular for digital cameras those are defects and deviations within tolerances of separate photosensitive elements, defects of elements of the unit of a photosensitive matrix [16, 20].
2. Features of postprocessing algorithms. The digital image received at output of ADC of digital cameras is then further processed. In digital cameras algorithms of the postprocessing that make the greatest impact on the resulted image are algorithms of image recovery from a mosaic (Bayer) structure of a sensor [17], algorithms of increasing contour sharpness and noise reduction. In the majority of the most widespread photocameras of the lower price segment algorithms of postprocessing can not be switched off and the only image formats accessible outside the camera are JPEG or processed TIFF.

In view of that algorithms of postprocessing are the general sometimes for all models of one vendor [16, 23], for detection by sample-unique features it is necessary to take identification on parameters of an analog section, i.e. on the first class of features.

1.4 Methods of matrix data-to-image conversion

Let's consider used algorithmic primitives of interpolation the colors applied to form the color image in digital photographic cameras.

Let light filters of primary colors are allocated in Bayer's grid according to a picture 1.

The algorithms used for recovery of missing color components, are represent "know-how" of vendors and, as a rule, vary depending on model of the camera and type of a photosensitive matrix. However most often they are constructed on the basis of linear and median filtrations primitives, threshold gradients and persistence of color tone.

r(1,1)	g(1,2)	r(1,3)	g(1,4)	r(1,5)	g(1,6)	...
g(2,1)	b(2,2)	g(2,3)	b(2,4)	g(2,5)	b(2,6)	...
r(3,1)	g(3,2)	r(3,3)	g(3,4)	r(3,5)	g(3,6)	...
g(4,1)	b(4,2)	g(4,3)	b(4,4)	g(4,5)	b(4,6)	...
r(5,1)	g(5,2)	r(5,3)	g(5,4)	r(5,5)	g(5,6)	...
g(6,1)	b(6,2)	g(6,3)	b(6,4)	g(6,5)	b(6,6)	...
...	...	...	...	...	...	...

Fig. 1. Color filter array in the Bayer structure

1.5 Interpolation based on linear filtering

The elementary primitive of color interpolation is the algorithm of a bilinear filtration which is applied to each channel independently. For channel G ("green") the filter kernel represents:

$$k = \frac{1}{4}\begin{bmatrix} 0 & 1 & 0 \\ 1 & 4 & 1 \\ 0 & 1 & 0 \end{bmatrix},$$

And for channels "red" and "blue" accordingly:

$$k = \frac{1}{4}\begin{bmatrix} 1 & 2 & 1 \\ 2 & 4 & 2 \\ 1 & 2 & 1 \end{bmatrix}.$$

Other algorithm of the general application is bicubic interpolation, at which kernels for channels of the primary colors are the following:

$$k_G = \frac{1}{256}\begin{bmatrix} 0 & 0 & 0 & 1 & 0 & 0 & 0 \\ 0 & 0 & -9 & 0 & -9 & 0 & 0 \\ 0 & -9 & 0 & 81 & 0 & -9 & 0 \\ 1 & 0 & 81 & 256 & 81 & 0 & 1 \\ 0 & -9 & 0 & 81 & 0 & -9 & 0 \\ 0 & 0 & -9 & 0 & -9 & 0 & 0 \\ 0 & 0 & 0 & 1 & 0 & 0 & 0 \end{bmatrix},$$

$$k_{R,B} = \frac{1}{256}\begin{bmatrix} 1 & 0 & -9 & -16 & -9 & 0 & 1 \\ 0 & 0 & 0 & 0 & 0 & 0 & 0 \\ -9 & 0 & 81 & 144 & 81 & 0 & -9 \\ -16 & 0 & 144 & 256 & 81 & 0 & -16 \\ -9 & 0 & 81 & 144 & 81 & 0 & -9 \\ 0 & 0 & 0 & 0 & 0 & 0 & 0 \\ 1 & 0 & -9 & -16 & 9 & 0 & 1 \end{bmatrix}.$$

1.6 Interpolation based on color hue constance

Color interpolation can be led also on the basis of assumptions of persistence of color tone in localized areas. Generally, selection of a color tone constant is possible considering property of orderliness of colors within a color circle. Interpolation of a constant of the color tone, offered in [7], is one of the most widespread methods used up to professional cameras. The constant of color tone is defined as a difference between the main color components. At the first stage the algorithm interpolates green channel G, using the bilinear method considered earlier. For an estimation of an error of "red" pixels bilinear interpolation of a difference $R^*(\bullet) - G(\bullet)$, which then incremented by $G(\bullet)$. The channel "blue" is recovered similarly.

1.7 Interpolation based on median filtering

Color interpolation can also be performed by median filtration. Application of the median filter offered in [8] is carried out in two stages. On the first by bilinear interpolation components $R^*(\bullet)$, $G^*(\bullet)$ and $B^*(\bullet)$ are calculated, and then the difference between channels with the subsequent median filtration. Let $M_{rg}(\bullet)$, $M_{gb}(\bullet)$, $M_{rb}(\bullet)$ designate differences after a median filtration. For each pixel sampling of missing colors is estimated as a difference between current value of a component and an appropriate difference after a median filtration.

Recovery of colors can be performed also by the gradient method offered in [9] and for the first time used in photocamera of Kodak DCS 200. The method is based on three-stage process which saves boundaries at interpolation led in a direction, perpendicular their orientation. In the beginning the G-channel along boundaries is interpolated. For example, in case of interpolation of "green" pixel in a position (4,4) horizontal and vertical gradients for "blue" are calculated:

$$H_{4,4} = \left|(b_{4,2} + b_{4,6}) / 2 - b_{4,4}\right|,$$
$$V_{4,4} = \left|b_{2,4} + b_{6.4}\right| / 2 - b_{4,4}.$$

If horizontal gradient $H4,4$ greater than vertical gradient $V4,4$, it specifies to possible boundary in a horizontal direction and then interpolation of value of "green" pixel is performed only in a vertical direction:

$$G(4,4) = (g_{3,4} + g_{5,4}) / 2.$$

And on the contrary. If horizontal and vertical gradients are equal, values of pixels of the "green" channel calculated by averaging four adjacent pixels:

$$G(4,4) = (g_{3,4} + g_{4,3} + g_{4,5} + g_{5,4}) / 4.$$

Missing $R(\bullet)$ and $B(\bullet)$ channels are recovered by interpolation on the basis of constant color tone. For example, the missing blue component of pixels with coordinates (3,4) and (4,4) according to [10] is interpolated by following expressions:

$$B(3,4) = b_{3,3} - G(3,3) + b_{3,5} - G(3,5)) / 2 + G(3,4),$$

$$B(4,4) = (b_{3,3} - G(3,3) + b_{3,5} - G(3,5) + b_{5,3} - G(5,3) + b_{5,5} - G(5,5)) / 4 + G(4,4).$$

1.8 Interpolation based on variable threshold gradients

The method on the basis of the variable gradients activated on a threshold (Threshold Based Variable Number of Gradient) is based on a variable amount of the gradients which usage is controlled by exceeding of threshold values. In the given primitive possibility to use gradients on all eight directions, namely in two horizontal, two vertical (N, S, E and W accordingly) and four diagonal NW, SW, NE and SE are added.

In each direction on a matrix of pixels the gradient for the selected point on the basis of an array of 5x5 adjacent pixels is calculated. The choice of a configuration of a neighborhood is

done by empirically detected feeble dependence of a difference of the calculated gradient from colors and considered pixels.
For example, vertical, horizontal and diagonal gradients for the "red" pixel allocated in a point (3,3) will be equal accordingly:

$$N=|g_{2,3}-g_{4,3}|+|r_{1,3}-r_{3,3}|+|b_{2,2}-b_{4,2}|/2+|b_{2,4}-b_{4,4}|/2+|g_{1,2}-g_{3,2}|/2+|g_{1,4}-g_{3,4}|/2$$
$$E=|g_{3,2}-g_{3,4}|+|r_{3,3}-r_{3,5}|+|b_{2,2}-b_{4,2}|/2+|b_{4,2}-b_{3,4}|/2+|g_{2,3}-g_{2,5}|/2+|g_{4,3}-g_{4,5}|/2$$
$$SW=|b_{2,4}-b_{4,2}|+|r_{5,1}-r_{3,3}|+|g_{2,3}-g_{3,2}|/2+|g_{3,4}-g_{4,3}|/2+|g_{3,2}-g_{4,1}|/2+|g_{4,3}+g_{5,2}|/2.$$

On the basis of a set containing 8 gradients, threshold T is calculated, allowing to define, what directions were used. T it is defined as $T=k_1\cdot min+k_2\cdot(max-min)$, where min and max are the minimum and maximum gradients accordingly, and $k1$ and $k2$ constants. Author's values are $k1=1,5$ and $k2=0,5$. Those directions which gradient is less than a threshold are selected, and for each selected direction mean values for "blue", "red" and "green" are calculated. For example, at coordinates (3,3) mean values for directions N, E, SW are the following:

$$R^N=(r_{1,3}+r_{3,3})/2,\ G^N=g_{2,3},\ B^N=(b_{2,2}+b_{2,4})/2,$$

$$R^E=(r_{3,3}+r_{3,5})/2,\ G^E=g_{3,4},\ B^E=(b_{2,4}+b_{4,4})/2,$$

$$R^{SW}=(r_{3,3}+r_{5,1})/2,\ G^{SW}=(g_{3,2}+g_{4,1}+g_{4,3}+g_{5,2})/4,\ B^{SW}=b_{4,2}.$$

Let's designate mean values red, blue and green as $Ravg$, $Gavg$ $Bavg$ accordingly. Then for the selected pixel mean averaging values for red, dark blue and green in the selected directions will be: $Ravg = (Rs+RE+Rse)/3$, $Gavg = (Gs+GE+GSE)/3$, $Bavg = (BS+BE+BSE)$ (for pixel pixel (3,3) and directions S, E, SE). A final estimation of missing color components levels are: G (3,3) $=r3,3 + (Gavg-Ravg)$ and B (3,3) $=r3,3 + (Bavg+Ravg)$ [11].

2. Cameras identification techniques

2.1 Camera identification based on artifacts of color interpolation

There are several approaches to the implementation of identification systems for digital cameras based on the above characteristics.
In [12] cameras identification is done based on color interpolation features. The recognition process involves the following steps.
Designating $I(\cdot)$ as one of $R(\cdot)$ $G(\cdot)$, $B(\cdot)$ channels provided that the pixel in coordinates (x,y) is correlated linearly with other pixels, it is possible to express value of brightness of a color component as the weighed total of brightness of components of adjacent pixels:

$$I(x,y)=\sum_{i=1}^{N}\alpha_i I(x+\Delta x_i,y+\Delta y_i),\tag{1}$$

Where N is a number of correlated pixels, α_i, Δx_j, Δy_j - weight and offset on an axis x and an axis y of the pixel correlated from i th pixel accordingly. The set of such coordinates Δx_i, Δy_i

allocated between $1 \le i \le N$ is considered as a set of the pixels correlated with adjacent pixels. Considering periodic layout of filters (lattice filters - color filter array (CFA) the given correlations will show periodicity. Being based on it in considered article the assumption about identity of scales of pixel sets with different x and y that a set of the correlated pixels, and according to their weight for each pixel in $I(\bullet)$ are identical.
Let's consider the right member of equation (1.1) as *filter* F applied to $I(\bullet)$, designating operation of a filtration $F(I(\bullet))$ as well as summed averaged square errors from both sides from $I(\bullet)$, we receive:

$$MSE(F(I(\bullet))) = \frac{1}{WH}\sum_{x=1}^{W}\sum_{y=1}^{H}\left|\sum_{i=1}^{N}\alpha_i I(x+\Delta x_i, y+\Delta y_i) - I(x,y)\right|^2 \quad (2)$$

Where H and W - height and width of an image accordingly. Adding the virtual correlated pixel $\alpha_{N+1}=-1$, $\Delta x_{N+1} = \Delta y_{N+1}=0$, the equation (1.2) assumes more arranged air:

$$MSE(F(I(\bullet))) = \frac{1}{WH}\sum_{x=1}^{W}\sum_{y=1}^{H}\left|\sum_{i=1}^{N}\alpha_i I(x+\Delta x_i, y+\Delta y_i)\right|^2 . \quad (3)$$

The extension of the equation (1.3) gives the square form rather $X = \{\alpha_1, \alpha_2, \ldots, \alpha_{N+1}\}^T$:

$$MSE(F(I(\bullet)), I(\bullet)) = X^T A X ,$$

Where

$$A(i,j) = \frac{1}{WH}\sum_{x=1}^{W}\sum_{y=1}^{H} I(x+\Delta x_i, y+\Delta y_i)\cdot I(x+\Delta x_j, y+\Delta y_j),\ 1 \le i,j \le N+1 .$$

The coefficient of a matrix A contains the full information for determination of variable vector X, however, in article obtaining X optionally and for the further analysis enough matrix affirms that A.
It was empirically revealed that the correlated pixels mask shown in a figure 2, yields good result (N=12).
On a following step the analysis of principal components is done.

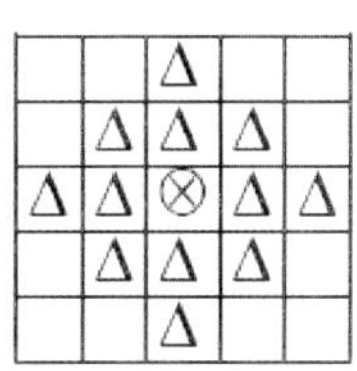

Fig. 2. Correlated pixels mask, where $\otimes$ — is a center and Δ — correlated pixels

Numerical values of elements A after obtaining are normalized:

$$A^*(i,j) = \left[A(i,j) - \overline{A}\right] / \overline{A},$$
$$(1 \le i,j \le N+1),$$

Where $$\bar{A} = \frac{1}{(N+1)^2} \sum_{1 \le i,j \le N+1} A(i,j).$$

Let A^* it is N^2 a-dimensional vector of signs β. Accepting total number of vectors for training of neural network as L $\{\beta_1, \beta_2, ..., \beta_L\}$ and their average according:

$$\bar{\beta} = \frac{1}{L}\sum_{i=1}^{L}\beta_i ,$$

$$\beta^*_i = \beta_i - \bar{\beta},$$
$$i = 1,2,...L.$$

The covariace matrix will be:

$$C = [\beta^*_1, \beta^*_2, ..., \beta^*_L][\beta^*_1, \beta^*_2, ..., \beta^*_L]^T / (L-1).$$

Let eigenvalues and eigenvectors C - $\{\lambda_1, \lambda_2, ... \lambda_L\}$ and $\{\xi_1, \xi_2, ... \xi_L\}$, $\{\zeta_1 \le \zeta_2 \le ...\zeta_{L-1} \le \zeta_L\}$ accordingly.
Eigenvectors corresponding M greatest eigenvalues, form a vector of features $V = [\lambda_1, \lambda_2, ... \lambda_M]^T$. The experiments led by authors, shown that $M = 15$ is enough. β_i^* as a result transforms to $\Gamma_i = V\beta_i^*$ with dimensionality reduction. Recognition of the image belonging to the specific camera was carried out by trained neural network of direct propagation with 15 input neurons, 50 neurons in the hidden layer (with tangential activation function) and one output neuron (sigmoidal activation function). If we denote a set of color interpolation algorithms D: $D^* = D \cup \{\varnothing\}$ where $\varnothing$ is the empty set corresponding initial $I(\cdot)$. Identification by color interpolation consists in defining conversion $d \in D^*$ which with the greatest probability has been fulfilled over $I(\cdot)$, i.e. the purpose is to reference available $I(\cdot)$ to one of classes $|D^*|$ of the learning images set nearest to $I(\cdot)$, in space of conversion characteristics of debayerization. Thus, to each class of images one neural network should be set in correspondence. To select total number of neural networks the following conditions according to authors is necessary to consider. For the different d_i it is necessary to use different neural networs $(d_1, d_2 \in D^*, d_1 \ne d_2)$ considering essential difference of debayerization operators, applied to the channel of "green" and channels "red"/"blue" should use different neural networks for each color component. By authors of a method the best result were shown when three networks was used, one for each channel. Thus the total of neural networks makes $|D^*| = |D| + 1$ for each channel and $3(|D| + 1)$ totally [12]. The given method does not depend on color channels used for identification, on each channel the independent decision which is pseudo-independent, as channels are mutually correlated as shown in [13-14]. Accuracy of identification has been checked by authors on learning and test samplings on 100 images [13]. Accuracy of recognition of 7 algorithms of the interpolation was 100 % (errors of the first and second type are equal to zero). Accuracy of classification by offered methods ot real photocameras made 95-100 % depending on a

photocamera. The method also can be used for check on authenticity of pictures since in those areas at which there are signs of editing the responses of neural network increase by 2-3 times. However, usage of replaced fragments of the image from the same photocamera as a background makes detection impossible [15]. Mehdi Kharrazi, etc. in [16] proposed the method of photocameras identification on the basis of image features. The task of determination of the camera with which help the analyzable picture has been received was thus considered. Proceeding from known sequence of information processing from a photosensitive matrix, it is possible to select two stages, importing the most essential distortions: a stage of debayerization, i.e. full-color image restoration and a postprocessing stage. Totally authors select 34 signs of classification, among them:

- Cross-channel correlation R-G, R-B, B-G (3 scalar features).
- Center of mass for histograms of differences number of pixels with i , $i+1$ and $i-1$ values (3 scalar features).
- Channelwise power channel wise ratio of color components:

$$E_1 = \frac{|G|^2}{|B|^2},\ E_1 = \frac{|G|^2}{|R|^2},\ E_1 = \frac{|B|^2}{|R|^2}.$$

- statistics of wavelet transform (subspace decomposition by quadrature mirror filters and averaging each sub-band) (9 features).

Along with enumerated features metrics of image quality proposed in [16] has been used. All used metrics can be divided into following groups:

- pixelwise difference metrics (MSE, AMSE).
- correlation metrics (for example normalized mutual correlation).
- spectral difference metrics.

To classify vectors the SVM-based classifier has been used. At learning stage 120 of 300 images were used, with 180 at test stage. An average accuracy of camera identification in "1 out of 2" were 98,73% with 88,02% when images were regular photos.

In [17] an identification method based on proprietary interpolation algorithms used in camera. The basis of algorithm is pixel correlation estimation listed in [18] with two estimations: estimation of pixel value by adjacent pixels' values and demosaic kernel used for raw data processing. As precise configuration of area used for interpolation is uknown, several different configurations were used, with additional assumbtion about different interpolation algorithms used in gradient and texturized areas. Camera identification experiments were done on a basis of two cameras: Sony DSC-P51 и Nikon E-2100. It has been acknowledged that filter kernel increase leads to accuracy increase (for kernels 3x3, 4x4, 5x5, accuracies were from 89.3 to 95.7%).

2.2 Camera identification based on matrix defects

Camera identification based on postprocessing algorithms features posess several disadvantages, the most fundamental is impossibility of practical use for one-model camera identification, even in "1 ot of 2" case.

In [19] camera identification method based on defective ("hot" and "dead" pixels) are presented but its effectiveness is limited for cameras without build-in pixel defects correction and "dark frame" subtraction.

Camera identification based on dark frame correction along with obvious advantage of identification of concrete camera sample inherent critical disadvantage namely requirement of dark frames to identify cameras, which makes this method nearly completely useless in practical sense.

In [20] the camera identification method based on non-uniformity of pixel matrix namely different photosensitivity of pixels.

There are many sources of defects and noises which are generated at different image processing stages. Even if sensors form several images of absolutely static scene, the resulted digital representations may posess insignificant alterations of intensity between "same pixel" of image. It appears partly from shot noise [14,15] which is random, and partially because of structure non-uniformity, which is deterministic and slowly changed across even very large sets of image for similar conditions.

Structural non-uniformity presented in every image and can be used for camera identification. Due to similarity of non-uniformity's nature and random noise it is frequently named structural noise.

By averaging multiple images context impact is reduced and structural noises are separated structural matrix noise can be viewed as two components – fixed pattern noise (FPN) and photo-response non-uniform noise (PRNU). Fixed pattern noise is induced by dark currents and defined primarily by pixels non-uniformity in absence of light on sensitive sensor area. Due to additive nature of FPN, modern digital cameras suppress it automatically by subtracting the dark frame from every image [14]. FPN depends on matrix temperature and time of exposure. Natural images primary structural noise component is PRNU. It is caused by pixels non-uniformity (PNU), primarily non-uniform photosensitivity due to non-homogeneity of silicon wafers and random fluctuations in sensor manufacturing process. Source and character of noise induced by pixels non-uniformities make correlation of noise extracted from two even identical matrixes small. Also temperature and humidity don't render influence to PNU-noise. Light refraction on dust particles and optical system also also induced its contribution to PRNU-noise, but these effects are not stable (dust can migrate over the matrix surface, vignette type changes with focal length or lens change) hence, can't be used for reliable identification.

The model of image obtaining process is the following. Let absolute photon number on pixel's area with coordinates (i,j) corresponds x_{ij}, where $i=1..m, j=1..n$, $m \times n$ – photosensitive matrix resolution. If we designate shooting noise as η_{ij}, additive noise due to reading and other noises as ε_{ij}, dark currents as c_{ij}. Then sensor's output y_{ij} is:

$$y_{ij} = f_{ij}(x_{ij} + \eta_{ij}) + c_{ij} + \varepsilon_{ij} . \qquad (*)$$

Here f_{ij} is almost 1 and is multiplicative PRNU-noise.

Final image pixels p_{ij} are completely formed after multiple-stage processing of y_{ij} including, interpolation over adjacent pixels, color correction and image filtering. Many of that operations are non-linear like gamma correction white balance estimation, adaptive Bayer structure interpolation based on strategies for missing color recoveries. So:

$$p_{ij} = P(y_{ij}, N(y_{ij}), i, j),$$

where P is a non-linear function of pixel's value, its coordinates and its neighborhood $N(y_{ij})$.

Structure noise can be suppressed by subtracting additive noise c_{ij}, then dividing pixels value by normalized frame's values:

$$x'_{ij} = (y_{ij} - c_{ij}) / f_{ij}',$$

where x_{ij}' is a corrected pixels value, f_{ij}' is an approximation of f_{ij} by averaging multiple flat-exposure frames $f_{ij}^{(k)}$, $k = 1..K$:

$$f_{ij}' = \frac{\sum_k f_{ij}^{(k)}}{\sum_{i,j,k} f_{ij}^{(k)}} m \cdot n .$$

This operation cannot be done on p_{ij}, only over raw data from photosensitive matrix y_{ij} prior successive image processing.

Properties of pixel non-uniformity noise

To get better understanding influence of structural noise onto resulted images and determine its characteristics in the following experiments were done:

Using ambient light 118 images were made on Canon camera with automatic exposure and focused on infinity. White balance was set to create neutral gray images.

All obtained images possessed pronounced brightness gradient (vignetting). To eliminate that low-frequency distortion the HF-filter with cutoff frequency at $(150/1136)\,\pi$. Then images were averaged thus random noise was suppressed and structural noise summed. Spectrum of the signal resembles white spectrum with decrease of HF-components area, which is explainable as consequences of color interpolation over pixel neighborhood. PNU-noises are not presented in saturated and completely dark areas where FPN prevails. Owing to noise-like of the PNU-components of matrix noise, it is natural to use correlation method for its detection [16].

2.3 Identification based on non-uniformity of pixels sensitivity

In the absence of access in consumer-grade cameras to sensors output y_{ij}, usually it is impossible to extract PNU from gray-frame. However it is possible to approximate noise by averaging multiple images *p(k) k= 1,…,Np*. Process speed-up is performed by filtering and averaging of residual noise *n(k)*:

$$n^{(k)} = p^{(k)} - F(p^{(k)}).$$

Other advantage of operation with residual noise that low-frequency component of PRNU is automatically suppressed. It is obvious that, the more the number of images (N> 50), the less influence of the single source image will take place. Originally, the filter based on wavelet transform was used. So advantages of this method are:

- No access to internals of camera is required;
- Applicable to all cameras built on the basis of photosensitive matrixes.

2.4 Detection based on correlation coefficient

To detect image p belonging to specific camera C it is possible to calculate correlation ρC between residual and structural noise $n=p-F(p)$ for the camera:

$$\rho_C = corr(n, P_C) = \frac{(n-\bar{n})(P_C - \bar{P}_C)}{\|n-\bar{n}\|\|P_C - \bar{P}_C\|}.$$

Now it is possible to define distribution $\rho C\ (q)$ for different images q made by the camera C and distribution $\rho C\ (q\ ')$ for images $q\ '$ made not by the camera C. *Based on* Neumann-Pirson approach and minimizing error rate the reached accuracy of classification made from 78 % to 95 % on 320 images from 9 digital cameras.

2.5 Identification technique of digital recording devices based on correlation of digital images

For development of a technique of identification of photocameras under images it is necessary to consider architecture of prospective system of identification. The system includes units:

- Input format converter;
- Detector of container modifying;
- Feature vector former;
- Feature vector saving;
- Feature vector search and extraction;
- Device identification.

An input format for identification system should be lossless format like full-color BMP to which all images and video streams are convertible. Typical output formats of modern cameras are JPEG and TIFF. In the feature vector former, digital image is converted to the feature vector represents an image for identification an storage purposes.

In the unit of device identification the estimation of likeness of two or more vectors is estimated allowing to accept or reject device similarity hypothesis.

2.5.1 Feature vector forming for digital cameras identification

Feature vector former is based on photosensitive matrix identification techniques, namely PRNU-features. As there will always be both signal and noise (PRNU-components and image context and (or) other noises) it is preferable to use filters to increase signal-noise ratio. To select HF-components, which represent PRNU can be done by Wiener filtering:

$$\mu = \frac{1}{NM} \sum_{n_1, n_2 \in \eta} a(n_1, n_2)$$

$$\sigma^2 = \frac{1}{NM} \sum_{n_1, n_2 \in \eta} a^2(n_1, n_2) - \mu^2$$

$$b(n_1, n_2) = \mu + \frac{\sigma^2 - v^2}{\sigma^2}(a(n_1, n_2) - \mu),$$

where N and M are number of pixels of neighborhood by y and x axis respectively.

$a(n_1, n_2)$ – is a value of pixel with (n_1, n_2) coordinates.

Thus averaged values for specific matrix is:

$$W_{apprx} = \frac{\sum_{N}(I - F(I))}{N},$$

where $F(I)$ is a filter operation.

The best results were achieved by 5×5 mask. It has been shown that Wiener filter provides better separation, comparing with wavelet transform filter in [21]. The offered identification technique has been researched for identification possibility of 13 cameras [22-27], each with 100 images. Images from every camera were divided into 2 sets - training set used for camera fingerprinting and the test set, used for identity check [21]. The central crop of an image with 1024x1024 pixels size was used for identification purposes.

To process an image I for fingerprint creation or identification the color-to-grayscale conversion has been applied. Fingerprint is an averaged sum of all HF-components, forming W_{apprx} value. To check identity of an image I_q, the correlation coefficient is evaluated:

$$p = cc(F(I_q), W_{apprx}),$$

where p — is a correlation coefficient, and cc — cross correlation.

	1	2	3	4	5	6	7	8	9	10	11	12	13
1	**0.0908**	0.0010	0.0042	0.0004	0.0037	0.0026	0.0028	0.0020	0.0040	0.0033	0.0031	0.0032	0.0035
2	0.0006	**0.1494**	-0.0001	0.0015	0.0005	0.0006	-0.0001	0.0000	0.0005	0.0003	0.0004	0.0001	0.0007
3	0.0028	-0.0001	**0.1364**	0.0003	0.0018	0.0004	0.0017	0.0020	0.0030	0.0023	0.0017	0.0024	0.0016
4	-0.0001	0.0009	-0.0004	**0.1889**	0.0002	-0.0007	-0.0004	-0.0009	-0.0000	-0.0003	-0.0007	-0.0001	-0.0005
5	0.0054	0.0019	0.0035	0.0000	**0.0727**	0.0025	0.0042	0.0024	0.0044	0.0022	0.0033	0.0029	0.0038
6	0.0022	0.0004	0.0004	0.0000	0.0015	**0.1423**	0.0006	0.0001	0.0016	0.0009	0.0013	0.0007	0.0036
7	0.0010	-0.0007	0.0005	-0.0001	0.0010	0.0001	**0.2645**	-0.0012	0.0010	0.0008	0.0012	0.0017	0.0014
8	0.0003	0.0008	0.0009	-0.0001	0.0004	-0.0001	-0.0005	**0.7079**	0.0000	0.0012	-0.0001	0.0009	-0.0005
9	0.0049	0.0004	0.0046	0.0001	0.0031	0.0018	0.0027	0.0029	**0.1038**	0.0017	0.0033	0.0036	0.0019
10	0.0027	0.0013	0.0031	0.0002	0.0023	0.0015	0.0017	0.0032	0.0025	**0.1005**	0.0090	0.0030	0.0014
11	0.0011	0.0007	0.0020	-0.0010	0.0013	0.0004	0.0012	-0.0002	0.0006	0.0013	**0.3776**	0.0006	0.0007
12	0.0025	0.0002	0.0023	-0.0006	0.0018	0.0004	0.0023	0.0016	0.0028	0.0016	0.0020	**0.1294**	0.0016
13	0.0015	0.0001	0.0010	-0.0006	0.0017	0.0022	0.0009	-0.0006	0.0008	0.0009	0.0009	0.0003	**0.2747**

Table 1. An averaged correlation coefficients for 13 cameras.

On intersection of columns and lines with identical indexes there are correlation coefficients of images and a fingerprint, received by the same camera. Thus, at matrix coincidence, correlation value is 0.1 - 0.7 and for incoincident cameras is 0.001 - 0.054.

2.6 Image rotation detection based on Radon transform

Photosensitive matrix of a modern digital camera naturally possesses non-uniformity of its elements, both photosensitive and signal amplifiers. As the charge is transferred by columns, the well-known phenomena called banding occurs, resulting high-frequency noise. After image reconstruction process [3] and subjective quality improvements completing, the resulted image is compressed, usually according to JPEG standard, which introduces blocking effect, and contributes regular pattern to rows and columns as well.

This phenomena can be used to detect angle of rotation. To detect an angle of image rotation the Radon transform of its fragment could be performed with successive analysis of Radon projections with Fourier transform. Radon transform can be defined as follows: Let function $f(x, y)$, is defined in D. We will consider some straight line L on a plane xy, crossing area D. Then, integrating function $f(x, y)$ along line L, we receive a projection or linear integral of function f. Integration along all possible lines L on a plane allows to define Radon transform:

$$f^* = Rf = \int_L f(x,y)ds,$$

where ds - an increment of length along L.

For minimization of edge effects impact of analyzed area on high-frequency part of an image it is advisable to apply Radon transform over circular fragment with smoothed borders. Selection of a fragment from the image and smoothing of its borders were done [4] by normalized two-dimensional Gaussian window

$$h(t) = \frac{\exp\left(\frac{-t^2}{2\delta^2}\right)}{\sqrt{2\pi}\cdot\delta}$$

$$\delta = \frac{\sqrt{ln(2)}}{2\pi BT}$$

shown in figure 3.

Refinement to an angle which 90° degrees multiple is possible to make due to uncompensated banding traces, which are consequences of non-uniformity of image brightness component obtained from CCD or CMOS matrixes [2] and traces of compression artifacts. A consequence of the given phenomenon will be unequal level of maxima of a Fourier spectrum obtained from result of Radon transform that allows to select only 2 or (in some cases) 4 angles. Examples of columns spectrograms for a matrix of Radon transformed image fragment 1024x1024 pixel size are represented in figure 4.

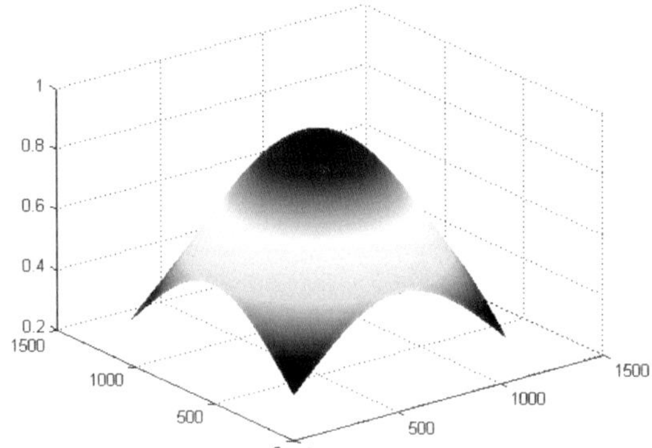

Fig. 3. Two-dimensional normalized Gaussian window used to select an image fragment

In drawing there are maxima at values of rotating angle with added 90° multiples.. At transition from a corner 89° to a corner 90° occurrence of maxima in a peak spectrum is observed. Similar change of character of a peak spectrum gives a possibility to establish value of a image rotation degree.

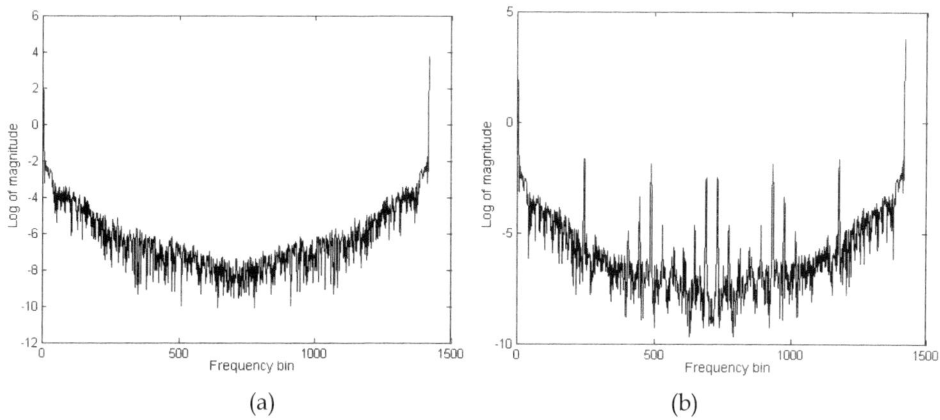

Fig. 4. Spectrums of the projections corresponding to angles 89° and 90° (a-b) for 1024x1024 pixels image fragment

Result (an average of a spectrum for the Radon-transformed image for projection angles from 0° to 360° with 10° step) is presented in figure 5 and a dissection of it – an average of Radon projection spectrograms for image fragment 1024x1024 pixels in figure 5. Local maximums at 10°, 100°, 190°, 280° in figure 6 correspond diagonally-placed maximums in figure 6. To determine the influence of image size change (resize operation), defined as the relation of the linear sizes of the initial image to resulted one on possibility of rotation detection by Radon transform, different scales of original image has also been investigated.

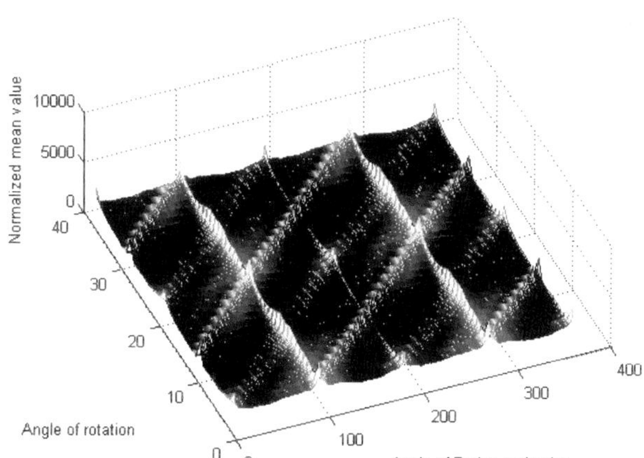

Fig. 5. An average of a spectrum of capacity of transformation of Radon (corners from 0° to 360°) at corners of turn from 0° to 360° with step 10°

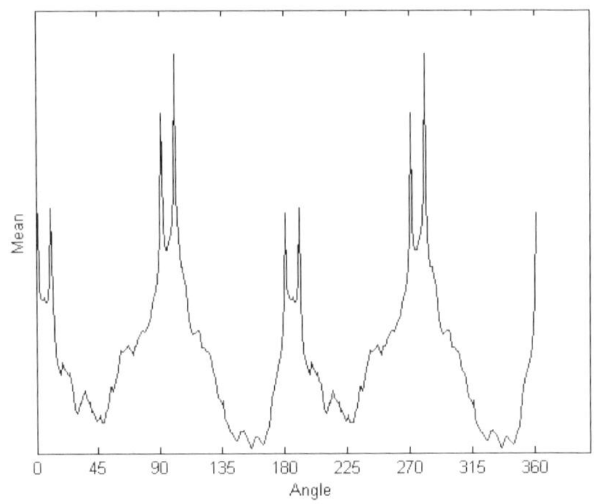

Fig. 6. An average of the spectrogram of projections of Radon of a fragment of the image dimension 1024x1024 pixel for corners from 0 to 360°

In figure 7 the two-dimensional dependence diagram of a magnitude average calculated on a set of image Radon transforms is shown where peak of normalized averaged spectrum located at 10° of Radon transform for angles from [5°..15°], applied to the image with an initial rotation angle of 10° and image scaling factor varied from 1 to 0,1 by 0,1 step. Even at 0.2 scale coefficient the maximum, which corresponds to correct rotation angle, is visible, so rotation operation can be undone. In figure 6 values of spectrogram average of Radon transform (rotation angles from (0°..20°)) for 80 images obtained from one camera and rotated by 10° (a) with histograms (b) are shown.

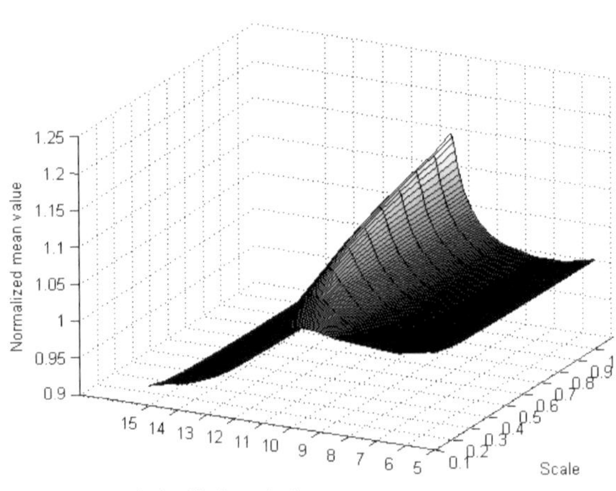

Fig. 7. Dependence of a normalized mean value of averaged spectrum on scaling factor and angle of a Radon projection

3. Conclusion

The methods of digital cameras identification allow defining the fact of origin of digital images from the specific camera. In comparison with artificial digital watermarks embedded by either the special software, or device modification, identification based on innate difference of every single camera allows to identify cameras by the analysis of statistical regularities in digital media. Explicit advantages of the given identification methods are their applicability to images of consumer cameras without necessity of internals access or camera firmware modification. Methods of cameras identification on the basis of processing differences allow to identify cameras vendors. Methods of identification based on non-uniformities of record path allow to identify separate copies of one model of camera. Essential hindrance for correct identification of cameras is scaling and rotation of the images which are exposed to identification process. To ascertain the fact of rotation and its reversing the Radon transform with the subsequent projections processing by Fourier transform can be used.

4. References

[1] Grubunin V. G., Okov I. N., Turintsev I. V. Tsifrovaya steganografiya. - M.: Solon-Press, 2002. - 272 p.
[2] Osborne C., van Schyndel R., Tirkel A. A digital watermark. - IEEE International Conference on Image Processing, 1994. - P. 86-90
[3] Ramkumar M. Data Hiding in Multimedia. PhD Thesis. - New Jersey Institute of Technology,1999. - 72 p.
[4] Hartung F., Su J., Girod B. Spread Spectrum Watermarking: Malicious Attacks and Counterattacks.
[5] Petitcolas F., Anderson R., Kuhn M. Attacks on Copyright Marking Systems. - Lecture Notes in Computer Science, 1998. - P. 218-238.
[6] Langelaar G., Lagendijk R., Biemond J. Removing spatial spread spectrum watermarks by non-linear filtering. - Proceedings EUSIPCO-98, 1998.
[7] B. E. Bayer, Color imaging array, U.S. Patent, No. 3,971,065, 1976
[8] D. R. Cok, Signal processing method and apparatus for producing interpolated chrominance values in a sampled color image signal, U.S.Patent, No. 4,642,678, 1986.
[9] W. T. Freeman, Median filter for reconstructing missing color samples, U.S. Patent, No. 4,724,395, 1988.
[10] C. A. Laroche and M. A. Prescott, Apparatus and method for adaptively interpolating a full color image utilizing chrominance gradients, U.S. Patent, No. 5,373,322, 1994.
[11] Yangjing Long Yizhen Huang Image Based Source Camera Identification using Demosaicking
[12] B. K. Gunturk, Y. Altunbasak and R. M. Mersereau, Color plane interpolation using alternating projections, IEEE Transactions on Image Processing, 11(9):997-1013, 2002.
[13] L. Lam and C.Y. Suen, Application of majority voting to pattern recognition: An analysis of its behavior and performance, IEEE Transactions on Systems, Man and Cybernetics, 27(5):553–568, 1997.

[14] Holst, G. C.: CCD Arrays, Cameras, and Displays, 2nd edition, JCD Publishing & SPIE Pres, USA, 1998.

[15] Janesick, J. R.: Scientific Charge-Coupled Devices, SPIE PRESS Monograph vol. PM83, SPIE–The International Society for Optical Engineering, January, 2001.

[16] Mehdi, K.L. Sencar, H.T. Memon, N. Blind source camera identification. International Conference on Image Processing, 2004, Vol. 1, pp. 709- 712.

[17] Kharrazi, M., Sencar, H. T., and Memon, N.: "Blind Source Camera Identification", Proc. ICIP' 04, Singapore, October 24–27, 2004. pp. 312-317.

[18] A. C. Popescu and H. Farid, "Exposing Digital Forgeries in Color Filter Array Interpolated Images," IEEE Transactions on Signal Processing, Vol. 53, No. 10, part 2, pp. 3948–3959, Oct 2005.

[19] Geradts, Z., Bijhold, J., Kieft, M., Kurosawa, K., Kuroki, K., and Saitoh, N.: "Methods for Identification of Images Acquired with Digital Cameras," Proc. of SPIE, Enabling Technologies for Law Enforcement and Security, vol. 4232, pp. 505–512, February 2001.

[20] Jan Lukás, Jessica J. Fridrich, Miroslav Goljan: Digital camera identification from sensor pattern noise. IEEE Transactions on Information Forensics and Security 1(2): 205-214 (2006)

[21] D.P. Rublev, V.M. Fedorov, A.B. Chumachenko, O.B. Makarevich Identifikatsiya ustroisv tsifrovoi zapisi po osobennostiam sozdavaemykh imi obrazov, Vserossiyskaya konferetsiнa s mezhdunarodnym uchastiem «Problemy informatisatsii obschestva», Nalchik, 2008, p 132-135.

[22] D.P. Rublev, A.B.Chumachenko Identifikaciya cifrovyh fotokamer po karte svetochuvstvitel'nosti matricy. HIII Vserossiiskaya nauchno-prakticheskaya konferenciya "Problemy informacionnoi bezopasnosti v sisteme vysshei shkoly", MIFI, 2007, s 78-79.

[23] Rublev D. P., Fedorov V.M., Chumachenko A.B., Makarevich O.B.; Identifikaciya fotokamer i skanerov po neodnorodnostyam cifrovyh obrazov; Materialy H Mejdunarodnoi nauchno-prakticheskoi konferencii "Informacionnaya bezopasnost'" Taganrog, 2008, 1, s. 238-244

[24] Rublev D.P., Fedorov V.M., Makarevich O.B. Arhitektura setevoi sistemy obnarujeniya vnedrennyh steganograficheskim metodom dannyh v rechevyh soobscheniyah i izobrajeniyah, VII Mejdunarodnaya nauchno-prakticheskaya konferenciya "Informacionnaya bezopasnost'"-2007.

[25] Rublev D. P. Fedorov V.M., Chumachenko A.B., Makarevich O.B., Metody identifikacii cifrovoi apparatury zapisi po ee vyhodnym dannym, Tret'ya mejdunarodnaya nauchno-tehnicheskaya konferenciya "Informacionnye tehnologii v nauke, proizvodstve i obrazovanii", Stavropol', 2008,s. 178-183.

[26] Rublev D. P. Fedorov V.M., Chumachenko A.B., Makarevich O.B., Ustanovlenie avtorskih prav po neodnorodnostyam cifrovyh obrazov, stat'ya v jurnale, Taganrog, Izvestiya YuFU. Tehnicheskie nauki. Tematicheskii vypusk. "Informacionnaya bezopasnost'"2008, 8 (85), s. 141-147.

[27] Rublev D. P., Makarevich O.B., Chumachenko A.B., Fedorov V.M., Ufa, Methods of Digital Recording Device Identification based on Created Records,статья в сборнике ,Proceedings of the 10 International Workshop on Computer Science and Information Technologies, 2008,1,c. 97-100.

16

Study of the Reverse Converters for the Large Dynamic Range Four-Moduli Sets

Amir Sabbagh Molahosseini[1] and Keivan Navi[2]
[1]*Kerman Branch, Islamic Azad University*
[2]*Shahid Beheshti University*
Iran

1. Introduction

The Residue Number System (RNS) is an efficient alternative number system which has been attracted researchers for over three decades. In RNS, arithmetic operations such as addition and multiplication can be performed on residues without carry-propagation between them; resulting in parallel arithmetic and high-speed hardware implementations (Parhami, 2000; Mohan, 2002; Omondi & Premkumar, 2007). Due to this feature, many Digital Signal Processing architectures based on RNS have been introduced in the literature (Soderstrand et al., 1986; Diclaudio et al., 1995; Chaves et al., 2004). In particular, RNS is an efficient method for the implementation of high-speed finite-impulse response (FIR) filters, where dominant operations are addition and multiplication. Implementation issues of RNS-based FIR filters show that performance can be considerably increased, in comparison with traditional two's complement binary number system (Jenkins et al., 1977; Conway et al., 2004; Cardarilli et al., 2007).
As described in (Navi et al., 2011) a typical RNS system is based on a moduli set which is included some pair-wise relatively prime integers. The product of the moduli is defined as the dynamic range, and it denotes the interval of integers which can be distinctively represented in RNS. The main components of an RNS system are a forward converter, parallel arithmetic channels and a reverse converter. The forward converter encodes a weighted binary number into a residue represented number, with regard to the moduli set; where it can be easily realized using modular adders or look-up tables. Each arithmetic channel includes modular adder, subtractor and multiplier for each modulo of set. The reverse converter decodes a residue represented number into its equivalent weighted binary number. The arithmetic channels are working in a completely parallel architecture without any dependency, and this results in a considerable speed enhancement. However; the overhead of forward and reverse converters can counteract this speed gain, if they are not designed efficiently. The forward converters can be designed using efficient methods. In contrast, design of reverse converters have many complexities with many important factors such as conversion algorithm, type and number of moduli.
An efficient moduli set with moduli of the form of powers of two can greatly reduce the complexity of the reverse converter as well as arithmetic channels. Due to this, many different moduli sets have been proposed for RNS which can be categorized based on their

dynamic range. The most well-known $3n$-bit dynamic range moduli set is $\{2^n-1, 2^n, 2^n+1\}$ (Gallaher et al., 1997; Bhardwaj et al., 1998; Wang et al., 2000; Wang et al., 2002). The main reasons for the popularity of this set are its well-form and balanced moduli. However, the modulo 2^n+1 has lower performance than the other two moduli. Hence, some efforts have been done to substitute the modulo 2^n+1 with other well-form RNS moduli, and the resulted moduli sets are $\{2^n-1, 2^n, 2^{n-1}-1\}$ (Hiasat & Abdel-Aty-Zohdy, 1998; Wang et al., 2000b), $\{2^n-1, 2^n, 2^{n+1}-1\}$ (Mohan, 2007; Lin et al., 2008).
The dynamic ranges provided by these three moduli sets are not adequate for recent applications which require higher performance. Two approaches have been proposed to solve this problem. First, using three-moduli sets to provide large dynamic range with some specific forms like $\{2^{\alpha}, 2^{\beta} - 1, 2^{\beta} + 1\}$ where $\alpha<\beta$ (Molahosseini et al., 2008) and $\{2^{2n}, 2^n-1, 2^{n+1}-1\}$ (Molahosseini et al., 2009). Second, using four and five moduli sets to increase dynamic range and parallelism in RNS arithmetic unit. The $4n$-bit dynamic range four-moduli sets are $\{2^n-1, 2^n, 2^n+1, 2^{n+1}+1\}$ (Bhardwaj et al., 1999; Mohan & Premkumar, 2007) and $\{2^n-1, 2^n, 2^n+1, 2^{n+1}-1\}$ (Vinod et al., 2000; Mohan & Premkumar, 2007). Although, these four-moduli sets include relatively balanced moduli, their multiplicative inverses are very complicated, and this results in low-performance reverse converters. Furthermore, some recent applications require even more dynamic range than $4n$-bit. This demand results in introducing new class of moduli sets which have been called *large dynamic range four-moduli sets*. The first one is the $5n$-bit dynamic range moduli set $\{2^n-1, 2^n, 2^n+1, 2^{2n}+1\}$ that was proposed by (Cao et al., 2003). Next, (Zhang et al., 2008) enhanced the dynamic range to $6n$-bit, and introduced the set $\{2^n - 1, 2^n +1, 2^{2n}-2, 2^{2n+1}-3\}$. Moreover, (Molahosseini et al., 2010) proposed the four-moduli sets $\{2^n-1, 2^n, 2^n+1, 2^{2n+1}-1\}$ and $\{2^n-1, 2^n+1, 2^{2n}, 2^{2n}+1\}$ in $5n$ and $6n$-bit dynamic range, respectively.
In this chapter, after an introduction about RNS and reverse conversion algorithms, the architecture of the state-of-the-art reverse cnverters which have been designed for the efficient large dynamic range four-moduli sets $\{2^n-1, 2^n, 2^n+1, 2^{2n}+1\}$, $\{2^n-1, 2^n+1, 2^{2n}, 2^{2n}+1\}$ and $\{2^n-1, 2^n, 2^n+1, 2^{2n+1}-1\}$ will be investigated. Furthermore, a recent contribution about modified version of the four-moduli set $\{2^n-1, 2^n, 2^n+1, 2^{2n+1}-1\}$ that is $\{2^n-1, 2^n+1, 2^{2n}, 2^{2n+1}-1\}$ will be studied. Finally, we present performance comparison in terms of hardware requirements and conversion delays, between the investigated reverse converters.

2. Background

The fundamental part of RNS (Omondi & Premkumar, 2007) is the moduli set $\{P_1,P_2, \ldots,P_n\}$ where numbers are relatively-prime, i.e. $\gcd(P_i,P_j)=1$ for $i\neq j$. The binary weighted number X can be represented as $X=(x_1,x_2, \ldots ,x_n)$, where

$$x_i = X \bmod P_i = |X|_{P_i}, 0 \le x_i < P_i \tag{1}$$

This representation is unique for any integer number X in the range $[0,M\text{-}1]$, where $M=P_1P_2\ldots P_n$ is the dynamic range of the moduli set $\{P_1,P_2, \ldots,P_n\}$ (Taylor, 1984). Addition (subtraction) and multiplication on RNS numbers can be performed in parallel due to the absence of carry propagation between residues.
The famous algorithms for performing reverse conversion are Chinese remainder theorem (CRT), mixed-radix conversion (MRC) and new Chinese remainder theorems (New CRTs).

In order to design a reverse converter, we have to select appropriate moduli set with considering the required parallelism and dynamic range requirements. Next, the moduli should be substituted in one of mentioned conversion algorithm formulas, and the resulted conversion equations should be simplified using some modulo arithmetic properties to reduce hardware complexity. Finally, hardware implementation of the simplified equations can be done using binary hardware's such as full adders, half adders, logic gates or lock-up tables. In the following, we briefly review the formulas of reverse conversion algorithms for four-moduli RNSs. Hence, consider the moduli set (P_1, P_2, P_3, P_4) with corresponding RNS number (x_1, x_2, x_3, x_4).
By CRT (Parhami, 2000) the weighted number X can be calculated by

$$X = \left| \sum_{i=1}^{4} |x_i N_i|_{P_i} M_i \right|_M \tag{2}$$

Where

$$M = P_1 P_2 P_3 P_4 \tag{3}$$

$$M_i = M/P_i \tag{4}$$

$$N_i = |M_i^{-1}|_{P_i} \tag{5}$$

The CRT has capability of parallel implementation; however its final big modulo adder results in inefficient hardware realization if it is considered in direct form.
By MRC (Koc, 1989) the conversion can be done using the following equation:

$$X = v_4 P_3 P_2 P_1 + v_3 P_2 P_1 + v_2 P_1 + v_1 \tag{6}$$

The v_i's coefficients are as follows

$$v_1 = x_1 \tag{7}$$

$$v_2 = \left| (x_2 - v_1) \left| P_1^{-1} \right|_{P_2} \right|_{P_2} \tag{8}$$

$$v_3 = \left| ((x_3 - v_1) \left| P_1^{-1} \right|_{P_3} - v_2) \left| P_2^{-1} \right|_{P_3} \right|_{P_3} \tag{9}$$

$$v_4 = \left| (((x_4 - v_1) \left| P_1^{-1} \right|_{P_4} - v_2) \left| P_2^{-1} \right|_{P_4} - v_3) \left| P_3^{-1} \right|_{P_4} \right|_{P_4} \tag{10}$$

Although MRC implies a sequential process, for two and three-moduli sets it can be lead to simple and efficient reverse conversion equations.
The New CRT-I (Wang, 2000; Molahosseini et al., 2010) uses a more efficient conversion formula

$$X = x_1 + P_1 \left| k_1 (x_2 - x_1) + k_2 P_2 (x_3 - x_2) + k_3 P_2 P_3 (x_4 - x_3) \right|_{P_2 P_3 P_4} \tag{11}$$

Where

$$\left|k_1 \times P_1\right|_{P_2 P_3 P_4} = 1 \tag{12}$$

$$\left|k_2 \times P_1 \times P_2\right|_{P_3 P_4} = 1 \tag{13}$$

$$\left|k_3 \times P_1 \times P_2 \times P_3\right|_{P_4} = 1 \tag{14}$$

Moreover, New CRT-II (Wang, 2000; Molahosseini et al., 2010) provides a tree-like architecture by using the following equations

$$X = Z + P_1 P_2 \left|k_1 (Y - Z)\right|_{P_3 P_4} \tag{15}$$

$$Z = x_1 + P_1 \left|k_2 (x_2 - x_1)\right|_{P_2} \tag{16}$$

$$Y = x_3 + P_3 \left|k_3 (x_4 - x_3)\right|_{P_4} \tag{17}$$

Where

$$\left|k_1 P_1 P_2\right|_{P_3 P_4} = 1 \tag{18}$$

$$\left|k_2 P_1\right|_{P_2} = 1 \tag{19}$$

$$\left|k_3 P_3\right|_{P_4} = 1 \tag{20}$$

The New CRTs have potentiality to create higher performance reverse converters than CRT and MRC particularly for some special four-moduli sets. Hence, many research have been done in the recent years to discover efficient four-moduli sets which can be fitted with properties of New CRTs. In the next sections, we investigate the reverse converters that are previously designed for these four-moduli sets.

3. Reverse converter for the moduli set $\{2^n-1, 2^n, 2^n+1, 2^{2n}+1\}$

The moduli set $\{2^n-1, 2^n, 2^n+1, 2^{2n}+1\}$ was introduced by (Cao et al., 2003). They have used New CRT-I to design a fully adder-based reverse converter. In the following, we briefly review the conversion formulas and hardware architecture of the converter of (Cao et al., 2003). First, consider the moduli set $\{2^n-1, 2^n, 2^n+1, 2^{2n}+1\}$ with corresponding residues (x_1, x_2, x_3, x_4). The residues can be represented in bit-level as below

$$x_1 = (\underbrace{x_{1,n-1} x_{1,n-2} \cdots x_{1,1} x_{1,0}}_{n\ bits})_2 \tag{21}$$

$$x_2 = (\underbrace{x_{2,n-1} x_{2,n-2} \cdots x_{2,1} x_{2,0}}_{n\ bits})_2 \tag{22}$$

$$x_3 = (\underbrace{x_{3,n}x_{3,n-1}\cdots x_{3,1}x_{3,0}}_{n+1\,bits})_2 \tag{23}$$

$$x_4 = (\underbrace{x_{4,2n}x_{4,2n-1}\cdots x_{4,1}x_{4,0}}_{2n+1\,bits})_2 \tag{24}$$

With substituting the required multiplicative inverses and values of moduli, i.e. $P_1=2^n$, $P_2=2^n+1$, $P_3=2^{2n}+1$ and $P_4=2^n-1$ in the New CRT-I formulas (11)-(14), we achieve the following conversion equation:

$$X = x_1 + 2^n \left| \begin{array}{l} 2^{3n}(x_3 - x_2) + (2^{3n-2} + 2^{2n-1} - 2^{n-2})(2^n+1)(x_4 - x_3) \\ +2^{n-2}(2^n+1)(2^{2n}+1)(x_1 - x_4) \end{array} \right|_{2^{4n}-1} \tag{25}$$

This main conversion equation can be simplified based on the following two well-known modulo (2^n-1) arithmetic properties.
Property 1: The residue of a negative residue number ($-v$) in modulo ($2^n - 1$) is the one's complement of v, where $0 \le v < 2^n - 1$ (Hariri et al. 2008).
Property 2: The multiplication of a residue number v by 2^P in modulo ($2^n - 1$) is carried out by P bit circular left shift, where P is a natural number (Hariri et al. 2008).
Now, (25) can be rewritten as follows

$$X = x_1 + 2^n Z \tag{26}$$

Where

$$Z = \left| v_1 + v_2 + v_{31} + v_{32} + v_{41} + v_{42} \right|_{2^{4n}-1} \tag{27}$$

Next, the binary vectors v_i's which have been simplified based on properties 1 and 2 are as below

$$v_1 = \underbrace{x_{1,1}x_{1,0}}_{2}\underbrace{x_{1,n-1}\cdots x_{1,1}x_{1,0}}_{n}\underbrace{x_{1,n-1}\cdots x_{1,1}x_{1,0}}_{n}\underbrace{x_{1,n-1}\cdots x_{1,1}x_{1,0}}_{n}\underbrace{x_{1,n-1}\cdots x_{1,3}x_{1,2}}_{n-2} \tag{28}$$

$$v_2 = \underbrace{\bar{x}_{2,n-1}\cdots \bar{x}_{2,1}\bar{x}_{2,0}}_{n}\underbrace{1\cdots 11}_{3n} \tag{29}$$

$$v_{31} = \underbrace{\bar{x}_{3,1}\bar{x}_{3,0}}_{2}\underbrace{1\cdots 11}_{n-1}\underbrace{\bar{x}_{3,n}\cdots \bar{x}_{3,1}\bar{x}_{3,0}}_{n+1}\underbrace{1\cdots 11}_{n-1}\underbrace{\bar{x}_{3,n}\cdots \bar{x}_{3,3}\bar{x}_{3,2}}_{n-1} \tag{30}$$

$$v_{32} = 0\underbrace{x_{3,n}\cdots x_{3,1}x_{3,0}}_{n+1}\underbrace{0\cdots 00}_{n-1}\underbrace{x_{3,n}\cdots x_{3,1}x_{3,0}}_{n+1}\underbrace{0\cdots 00}_{n-2} \tag{31}$$

$$v_{41} = \underbrace{x_{4,n}\cdots x_{4,1}x_{4,0}}_{n+1}\underbrace{0\cdots 00}_{2n-1}\underbrace{x_{4,2n}\cdots x_{4,n+2}x_{4,n+1}}_{n} \tag{32}$$

$$v_{42} = \underbrace{1\cdots 11}_{n}\underbrace{\bar{x}_{4,2n}\cdots \bar{x}_{4,1}\bar{x}_{4,0}}_{2n+1}\underbrace{1\cdots 11}_{n-1} \tag{33}$$

Therefore, these six operands should be added using a modulo (2^{4n}-1) multi-operand adder which can be realised by four carry-save adders (CSAs) with end-around carry (EAC) followed by a modulo (2^{4n}-1) carry propagate adder (CPA) with EAC (Piestrak, 1994, 1995). The hardware architecture of the resulted converter is shown in Fig. 1.

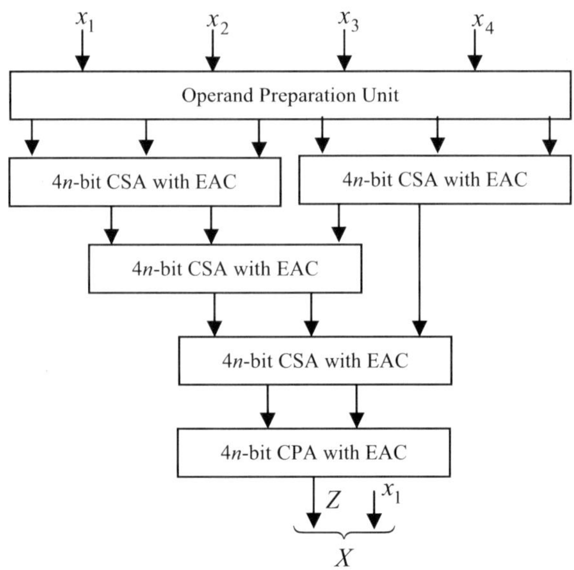

Fig. 1. The converter for moduli set $\{2^n-1, 2^n, 2^n+1, 2^{2n}+1\}$ (Cao et al., 2003)

4. Reverse converter for the moduli set $\{2^n-1, 2^n+1, 2^{2n}, 2^{2n}+1\}$

The moduli set $\{2^n-1, 2^n+1, 2^{2n}, 2^{2n}+1\}$ has been recently introduced by (Molahosseini et al., 2010) to provide large dyamic range (6n-bit), and high-speed reverse converter. Similar to (Cao et al., 2003), the New CRT-I has used to design converter but with different moduli order, i.e. $\{2^{2n}, 2^{2n}+1, 2^n+1, 2^n-1\}$. Therefore, by letting $P_1=2^{2n}$, $P_2=2^{2n}+1$, $P_3=2^n+1$ and $P_4=2^n-1$, and putting the multiplicative inverses in the New CRT-I formulas (11)-(14), we have the following main conversion equation (Molahosseini et al., 2010).

$$X = x_1 + 2^{2n}\left| \begin{array}{l} 2^{2n}(x_2 - x_1) + 2^{2n-1}(2^{2n}+1)(x_3 - x_2) \\ +2^{n-2}(2^{2n}+1)(2^n+1)(x_4 - x_3) \end{array} \right|_{2^{4n}-1} \quad (34)$$

Simplification of this equation can be done as follows

$$X = x_1 + 2^{2n}Z \quad (35)$$

Where

$$Z = \left| v_1 + v_2 + v_{31} + v_{32} + v_4 \right|_{2^{4n}-1} \quad (36)$$

$$v_1 = \underbrace{\bar{x}_{1,2n-1} \cdots \bar{x}_{1,1}\bar{x}_{1,0}}_{2n}\underbrace{\bar{x}_{2,2n} \cdots \bar{x}_{2,2}\bar{x}_{2,1}}_{2n} \quad (37)$$

$$v_2 = \underbrace{x_{2,2n} \cdots x_{2,1}x_{2,0}}_{2n+1}\underbrace{0 \cdots 00}_{2n-1} \quad (38)$$

$$v_{31} = x_{3,1}x_{3,0}\underbrace{0 \cdots 00}_{n-1}\underbrace{x_{3,n} \cdots x_{3,1}x_{3,0}}_{n+1}\underbrace{0 \cdots 00}_{n-1}\underbrace{x_{3,n} \cdots x_{3,3}x_{3,2}}_{n-1} \quad (39)$$

$$v_{32} = \bar{x}_{2,0}\underbrace{\bar{x}_{3,n} \cdots \bar{x}_{3,1}\bar{x}_{3,0}}_{n+1}\underbrace{1 \cdots 11}_{n-1}\underbrace{\bar{x}_{3,n} \cdots \bar{x}_{3,1}\bar{x}_{3,0}}_{n+1}\underbrace{1 \cdots 11}_{n-2} \quad (40)$$

$$v_4 = x_{4,1}x_{4,0}\underbrace{x_{4,n-1} \cdots x_{4,1}x_{4,0}}_{n}\underbrace{x_{4,n-1} \cdots x_{4,1}x_{4,0}}_{n}\underbrace{x_{4,n-1} \cdots x_{4,1}x_{4,0}}_{n}\underbrace{x_{4,n-1} \cdots x_{4,3}x_{4,2}}_{n-2} \quad (41)$$

Therefore, only five operands should be added using three CSAs with EAC followed by a CPA with EAC (Piestrak, 1994, 1995). Hence, in comparison with (Cao et al., 2003) which needed four CSAs, the (Molahosseini et al., 2010) results in reduction of one $4n$-bit CSA with EAC; while providing larger dynamic range. The Fig. 2 shows the hardware implementation of this converter.

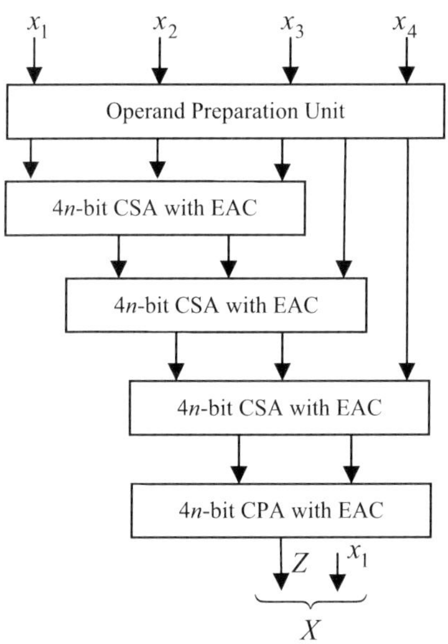

Fig. 2. The converter for moduli set $\{2^n-1, 2^n+1, 2^{2n}, 2^{2n}+1\}$ (Molahosseini et al., 2010)

5. The reverse converter for the moduli set $\{2^n-1, 2^n, 2^n+1, 2^{2n+1}-1\}$

The main disadvantage of the moduli sets $\{2^n-1, 2^n, 2^n+1, 2^{2n}+1\}$ and $\{2^n-1, 2^n+1, 2^{2n}, 2^{2n}+1\}$ is the modulo $2^{2n}+1$. Because, performance of modulo arithmetic circuits for $2^{2n}+1$ is much

lower than the moduli 2^n-1 and 2^n+1. Hence, (Molahosseini et al., 2010) have been substituted $2^{2n}+1$ with well-formed number $2^{2n+1}-1$ that results in introducing the large dynamic range four-moduli set $\{2^n-1, 2^n, 2^n+1, 2^{2n+1}-1\}$. Besides, they have used New CRT-II to design an efficient reverse converter for this moduli set as described below.
With considering $P_1=2^n$, $P_2=2^{2n+1}-1$, $P_3=2^n+1$, $P_4=2^n-1$, and the New CRT-II formulas (15)-(17), we have the following conversion equations (Molahosseini et al., 2010)

$$X = Z + 2^n(2^{2n+1}-1)\left|2^n(Y-Z)\right|_{2^{2n}-1} \tag{42}$$

Where

$$Z = x_1 + 2^n\left|2^{n+1}(x_2-x_1)\right|_{2^{2n+1}-1} \tag{43}$$

$$Y = x_3 + (2^n+1)\left|2^{n-1}(x_4-x_3)\right|_{2^n-1} \tag{44}$$

Simplified versions of these equations have been computed in (Molahosseini et al., 2010). Here, we briefly review the final simplified equations. First, (43) can be rewritten as

$$Z = x_1 + 2^n H \tag{45}$$

Where

$$H = \left|v_1 + v_2\right|_{2^{2n+1}-1} \tag{46}$$

$$v_1 = \underbrace{x_{2,n-1}\cdots x_{2,1}x_{2,0}}_{n}\underbrace{x_{2,2n}\cdots x_{2,n+1}x_{2,n}}_{n+1} \tag{47}$$

$$v_2 = \underbrace{\bar{x}_{1,n-1}\cdots\bar{x}_{1,1}\bar{x}_{1,0}}_{n}\underbrace{1\cdots 11}_{n+1} \tag{48}$$

Next, for simplifying (44) we have

$$Y = x_3 + (2^n+1)K \tag{49}$$

Where

$$K = \left|v_3 + v_4\right|_{2^n-1} \tag{50}$$

$$v_3 = \underbrace{x_{4,0}x_{4,n-1}\cdots x_{4,2}x_{4,1}}_{n} \tag{51}$$

$$v_4 = \begin{cases} \underbrace{\bar{x}_{3,0}\bar{x}_{3,n-1}\cdots\bar{x}_{3,2}\bar{x}_{3,1}}_{n} & \text{if } x_{3,n}=0 \\ 0\underbrace{1\cdots 11}_{n-1} & \text{if } x_{3,n}=1 \end{cases} \tag{52}$$

Eventually, (42) can be computed as below

$$X = Z + 2^n(2^{2n+1} - 1)T = x_1 + 2^n(P - T) \tag{53}$$

Where

$$P = H + 2^{2n+1}T = \underbrace{T_{2n-1} \cdots T_1 T_0}_{2n} \underbrace{H_{2n} \cdots H_1 H_0}_{2n+1} \tag{54}$$

$$T = \left| v_5 + v_6 + v_7 + v_8 \right|_{2^{2n}-1} \tag{55}$$

$$v_5 = \underbrace{x_{3,n-1} \cdots x_{3,1} x_{3,0}}_{n} \underbrace{0 \cdots 00}_{n-1} x_{3,n} \tag{56}$$

$$v_6 = \underbrace{K_{n-1} \cdots K_1 K_0}_{n} \underbrace{K_{n-1} \cdots K_1 K_0}_{n} \tag{57}$$

$$v_7 = \underbrace{\bar{x}_{1,n-1} \cdots \bar{x}_{1,1} \bar{x}_{1,0}}_{n} \underbrace{1 \cdots 11}_{n-1} \bar{H}_{2n} \tag{58}$$

$$v_8 = \underbrace{\bar{H}_{2n-1} \cdots \bar{H}_1 \bar{H}_0}_{2n} \tag{59}$$

Therefore, two modulo adders needed to realize (46) and (50). Moreover, (55) can be implemented using three CSAs with EAC followed by a CPA with EAC. Note that some of the full adders (FAs) of these CPAs and CSAs are simplified to XOR/AND or XNOR/OR pairs due to the constant bits of the inputs. The final result, i.e. (53) can be obtained by a ($4n$+1)-bit binaryadder with '1' carry-in. Fig. 3 presents the reverse converter for the moduli set $\{2^n-1, 2^n, 2^n+1, 2^{2n+1}-1\}$.

6. Reverse converter for the moduli set $\{2^n–1, 2^n+1, 2^{2n}, 2^{2n+1}–1\}$

The moduli set $\{2^n-1, 2^n, 2^n+1, 2^{2n+1}-1\}$ reduces the total delay of RNS arithmetic unit versus the moduli sets $\{2^n-1, 2^n, 2^n+1, 2^{2n}+1\}$ and $\{2^n-1, 2^n+1, 2^{2n}, 2^{2n}+1\}$. However, still the inter-channel delay of modulo $2^{2n+1}-1$ is larger than the other three moduli, i.e. 2^n-1, 2^n and 2^n+1. Due to this, the moduli set $\{2^n-1, 2^n+1, 2^{2n}, 2^{2n+1}-1\}$ has been recently proposed by (Molahosseini & Navi, 2010). The main advantage of this set is that it provides all of the merits of the moduli set $\{2^n-1, 2^n, 2^n+1, 2^{2n+1}-1\}$ while providing larger dynamic range ($6n$-bit). Because, enhancing modulo 2^n to 2^{2n} is not increasing the complexity of the reverse converter.
The converter of (Molahosseini & Navi, 2010) has a two-level architecutre. In other words, they have used a combinatorial conversion algorithm; consisting both CRT and MRC. First, the previous CRT-Based design of reverse converter for the subset $\{2^{2n}, 2^n-1, 2^n+1\}$ (Hiasat & Sweidan, 2004) is used to achieve the weighted equivalent of the residues (x_1, x_2, x_3) as below

$$Z = x_1 + 2^{2n}Y \tag{60}$$

Where

$$Y = \left| v_1 + v_2 + v_3 + v_4 \right|_{2^{2n}-1} \tag{61}$$

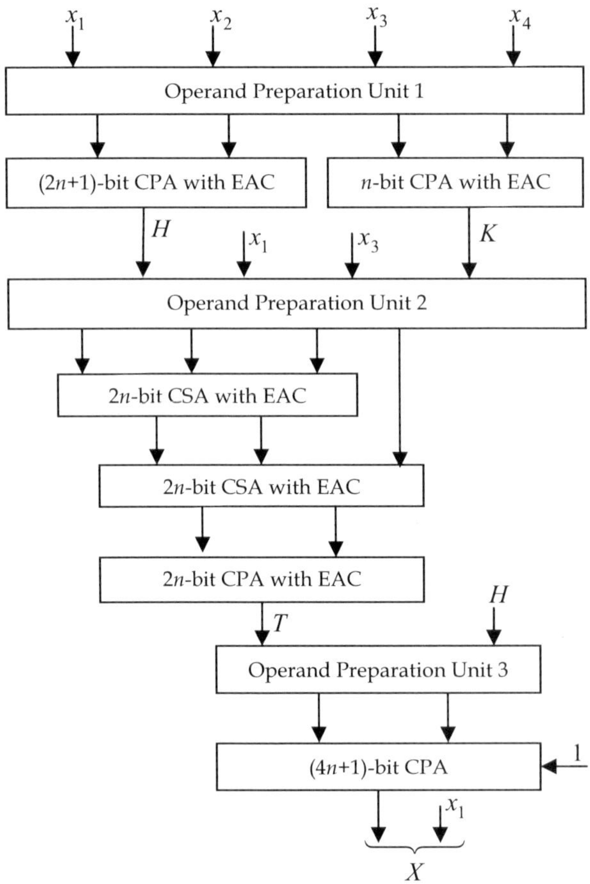

Fig. 3. The converter for moduli set $\{2^n-1, 2^n, 2^n+1, 2^{2n+1}-1\}$ (Molahosseini et al., 2010)

$$v_1 = \underbrace{\bar{x}_{1,2n-1} \cdots \bar{x}_{1,1}\bar{x}_{1,0}}_{2n} \tag{62}$$

$$v_2 = \underbrace{x_{2,0}x_{2,n-1} \cdots x_{2,1}}_{n}\underbrace{x_{2,0}x_{2,n-1} \cdots x_{2,1}}_{n} \tag{63}$$

$$v_3 = x_{3,0}\underbrace{\bar{x}_{3,n-1} \cdots \bar{x}_{3,1}\bar{x}_{3,0}}_{n}\underbrace{x_{3,n-1} \cdots x_{3,2}x_{3,1}}_{n-1} \tag{64}$$

$$v_4 = \bar{x}_{3,n}\underbrace{0 \cdots 00}_{n-1}x_{3,n}\underbrace{1 \cdots 11}_{n-1}. \tag{65}$$

Next, two-channel MRC (eqs. (6)-(8)) is used to derive the final result by considering the composite set $\{2^{2n}(2^{2n}-1), 2^{2n+1}-1\}$ and corresponding numbers (Z, x_4). The final conversion equation is as follows

$$X = Z + 2^{2n}(2^{2n}-1)\left|-2^{2n+3}(x_4 - Z)\right|_{2^{2n+1}-1} = x_1 + 2^{2n}(Y + 2^{2n}T - T) \tag{66}$$

Where

$$T = \left|v_5 + v_6 + v_7\right|_{2^{2n+1}-1} \tag{67}$$

$$v_5 = \underbrace{x_{1,2n-2} \cdots x_{1,1}x_{1,0}}_{2n-1} 0 x_{1,2n-1} \tag{68}$$

$$v_6 = \underbrace{Y_{2n-1} \cdots Y_1 Y_0}_{2n} 0 \tag{69}$$

$$v_7 = \underbrace{\bar{x}_{4,2n-2} \cdots \bar{x}_{4,1}\bar{x}_{4,0}}_{2n-1} \bar{x}_{4,2n}\bar{x}_{4,2n-1} \tag{70}$$

The hardware implementation of this converter relies on two modulo adders for realization of (61) and (67). In other words, (61) needed two $2n$-bit CSAs with EAC and a $2n$-bit CPA with EAC, and a $(2n+1)$-bit CPA with EAC is used to realize (67). Besides, (66) only requires one $(4n+1)$-bit regular binary adder; the required multiplications all can be done using shift and concatenation. The converter has been depicted in Fig. 4.

7. Complexity comparison

Table 1 presents the total hardware requirements and conversion delays of the reverse converters for the large dynamic range four-moduli sets in terms of logic gates and FAs. Note that A_{FA} and D_{FA} indicate the area and delay of one FA, respectively. It can be seen that the fastest converter is the converter for moduli set $\{2^n-1, 2^n+1, 2^{2n}, 2^{2n}+1\}$. Because, the dynamic range of this set is $6n$-bit while the dynamic range of moduli set $\{2^n-1, 2^n, 2^n+1, 2^{2n}+1\}$ is $5n$-bit. Therefore, for providing the same dynamic range, the value of n for the first

Moduli set	Hardware Requirements	Conversion Delay
$\{2^n-1, 2^n, 2^n+1, 2^{2n+1}-1\}$	$(8n+2)A_{FA} + (n-1)A_{XOR} + (n-1)A_{AND}$ $+ (4n+1)A_{XNOR} + (4n+1)A_{OR}$ $+ (7n+1)A_{NOT} + (n)A_{MUX2\times1}$	$(12n+5)D_{FA}+3D_{NOT}+D_{MUX}$
$\{2^n-1, 2^n+1, 2^{2n}, 2^{2n+1}-1\}$	$(10n+3)A_{FA} + (n+1)A_{XOR}$ $+ (n+1)A_{AND} + (3n-1)A_{XNOR}$ $+(3n-1)A_{OR} + (7n+3)A_{NOT}$	$(12n+6)D_{FA}+2D_{NOT}$
$\{2^n -1, 2^n, 2^n +1, 2^{2n} +1\}$	$(11n+6)A_{FA} + (2n-1)A_{XOR}$ $+ (2n-1)A_{AND} + (4n)A_{XNOR}$ $+(4n)A_{OR} + (5n+3)A_{NOT}$	$(8n+3)D_{FA}+D_{NOT}$
$\{2^n-1, 2^n+1, 2^{2n}, 2^{2n}+1\}$	$(10n+6)A_{FA} + (4n-3)A_{XOR}$ $+ (4n-3)A_{AND} + (2n-3)A_{XNOR}$ $+(2n-3)A_{OR} + (6n+3)A_{NOT}$	$(8n+3)D_{FA}+D_{NOT}$

Table 1. Hardware requirements and conversion delays of the reverse converters for the large dynamic range four-moduli sets

set is smaller than the second set. Furthermore, the reverse converter for the moduli set $\{2^n-1, 2^n, 2^n+1, 2^{2n+1}-1\}$ relies on less hardware requirements than others. From another side, the moduli sets $\{2^n-1, 2^n, 2^n+1, 2^{2n+1}-1\}$ and $\{2^n-1, 2^n+1, 2^{2n}, 2^{2n+1}-1\}$ results in faster RNS arithmetic units than the moduli sets $\{2^n-1, 2^n, 2^n+1, 2^{2n}+1\}$ and $\{2^n-1, 2^n+1, 2^{2n}, 2^{2n}+1\}$.

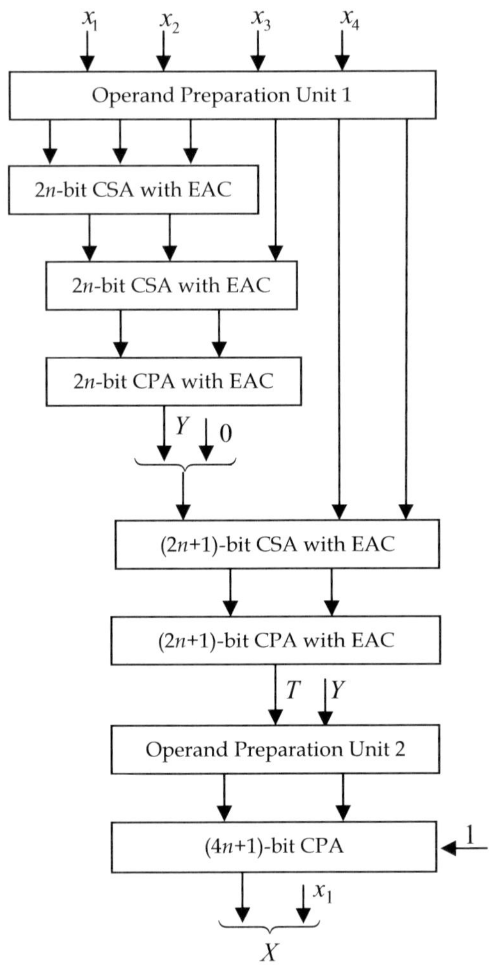

Fig. 4. The converter for moduli set $\{2^n-1, 2^n+1, 2^{2n}, 2^{2n+1}-1\}$ (Molahosseini & Navi, 2010)

8. Conclusion

The Residue Number System has been recognized as one of the efficient alternative number systems which can be used to high-speed hardware implementation of Digital Signal Processing computation algorithms. However, forward and reverse converters are needed to act as interfaces between RNS and the conventional binary digital systems. The overhead of these converters can frustrate the speed efficiency of RNS, and due to this a lot of research has been done to design efficient reverse converters. This chapter presents a study on the state-of-the-art reverse converters which have been designed for the recently introduced large dynamic range RNS four-moduli sets. We provide an overview about different reverse conversion algorithms, the recent four-moduli sets, and the reverse converter architectures.

9. References

Bhardwaj, M., Premkumar, A. B. & Srikanthan, T. (1998). Breaking the $2n$-Bit Carry Propagation Barrier in Residue to Binary Conversion for the [2^n-1, 2^n, 2^n+1] Modula Set. *IEEE Transactions on Circuits and Systems-I*, Vol. 45, No. 9, pp. 998-1002.

Bhardwaj, M., Srikanthan, T. & Clarke, C.T. (1999). A reverse converter for the 4-moduli superset {2^n-1, 2^n, 2^n+1, 2^{n+1}+1}, *Proceedings of IEEE Symposium on Computer Arithmetic.*

Cao, B., Chang, C.H. & Srikanthan, T. (2003). An Efficient Reverse Converter for the 4-Moduli Set {2^n-1, 2^n, 2^n+1, 2^{2n}+1} Based on the New Chinese Remainder Theorem. *IEEE Transactions on Circuits and Systems-I*, Vol. 50, No. 10, pp. 1296-1303.

Cardarilli, G.C., Nannarelli, A. and Re, M. (2007). Residue Number System for Low-Power DSP Applications, *Proceedings of Asilomar Conference on Signals, Systems, and Computers*, Asilomar, USA.

Chaves, R. & Sousa, L. (2003). RDSP: A RISC DSP based on residue number system, *Proceedings of Euromicro Symposium on digital system design: architectures, methods and tools.*

Conway, R. and Nelson, J. (2004). Improved RNS FIR Filter Architectures. *IEEE Transactions on Circuits and Systems-II*, Vol. 51, No. 1, pp. 26-28.

Diclaudio, E., Piazza F. & Orlandi, G. (1995). Fast combinatorial RNS processors for DSP applications. *IEEE Transactions on Computers*, Vol. 44, pp. 624-331.

Gallaher, D., Petry, F.E. & Srinivasan, P. (1997). The Digit Parallel Method for Fast RNS to Weighted Number System Conversion for Specific Moduli (2^k-1, 2^k, 2^k+1). *IEEE Transactions on Circuits and Systems-II*, Vol. 44, No. 1, pp. 53-57.

Hariri, A., Navi, K. & Rastegar, R. (2008). A new high dynamic range moduli set with efficient reverse converter. *Elsevier Journal of Computers and Mathematics with Applications*, Vol. 55, No. 4, pp. 660-668.

Hiasat A. & Abdel-Aty-Zohdy, H. S. (1998). Residue-to-binary arithmetic converter for the moduli set (2^k, 2^k-1, 2^{k-1}-1). *IEEE Transactions on Circuits and System-II*, Vol. 45, No. 2, pp. 204–208.

Hiasat, A. & Sweidan, A. (2004). Residue-to-binary decoder for an enhanced moduli set. *IEE Proc.-Comput. Digit. Tech.*, Vol. 151, No. 2, pp. 127-130.

Jenkins, W. K. & Leon, B. J. (1977). The use of residue number systems in the design of finite impulse response digital filters. *IEEE Transactions on Circuits and Systems*, Vol. CAS-24, pp. 191-201.

Koc, C.K. (1989). A fast algorithm for mixed-radix conversion in residue arithmetic, *Proceedings of IEEE International Conference on Computer Design: VLSI in Computers and Processors.*

Lin, S.H., Sheu, M.H. & Wang, C.H. (2008). Efficient VLSI Design of a Residue-to-Binary Converter for the moduli set (2^n, 2^{n+1}–1, 2^n–1). *IEICE Transactions on Information and Systems*, Vol. E91-D, No. 7, pp.2058-2060.

Mohan, P.V.A. (2002). *Residue Number Systems: Algorithms and Architectures,* Kluwer Academic.

Mohan, P. V. A. (2007). RNS-To-Binary Converter for a New Three-Moduli Set {2^{n+1}–1, 2^n, 2^n–1}. *IEEE Transactions on Circuits and Systems-II*, Vol. 54, No. 9, pp. 775-779.

Mohan P. V. A. & Premkumar, A. B. (2007). RNS-to-Binary Converters for Two Four-Moduli Set $\{2^n-1, 2^n, 2^n+1, 2^{n+1}-1\}$ and $\{2^n-1, 2^n, 2^n+1, 2^{n+1}+1\}$. *IEEE Transactions on Circuits and Systems-I*, Vol. 54, No. 6, pp. 1245-1254.

Molahosseini, A.S., Dadkhah, C., Navi, K. & Eshghi, M. (2009). Efficient MRC-Based Residue to Binary Converters for the New Moduli Sets $\{2^{2n}, 2^n-1, 2^{n+1}-1\}$ and $\{2^{2n}, 2^n-1, 2^{n-1}-1\}$. *IEICE Transactions on Information and Systems*, vol. E92-D, pp. 1628-1638.

Molahosseini, A.S. & Navi, K. (2010). A Reverse Converter for the Enhanced Moduli Set $\{2^n-1, 2^n+1, 2^{2n}, 2^{2n+1}-1\}$ Using CRT and MRC, *Proceedings of IEEE Computer Society Annual Symposium on VLSI (ISVLSI'10)*, Kefalonia, Greece, Jul. 13-15.

Molahosseini, A.S., Navi, K., Dadkhah, C., Kavehei, O. & Timarchi, S. (2010). Efficient Reverse Converter Designs for the New 4-Moduli Sets $\{2^n-1, 2^n, 2^n+1, 2^{2n+1}-1\}$ and $\{2^n-1, 2^n+1, 2^{2n}, 2^{2n}+1\}$ Based on New CRTs. *IEEE Transactions on Circuits and Systems-I*, vol. 57, no. 4, pp. 823-835.

Molahosseini, A.S., Navi, K., Hashemipour, O. & Jalali, A. (2008). An efficient architecture for designing reverse converters based on a general three-moduli set. *Elsevier Journal of Systems Architecture*, vol. 54, pp. 929-934.

Navi, K., Molahosseini, A.S. & Esmaeildoust, M. (2011). How to Teach Residue Number System to Computer Scientists and Engineers. *IEEE Transactions on Education*, vol. 54, pp. 156-163.

Omondi, A. & Premkumar, B. (2007). *Residue Number Systems: Theory and Implementations*, Imperial College Press, London.

Parhami, B. (2000). *Computer Arithmetic: Algorithms and Hardware Design*, Oxford University Press.

Piestrak, S. J. (1994). Design of residue generators and multioperand modular adders using carry-save adders. *IEEE Transactions on Computers*, Vol. 423, No. 1, pp. 68-77.

Piestrak, S.J. (1995). A high speed realization of a residue to binary converter. *IEEE Transactions on Circuits and Systems-II*, Vol. 42, pp. 661-663.

Soderstrand, M.A. & et al. (1986). *Residue number system arithmetic: modern applications in digital signal processing*, IEEE Press.

Stouratitis, T. & Paliouras, V. (2001). Considering the alternatives in lowpower design. *IEEE Circuits and Devices*, Vol. 7, pp. 23-29.

Taylor, F.J. (1984). Residue arithmetic: a tutorial with examples. *IEEE Computer*, Vol. 17, pp. 50–62.

Vinod, A.P. & Premkumar, A.B. (2000). A residue to binary converter for the 4-moduli superset $\{2^n-1, 2^n, 2^n+1, 2^{n+1}-1\}$. *Journal of Circuits, Systems and Computers*, Vol. 10, pp. 85-99.

Wang, Y. (2000). Residue-to-Binary Converters Based on New Chinese remainder theorems. *IEEE Transactions on Circuits and Systems-II*, Vol. 47, No. 3, pp. 197-205.

Wang, Z., Jullien, G. A. & Miller, W. C. (2000). An Improved Residue-to-Binary Converter. *IEEE Transactions on Circuits and Systems-I*, Vol. 47, No. 9, pp. 1437-1440.

Wang, Y., Song, X., Aboulhamid, M. & Shen, H. (2002). Adder based residue to binary numbers converters for $(2^n-1, 2^n, 2^n+1)$. *IEEE Transactions on Signal Processing*, Vol. 50, No. 7, pp. 1772-1779.

Wang, W., Swamy, M. N. S., Ahmad, M. O. & Wang, Y. (2000b). A high-speed residue-to-binary converter and a scheme of its VLSI implementation. *IEEE Transactions on Circuits and Systems-II*, Vol. 47, No. 12, pp. 1576–1581.

Zhang, W. & Siy, P. (2008). An efficient design of residue to binary converter for four moduli set $(2^n-1, 2^n+1, 2^{2n}-2, 2^{2n+1}-3)$ based on new CRT II. *Elsevier Journal of Information Sciences*, Vol. 178, No. 1, pp. 264-279.

17

A Description of Experimental Design on the Basis of an Orthonormal System

Yoshifumi Ukita[1] and Toshiyasu Matsushima[2]
[1]*Yokohama College of Commerce*
[2]*Waseda University*
Japan

1. Introduction

The Fourier series representation of a function is a classic representation which is widely used to approximate real functions (Stein & Shakarchi, 2003). In digital signal processing (Oppenheim & Schafer, 1975), the sampling theorem states that any real valued function f can be reconstructed from a sequence of values of f that are discretely sampled with a frequency at least twice as high as the maximum frequency of the spectrum of f. This theorem can also be applied to functions over finite domains (Stankovic & Astola, 2007; Takimoto & Maruoka, 1997). Then, the range of frequencies of f can be expressed in more detail by using a bounded set instead of the maximum frequency. A function whose range of frequencies is confined to a bounded set I is referred to as "bandlimited to I". Ukita et al. obtained a sampling theorem for bandlimited functions over Boolean (Ukita et al., 2003) and $GF(q)^n$ domains (Ukita et al., 2010a), where q is a prime power and $GF(q)$ is Galois field of order q. The sampling theorem can be applied in various fields as well as in digital signal processing, and one of the fields is the experimental design.

In most areas of scientific research, experimentation is a major tool for acquiring new knowledge or a better understanding of the target phenomenon. Experiments usually aim to study how changes in various factors affect the response variable of interest (Cochran & Cox, 1992; Toutenburg & Shalabh, 2009). Since the model used most often at present in experimental design is expressed through the effect of each factor, it is easy to understand how each factor affects the response variable. However, since the model contains redundant parameters and is not expressed in terms of an orthonormal system, a considerable amount of time is often necessary to implement the procedure for estimating the effects.

In this chapter, we propose that the model of experimental design be expressed as an orthonormal system, and show that the model contains no redundant parameters. Then, the model is expressed by using Fourier coefficients instead of the effect of each factor. As there is an abundance of software for calculating the Fourier transform, such a system allows for a straightforward implementation of the procedures for estimating the Fourier coefficients by using Fourier transform. In addition, the effect of each factor can be easily obtained from the Fourier coefficients (Ukita & Matsushima, 2011). Therefore, it is possible to implement easily the estimation procedures as well as to understand how each factor affects the response variable in a model based on an orthonormal system. Moreover, the analysis of variance can also be performed in a model based on an orthonormal system (Ukita et al., 2010b). Hence,

it is clear that two main procedures in the experimental design, that is, the estimation of the effects and the analysis of variance can be executed in a description of experimental design on the basis of an orthonormal system.
This chapter is organized as follows. In Section 2, we give preliminaries that are necessary for this study. In Section 3, we provide an introduction to experimental design and describe the characteristic of the previous model in .experimental design. In Section 4, we propose the new model of experimental design on the basis of an orthonormal system and clarify the characteristic of the model. Finally, Section 5 concludes this chapter.

2. Preliminaries

2.1 Fourier analysis on finite Abelian groups

Here, we provide a brief explanation of Fourier analysis on finite Abelian groups. Characters are important in the context of finite Fourier series.

2.1.1 Characters

Let G be a finite Abelian group (with additive notation), and let S^1 be the unit circle in the complex plane. A character on G is a complex-valued function $\mathcal{X} : G \to S^1$ that satisfies the condition

$$\mathcal{X}(x+x') = \mathcal{X}(x)\mathcal{X}(x') \quad \forall x, x' \in G. \tag{1}$$

In other words, a character is a homomorphism from G to the circle group.

2.1.2 Fourier transform

Let $G_i, i = 1, 2, \ldots, n$, be Abelian groups of respective orders $|G_i| = g_i, i = 1, 2, \ldots, n, g_1 \leq g_2 \leq \cdots \leq g_n$, and let

$$G = \times_{i=1}^{n} G_i \quad and \quad g = \prod_{i=1}^{n} g_i. \tag{2}$$

Since the character group of G is isomorphic to G, we can index the characters by the elements of G, that is, $\{\mathcal{X}_{\boldsymbol{a}}(\boldsymbol{x}) | \boldsymbol{a} \in G\}$ are the characters of G. Note that $\mathcal{X}_{\boldsymbol{0}}(\boldsymbol{x})$ is the principal character and identically equal to 1. The characters $\{\mathcal{X}_{\boldsymbol{a}}(\boldsymbol{x}) | \boldsymbol{a} \in G\}$ form an orthonormal system:

$$\frac{1}{g} \sum_{\boldsymbol{x} \in G} \mathcal{X}_{\boldsymbol{a}}(\boldsymbol{x}) \mathcal{X}_{\boldsymbol{b}}^{*}(\boldsymbol{x}) = \begin{cases} 1, & \boldsymbol{a} = \boldsymbol{b}, \\ 0, & \boldsymbol{a} \neq \boldsymbol{b}, \end{cases} \tag{3}$$

where $\mathcal{X}_{\boldsymbol{b}}^{*}(x)$ is the complex conjugate of $\mathcal{X}_{\boldsymbol{b}}(x)$.
Any function $f : G \to \mathbb{C}$, where $\mathbb{C}$ is the field of complex numbers, can be uniquely expressed as a linear combination of the following characters:

$$f(\boldsymbol{x}) = \sum_{\boldsymbol{a} \in G} f_{\boldsymbol{a}} \mathcal{X}_{\boldsymbol{a}}(\boldsymbol{x}), \tag{4}$$

where the complex number

$$f_{\boldsymbol{a}} = \frac{1}{g} \sum_{\boldsymbol{x} \in G} f(\boldsymbol{x}) \mathcal{X}_{\boldsymbol{a}}^{*}(\boldsymbol{x}) \tag{5}$$

is the $\boldsymbol{a}$-th *Fourier coefficient* of f.

2.2 Fourier analysis on $GF(q)^n$

Assume that q is a prime power. Let $GF(q)$ be a Galois field of order q which contains a finite number of elements. We also use $GF(q)^n$ to denote the set of all n-tuples with entries from $GF(q)$. The elements of $GF(q)^n$ are referred to as vectors.

Example 1. *Consider $GF(3) = \{0, 1, 2\}$. Addition and multiplication are defined as follows:*

+	0	1	2
0	0	1	2
1	1	2	0
2	2	0	1

·	0	1	2
0	0	0	0
1	0	1	2
2	0	2	1

Moreover, consider $n = 5$.

$$GF(3)^5 = \{00000, 10000, \cdots, 22222\}, \tag{6}$$

and $|GF(3)^5| = 243$. □

Specifying the group G in Section 2.1.2 to be the support group of $GF(q)^n$ and $g = q^n$, the relations (3), (4) and (5) also hold over the $GF(q)^n$ domain.

3. Experimental design

In this section, we provide a short introduction to experimental design.

3.1 Model in experimental design

Let $F_1, F_2, \ldots, F_n$ denote n factors to be included in an experiment. The levels of each factor can be represented by $GF(q)$, and the combinations of levels can be represented by the n-tuples $\boldsymbol{x} = (x_1, x_2, \ldots, x_n) \in GF(q)^n$.

Example 2. *Let Machine (F_1) and Worker (F_2) be factors that might influence the total amount of the product. Assume each factor has two levels.*
F_1 : new machine (level 0), old machine (level 1).
F_2 : skilled worker (level 0), unskilled worker (level 1).
For example, $\boldsymbol{x} = 01$ represents a combination of new machine and unskilled worker.
Then, the effect of the machine, averaged over both workers, is referred to as the effect of main factor F_1. Similarly, the effect of the worker, averaged over both machines, is referred to as the effect of main factor F_2. The difference between the effect of the machine for an unskilled worker and that for a skilled worker is referred to as the effect of the interaction of F_1 and F_2. □

Let the set $A \subseteq \{0, 1\}^n$ represent all factors that might influence the response of an experiment. The *Hamming weight* $w(\boldsymbol{a})$ of a vector $\boldsymbol{a} = (a_1, a_2, \ldots, a_n) \in A$ is defined as the number of nonzero components. The main factors are represented by $MF = \{l | a_l = 1, \boldsymbol{a} \in A_1\}$, where $A_1 = \{\boldsymbol{a} | w(\boldsymbol{a}) = 1, \boldsymbol{a} \in A\}$. The interactive factors are represented by $IF = \{\{l, m\} | a_l = 1, a_m = 1, \boldsymbol{a} \in A_2\}$, where $A_2 = \{\boldsymbol{a} | w(\boldsymbol{a}) = 2, \boldsymbol{a} \in A\}$.

Example 3. *Consider $A = \{000, 100, 010, 001, 110\}$. Then, $A_1 = \{100, 010, 001\}$ and $MF = \{1, 2, 3\}$, $A_2 = \{110\}$ and $IF = \{\{1, 2\}\}$.*
For example, $1 \in MF$ indicates the main factor F_1, and $\{1, 2\} \in IF$ indicates the interactive factors F_1 and F_2. □

It is usually assumed that the set A satisfies the following monotonicity condition (Okuno & Haga, 1969).

Definition 1. *Monotonicity*

$$\boldsymbol{a} \in A \rightarrow \boldsymbol{b} \in A \qquad \forall \boldsymbol{b}\ (\boldsymbol{b} \sqsubseteq \boldsymbol{a}), \tag{7}$$

where $(b_1, b_2, \ldots, b_n) \sqsubseteq (a_1, a_2, \ldots, a_n)$ *indicates that if* $a_i = 0$ *then* $b_i = 0, i = 1, 2, \ldots, n.$ □

Example 4. *Consider* $A = \{00000, 10000, 01000, 00100, 00010, 00001, 11000, 10100, 10010\}$. *Since the set A satisfies (7), A is monotonic.* □

Let $y(\boldsymbol{x})$ denote the response of the experiment with level combination $\boldsymbol{x}$. Assume the model

$$y(\boldsymbol{x}) = \mu + \sum_{l \in MF} \alpha_l(x_l) + \sum_{\{l,m\} \in IF} \beta_{l,m}(x_l, x_m) + \epsilon_{\boldsymbol{x}}, \tag{8}$$

where μ is the general mean, $\alpha_l(x_l)$ is the effect of the x_l-th level of Factor F_l, $\beta_{l,m}(x_l, x_m)$ is the effect of the interaction of the x_l-th level of Factor F_l and the x_m-th level of Factor F_m and $\epsilon_{\boldsymbol{x}}$ is a random error with a zero mean and a constant variance σ^2.

Since the model is expressed through the effect of each factor, it is easy to understand how each factor affects the response variable. However, because the constraints

$$\sum_{\varphi=0}^{q-1} \alpha_l(\varphi) = 0, \tag{9}$$

$$\sum_{\varphi=0}^{q-1} \beta_{l,m}(\varphi, \psi) = 0, \tag{10}$$

$$\sum_{\psi=0}^{q-1} \beta_{l,m}(\varphi, \psi) = 0, \tag{11}$$

are assumed, the model contains redundant parameters.

Example 5. *Consider* $q = 3, n = 5$ *and* $A = \{00000, 10000, 01000, 00100, 00010, 00001, 11000, 10100, 10010\}$. *Then,*

$\mu, \alpha_1(0), \alpha_1(1), \alpha_1(2), \alpha_2(0), \alpha_2(1), \alpha_2(2), \alpha_3(0), \alpha_3(1), \alpha_3(2), \alpha_4(0), \alpha_4(1), \alpha_4(2), \alpha_5(0), \alpha_5(1),$
$\alpha_5(2), \beta_{1,2}(0,0), \beta_{1,2}(0,1), \beta_{1,2}(0,2), \beta_{1,2}(1,0), \beta_{1,2}(1,1), \beta_{1,2}(1,2), \beta_{1,2}(2,0), \beta_{1,2}(2,1),$
$\beta_{1,2}(2,2),, \beta_{1,3}(0,0), \beta_{1,3}(0,1), \beta_{1,3}(0,2), \beta_{1,3}(1,0), \beta_{1,3}(1,1), \beta_{1,3}(1,2), \beta_{1,3}(2,0), \beta_{1,3}(2,1),$
$\beta_{1,3}(2,2), \beta_{1,4}(0,0), \beta_{1,4}(0,1), \beta_{1,4}(0,2), \beta_{1,4}(1,0), \beta_{1,4}(1,1), \beta_{1,4}(1,2), \beta_{1,4}(2,0), \beta_{1,4}(2,1),$
$\beta_{1,4}(2,2)$

are parameters. The number of parameters is 43, *but the number of the independent parameters is* 23 *by the constraints.* □

In experimental design, we are presented with a model of an experiment, which consists of a set $A \subseteq \{0,1\}^n$. First, we determine a set of level combinations $\boldsymbol{x} \in X, X \subseteq GF(q)^n$. The set X is referred to as a design. Then, we perform a set of experiments in accordance to the design X and estimate the effects from the obtained results $\{(\boldsymbol{x}, y(\boldsymbol{x})) | \boldsymbol{x} \in X\}$.

An important standard for evaluating experimental design is the maximum of the variances of the unbiased estimators of effects, as calculated from the results of the conducted experiments. It is known that, for a given number of experiments, this criterion is minimized when using orthogonal design (Takahashi, 1979). Hence, there has been extensive research focusing on orthogonal design (Hedayat et al., 1999; Takahashi, 1979; Ukita et al., 2003; 2010a;b; Ukita & Matsushima, 2011).

3.2 Orthogonal design

Definition 2. *(Orthogonal design)*
Define $v(\boldsymbol{a}) = \{i | a_i \neq 0, 1 \leq i \leq n\}$. *For* $A \subseteq \{0,1\}^n$, *let* H_A *be the* $k \times n$ *matrix*

$$H_A = \begin{bmatrix} h_{11} & h_{12} & \dots & h_{1n} \\ h_{21} & h_{22} & \dots & h_{2n} \\ \vdots & \vdots & \ddots & \vdots \\ h_{k1} & h_{k2} & \dots & h_{kn} \end{bmatrix}. \tag{12}$$

The components of this matrix, $h_{ij} \in GF(q)$ $(1 \leq i \leq k, 1 \leq j \leq n)$, *satisfy the following conditions.*

1. *The set* $\{\boldsymbol{h}_{\cdot j} | j \in v(\boldsymbol{a}' + \boldsymbol{a}'')\}$[1], *where* $\boldsymbol{h}_{\cdot j}$ *is the j-th column of* H_A, *is linearly independent over* $GF(q)$ *for any given* $\boldsymbol{a}', \boldsymbol{a}'' \in A$.
2. *The set* $\{\boldsymbol{h}_{i\cdot} | 1 \leq i \leq k\}$, *where* $\boldsymbol{h}_{i\cdot}$ *is the i-th row of* H_A, *is linearly independent over* $GF(q)$.

An orthogonal design $C^{\perp}$ *for main and interactive factors* $A \subseteq \{0,1\}^n$ *is defined as*

$$C^{\perp} = \{\boldsymbol{x} | \boldsymbol{x} = \boldsymbol{r}H_A, \ \boldsymbol{r} \in GF(q)^k\}, \tag{13}$$

and $|C^{\perp}| = q^k$. □

Example 6. *We consider the case* $q = 3, n = 5$ *and*

$$A = \{00000, 10000, 01000, 00100, 00010, 00001, 11000, 10100, 10010\}. \tag{14}$$

In this case,

$$H_A = \begin{bmatrix} 1 & 0 & 0 & 0 & 0 \\ 0 & 1 & 0 & 1 & 1 \\ 0 & 0 & 1 & 1 & 2 \end{bmatrix}, \tag{15}$$

satisfies the conditions in Definition 2. Therefore,

$$\begin{aligned} C^{\perp} = \{&00000, 00112, 00221, 01011, 01120, 01202, 02022, 02101, 02210, 10000, 10112, \\ &10221, 11011, 11120, 11202, 12022, 12101, 12210, 20000, 20111, 20221, 21011, \\ &21120, 21202, 22022, 22101, 22210\}, \end{aligned}$$

is an orthogonal design for A. □

Many algorithms for constructing H_A have been proposed (Hedayat et al., 1999; MacWilliams & Sloane, 1977; Takahashi, 1979; Ukita et al., 2003). However, it is still an extremely difficult problem to construct H_A when the number of factors n is large and a large number of interactions are included in the model. In this regard, algorithms for the construction of orthogonal design are not presented here since this falls outside the scope of this chapter.

---

[1] For $\boldsymbol{a}_1 = (a_{11}, a_{12}, \dots, a_{1n}), \boldsymbol{a}_2 = (a_{21}, a_{22}, \dots, a_{2n}) \in \{0,1\}^n$, the addition of vectors $\boldsymbol{a}_1$ and $\boldsymbol{a}_2$ is defined as $\boldsymbol{a}_1 + \boldsymbol{a}_2 = (a_{11} \oplus a_{21}, a_{12} \oplus a_{22}, \dots, a_{1n} \oplus a_{2n})$, where $\oplus$ is the *exclusive OR* operator.

3.3 Estimation of effects in experimental design

First, we adopt the following definitions.

$$Y = \sum_{\boldsymbol{x} \in C^{\perp}} y(\boldsymbol{x}), \tag{16}$$

where $|C^{\perp}| = q^k$.

$$Y_l(\varphi) = \sum_{\boldsymbol{x} \in C_l^{\perp}(\varphi)} y(\boldsymbol{x}), \tag{17}$$

where $C_l^{\perp}(\varphi) = \{\boldsymbol{x} | x_l = \varphi, \boldsymbol{x} \in C^{\perp}\}$ and $|C_l^{\perp}(\varphi)| = q^{k-1}$.

$$Y_{l,m}(\varphi, \psi) = \sum_{\boldsymbol{x} \in C_{l,m}^{\perp}(\varphi,\psi)} y(\boldsymbol{x}), \tag{18}$$

where $C_{l,m}^{\perp}(\varphi, \psi) = \{\boldsymbol{x} | x_l = \varphi, x_m = \psi, \boldsymbol{x} \in C^{\perp}\}$ and $|C_{l,m}^{\perp}(\varphi, \psi)| = q^{k-2}$.
Let $\bar{y} = \frac{1}{q^k} Y$, $\bar{y}_l(\varphi) = \frac{1}{q^{k-1}} Y_l(\varphi)$, $\bar{y}_{l,m}(\varphi, \psi) = \frac{1}{q^{k-2}} Y_{l,m}(\varphi, \psi)$. Then, the unbiased estimators of the parameters in (8) are given as

$$\hat{\mu} = \bar{y}, \tag{19}$$

$$\hat{\alpha}_l(\varphi) = \bar{y}_l(\varphi) - \hat{\mu}, \tag{20}$$

$$\hat{\beta}_{l,m}(\varphi, \psi) = \bar{y}_{l,m}(\varphi, \psi) - \hat{\alpha}_l(\varphi) - \hat{\alpha}_m(\psi) - \hat{\mu}. \tag{21}$$

Example 7. *Consider the case that a set A is given by (14) and the result of experiments is given by Table 1.*

$\boldsymbol{x}$	$y(\boldsymbol{x})$	$\boldsymbol{x}$	$y(\boldsymbol{x})$	$\boldsymbol{x}$	$y(\boldsymbol{x})$
00000	93	10000	99	20000	87
00112	97	10112	109	20111	86
00221	98	10221	112	20221	90
01011	90	11011	102	21011	85
01120	96	11120	111	21120	82
01202	102	11202	111	21202	94
02022	97	12022	105	22022	84
02101	95	12101	104	22101	88
02210	95	12210	101	22210	83

Table 1. Result of experiments

First, using (16)–(18),

$$\begin{array}{llll}
Y = 2596, & Y_1(0) = 863, & Y_1(1) = 954, & Y_1(2) = 779, \\
Y_2(0) = 871, & Y_2(1) = 873, & Y_2(2) = 852, & Y_3(0) = 842, \\
Y_3(1) = 868, & Y_3(2) = 886, & Y_4(0) = 873, & Y_4(1) = 848, \\
Y_4(2) = 875, & Y_5(0) = 847, & Y_5(1) = 864, & Y_5(2) = 885, \\
Y_{1,2}(0,0) = 288, & Y_{1,2}(0,1) = 288, & Y_{1,2}(0,2) = 287, & Y_{1,2}(1,0) = 320, \\
Y_{1,2}(1,1) = 324, & Y_{1,2}(1,2) = 310, & Y_{1,2}(2,0) = 263, & Y_{1,2}(2,1) = 261, \\
Y_{1,2}(2,2) = 255, & Y_{1,3}(0,0) = 280, & Y_{1,3}(0,1) = 288, & Y_{1,3}(0,2) = 295, \\
Y_{1,3}(1,0) = 306, & Y_{1,3}(1,1) = 324, & Y_{1,3}(1,2) = 324, & Y_{1,3}(2,0) = 256, \\
Y_{1,3}(2,1) = 256, & Y_{1,3}(2,2) = 267, & Y_{1,4}(0,0) = 290, & Y_{1,4}(0,1) = 282, \\
Y_{1,4}(0,2) = 291, & Y_{1,4}(1,0) = 314, & Y_{1,4}(1,1) = 312, & Y_{1,4}(1,2) = 328, \\
Y_{1,4}(2,0) = 269, & Y_{1,4}(2,1) = 254, & Y_{1,4}(2,2) = 256. &
\end{array}$$

Next, the following values are obtained.

$$\begin{array}{llll}
\bar{y} = 96.15, & \bar{y}_1(0) = 95.89, & \bar{y}_1(1) = 106.00, & \bar{y}_1(2) = 86.56, \\
\bar{y}_2(0) = 96.78, & \bar{y}_2(1) = 97.00, & \bar{y}_2(2) = 94.67, & \bar{y}_3(0) = 93.56, \\
\bar{y}_3(1) = 96.44, & \bar{y}_3(2) = 98.44, & \bar{y}_4(0) = 97.00, & \bar{y}_4(1) = 94.22, \\
\bar{y}_4(2) = 97.22, & \bar{y}_5(0) = 94.11, & \bar{y}_5(1) = 96.00, & \bar{y}_5(2) = 98.33, \\
\bar{y}_{1,2}(0,0) = 96.00, & \bar{y}_{1,2}(0,1) = 96.00, & \bar{y}_{1,2}(0,2) = 95.67, & \bar{y}_{1,2}(1,0) = 106.67, \\
\bar{y}_{1,2}(1,1) = 108.00, & \bar{y}_{1,2}(1,2) = 103.33, & \bar{y}_{1,2}(2,0) = 87.67, & \bar{y}_{1,2}(2,1) = 87.00, \\
\bar{y}_{1,2}(2,2) = 85.00, & \bar{y}_{1,3}(0,0) = 93.33, & \bar{y}_{1,3}(0,1) = 96.00, & \bar{y}_{1,3}(0,2) = 98.33, \\
\bar{y}_{1,3}(1,0) = 102.00, & \bar{y}_{1,3}(1,1) = 108.00, & \bar{y}_{1,3}(1,2) = 108.00, & \bar{y}_{1,3}(2,0) = 85.33, \\
\bar{y}_{1,3}(2,1) = 85.33, & \bar{y}_{1,3}(2,2) = 89.00, & \bar{y}_{1,4}(0,0) = 96.67, & \bar{y}_{1,4}(0,1) = 94.00, \\
\bar{y}_{1,4}(0,2) = 97.00, & \bar{y}_{1,4}(1,0) = 104.67, & \bar{y}_{1,4}(1,1) = 104.00, & \bar{y}_{1,4}(1,2) = 109.33, \\
\bar{y}_{1,4}(2,0) = 89.67, & \bar{y}_{1,4}(2,1) = 84.67, & \bar{y}_{1,4}(2,2) = 85.33. &
\end{array}$$

Last, by using (19)–(21),

$$\begin{array}{llll}
\hat{\mu} = 96.15, & \hat{\alpha}_1(0) = -0.26, & \hat{\alpha}_1(1) = 9.85, & \hat{\alpha}_1(2) = -9.59, \\
\hat{\alpha}_2(0) = 0.63, & \hat{\alpha}_2(1) = 0.85, & \hat{\alpha}_2(2) = -1.48, & \hat{\alpha}_3(0) = -2.59, \\
\hat{\alpha}_3(1) = 0.30, & \hat{\alpha}_3(2) = 2.30, & \hat{\alpha}_4(0) = 0.85, & \hat{\alpha}_4(1) = -1.93, \\
\hat{\alpha}_4(2) = 1.07, & \hat{\alpha}_5(0) = -2.04, & \hat{\alpha}_5(1) = -0.15, & \hat{\alpha}_5(2) = 2.19, \\
\hat{\beta}_{1,2}(0,0) = -0.52, & \hat{\beta}_{1,2}(0,1) = -0.74, & \hat{\beta}_{1,2}(0,2) = 1.26, & \hat{\beta}_{1,2}(1,0) = 0.04, \\
\hat{\beta}_{1,2}(1,1) = 1.15, & \hat{\beta}_{1,2}(1,2) = -1.19, & \hat{\beta}_{1,2}(2,0) = 0.48, & \hat{\beta}_{1,2}(2,1) = -0.41, \\
\hat{\beta}_{1,2}(2,2) = -0.07, & \hat{\beta}_{1,3}(0,0) = 0.04, & \hat{\beta}_{1,3}(0,1) = -0.19, & \hat{\beta}_{1,3}(0,2) = 0.15, \\
\hat{\beta}_{1,3}(1,0) = -1.41, & \hat{\beta}_{1,3}(1,1) = 1.70, & \hat{\beta}_{1,3}(1,2) = -0.30, & \hat{\beta}_{1,3}(2,0) = 1.37, \\
\hat{\beta}_{1,3}(2,1) = -1.52, & \hat{\beta}_{1,3}(2,2) = 0.15, & \hat{\beta}_{1,4}(0,0) = -0.07, & \hat{\beta}_{1,4}(0,1) = 0.04, \\
\hat{\beta}_{1,4}(0,2) = 0.04, & \hat{\beta}_{1,4}(1,0) = -2.19, & \hat{\beta}_{1,4}(1,1) = -0.07, & \hat{\beta}_{1,4}(1,2) = 2.26, \\
\hat{\beta}_{1,4}(2,0) = 2.26, & \hat{\beta}_{1,4}(2,1) = 0.04, & \hat{\beta}_{1,4}(2,2) = -2.30. &
\end{array}$$

□

Although there are software packages that can be used to estimate the effects on the basis of (19)–(21), as yet no software can be used for an arbitrary monotonic set A. Therefore, it is often necessary to implement the procedure for estimating the effects, which requires a considerable amount of time.

3.4 Analysis of variance

When there are many factors, a comprehensive view of whether an interaction in A can be disregarded is needed. The test procedure involves an analysis of variance. For a detailed explanation of analysis of variance, refer to (Toutenburg & Shalabh, 2009).

The statistics needed in analysis of variance are the following. SS_{Mean} is the correction term (the sum of squares due to the mean), SS_{F_l} is the sum of squares due to the effect of F_l, $SS_{F_l \times F_m}$ is the sum of squares due to the interaction effect of $F_l \times F_m$, and SS_{Error} is the sum of squares due to error. These can be computed as follows.

$$SS_{Mean} = \frac{1}{q^k} Y^2, \tag{22}$$

$$SS_{F_l} = \frac{1}{q^{k-1}} \sum_{\varphi=0}^{q-1} Y_l^2(\varphi) - SS_{Mean}, \tag{23}$$

$$SS_{F_l \times F_m} = \frac{1}{q^{k-2}} \sum_{\varphi=0}^{q-1} \sum_{\psi=0}^{q-1} Y_{l,m}^2(\varphi, \psi) - SS_{F_l} - SS_{F_m} - SS_{Mean}, \tag{24}$$

$$SS_{Error} = \sum_{\boldsymbol{x} \in C^{\perp}} y^2(\boldsymbol{x}) - SS_{Mean} - \sum_{l \in MF} SS_{F_l} - \sum_{\{l,m\} \in IF} SS_{F_l \times F_m}. \tag{25}$$

Example 8. *Consider the case that a set A is given by (14) and the result of experiments is given by Table 1. Then, using (22)–(25),*

$$\begin{aligned} SS_{Mean} &= 249600.6, & SS_{F_1} &= 1702.3, & SS_{F_2} &= 29.9, & SS_{F_3} &= 108.7, \\ SS_{F_4} &= 50.3, & SS_{F_5} &= 80.5, & SS_{F_1 \times F_2} &= 16.6, & SS_{F_1 \times F_3} &= 27.7, \\ SS_{F_1 \times F_4} &= 60.8, & SS_{Error} &= 16.6. \end{aligned}$$

□

4. Description of experimental design on the basis of an orthonormal system

In this section, we propose the model of experimental design on the basis of an orthonormal system.

4.1 Model on the basis of an orthonormal system in experimental design

We use $y(\boldsymbol{x})$ to denote the response of an experiment with a level combination $\boldsymbol{x}$, and assume the following model:

$$y(\boldsymbol{x}) = \sum_{\boldsymbol{a} \in I_A} f_{\boldsymbol{a}} \mathcal{X}_{\boldsymbol{a}}(\boldsymbol{x}) + \epsilon_{\boldsymbol{x}}, \tag{26}$$

where $I_A = \{(b_1 a_1, \ldots, b_n a_n) | \boldsymbol{a} \in A, b_i \in GF(q)\}$ and $\epsilon_{\boldsymbol{x}}$ is a random error with a zero mean and a constant variance.

Then, the model is expressed by using Fourier coefficients instead of the effect of each factor. The effects are represented by the parameters $\{f_{\boldsymbol{a}} | \boldsymbol{a} \in I_A\}$. In addition, there are no constraints between the parameters, and the parameters are independent. Hence, it is clear that the model contains no redundant parameters.

Example 9. *Consider* $q = 3, n = 5$ *and* $A = \{00000, 10000, 01000, 00100, 00010, 00001, 11000, 10100, 10010\}$. *Then,* I_A *is given by*

$$\begin{aligned} I_A = \{&00000, 10000, 20000, 01000, 02000, 00100, 00200, 00010, 00020, 00001, 00002, 11000, \\ &12000, 21000, 22000, 10100, 10200, 20100, 20200, 10010, 10020, 20010, 20020\}, \end{aligned}$$

and Fourier coefficients
$f_{00000}, f_{10000}, f_{20000}, f_{01000}, f_{02000}, f_{00100}, f_{00200}, f_{00010}, f_{00020}, f_{00001}, f_{00002}, f_{11000}, f_{12000}, f_{21000}, f_{22000}, f_{10100}, f_{10200}, f_{20100}, f_{20200}, f_{10010}, f_{10020}, f_{20010}, f_{20020}$
are parameters. The number of parameters is 23, and these parameters are independent. □

4.2 Estimation of Fourier coefficients in experimental design

First, we present the following theorem (Ukita et al., 2010a).

Theorem 1. *Sampling Theorem for Bandlimited Functions over a $GF(q)^n$ Domain*
Assume that $A \subseteq \{0,1\}^n$ is monotonic and

$$f(\boldsymbol{x}) = \sum_{\boldsymbol{a} \in I_A} f_{\boldsymbol{a}} \mathcal{X}_{\boldsymbol{a}}(\boldsymbol{x}), \tag{27}$$

where $I_A = \{(b_1 a_1, \ldots, b_n a_n) | \boldsymbol{a} \in A, b_i \in GF(q)\}$. Then, the Fourier coefficients can be computed as follows:

$$f_{\boldsymbol{a}} = \frac{1}{q^k} \sum_{\boldsymbol{x} \in C^{\perp}} f(\boldsymbol{x}) \mathcal{X}^*_{\boldsymbol{a}}(\boldsymbol{x}), \tag{28}$$

where $C^{\perp}$ is an orthogonal design for A ($|C^{\perp}| = q^k$). □

When an experiment is conducted in accordance to the orthogonal design $C^{\perp}$, unbiased estimators of $f_{\boldsymbol{a}}$ in (26) can be obtained by using Theorem 1 and assuming that $E(\epsilon_{\boldsymbol{x}}) = 0$:

$$\hat{f}_{\boldsymbol{a}} = \frac{1}{q^k} \sum_{\boldsymbol{x} \in C^{\perp}} y(\boldsymbol{x}) \mathcal{X}^*_{\boldsymbol{a}}(\boldsymbol{x}). \tag{29}$$

Then, the Fourier coefficients can be easily estimated by using Fourier transform. There are a number of software packages for Fourier transform, which can be used to calculate (29) for any monotonic set A.

Example 10. *Consider the case that a set A is given by (14) and the result of experiments is given by Table 1. Then,*

$$\mathcal{X}^*_{\boldsymbol{a}}(\boldsymbol{x}) = e^{-2\pi i(a_1x_1 + a_2x_2 + a_3x_3 + a_4x_4 + a_5x_5)/3}. \tag{30}$$

Using (29), (30) and $e^{2\pi i k} = 1$ for any integer k,

$$\begin{aligned}
\hat{f}_{00000} &= 2596/27, \\
\hat{f}_{10000} &= (863 + 954e^{-2\pi i/3} + 779e^{-4\pi i/3})/27, \\
\hat{f}_{20000} &= (863 + 779e^{-2\pi i/3} + 954e^{-4\pi i/3})/27, \\
\hat{f}_{01000} &= (871 + 873e^{-2\pi i/3} + 852e^{-4\pi i/3})/27, \\
\hat{f}_{02000} &= (871 + 852e^{-2\pi i/3} + 873e^{-4\pi i/3})/27, \\
\hat{f}_{00100} &= (842 + 868e^{-2\pi i/3} + 886e^{-4\pi i/3})/27, \\
\hat{f}_{00200} &= (842 + 886e^{-2\pi i/3} + 868e^{-4\pi i/3})/27, \\
\hat{f}_{00010} &= (873 + 848e^{-2\pi i/3} + 875e^{-4\pi i/3})/27, \\
\hat{f}_{00020} &= (873 + 875e^{-2\pi i/3} + 848e^{-4\pi i/3})/27, \\
\hat{f}_{00001} &= (847 + 864e^{-2\pi i/3} + 885e^{-4\pi i/3})/27, \\
\hat{f}_{00002} &= (847 + 885e^{-2\pi i/3} + 864e^{-4\pi i/3})/27, \\
\hat{f}_{11000} &= (859 + 863e^{-2\pi i/3} + 874e^{-4\pi i/3})/27, \\
\hat{f}_{12000} &= (867 + 868e^{-2\pi i/3} + 861e^{-4\pi i/3})/27, \\
\hat{f}_{21000} &= (867 + 861e^{-2\pi i/3} + 868e^{-4\pi i/3})/27, \\
\hat{f}_{22000} &= (859 + 874e^{-2\pi i/3} + 863e^{-4\pi i/3})/27, \\
\hat{f}_{10100} &= (860 + 861e^{-2\pi i/3} + 875e^{-4\pi i/3})/27, \\
\hat{f}_{10200} &= (871 + 857e^{-2\pi i/3} + 868e^{-4\pi i/3})/27,
\end{aligned}$$

$$\begin{aligned}
\hat{f}_{20100} &= (871 + 868e^{-2\pi i/3} + 857e^{-4\pi i/3})/27, \\
\hat{f}_{20200} &= (860 + 875e^{-2\pi i/3} + 861e^{-4\pi i/3})/27, \\
\hat{f}_{10010} &= (872 + 852e^{-2\pi i/3} + 872e^{-4\pi i/3})/27, \\
\hat{f}_{10020} &= (858 + 859e^{-2\pi i/3} + 879e^{-4\pi i/3})/27, \\
\hat{f}_{20010} &= (858 + 879e^{-2\pi i/3} + 859e^{-4\pi i/3})/27, \\
\hat{f}_{20020} &= (872 + 872e^{-2\pi i/3} + 852e^{-4\pi i/3})/27.
\end{aligned}$$

□

In particular, when $q = 2^m$, where m is an integer and $m \geq 1$, it is possible to use the vector-radix fast Fourier transform (FFT), which is a multidimensional implementation of the FFT algorithm, for calculating (29) for all $\boldsymbol{a} \in I_A$. The complexity of the vector-radix FFT is $O(q^k \log q^k)$. In addition, it can be shown that the Yates' Method (Yates, 1937) for efficient calculation of (19)–(21) in the case of $q = 2$ is equivalent to the vector-radix FFT for calculation of (29).

4.3 The relation between the Fourier coefficients and the effect of each factor

In a description of experimental design on the basis of an orthonormal system, the model is expressed by using Fourier coefficients. Fourier coefficients themselves do not provide a direct representation of the effect of each factor.

On the other hand, since the previous model in experimental design is expressed through the effect of each factor, it is easy to understand how each factor affects the response variable.

In this section, we present three theorems of the relation between the Fourier coefficients and the effect of each factor (Ukita & Matsushima, 2011).

First, we present a theorem of the relation between the Fourier coefficient and the general mean.

Theorem 2. *Let $\hat{\mu}$ be the unbiased estimator of the general mean μ in the model of Sect.3.1, and let $\hat{f}_{0\dots0}$ be that of the Fourier coefficient $f_{0\dots0}$ in the model of Sect.4.1.*
Then, the following equation holds:

$$\hat{\mu} = \hat{f}_{0\dots0}. \tag{31}$$

□

Next, we present a theorem of the relation between the Fourier coefficients and the effect of the main factor.

Theorem 3. *Let $\hat{\alpha}_l(\varphi)$ be the unbiased estimator of the effect of the main factor $\alpha_l(\varphi)$ in the model of Sect.3.1, and let $\hat{f}_{0\dots0a_l0\dots0}$ be that of the Fourier coefficient $f_{0\dots0a_l0\dots0}$ in the model of Sect.4.1.*
Then, the following equation holds:

$$\hat{\alpha}_l(\varphi) = \sum_{\substack{a_l \in GF(q) \\ a_l \neq 0}} \mathcal{X}_{a_l}(\varphi) \hat{f}_{0\dots0a_l0\dots0}. \tag{32}$$

□

Last, we present a theorem of the relation between the Fourier coefficients and the effect of the interaction.

Theorem 4. *Let $\hat{\beta}_{l,m}(\varphi,\psi)$ be the unbiased estimator of the effect of the interaction $\beta_{l,m}(\varphi,\psi)$ in the model of Sect.3.1, and let $\hat{f}_{0\ldots0a_l0\ldots0a_m0\ldots0}$ be that of the Fourier coefficient $f_{0\ldots0a_l0\ldots0a_m0\ldots0}$ in the model of Sect.4.1.*
Then, the following equation holds:

$$\hat{\beta}_{l,m}(\varphi,\psi) = \sum_{\substack{a_l\in GF(q)\\ a_l\neq 0}} \sum_{\substack{a_m\in GF(q)\\ a_m\neq 0}} \mathcal{X}_{a_l}(\varphi)\mathcal{X}_{a_m}(\psi)\hat{f}_{0\ldots0a_l0\ldots0a_m0\ldots0}. \tag{33}$$

□

From these theorems, the effect of each factor can be easily obtained from the computed Fourier coefficients.

Example 11. *Let $q = 3$ and $n = 5$. Consider the general mean, the effect of main factor F_1, and the effect of the interaction of F_1 and F_2. Then,*

$$\mathcal{X}_l(k) = e^{2\pi ilk/3}. \tag{34}$$

First, using (31), $\hat{\mu} = \hat{f}_{00000}$ holds.
Next, using (32) and (34), the following equations

$$\begin{aligned}
\hat{\alpha}_1(0) &= \hat{f}_{10000} + \hat{f}_{20000},\\
\hat{\alpha}_1(1) &= e^{2\pi i/3}\hat{f}_{10000} + e^{4\pi i/3}\hat{f}_{20000},\\
\hat{\alpha}_1(2) &= e^{4\pi i/3}\hat{f}_{10000} + e^{2\pi i/3}\hat{f}_{20000},
\end{aligned}$$

hold. Hence, it is clear that the effects of main factor F_1 (3 parameters) can be obtained from the computed Fourier coefficients (2 parameters).
Last, using (33) and (34), the following equations

$$\begin{aligned}
\hat{\beta}_{1,2}(0,0) &= \hat{f}_{11000} + \hat{f}_{12000} + \hat{f}_{21000} + \hat{f}_{22000},\\
\hat{\beta}_{1,2}(0,1) &= e^{2\pi i/3}\hat{f}_{11000} + e^{4\pi i/3}\hat{f}_{12000} + e^{2\pi i/3}\hat{f}_{21000} + e^{4\pi i/3}\hat{f}_{22000},\\
\hat{\beta}_{1,2}(0,2) &= e^{4\pi i/3}\hat{f}_{11000} + e^{2\pi i/3}\hat{f}_{12000} + e^{4\pi i/3}\hat{f}_{21000} + e^{2\pi i/3}\hat{f}_{22000},\\
\hat{\beta}_{1,2}(1,0) &= e^{2\pi i/3}\hat{f}_{11000} + e^{2\pi i/3}\hat{f}_{12000} + e^{4\pi i/3}\hat{f}_{21000} + e^{4\pi i/3}\hat{f}_{22000},\\
\hat{\beta}_{1,2}(1,1) &= e^{4\pi i/3}\hat{f}_{11000} + \hat{f}_{12000} + \hat{f}_{21000} + e^{2\pi i/3}\hat{f}_{22000},\\
\hat{\beta}_{1,2}(1,2) &= \hat{f}_{11000} + e^{4\pi i/3}\hat{f}_{12000} + e^{2\pi i/3}\hat{f}_{21000} + \hat{f}_{22000},\\
\hat{\beta}_{1,2}(2,0) &= e^{4\pi i/3}\hat{f}_{11000} + e^{4\pi i/3}\hat{f}_{12000} + e^{2\pi i/3}\hat{f}_{21000} + e^{2\pi i/3}\hat{f}_{22000},\\
\hat{\beta}_{1,2}(2,1) &= \hat{f}_{11000} + e^{2\pi i/3}\hat{f}_{12000} + e^{4\pi i/3}\hat{f}_{21000} + \hat{f}_{22000},\\
\hat{\beta}_{1,2}(2,2) &= e^{2\pi i/3}\hat{f}_{11000} + \hat{f}_{12000} + \hat{f}_{21000} + e^{4\pi i/3}\hat{f}_{22000},
\end{aligned}$$

hold. Hence, it is clear that the effects of the interaction of F_1 and F_2 (9 parameters) can be obtained from the computed Fourier coefficients (4 parameters). □

From these theorems, the effect of each factor can be easily obtained from the Fourier coefficients. Therefore, it is possible to implement easily the estimation procedures as well as to understand how each factor affects the response variable in a model based on an orthonormal system.

4.4 Analysis of variance in experimental design

On the other hand, it is already shown that the analysis of variance can also be performed in the model of experimental design on the basis of an orthonormal system (Ukita et al., 2010b). We present three theorems with respect to the sum of squares needed in analysis of variance.

Theorem 5. *Let SS_{Mean} be the sum of squares due to the mean in Sect.3.4, and let $\hat{f}_{0\ldots0}$ be the unbiased estimator of the Fourier coefficient $f_{0\ldots0}$ in the model of Sect.4.1. Then,*

$$q^k|\hat{f}_{0\ldots0}|^2 = SS_{Mean}, \tag{35}$$

where

$$\hat{f}_{0\ldots0} = \frac{1}{q^k}\sum_{\boldsymbol{x}\in C^{\perp}} y(\boldsymbol{x})\mathcal{X}^*_{0\ldots0}(\boldsymbol{x}). \tag{36}$$

□

Theorem 6. *Let SS_{F_l} be the sum of squares due to the effect of F_l in Sect.3.4, and let $\hat{f}_{0\ldots0a_l0\ldots0}$ be the unbiased estimator of the Fourier coefficient $f_{0\ldots0a_l0\ldots0}$ in the model of Sect.4.1. Then,*

$$\sum_{\substack{a_l\in GF(q)\\ a_l\neq 0}} q^k|\hat{f}_{0\ldots0a_l0\ldots0}|^2 = SS_{F_l},\ \ l=1,2,\cdots,n, \tag{37}$$

where

$$\hat{f}_{0\ldots0a_l0\ldots0} = \frac{1}{q^k}\sum_{\boldsymbol{x}\in C^{\perp}} y(\boldsymbol{x})\mathcal{X}^*_{0\ldots0a_l0\ldots0}(\boldsymbol{x}). \tag{38}$$

□

Theorem 7. *Let $SS_{F_l\times F_m}$ be the sum of squares due to the interaction effect of $F_l \times F_m$ in Sect.3.4, and let $\hat{f}_{0\ldots0a_l0\ldots0a_m0\ldots0}$ be the unbiased estimator of the Fourier coefficient $f_{0\ldots0a_l0\ldots0a_m0\ldots0}$ in the model of Sect.4.1. Then,*

$$\sum_{a_l\neq 0}\sum_{a_m\neq 0} q^k|\hat{f}_{0\ldots0a_l0\ldots0a_m0\ldots0}|^2 = SS_{F_l\times F_m},$$
$$l,m=1,2,\cdots,n,(l<m), \tag{39}$$

where the sums are taken over $a_l, a_m \in GF(q)$ and

$$\hat{f}_{0\ldots0a_l0\ldots0a_m0\ldots0} = \frac{1}{q^k}\sum_{\boldsymbol{x}\in C^{\perp}} y(\boldsymbol{x})\mathcal{X}^*_{0\ldots0a_l0\ldots0a_m0\ldots0}(\boldsymbol{x}). \tag{40}$$

□

By these theorems, SS_{Mean}, SS_{F_l} and $SS_{F_l\times F_m}$ can be obtained in the proposed description of experimental design. In addition, using the Parseval-Plancherel formula and these theorems, SS_{Error} can be computed as follows.

$$SS_{Error} = \sum_{\boldsymbol{x}\in C^{\perp}} y^2(\boldsymbol{x}) - SS_{Mean} - \sum_{l\in MF} SS_{F_l} - \sum_{\{l,m\}\in IF} SS_{F_l\times F_m}. \tag{41}$$

Example 12. *Consider the case that a set A is given by (14) and the result of experiments is given by Table 1. Then, using the result of Example10, $\sum_{k=0}^{2} e^{2\pi ik/3} = 0$ and $e^{2\pi i} = 1$,*

$$|\hat{f}_{00000}|^2 = \hat{f}_{00000}\hat{f}^*_{00000} = (2596/27)(2596/27) = 9244.466, \tag{42}$$

$$\begin{aligned} |\hat{f}_{10000}|^2 &= \hat{f}_{10000}\hat{f}^*_{10000} \\ &= (863 + 954e^{-2\pi i/3} + 779e^{-4\pi i/3})(863 + 954e^{2\pi i/3} + 779e^{4\pi i/3})/27^2 \\ &= (863^2 + 954^2 + 779^2 - 863 \cdot 954 - 863 \cdot 779 - 954 \cdot 779)/27^2 \\ &= 31.52401. \end{aligned} \tag{43}$$

Similarly,

$$\begin{aligned} &|\hat{f}_{20000}|^2 = 31.52401, \\ &|\hat{f}_{01000}|^2 = |\hat{f}_{02000}|^2 = 0.552812, \ |\hat{f}_{00100}|^2 = |\hat{f}_{00200}|^2 = 2.013717, \\ &|\hat{f}_{00010}|^2 = |\hat{f}_{00020}|^2 = 0.931413, \ |\hat{f}_{00001}|^2 = |\hat{f}_{00002}|^2 = 1.491084, \\ &|\hat{f}_{11000}|^2 = |\hat{f}_{22000}|^2 = 0.248285, \ |\hat{f}_{12000}|^2 = |\hat{f}_{21000}|^2 = 0.058985, \\ &|\hat{f}_{10100}|^2 = |\hat{f}_{20200}|^2 = 0.289438, \ |\hat{f}_{10200}|^2 = |\hat{f}_{20100}|^2 = 0.223594, \\ &|\hat{f}_{10010}|^2 = |\hat{f}_{20020}|^2 = 0.548697, \ |\hat{f}_{10020}|^2 = |\hat{f}_{20010}|^2 = 0.577503. \end{aligned}$$

Hence, using Theorem 5–7 and (41),

$$\begin{aligned} &SS_{Mean} = 27|\hat{f}_{00000}|^2 = 249600.6, \\ &SS_{F_1} = 27(|\hat{f}_{10000}|^2 + |\hat{f}_{20000}|^2) = 1702.3, \\ &SS_{F_2} = 27(|\hat{f}_{01000}|^2 + |\hat{f}_{02000}|^2) = 29.9, \\ &SS_{F_3} = 27(|\hat{f}_{00100}|^2 + |\hat{f}_{00200}|^2) = 108.7, \\ &SS_{F_4} = 27(|\hat{f}_{00010}|^2 + |\hat{f}_{00020}|^2) = 50.3, \\ &SS_{F_5} = 27(|\hat{f}_{00001}|^2 + |\hat{f}_{00002}|^2) = 80.5, \\ &SS_{F_1 \times F_2} = 27(|\hat{f}_{11000}|^2 + |\hat{f}_{12000}|^2 + |\hat{f}_{21000}|^2 + |\hat{f}_{22000}|^2) = 16.6, \\ &SS_{F_1 \times F_3} = 27(|\hat{f}_{10100}|^2 + |\hat{f}_{10200}|^2 + |\hat{f}_{20100}|^2 + |\hat{f}_{20200}|^2) = 27.7, \\ &SS_{F_1 \times F_4} = 27(|\hat{f}_{10010}|^2 + |\hat{f}_{10020}|^2 + |\hat{f}_{20010}|^2 + |\hat{f}_{20020}|^2) = 60.8, \\ &SS_{Error} = 16.6. \end{aligned}$$

□

Therefore, the analysis of variance can be executed in the proposed description of experimental design.
Hence, it is clear that two main procedures in the experimental design, that is, the estimation of the effects and the analysis of variance can be executed in a description of experimental design on the basis of an orthonormal system.

5. Conclusion

In this chapter, we have proposed that the model of experimental design be expressed as an orthonormal system, and shown that the model contains no redundant parameters. Then, the model is expressed by using Fourier coefficients instead of the effect of each factor. As there is an abundance of software for calculating the Fourier transform, such a system allows for a straightforward implementation of the procedures for estimating the Fourier coefficients by using Fourier transform. In addition, the effect of each factor can be easily obtained from the Fourier coefficients. Therefore, it is possible to implement easily the estimation procedures as well as to understand how each factor affects the response variable in a model based on an orthonormal system. Moreover, it is already shown that the analysis of variance can also be performed in a model based on an orthonormal system. Hence, it is clear that two main procedures in the experimental design, that is, the estimation of the effects and the analysis of variance can be executed in a description of experimental design on the basis of an orthonormal system.

6. References

Cochran, W.G. & Cox, G.M. (1992). *Experimental Designs* (Second Edition), Wiley Classics library.

Hedayat, A.S., Sloane, N.J.A. & Stufken, J. (1999). *Orthogonal Arrays: Theory and Applications*, Springer, New York.

MacWilliams, F.J. & Sloane, N.J.A. (1977). *The theory of error-correcting codes*, North-Holland Mathematical Library.

Okuno, T. & Haga, T. (1969). *Experimental Designs*, Baifukan, Tokyo.

Oppenheim, A.V. & Schafer, R.W. (1975). *Digital Signal Processing*, Prentice-Hall.

Stankovic, R.S. & Astola, J. (2007). Reading the Sampling Theorem in Multiple-Valued Logic: A journey from the (Shannon) sampling theorem to the Shannon decomposition rule, *Proc. 37th Int. Symp. on Multiple-Valued Logic*, Oslo, Norway.

Stein, E.M. & Shakarchi, R. (2003). *Fourier Analysis: An Introduction*, Princeton University Press.

Takahashi, I. (1979). *Combinatorial Theory and its Application*, Iwanami Syoten, Tokyo.

Takimoto, E. & Maruoka, A. (1997). A Sampling Theorem for Functions over Boolean Domain, *IEICE Technical Report*, Vol.COMP97–56, pp.87-93.

Toutenburg, H. & Shalabh, (2009). *Statistical Analysis of Designed Experiments* (Third Edition), Springer, New York.

Ukita, Y., Matsushima, T. & Hirasawa, S. (2003). A Note on Learning Boolean Functions Using Orthogonal Designs, *IEICE Trans. Fundamentals*, Vol.J86-A, No.4, pp.482–490.

Ukita, Y., Saito, T., Matsushima, T. & Hirasawa, S. (2010a). A Note on a Sampling Theorem for Functions over $GF(q)^n$ Domain, *IEICE Trans. Fundamentals*, Vol.E93-A, No.6, pp.1024–1031.

Ukita, Y., Matsushima, T. & Hirasawa, S. (2010b). Estimation of the Effects in the Experimental Design using Fourier Transforms, *IEICE Trans. Fundamentals*, Vol.E93-A, No.11, pp.2077–2082.

Ukita, Y. & Matsushima, T. (2011). A Note on Relation between the Fourier Coefficients and the Effects in the Experimental Design, (submitted paper).

Yates, F. (1937). The design and analysis of factorial experiments, *Imperial Bureau of Soil Science, Technical Communication*, No.35.

18

Entropic Complexity Measured in Context Switching

Paul Pukite and Steven Bankes
BAE Systems
USA

1. Introduction

Abstract: A complexity metric for concurrent software controlled systems is defined and derived. It is equivalent or comparable to the Shannon information metric, which essentially measures entropy of a system, but uses a novel and efficient technique based on a FFT to calculate its value. This can be extended to other temporal realizations of behaviour.

For concurrent software, the amount of context switching that occurs indicates the level of complexity of the program. If the program consists of a fixed period cyclic scheduler, context switches will occur very predictably at those same fixed intervals. However, if the program consists of a mix of periods, some long, some short, and some perhaps sporadically aperiodic, the complexity of such a program will be much greater. Further, the greater the spread between short periods and long periods will indicate that the program will be much harder to verify, as the shorter cycles will accumulate more testing coverage than the longer or sporadic interval context switching.

In some sense, this is what makes clock-driven synchronous logic much easier to test. The state space of possible events is reduced in direct relation to the reduction of available slots for computation. As the context switching and potential interactions between threads occurs only at these slots, a simplification of the temporal behaviour here will result in a better chance to verify the correctness of its execution.

By the same token, any concurrent program will show greater non-determinism than an equivalent sequential program. The benefits of making a concurrent program more synchronous grants a greater predictability in its execution semantics. In terms of meeting hard real-time constraints, a synchronously designed program will have predictable points for schedulability, allowing for techniques such as rate-monotonic scheduling (Klein, 1993) to meet strict timing deadlines. These categories of techniques grant a concurrent program the same possibilities for verification as a sequential one.

Even though these strict scheduling programs are effective for their problem domains, they are difficult to maintain and do not scale that well for the large software systems required. In fact most interactive programs and other soft real-time systems use the concept of event-driven semantics, which can allow interruption at any point in time. These have the benefit of dealing with interactions only upon demand, and so scale better, especially in terms of not placing a huge computational or communication load on the system when not needed.

New events can also be handled well, either by adding a new thread or task or interrupt to watch for their occurrence or by adding an event handler to a queue.

Unfortunately these benefits of scalability and flexibility detract from being able to reason about the execution semantics of a program, thus leading to an element of non-determinism. One can thus consider the obvious potential for introducing stochastic ideas to qualitatively and quantitatively understand how complex a design has become. If we can attach a simple and fundamental measure to this complexity and thus generate a useful metric, it has the capacity for comparing software designs or of providing a continuous benchmark of growing potential complexity. That leads to the notion of supplying an entropy-based measure to characterize the behaviour of a concurrent software system.

2. Approach – The context-switching entropy measure

The premise is to evaluate the complexity of a concurrently executing program. A novel approach of creating such a complexity metric involves the analysis of the context-switching frequency spectrum. We take a Fourier transform (*FT*) of the temporally distributed context switching events, *c(t),* and treat that as a probability density function (Feller, 1957) in frequency space (i.e. a normalized power spectrum).

$$p(f) = \left|FT\big(c(t)\big)\right| = \left|\int_{-\infty}^{\infty} c(t) \cdot e^{-i2\pi ft}\, dt\right| \tag{1}$$

Then the Shannon information entropy (*S*) of *p(f)* (Leon-Garcia, 2008) will generate a simple complexity measure.

$$S = -\int_0^{\infty} p(f) \cdot \ln\big(p(f)\big)\, df \tag{2}$$

The entropy of a spectrum has meaning because the power spectrum itself transforms the autocorrelation of the signal in the time domain. The autocorrelation is nothing more than the probability of a time-shift occurring between points of interest. Thus by transforming the time domain information to the frequency domain we do not lose the essential information pertaining to order and disorder in the signal, in particular when it comes to comparing one waveform against another. The information containing the essential order is largely retained across the transformation.

$$prob(t) \approx autocorrelation\big(c(t)\big) = \int_0^{\infty} c(t) \cdot c(t+\tau) d\tau \tag{3}$$

So the Fourier transform represents the same probabilities as that of the autocorrelation but computed in the frequency domain. For any captured sequence of events, the Fourier transform of the autocorrelation is practically obtained as the magnitude of the Fourier transform of the waveform.

$$p(f) = FT\big(prob(t)\big) = |FT(c(t))| \tag{4}$$

For a given mean frequency $\langle f \rangle$, the value of *S* will be the greatest when the probability spread of the context switching frequency components is maximized. In this case, complexity tracks entropy and the maximum entropy principle (Jaynes, 2003) can be applied to understand the trend toward greater complexity. By Parseval's theorem, we do not lose the integrated signal when we take the Fourier transform.

This general approach is also known as Frequency Domain Entropy (Bao, 2004).

	Signal	Entropy	FT(Signal)	Entropy
Periodic Impulse Train	$\sum_{n=-\infty}^{\infty} \delta(t-nT)$	low	$\frac{2\pi}{T}\sum_{k=-\infty}^{\infty} \delta\left(\omega - k\frac{2\pi}{T}\right)$	low
Disordered Autocorrelation with Mean	$\alpha \cdot e^{-\alpha t}$	1+ln(α)	$\frac{\alpha}{\alpha^2+\omega^2}$	~ 2+ln(α)
Noise	Random(u(t))	Maximum entropy over time interval	Random in frequency domain	Maximum entropy over frequency range

Table 1. Mapping of signal entropies to their corresponding frequency domain entropies. The trend of increasing entropies with increasing disorder is maintained through the transformation process.

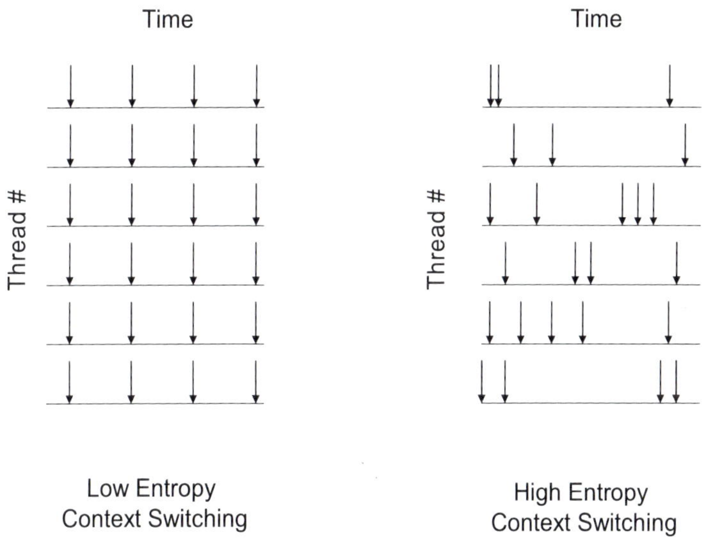

Fig. 1. Example of context-switching waveforms with low entropy (left) and high entropy (right). The low entropy signals are more regular.

In practice using a discrete Fourier transform such as the Fast Fourier Transform (FFT) (Brigham & Morrow, 1967) allows us to then use the summation variant of the entropy, where N is the order of the FFT.

$$S = -\sum_{i=1}^{N} p(f_i) \cdot \ln\left(p(f_i)\right) \quad (5)$$

The utility of the frequency domain approach is apparent when we consider that the frequency domain is partitioned into FFT bins and we have a very fast way of computing the metric, and in many cases more convenient than by calculating the autocorrelation, or by traditional periodogram (Schuster, 1898). If we did have the extra processing time, it may make sense to calculate the entropy off the autocorrelation as that would definitely retain more of the information measure, and thus not losing information due to the transformation process. In practice, however, many times we want to know only *relative* entropy measures, so the loss in precision is not as important as long as we maintain the relative ordering or rank.

As an intuitive interpretation of this measure consider that a value of S near 0 (if <*f*> normalized to 1) indicates that the program is very specialized and is only trying to a few things at once. As S approaches 1 it becomes a general purpose application that attempts to control many different behaviours or responds to events of a more random nature. If S goes much greater than 1, then the temporal complexity turns even more random, even approaching fractal. It becomes progressively harder to test such systems because of the diverging spread in time scales.

This turns into a useful metric since any sufficiently capable controller program or event-driven system will try to accommodate as much real-world functionality as possible, and since the real world tends to disorder and to maximize entropy, this will track an increasingly complex scenario. So in practice, we will have fast cycles interspersed with slower cycles corresponding perhaps to human events or sporadic signals.

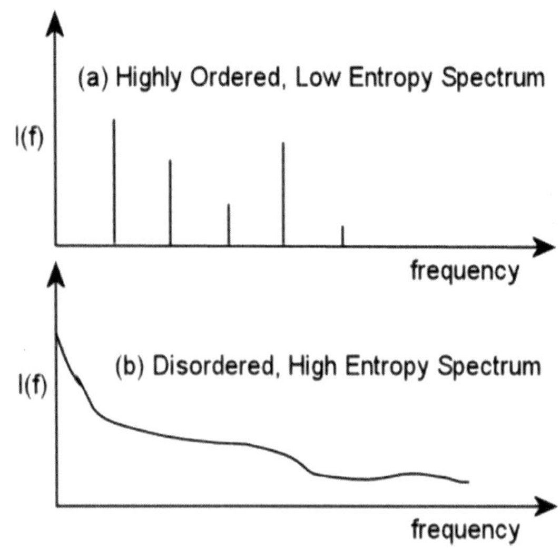

Fig. 2. Variation of power spectrum with increasing disorder. (a) Ordered signal with harmonics generates a low entropy signal. (b) Disordered signal generates a larger intropy.

In further practical terms, to apply the metric we need to run a simulation or collect data from a real or prototype system to gather the statistics. It may be possible to evaluate a program statically but that would require an introspective analysis of a program's structure, a task much more involved than collecting the statistics during typical use or as a test scenario.

The possibility of using this approach on non-concurrent programs such as a single-tasking event-driven program exists but these have less inherent complexity due to their sequential

nature. When the concurrency becomes an integral aspect of the system and in particular when threads must interact with one another does the complexity start to increase quickly. That becomes the essence of the disorder or entropy that we try to capture.
System designers will use tools such as state diagrams, activity diagrams, and Petri nets (Peterson 1981) to architect concurrency behaviors. A typical Petri net is shown below which illustrates the parallel activities and synchronous behaviors. These have various levels of contol, in that they can abstract some of the temporal scale differences by encapsulation of behaviours (Pukite & Ludwig, 2007). However, the actual synchronization will always reveal itself in the realized system.

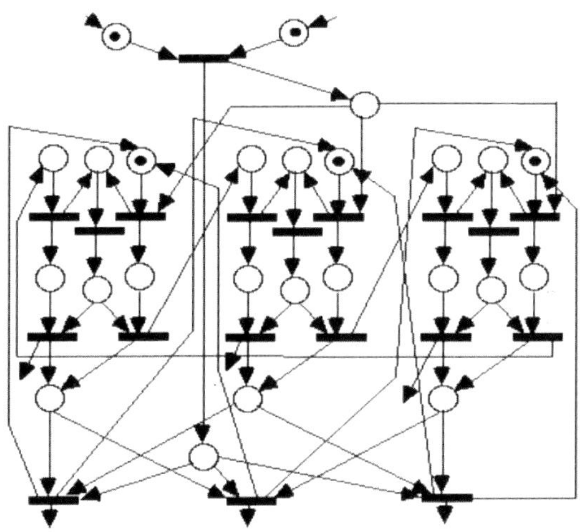

Fig. 3. A typical concurrent automata represented by a Petri net. Multiple threads are represented by filled tokens (places), and context switches occur when threads meet at synchronization points (bars). For large systems, the complexity of interactions canl grow well beyond this level.

2.1 Application

To effectively put the temporal complexity measure into practice, the program or simulation implementation will need a way to time stamp and data log context switches during execution. An implementation of a discrete event task scheduler such as the public source Degas (Ludwig & Pukite, 2006) provides such an instrumentation facility.
The simplest input stream is a list of events signified by time-stamps, which will serve to capture the *c(t)* signal. The values for *c(t)* can be as simple as delta functions with a value of unity indicating a context switch.

2.2 Computation

This algorithm needs a FFT routine. The following pseudo-code guides the calculation calling a generic FFT routine named **Fourier8**. This assumes that each line contains the number of events that occur in the time step and is read through standard input. (This is set to read in exactly 2N lines, where N is an order 15 FFT) If sparse input is provided with time stamps then the data must be transferred into a discrete set of lines with either 0 or N events on each line, due to the nature of the discrete time FFT.

```
with Fourier8;
with Text_IO;
with Ada.Numerics.Generic_Elementary_Functions;
procedure Context_Switching_Entropy is
   subtype Real is Long_Float;
   package mth is new
     Ada.Numerics.Generic_Elementary_Functions (Real);
   use mth;
   Log_Of_Max_Data_Length : constant := 21;
   type Array_Index is range 0..2**(Log_Of_Max_Data_Length+1)-1;
   type Data_Array is array(Array_Index) of Real;
   package fft8 is new Fourier8
     (Real, Array_Index, Data_Array, Log_Of_Max_Data_Length);
   use fft8;
   D_Re, D_Im : Data_Array;
   N : Data_Index;
   Transformed_Data_Last : Data_Index;
   Exp_Table : Exp_Storage;
   Val : Real;
   Sum : Real := 0.0;
   Prob : Real;
   Entropy : Real := 0.0;
begin
    N := Data_Index (2**15 - 1);
    for I in 0..N loop
      -- Read from standard input
      Val := Real'Value(Text_IO.Get_Line);
      D_Re(I) := Val;
      D_Im(I) := 0.0;
   end loop;
    FFT
     (Data_Re                   => D_Re,
      Data_Im                   => D_Im,
      Transformed_Data_Last     => Transformed_Data_Last,
      Input_Data_Last           => N,
      Exp_Table                 => Exp_Table,
      Inverse_FFT_Desired       => False,
      Normalized_Data_Desired   => True);
   for I in Data_Index range 1 .. N-1 loop
      Val := sqrt(D_Re(I)*D_Re(I) + D_Im(I)*D_Im(I));
      Sum := Sum + Val;
   end loop;
   for I in Data_Index range 1..N-1 loop
      Val := sqrt(D_Re(I)*D_Re(I) + D_Im(I)*D_Im(I));
      Prob := Val / Sum;
      if Prob /= 0.0 then
         Entropy:= Entropy - Prob * Log(Prob, 2.0);
      end if;
   end loop;
   Text_IO.Put_Line ("Entropy =" & Entropy'Img);
 end Context_Switching_Entropy;
```

Fig. 4. Pseudo-code algorithm for calculating the context-switching entropy using a library package FFT routine

Note that for a running simulator, if the data is provided continuously through an output port or pipe, then a suitably fast FFT can report a context switching metric in real time. This is useful for diagnostics, for example correlating the detection of errors with high complexity regimes.

2.3 Example of use

We evaluated hypothetical systems with 15 concurrent tasks. One system featured asynchronous events. The other system had tasks cyclically scheduled at different periods. The asynchronous system always had higher switching entropy. The context switching metric distinguishes complicated but synchronized architectures from those with complex temporal behaviour

For this metric, by computing the entropy of the phase spectrum in addition to the amplitude may help in discriminating between complexity and noise. That would be straightforward to include because right now we discard the phase information. Any stationary signal that shows asymmetry must have some peculiar phase relationships going on. So it might be easier to discount randomness in favour of more complex phase relationships if we were to include both the amplitude spectrum and the phase spectrum in the final metric.

Another simple alternative that works on the timing alone is the **gzip** program. This looks at the distribution complexity in times and calculates the entropy metric. The usage of this application is trivial. Take the output file, if a sequence of discrete inputs (zeros and number of values), run it through **gzip** and record the size (Benedetto, et al, 2003). Comparisons of two different sequences of identical length will suggest that the less complex program is the one of the smaller zipped size.

3. Complementary approach – the multi-scale entropy measure

An alternate approach to measuring the temporal complexity involves the application of the multi-scale entropy metric suggested by (Costa, 2005). This differs from the just described context-witching metric in that it measures the complexity of a temporal behaviour or signal over a wide range of fundamental periods. Whereas the single-scale metric works best over a single-decade frequency spectrum scale, a multi-scale metric offers up the possibility of looking at complexities at a variety of time scales, ranging over potentially orders of magnitude. This is definitely useful but it can't be boiled down into a single complexity metric; instead we will need to depict this graphically over what amounts to a logarithmic frequency scale.

The basis for the multi-scale entropy metric is that many real-world behaviours often occur over time scales of varying dynamic ranges. Costa originally applied this to a biomedical application, trying to extract the temporal complexities of cardiac-driven circulatory systems. When such a pulsed cardiac signal is multi-scale, it is actually composed of a fundamental pulse and various arrhythmias, leading to a complicated spectrum of events. The signal actually appears buried amongst competing behavioural periodicities at different time scales and so it becomes that much harder to extract the information.

At first glance, we would imagine that a Fourier transform would work well to extract the periods but in fact a typical FFT algorithm actually works best over a limited dynamic range. By expanding the scope to a multi-scale level, Costa showed that this complexity measure has use in a real-world application and we contend that it may also prove useful for cyber-physical applications such as a complex event-driven system. In this case, the time scales can range from fast interrupt-processing, to human-scale interactivity, to the even more sporadic environmental influences.

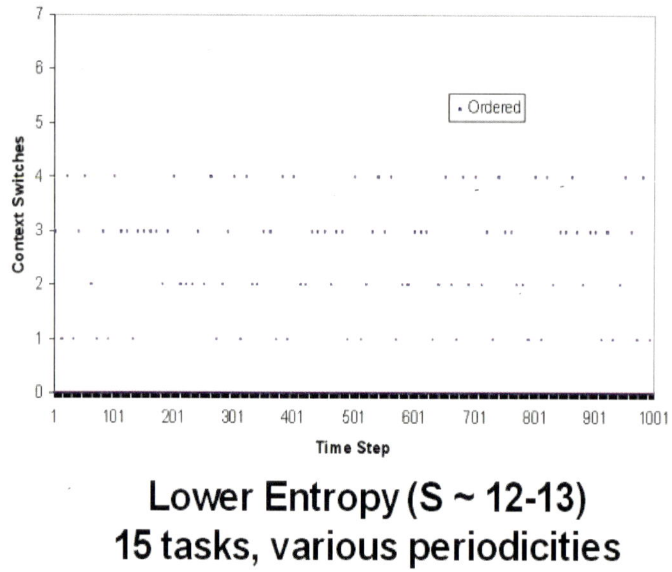

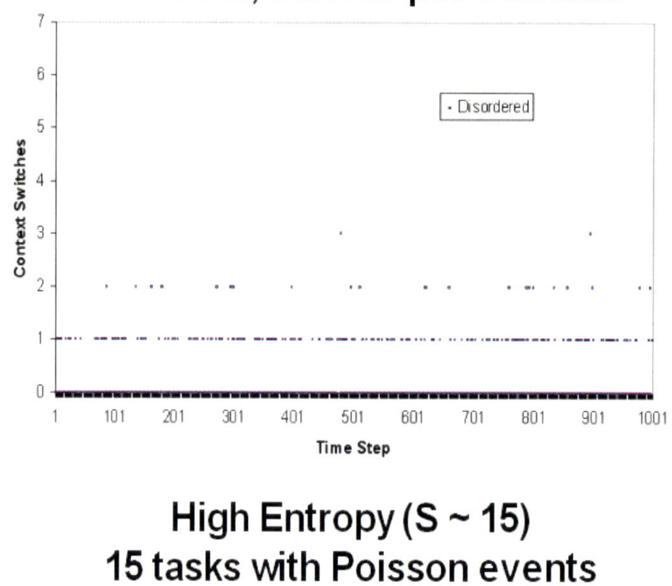

Fig. 5. Context switching metric distinguishes complicated but synchronized architectures from those with complex temporal behaviour.
(top) High Entropy (S ~ 15) 15 tasks with Poisson events
(bottom) Lower Entropy (S ~ 12-13) 15 tasks, randomized events

The multi-scale aspect senses the different temporal frequencies in the underlying signal, comparable to what the contest-switching metric does, but instead pairs or groups the data points to measure a different coarse graining effect. This works out fairly straightforwardly in terms of a autocorrelation. The essential algorithm groups adjacent time samples together in a window of length *Scale* as the coarse graining or moving average measure. Then it counts the number of times, n, that the amplitude will change from one coarse-grained time step to the next.

If the amplitudes don't change for a given coarse-grain then it is predictable and the entropy will be low. To calculate the sample entropy they calculate

$$complexity(Scale) = -\log\left(\frac{n(Scale+1)}{n(Scale)}\right) \quad (6)$$

over each of the scale factors $Scale = 1 .. maxScaleFactor$.

3.1 Usage domains

A graph of the multi-scale entropy will appear flat if it is measuring "1/f" (van der Ziel, 1950) or the so-called pink noise as the underlying behaviour. Pink noise shows a predictable constant change of amplitude density per scale factor; in other words it has a constant energy per frequency doubling while white noise shows constant per frequency interval (Montroll & Shlesinger, 2002).

In comparison to the structure-less noise, if structure does exist in the signal, you will see observable changes in the entropy from one scale factor to the next. For example a superimposed sine wave would show a spike downward in sample entropy when it crossed a harmonic in the scale factor.

A simple interpretation suggests that we scale the measured results relative to the *1/f* noise part of the signal. The *1/f* noise includes the greatest variety of frequencies of any behaviour known and therefore the highest entropy (Milotti, 2002). So by providing a good visualization or graph that plots the *1/f* asymptotic value we can immediately gauge the complexity of a signal. Costa *et al* discuss the difficulty of distinguishing between randomness and increasing complexity, which has importance in the realm of event-driven systems.

> *"In fact, entropy-based metrics are maximized for random sequences, although it is generally accepted that both perfectly ordered and maximally disordered systems possess no complex structures. A meaningful physiologic complexity measure, therefore, should vanish for these two extreme states."*

This is a key insight and one echoed by researchers in complexity theory (Gell-Mann, 1994) in that the most interesting and challenging complexity measures occupy the middle range of the complexity scale. In other words, the most ordered signals can be described by a few harmonic periods and at the other extreme, the complexity reduces to simple stochastic measures akin to statistical mechanics (Reif, 1965). In between these extremes, we require a different level of sophistication.

3.2 Comparison to single-scale metric

The same inputs as that for the Context-Switching Metric described earlier result in the following *Figure 6*. Note that in this case as well, the sample entropy is always higher for the disordered signal than for the ordered signal. The reference *1/f* noise level is shown on the plot to indicate the asymptotic maximum entropy level achievable.

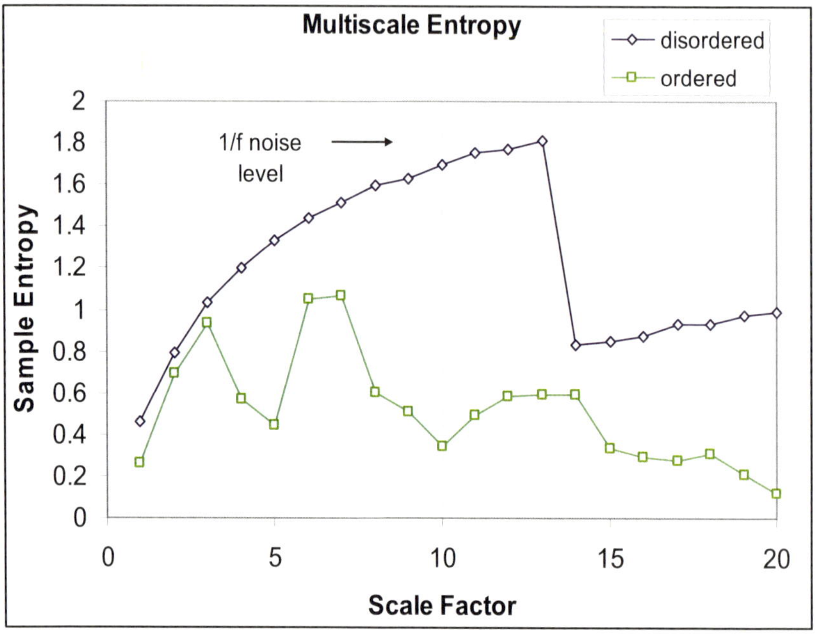

Fig. 6. For the same pair of inputs we used on the context-switching metric, the multiscale entropy appears as the following graph. It shows greater variety than the context-switching metric over the time scales because the metric compares at different levels of resolution.

In practice, the multi-scale algorithm requires only a basic periodogram method invoked over different time scales. The output is one value per temporal scale factor so the results are best displayed as a graph, via a spreadsheet or bar-charting software for example. The calculation is somewhat more brute force compared to the FFT, with complexity $o(n^2)$ versus $o(n*ln(n))$. The context-switching metric operates over a narrower time scale so gets rolled into a single value, simplifying the presentation into a classical scalar metric.

4. Discussion

Interacting concurrent activities can produce behaviour that is difficult to anticipate. The combinatorial capacities of just the ordering of parallel thread execution will exhaust any brute force attempt at testing the possible permutations. Moreover, the possibilities of rare anomalies attributed to corner cases of executions may take non-determinant times to reveal, or worse, are not easily repeatable. The sporadic defect that occurs rarely in the lab usually has a real origin and should never be ignored **(Kuhn, et al, 2004).**

The figure below shows a representative timeline trace of the interacting subsystems on a typical automated ground vehicle. The task interactions are interleaved and pipelined.

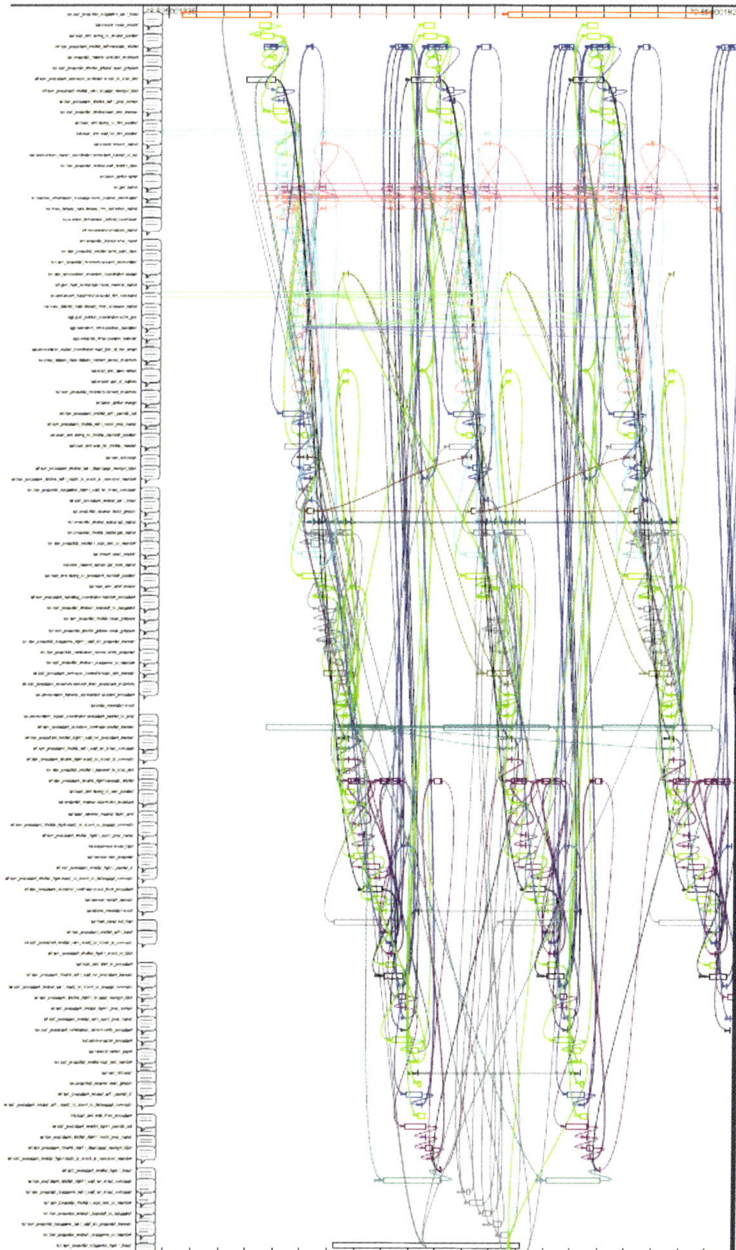

Fig. 7. GANTT chart of a typical vehicle system execution trace showing interacting threads. Time proceeds left to right, and one thread exists per horizontal entry. The lines indicate thread synchronization points. This diagram is only meant to give a notional idea of complexity, and the text description along the left edge is irrelevant to the discussion.

4.1 Applications to design

The difficulty in designing and debugging of such systems necessitates practical simulation tools and simplified metrics such as the entropy measures described here. Both of these entropy metrics are potentially useful during system development as an *analysis of alternatives* or *design space exploration* tool. If several concurrent options are available for a design, either of these metric could provide criteria to establish the least complex design. In that sense, it serves as a simple utility function to measure unwanted or creeping complexity much like a duty-cycle utilization measures processor contention.

4.2 Applications to test and verification

A close connection exists between the entropy metrics and usage modelling (Whittaker, 1994) for program verification. Most non-exhaustive testing requires a mix of tests taken during nominal conditions along with tests sampled according to potentially rare conditions. This consideration takes into account the number of test vectors and the path coverage for testing. Any characterization at this stage will provide useful inputs to generating a stochastic measure of reliability. This could incorporate stochastic usage models and a log-likelihood metric is often used to compare between two state space probabilities. Between an entropic measure and a usage model, we can cover the temporal and path dimensions of a program's execution and its programmatic complexity.

4.3 Applications to diagnostics

As a diagnostic tool, the context switching metric can also detect potential complexities during execution. Since the FFT can easily compute in real-time for typical sample sizes N, parallel execution of the entropy algorithm with the context switching data can reveal deviations from expected operation. For example, if an execution profile shows a high regularity of frequent context switches during some interval and then transitions to a more irregular sequence of switches with the same overall density, the expected entropy measure will definitely increase. In that sense, the entropy metric measures an intrinsic property of the signal, and that strictly speaking, density fluctuations such as expected increases in the rate of context switches will not influence the measure. In other words, density alone does not affect the complexity.

By the same token, the multi-scale metric has obvious benefit for detecting long term complexity changes or short-term bursts buried in a nominally sampled signal. The idea of using frequency domain entropy for diagnostics of complex machinery is further explored in (Shen, 2000).

5. Conclusion

We described a complexity metric for concurrent software controlled systems or concurrent realizations of behaviour. The novel approach of creating a complexity metric for context switching involves the analysis of the switching frequency spectrum. We take a Fourier transform of the temporally distributed context switching events ($c(t)$) and treat that as a probability density function in frequency space (i.e. a normalized power spectrum). Then the entropy (S) of $p(f)$ will generate a simple complexity measure.

The context switching metric can be used during system development as an analysis of alternative utility function. If several design options or algorithms are available, the context-

switching metric can be used as selection criteria to minimize inherent algorithmic complexity. It is comparable or equivalent to the Shannon information metric, which essentially measures entropy of a system.
As an alternative approach we compared this against a multi-scale entropy measure. Although more involved in construction, the multi-scale entropy can be used as an orthogonal metric, perhaps more useful for measuring temporal behaviours of a wide dynamic range or as a more detailed diagnostic tool. This will reveal finer structures in complexity than the single-scale metric can.

6. Acknowledgment

We wish to thank the DARPA META program for providing encouragement to develop complexity metrics for cyber-physical system development.

7. References

Bao, L. Intille, S. S., (2004). Activity Recognition from User-Annotated Acceleration Data, Pervasive Computing, Lecture Notes in Computer Science, Vol.3001, pages 1-17, Springer-Verlag, ISSN 0302-9743

Benedetto, D. Caglioti, E. Loreto, V., (2002). Language Trees and Zipping, *Physical Review Letters,* Vol.88, No.4, ISSN 0031-9007.

Brigham, E. O. and Morrow, R. E., (1967). The Fast Fourier Transform, IEEE Spectrum, Vol.4, No.12, p.63-70, ISSN: 0018-9235 .

Costa, M., Goldberger, A.L., and Peng, C.-K. (2005). Multiscale entropy analysis of biological signals, *Physical Review E,* Vol.71, No.2.

Feller, W., (1957). *An Introduction to Probability Theory and Its Applications,* Volume I and II, John Wiley, 1957

Gell-Mann, M., (1994). *The Quark and the Jaguar: Adventures in the Simple and the Complex,* W.H. Freeman and Co, ISBN0-7167-2725-0.

Jaynes, E. T., (2003). *Probability Theory: The Logic of Science,* Cambridge University Press, ISBN 052159271.

Klein, M., (1993). *A Practitioner's handbook for real-time analysis: Guide to rate monotonic analysis for real-time systems,* Kluwer Academic Publishers, ISBN 0792393619 .

Kuhn, D.R., Wallace, D.R., Gallo, A.M., Jr., (2004). Software fault interactions and implications for software testing, *IEEE Transactions on Software Engineering,* Vol.30 No.6, p.418-421, ISSN: 0098-5589

Leon-Garcia, A., (2008). *Probability and Random Processes for Electrical Engineering,* Prentice-Hall, p.273, (2008).

Ludwig, L. and Pukite, P., (2006). DEGAS: discrete event Gnu advanced scheduler, *Proceedings of the 2006 annual ACM international conference on SIGAda,* ACM SIGAda Ada Letters, Vol XXVI Issue 3 , ISBN 1-59593-563-0.

Milotti, E., (2002). 1/f Noise: A Pedagogical Review, invited talk to E-GLEA-2, Buenos Aires, arXiv:physics/0204033v1

Montroll, E.W. and M.F. Shlesinger, (1982) .On 1/f noise and other distributions with long tails, Proceedings of the National Academy of Sciences, 79/10, 1982.

Peterson, J. L., (1981). Petri Net Theory and the Modeling of Systems, Prentice-Hall, Englewood Cliffs, NJ.

Pukite, P., and Ludwig, L., (2007). Generic discrete event simulations using *DEGAS*:: application to logic design and digital signal processing, *Proceedings of the 2007 ACM international conference on SIGAda,* ACM SIGAda Ada Letters, Vol XXVII Issue 3, ISBN 978-1-59593-876-3

Reif, F., (1965). *Statistical and Thermal Physics*, McGraw-Hill.

Schuster, A., (1898). On the investigation of hidden periodicities with application to a supposed 26 day period of meteorological phenomena, *Terrestrial Magnetism and Atmospheric Electricity*, Vol. 3, p.13-41.

Shen, L., Tay F.E.H., Qu. L., Shen Y., (2000). Fault diagnosis using Rough Sets Theory, *Computers in Industry*, Vol.43.

van der Ziel, A. (1950). On the noise spectra of semi-conductor noise and of flicker effect, 1950, Physica 16, 359-372.

Whittaker, M., (1994). A Markov chain model for statistical software testing, *IEEE Transactions on Software Engineering*.

19

An Optimization of 16-Point Discrete Cosine Transform Implemented into a FPGA as a Design for a Spectral First Level Surface Detector Trigger in Extensive Air Shower Experiments

Zbigniew Szadkowski
University of Łódź
Department of Physics and Applied Informatics,
Faculty of High Energy Astrophysics, Łódź
Poland

1. Introduction

The Pierre Auger Observatory is a ground based detector located in Malargue (Argentina) (Auger South) at 1400 m above the sea level and dedicated to the detection of ultra high-energy cosmic rays with energies above 10^{18} eV with unprecedented statistical and systematical accuracy. The main goal of cosmic rays investigation in this energy range is to determine the origin and nature of particles produced at these enormous energies as well as their energy spectrum. These cosmic particles carry information complementary to neutrinos and photons and even gravitational waves. They also provide an extremely energetic stream for the study of particle interactions at energies orders of magnitude above energies reached at terrestrial accelerators (Abraham J. et al., 2004).
The flux of cosmic rays above 10^{19} eV is extraordinarily low: on the order of one event per square-kilometer per century. Only detectors of exceptional size, thousands of square-kilometers, may acquire a significant number of events. The nature of the primary particles must be inferred from properties of the associated extensive air showers (EAS).
The Pierre Auger Observatory consists of a surface detectors (SD) array spread over 3000 km^2 for measuring the charged particles of EAS and their lateral density profile of muon and electromagnetic components in the shower front at ground, and of 24 wide-angle Schmidt telescopes installed at 4 locations at the boundary of the ground array measuring the fluorescence light associated with the evolution of air showers: the growth and subsequent deterioration during a development. Such a "hybrid" measurements allow cross-calibrations between different experimental techniques, controlling and reducing the systematic uncertainties.
Very inclined showers are different from the ordinary vertical ones. At large zenith angles the slant atmospheric depth to ground level is enough to absorb the part of the shower that follows from the standard cascading interactions, both of electromagnetic and hadronic type. Only penetrating particles such as muons and neutrinos can traverse the atmosphere at large zenith angles to reach the ground or to induce secondary showers deep in the atmosphere and close to an air shower detector.

The ability to analyze inclined showers with zenith angles larger than 60° induced by neutrinos or photons essentially increases the acceptance of the surface array and opens a part of the sky that was previously inaccessible to the detector. These showers provide a new tool for ultra high energy cosmic rays interpretation because they are probing muons of significantly higher energies than vertical showers. Spectral triggers offering a pattern recognition in a frequency domain may improve a standard detection technique based on the signal coincidences from many PMT channels above some thresholds in the time domain. The "old" muon shower fronts have only a small longitudinal extension, which is leading to short detector signals also in time. To identify these showers at the presence of "young" showers with a large electromagnetic component one may need a very good spectral sensitivity to the fast muon component in the trigger.

The main advantage of the spectral trigger is the scaling feature. The set of the DCT coefficients depends only on the shape of signals, not on their amplitudes. Triggers sensitive on the shape of FADC traces may detect events with expected characteristics i.e. the fast attenuated, very short peaks related to the muonic, flat fronts coming from very inclined showers. Independence of the amplitude is especially promising for the Auger North, where due to a single PMT in the surface detectors the coincidence technique cannot be used. In order to keep reasonable trigger rate for the 1st level trigger (ca. 100 Hz), the threshold for the 1st trigger should be much higher than for example in the Pierre Auger Observatory, where 3-fold coincidences attenuated a noise.

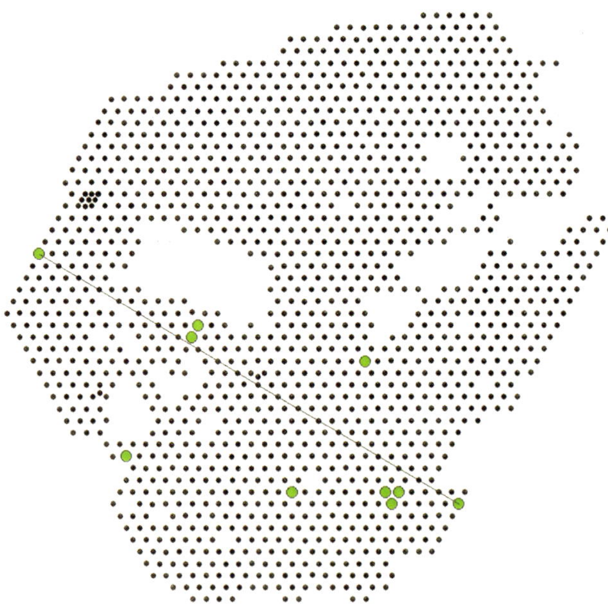

Fig. 1. Position of triggered surface detectors on the Auger array for the very inclined shower ($\theta = 83.5°$) nr 1155555. Muons triggered only few surface detectors, although they crossed several hundred detectors. A distance between opposite detectors is 54 km.

2. Triggers

Two different triggers are currently implemented at the 1st level. The first is a single-bin trigger generated as 3-fold coincidence of the 3 PMTs at a threshold equivalent to 1.75 vertical emitted muons. The estimated current for a Vertical Equivalent Muon (I_{VEM}) is the reference unit for the calibration of FADC traces signals and corresponds to ca. 50 ADC-counts. This trigger has a rate of about 100 Hz. It is used mainly to detect fast signals, which correspond also to the muonic component generated by horizontal showers. The single bin trigger is generated when the input signal is above the fixed thresholds calculated in the micro-controller during the calibration process. It is the simplest trigger useful for high-level signals. The second trigger is the Time over Threshold (ToT) trigger that requires at least 13 time bins above a threshold of 0.2 I_{VEM}. A pre-trigger ("fired" time bin) is generated if in a sliding time window of 120 $\times$ 25 ns length a coincidence of any two channels appears. This trigger has a relatively low rate of about 1.6 Hz, which is the expected rate for two muons crossing the Auger surface detector. It is designed mainly for selecting small but spread-in-time signals, typical for high energy distant EAS or for low energy showers, while ignoring the single muon background (Abraham J. et al., 2010).

Cherenkov light generated by very inclined showers crossing the Auger surface detector can reach the PMT directly without reflections on Tyvec liners. Especially for "old" showers the muonic front is very flat. This together corresponds to very short direct light pulse falling on the PMT and in consequence very short rise time of the PMT response. For vertical or weakly inclined showers, where the geometry does not allow reaching the Cherenkov light directly on the PMT, the light pulse is collected from many reflections on the tank walls. Additionally, the shower developed for not so high slant depth are relatively thick. These give a signal from a PMT as spread in time and relatively slow increasing.

Hadron induced showers with dominant muon component give an early peak with a typical rise time mostly from 1 to 2 time bins (by 40 MHz sampling) and decay time of the order of 80 ns (Aglietta et al., 2005). The estimation of the rise time for the front on the base of one or two time bins is rather rough. The rise time calculated as for two time bins may be overestimated due to a low sampling rate and an error in a quantization in time. Higher time resolution would be favorable. The expected shape of FADC traces suggests to use a spectral trigger, instead of a pure threshold analysis in order to recognize the shape of the FADC traces characteristic for the traces of very inclined showers. The monitoring of the shape would include both the analysis of the rising edge and the exponentially attenuated tail. A very short rise time together with a relatively fast attenuated tail could be a signature of very inclined showers. We observe numerous very inclined showers crossing the full array but which "fire" only few surface detectors (Fig. 1). For that showers much more detectors should have been hit. Muonic front probably produces PMT signals not high enough to generate 3-fold coincidences, some of signals are below of thresholds (see Fig. 2). This may be a reason of "gaps" in the array of activated surface detectors.

3. Discrete Fourier Transform vs. Discrete Cosine Transform

There are several variants of the DCT with slightly modified definitions. The DCT-I is exactly equivalent (up to an overall scale factor of 2), to a DFT of 2N - 2 real numbers with even symmetry. The most commonly used form of the Discrete Cosine Transform is DCT-II.

$$\bar{X}_k = \alpha_k \sum_{n=0}^{N-1} x_n cos\left(\frac{\pi}{N}(n+\frac{1}{2})k\right) \tag{1}$$

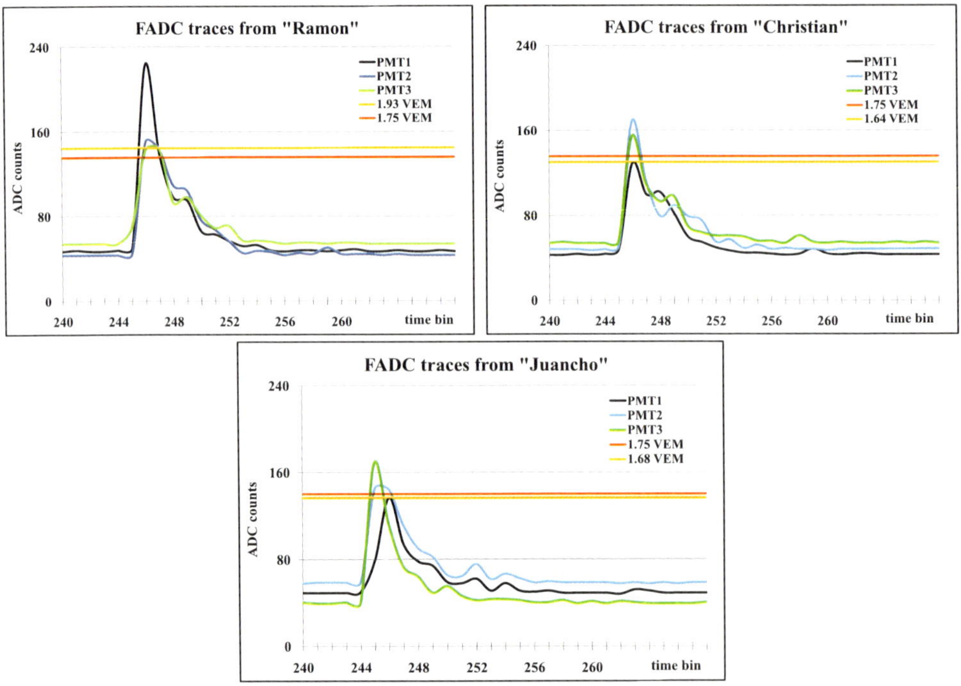

Fig. 2. FADC traces (in ADC-counts) of a horizontal shower (no. 01145055: $\theta = 83.3^{\circ}$) registered in three detectors: Ramon, Christian and Juancho, respectively, and shown for the range of (240 - 265) time bins. Only the signal in the Ramon detector (1.93 I_{VEM}) is above the standard threshold of 1.75 I_{VEM}. Signals in Cristian (1.64 I_{VEM}) and Juancho (1.68 I_{VEM}) detectors are below the standard thresholds and they are detected by chance (compare a registration efficiency for a similar event shown in Fig. 1). For all very inclined showers the rising edge corresponds to one or two time bins.

where $\alpha_0 = \frac{1}{\sqrt{N}}$ and $\alpha_k = \frac{2}{\sqrt{N}}$ for $k \geq 1$.
The DCT-III form is sometimes simply referred to as "the inverse DCT" (IDCT). A variant of the DCT-IV, where data from different transforms are overlapped, is called the Modified Discrete Cosine Transform (MDCT). The DCT is a Fourier-related transform similar to the DFT, but using only real numbers. DCT are equivalent to DFT of roughly twice the length, operating on real data with even symmetry (since the Fourier transform of a real and even function is real and even), where in some variants the input and/or output data are shifted by half a sample. The DCT-II and DCT-IV are considered as the alternative approach to the FFT. In fact, the FFT routine can be supplied in an interleaving mode, even samples treated as real data, odd samples as imaginary data. A trigger based on Discrete Fourier Transform (DFT) (Radix-2 FFT) (Szadkowski, 2006) has already been implemented in the 3rd generation of the Front FEB based on Cyclone™ Altera® chip (Szadkowski, 2005b). However, for real signal x_n

$$\bar{X}_{\frac{N}{2}+k} = \sum_{n=0}^{N-1} x_n e^{-j\frac{2\pi}{N}\left(\frac{N}{2}+k\right)n} = \sum_{n=0}^{N-1} x_n (-1)^n \left[e^{j\frac{2\pi}{N}kn}\right]^* = \bar{X}^*_{\frac{N}{2}-k} \quad (2)$$

and $\left(\frac{N}{2}^{th}\right)$ spectral line of $\bar{X}_k$, k = 0,1,...,N-1 is lying on a symmetry axis: the real part is symmetric, the imaginary part is asymmetric. The useful information is contained only in 1st $\left(\frac{N}{2}+1\right)$ spectral lines for k = 0,1,...,N/2 corresponding to frequencies $f_k = k \cdot f_0 = k\frac{1}{N\Delta t}$, changing from zero to $\frac{f_{smpl}}{2}$ with $\frac{f_{smpl}}{N}$ grid.

3.1 Pedestal independence

The analog section of the FEB has been designed to have a pedestal of ca. 10 % of the full FADC range in order to investigate undershoots. However, the pedestal is relatively sensitive on the temperature. Daily variation of the pedestal may reach 5 ADC-counts. The trigger pedestal-independent is very welcome. Let us consider signal with a constant pedestal:

$$\bar{X}_k(ped) = \sum_{n=0}^{N-1}(x_n + ped)F(k,n) = \bar{X}_k + ped\sum_{n=0}^{N-1}F(k,n) = \bar{X}_k + ped \times W \tag{3}$$

$$F(k,n) = cos\left(\frac{k\pi}{N}(n+\frac{1}{2})\right) \tag{4}$$

Due to symmetry and parity of the cosine, we get for odd and even indices respectively:

$$W = 2\sum_{n=0}^{\frac{N}{2}-1}\left\{cos\left(\frac{k\pi}{2}\right)cos\left[\frac{\pi}{N}\left(n+\frac{1}{2}\right)k - \frac{k\pi}{2}\right]\right\} = \begin{cases} 0 & , k-odd \\ 2\sum_{n=0}^{\frac{N}{2}-1}F(k,n) & , k-even \end{cases} \tag{5}$$

By a recursion, repeating (5) we get finally $\frac{N}{2} = 2$ and $k = 0, \frac{N}{2}$. For $k = \frac{N}{2}$

$$\sum_{n=0}^{1} cos\left(\frac{\pi}{2}\left(n+\frac{1}{2}\right)\right) = 0 \tag{6}$$

In a consequence for k > 0 the DCT coefficients are independent of the pedestal.

3.2 Scaling

The DCT algorithm has a significant advantage in comparison to the FFT one. The structure of DCT coefficients is much simpler for interpretation and for a trigger implementation than the structure of the FFT real and imaginary coefficients (compare 4th of the FFT data vs. 2nd row for the DCT coefficients in Fig. 3). For the exponentially attenuated signals from the PMTs higher DCT coefficients (scaled to the 1st harmonics)

$$\xi_k = \frac{\bar{X}_k}{\bar{X}_1} \tag{7}$$

are almost negligible, while both real and imaginary parts of the FFT (scaled to the module of the 1st harmonics) give relatively significant contributions and are not relevant for triggering. When a peak appears in the pure attenuated signal (last column in Fig. 3) the structure of the DCT dramatically changes and trigger condition immediately expires, while modules of FFT components almost do not change. The structure of FFT harmonics for the last graph in Fig. 3 would be more suitable for a trigger (almost negligible imaginary part for higher harmonics and also relatively low real harmonics), however it corresponds just to situation, when the

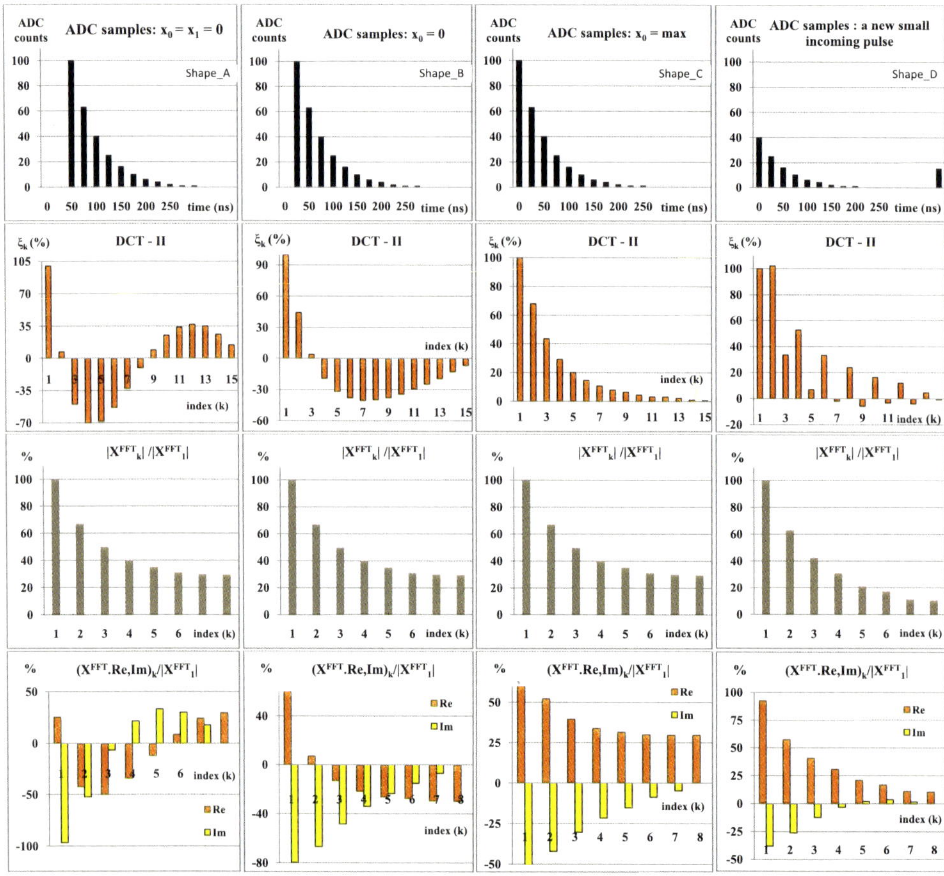

Fig. 3. A propagation of the pulse (1st row) through the shift register, DCT-II coefficients (2nd row), absolute values of the DFT (3rd row) and corresponding real (Re), imaginary parts (Im) (4th row). The 1st column shows the pulse (shape A), when two time bins are on the pedestal level, the 2nd one (shape B), when only the one time bin is still on the pedestal level, while the 3rd one (shape C) shows the pulse fully fulfilled the range of investigating shift registers. For a signal shape related to the exponential attenuation (shape C), the contribution of higher DCT coefficients is small and suitable for a trigger. When a peak appears in the declining signal (last column - shape D), the DCT coefficients immediately excesses assumed relatively narrow acceptance range for triggers. The DFT coefficients (Re and Im in 4th row) have similar structure as the DCT, however for the pure exponentially declining signal the higher real DFT harmonics have relatively high values and they are not suitable for triggering. Absolute values of DFT components (3rd row) are clearly insensitive on discussed conditions.

pure attenuated signal is distorted by some peak on the tail and a trigger condition has been violated.
The plot in the 4th row and 3rd column on Fig. 3 shows a contribution of the DFT vs. the absolute value of the 1st harmonic. For an exponential attenuated signal (with the attenuation factor = β) the contribution of both real and imaginary coefficients decreases monotonically with a significant value for all real coefficients. From the DFT definition we get:

$$\bar{X}_k = A \sum_{n=0}^{N-1} e^{\beta n} e^{-j\frac{2\pi k}{N} n} = A \frac{1 - e^{-N\beta}}{1 - e^{\beta + j\frac{2\pi}{N} k}} \tag{8}$$

$$\xi = \frac{Re(\bar{X}_k)}{\|\bar{X}_1\|} = \frac{(1 + e^{-\beta N})}{\sqrt{(1 - e^{-\beta} cos\phi)^2 + e^{-2\beta} sin^2\phi}} \tag{9}$$

where $\phi = \frac{2\pi k}{N}$.
Calculating (8) for boundary factors β = (0.28, 0.42) (from the Auger database) and for k = N/2 (as the lowest in a monotonically decreasing chain), we obtain for N = 16: ξ = 24% and 28%, respectively. These values are too large to be use for triggering. Even an extension of the DFT size does not help very much. For N = 32: we get still large values: ξ = 17% and 23%. Almost vanishing higher DCT coefficients provide much natural trigger conditions. 32-point FFT (roughly equivalent to 16-point DCT) does not offer better stability.

4. Genaral DCT algorithm

The DCT for real signal x_n gives independent spectral coefficients for k = 0,1,...,N-1, changing f_k also from zero to $\frac{f_{sampl}}{2}$ but with $\frac{f_{sampl}}{2N}$ grid. DCT vs. DFT gives twice better resolution. Splitting the sum (1) and redefine the indices we get:

$$\bar{X}_k = \alpha_k \left\{ \sum_{n=0}^{\frac{N}{2}-1} \left(x_n cos\left[\frac{\pi}{N}\left(n + \frac{1}{2}\right)k\right] - x_{N-1-n} cos\left[\frac{\pi}{N}\left(N - \left(n + \frac{1}{2}\right)\right)k\right]\right)\right\} \tag{10}$$

Due to symmetry of the cosine function

$$\bar{X}_k = \alpha_k \sum_{n=0}^{\frac{N}{2}-1} cos\left[\frac{\pi}{N}\left(n + \frac{1}{2}\right)k\right] \times \left(x_n + (-1)^k x_{N-1-n}\right) \tag{11}$$

We can introduce the new set of variables:

$$A_n = \begin{cases} x_n + x_{N-1-n} & (n = 0, ..., \frac{N}{2} - 1) \\ x_{N-1-n} - x_n & (n = \frac{N}{2}, ..., N - 1) \end{cases} \tag{12}$$

DCT coefficients can be separated for even and odd indices respectively:

$$\bar{X}_{\binom{even}{odd}} = \alpha_k \sum_{n=\binom{0}{N-1}}^{\frac{N}{2}-\binom{1}{0}} A_n cos\left[\frac{\pi}{N}\left(n + \frac{1}{2}\right)k\right] \tag{13}$$

Let us notice that (13) for even indices has the same structure as (1) with only shorter range of indices. Recurrently we can introduce new sets of variables for the set of indices k = 2p, where p is integer, till k < N. In order to use symmetry of trigonometric functions in a maximal way, N should be a power of 2, similarly to Radix-2 approach used in FFT algorithm. If N = 2q, recurrent minimization is possible till p = q. The twiddle factors for successive minimization steps m equal to $cos\left(\frac{2\pi}{2^q}\frac{2^{p+m}}{2}\right) = -1$, because the sum of step index m and range factor p is constant and equals to q. For the rest of indices twiddle factor depends on fractional angle $\alpha = \frac{\pi 2^{q-m-1}}{N}$.

After the 1st step of minimization, the terms of the sum (13) for odd indices depends only on the odd multiplicity of the fractional angle

$$\bar{X}_k = \alpha_k \sum_{n=N-1}^{\frac{N}{2}-1} A_n cos\left[\frac{\pi}{2N}(2n+1)k\right] \tag{14}$$

Using a following trigonometric identity

$$cos\alpha = \frac{1}{2cos\beta}(cos(\alpha+\beta) + cos(\alpha-\beta)) \tag{15}$$

the fractional angles can be increased by the factor of 2 for $\beta = \frac{k\pi}{2N}$. Thus:

$$\bar{X}_k = \frac{\alpha_k}{2cos\left(\frac{k\pi}{2N}\right)} \sum_{n=N-1}^{N/2} A_n \times \left[cos\left(\frac{k\pi}{N}(n+1)\right) + cos\left(\frac{k\pi}{N}n\right)\right] \tag{16}$$

Let us notice that:

1). $cos(k\pi) = (-1)^k$, for n = N-1, hence pure A_n coefficient survives,

2). $cos(\frac{k\pi}{2}) = 0$, for $n = \frac{N}{2}$ because of odd k,

3). the rest of indices appear in cosine terms twice in A_{n+1} and A_n coefficients, which allows introducing the new set of variables

$$\begin{aligned} B_{N-1} &= A_{N-1} \\ B_{N-1-n} &= A_{N-n} + A_{N-1-n} \end{aligned} \tag{17}$$

The range of B_n indices is continuous and can be split again on even and odd parts. The above procedure can be repeated in recurrence.

5. 8-point DCT algorithm

For N = 8 according to formulae (12) and (17) we get :

$$A_{0,1,2,3} = x_{0,1,2,3} + x_{7,6,5,4} \qquad A_{7,6,5,4} = x_{0,1,2,3} - x_{7,6,5,4} \tag{18}$$

$$B_{0,1} = A_{0,1} + A_{3,2} \qquad B_{2,3} = A_{0,1} - A_{3,2} \qquad B_{6,5,4} = A_{7,6,5} + A_{6,5,4} \qquad B_7 = A_7 \tag{19}$$

For even indices the DCT coefficients are expressed as follows:

$$2\sqrt{2}\begin{bmatrix}\bar{X}_0\\ \bar{X}_4\end{bmatrix} = \begin{bmatrix}1 & 1\\ 1 & -1\end{bmatrix}\begin{bmatrix}B_0\\ B_1\end{bmatrix} \qquad 4\begin{bmatrix}S_2\bar{X}_2\\ S_6\bar{X}_6\end{bmatrix} = B_3 + S_4\begin{bmatrix}1 & 1\\ -1 & -1\end{bmatrix}\begin{bmatrix}B_3\\ B_2\end{bmatrix} \tag{20}$$

where

$$S_k = cos\left(\frac{k\pi}{16}\right) \tag{21}$$

For odd indices with a support of (15) we get:

$$4\begin{bmatrix} S_1\bar{X}_1 \\ S_7\bar{X}_7 \\ S_3\bar{X}_3 \\ S_5\bar{X}_5 \end{bmatrix} = \begin{bmatrix} 1 & 0 & 1 & 1 & 1 & 0 \\ 1 & 0 & -1 & 1 & -1 & 0 \\ 1 & -1 & 0 & -1 & 0 & 1 \\ 1 & 1 & 0 & -1 & 0 & -1 \end{bmatrix} \begin{bmatrix} B_7 \\ S_2B_4 \\ S_6B_4 \\ S_4B_5 \\ S_2B_6 \\ S_6B_6 \end{bmatrix} \tag{22}$$

A direct approach from the classical definition requires: a single multiplication for even indices (20) and 5 multiplications for odd indices (22). The scaled coefficients $S_{1,7,3,5}\bar{X}_{1,7,3,5}$ in (22) can be expressed in an equivalent way introduced by Arai, Agui, Nakajima (AAN, 1988)., which allows reducing an amount of multiplications from 5 to 4 only.

$$4\begin{bmatrix} S_1\bar{X}_1 \\ S_7\bar{X}_7 \\ S_3\bar{X}_3 \\ S_5\bar{X}_5 \end{bmatrix} = \begin{bmatrix} 1 & 1 & 1 & 0 & -1 \\ 1 & 1 & -1 & 0 & 1 \\ 1 & -1 & 0 & -1 & 1 \\ 1 & -1 & 0 & 1 & -1 \end{bmatrix} \begin{bmatrix} B_7 \\ S_4B_5 \\ (S_2+S_6)B_6 \\ (S_2-S_6)B_4 \\ S_6(B_6-B_4) \end{bmatrix} \tag{23}$$

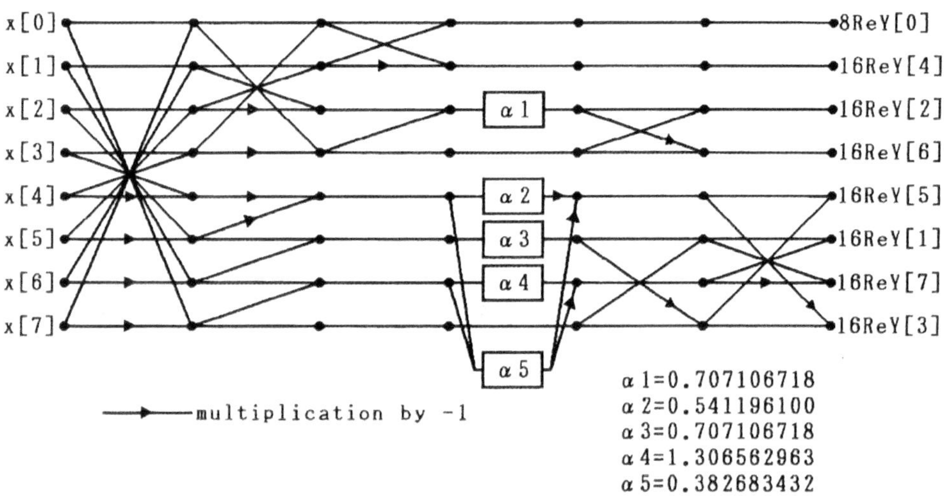

Fig. 4. A fast DCT algorithm developed in 1988 by Arai, Agui and Nakajima

A minimization of multiplications amounts is one of a fundamental goal in long-term numerical calculations. Reduction of product terms significantly speed up sophisticated calculations, because a single multiplication requires several clock cycles of processor. Multiplications in powerful FPGA chips can be however performed in very fast dedicated DSP blocks in a single clock cycle. Signals processed in parallel threads in a hardware implementation of a pipeline design have to be synchronized to each other. Pipeline approach requires additional shift registers for synchronization also for signal currently not being processed. However, such synchronization needs additional resources. Fig. 5 shows the part of pipeline chain corresponding to odd indices of DCT coefficients (lower part in Fig. 4).

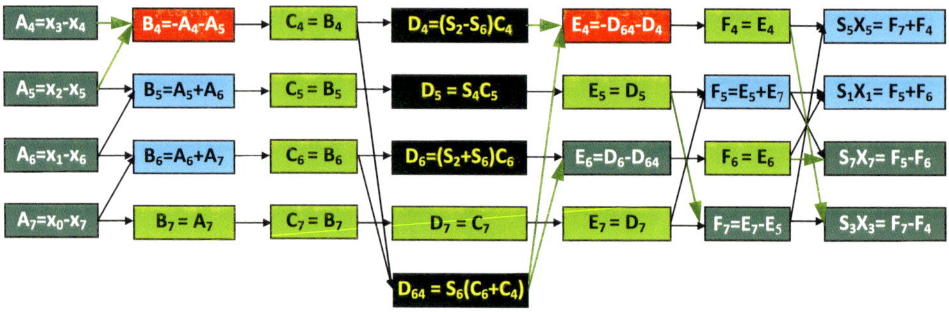

Fig. 5. The AAN algorithm limited to indices 4 - 7 only with a time-oriented structure. Adders, sub-tractors, multipliers and shift registers are marked by the following colours: blue, gray, black and green, respectively. Red colour corresponds to routines requiring a cascade processes.

A direct implementation of the pure AAN algorithm requires 7 pipeline stages, which utilize additional resources of shift registers for synchronization for operations like: X(t+1) = X(t). In a numerical calculation in processors data are simply waiting for a next performance cycle. The D_{64} block contains a cascade of the sum and the multiplication. An implementation of the cascade in a single clock FPGA logic block significantly reduce a speed. Additionally, the *lpm_add_sub* mega-function from the Altera® library of parameterized modules (LPM) does not support an inversion of a sum i.e. $B_4 = -(A_4 + A_5)$ or $E_4 = -(D_{64} + D_4)$. These operations would have to be performed in a cascade way by an adder and a sign inversion. Cascade operations performed in the same clock cycle significantly slow down a global registered performance.

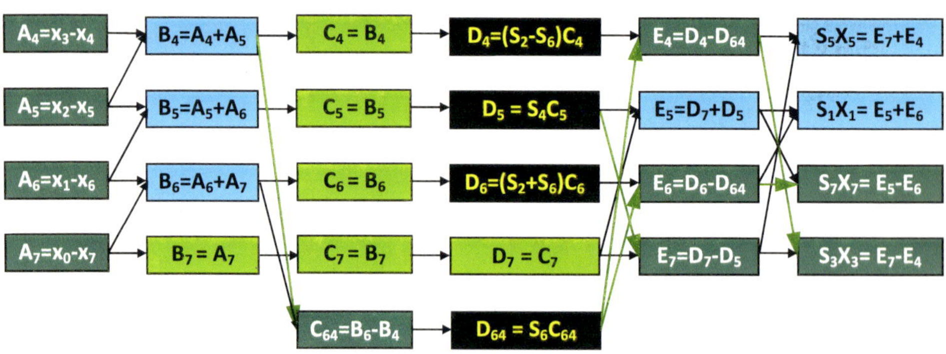

Fig. 6. Optimized AAN algorithm for indices 4 - 7. A redefinition and splitting of variables allowed a reduction of the chain length.

A simple redefinition of nodes removes difficulties mentioned above. The B_4 node defined as the sum of $A_{4,5}$ nodes requires a simple *lpm_add_sub* mega-function. The D_4 node with currently inverted sign allows using *lpm_add_sub* in E_4 performing a subtraction. The D_{64} node from Fig. 5 can be split into the subtraction C_{64} and the multiplication D_{64} in the next clock cycle (Fig. 6).

Routines E and F from Fig. 5 have been merged into single routine E (Fig. 6) to short an amount of pipeline stages and remove unnecessary shift registers.

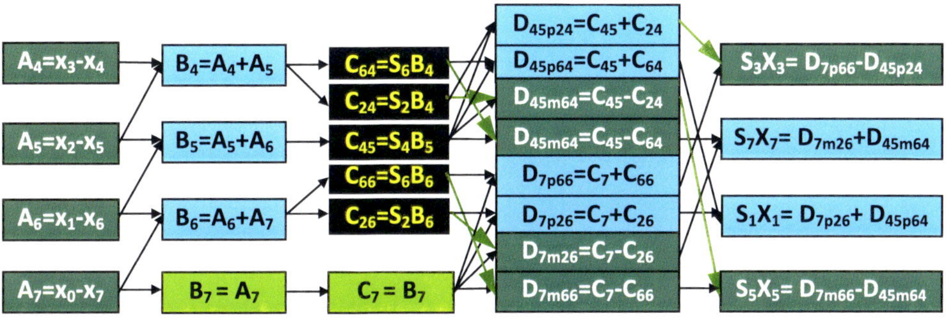

Fig. 7. Optimized, shorter pipeline chain based on the classical approach. The reduction of the length of the chain at the cost of an additional multiplier.

A classical approach reduces a length of the chain from 6 to 5 stages only, at the cost of one additional multipliers. An abridgement of the pipeline chain and in a consequence a reduction of the shift registers needed for synchronization allows saving significant amount of logic blocks, especially for wide data bus. In order to reduce an approximation errors, the data bus in the intermediate stages is enlarged.

6. 16-point DCT algorithm

The 16-point DCT algorithm will be implemented according to the classical approach with an optimization of the number of pipeline stages at the cost of an utilization of embedded multipliers (Szadkowski, 2009). The 1st and the 2nd pipeline stages utilize the set of variables (12) and (17) respectively. For N = 16 the fractional angle of the twiddle factor in the 1st step of minimization equals to $\beta = \pi$. The same fractional angle corresponds to the 2nd step of minimization for even indices corresponded to A_n.

$$B_{0,1,2,3} = A_{0,1,2,3} + A_{7,6,5,4} \qquad B_{4,5,6,7} = A_{3,2,1,0} - A_{4,5,6,7} \tag{24}$$

The scaling procedure used for odd indices of $\bar{X}_k$ with the fractional angles $\beta = \frac{k\pi}{32}$ gives:

$$B_{15} = A_{15} \qquad B_{14,\dots,8} = A_{15,\dots,9} + A_{14,\dots,8} \tag{25}$$

Coefficients $\bar{X}_k$ for even indices can be expressed by variables (24) and scaling factor (21)

$$\begin{bmatrix} \bar{X}_0 \\ \bar{X}_8 \\ \bar{X}_4 \\ \bar{X}_{12} \end{bmatrix} = \frac{1}{2\sqrt{2}} \begin{bmatrix} S_4 & S_4 & S_4 & S_4 \\ S_4 & -S_4 & -S_4 & S_4 \\ S_2 & S_6 & -S_6 & -S_2 \\ S_6 & -S_2 & S_2 & -S_6 \end{bmatrix} \begin{bmatrix} B_0 \\ B_1 \\ B_2 \\ B_3 \end{bmatrix} \tag{26}$$

$$\begin{bmatrix} \bar{X}_2 \\ \bar{X}_{14} \\ \bar{X}_6 \\ \bar{X}_{10} \end{bmatrix} = \frac{1}{2\sqrt{2}} \begin{bmatrix} S_7 & S_5 & S_3 & S_1 \\ -S_1 & S_3 & -S_5 & S_7 \\ -S_5 & -S_1 & -S_7 & S_3 \\ S_3 & S_7 & -S_1 & S_5 \end{bmatrix} \begin{bmatrix} B_4 \\ B_5 \\ B_6 \\ B_7 \end{bmatrix} \tag{27}$$

After a scaling according to (15) we can introduce the new set of variables for the 3rd pipeline stage:

$$4\begin{bmatrix}\bar{X}_0\\ \bar{X}_8\end{bmatrix}=\begin{bmatrix}1 & 1\\ 1 & -1\end{bmatrix}\begin{bmatrix}C_0\\ C_1\end{bmatrix}\qquad 4\sqrt{2}\begin{bmatrix}S_2\bar{X}_4\\ S_6\bar{X}_{12}\end{bmatrix}=\begin{bmatrix}1+S_4 & S_4\\ 1-S_4 & -S_4\end{bmatrix}\begin{bmatrix}C_3\\ C_2\end{bmatrix} \tag{28}$$

$$4\sqrt{2}\begin{bmatrix}S_1\bar{X}_2\\ S_7\bar{X}_{14}\\ S_3\bar{X}_6\\ S_5\bar{X}_{10}\end{bmatrix}=\begin{bmatrix}1 & S_4 & S_2 & S_6\\ 1 & -S_4 & S_6 & -S_2\\ 1 & -S_4 & -S_6 & S_2\\ 1 & S_4 & -S_2 & -S_6\end{bmatrix}\begin{bmatrix}C_7\\ C_5\\ C_6\\ C_4\end{bmatrix}=\begin{bmatrix}1 & 0 & 1 & 1 & 1 & 0\\ 1 & -1 & 0 & -1 & 0 & 1\\ 1 & 1 & 0 & -1 & 0 & -1\\ 1 & 0 & -1 & 1 & -1 & 0\end{bmatrix}\begin{bmatrix}C_7\\ S_2C_4\\ S_6C_4\\ S_4C_5\\ S_2C_6\\ S_6C_6\end{bmatrix} \tag{29}$$

$$\begin{aligned} C_{0,1} &= B_{0,1}+B_{3,2} & C_{3,2} &= B_{0,1}-B_{3,2}\\ C_{4,5,6} &= B_{4,5,6}+B_{5,6,7} & C_7 &= B_7 \end{aligned} \tag{30}$$

Let us notice that the structure of the right vector in (29) is exactly the same as in (22), but the structures of the 6x4 matrices are different. In (22) the matrix comes from a transformation for the odd indices supported by (21), while in (29) the matrix comes from a transformation of even indices.

Scaled coefficients corresponding to odd indices

$$\bar{Z}_k = 4\sqrt{2}\bar{X}_k cos\left(\frac{k\pi}{32}\right) \tag{31}$$

can be expressed by variables (25) and scaling factors (21) as follows:

$$\begin{bmatrix}\bar{Z}_{1,15}\\ \bar{Z}_{3,13}\\ \bar{Z}_{5,11}\\ \bar{Z}_{7,9}\end{bmatrix}=\begin{bmatrix}1 & S_4 & S_2 & S_6\\ 1 & -S_4 & S_6 & -S_2\\ 1 & -S_4 & -S_6 & S_2\\ 1 & S_4 & -S_2 & -S_6\end{bmatrix}\begin{bmatrix}B_{15}\\ B_{11}\\ B_{13}\\ B_9\end{bmatrix}\pm\begin{bmatrix}S_1 & S_3 & S_5 & S_7\\ S_3 & -S_7 & -S_1 & -S_5\\ S_5 & -S_1 & S_7 & S_3\\ S_7 & -S_5 & S_3 & -S_1\end{bmatrix}\begin{bmatrix}B_{14}\\ B_{12}\\ B_{10}\\ B_8\end{bmatrix} \tag{32}$$

Matrix (32) can be factorized as follows:

$$\begin{bmatrix}\bar{Z}_{1,15}\\ \bar{Z}_{7,9}\\ \bar{Z}_{5,11}\\ \bar{Z}_{3,13}\end{bmatrix}=\begin{bmatrix}(C_{15}+C_{11})+(C_{13}^2+C_9^6)\\ (C_{15}+C_{11})-(C_{13}^2+C_9^6)\\ (C_{15}-C_{11})-(C_{13}^6-C_9^2)\\ (C_{15}-C_{11})+(C_{13}^6-C_9^2)\end{bmatrix}\pm\begin{bmatrix}\frac{1}{2S_1} & 0 & 0 & 0\\ 0 & \frac{1}{2S_3} & 0 & 0\\ 0 & 0 & \frac{1}{2S_5} & 0\\ 0 & 0 & 0 & \frac{1}{2S_7}\end{bmatrix}\begin{bmatrix}1 & S_4 & S_2 & S_6\\ 1 & -S_4 & S_6 & -S_2\\ 1 & -S_4 & -S_6 & S_2\\ 1 & S_4 & -S_2 & -S_6\end{bmatrix}\begin{bmatrix}C_{14}\\ C_{10}\\ C_{12}\\ C_8\end{bmatrix} \tag{33}$$

where:

$$C_{8,10,12} = B_{8,10,12}+B_{10,12,14} \qquad C_{14,15} = B_{14,15} \qquad C_{9,13}^{2,6} = B_{9,13}S_{2,6} \qquad C_{11} = B_{11}S_4 \tag{34}$$

In the 4th pipeline step directly from (32) we can introduce new variables:

$$D_{15,11} = C_{15}\pm C_{11} \qquad D_{13} = C_{13}^2+C_9^6 \qquad D_9 = C_{13}^6-C_9^2 \tag{35}$$

The rest of variables require 10 next multipliers, 3 adders/sub-tractors and 3 shift registers:

$$D_{5,10} = S_4C_{5,10} \qquad D_{4,6,8,12}^{2,6} = S_{2,6}C_{4,6,8,12} \qquad D_{3.7.14} = C_{3,7,14}$$

$$D_{0,1} = C_0 \pm C_1 \qquad D_2 = C_2 + C_3 \tag{36}$$

However, the 5th pipeline stage requires only a single multiplier for the E_2 variable:

$$E_2 = S_4 D_2 \qquad E_{0,1,3} = D_{0,1,3} \qquad E_4 = D_6^6 - D_4^2 \qquad E_6 = D_6^2 + D_4^6 \tag{37}$$

$$E_{\substack{7,11,15\\5,9,13}} = D_{7,11,15} \pm D_{5,9,13} \quad E_{14,10} = D_{14} \pm D_{10} \quad E_{12} = D_{12}^2 + D_8^6 \quad E_8 = D_{12}^6 - D_8^2 \tag{38}$$

The 6th stage does not require any multiplier, only 10 adders/sub-tractors and 6 shift registers for synchronization:

$$F_{\substack{3,5,7,9,13\\2.4.6.8.12}} = E_{3,5,7,9,13} \pm E_{2,4,6,8,12} \qquad F_{0,1.9.11.13.15} = E_{0,1.9.11.13.15} \tag{39}$$

In the 7th pipeline stage 12 signals are delayed only for synchronization and 4 are scaled for the following (n,k) pairs: (14,1),(12,7),(10,3),(8,5):

$$G_n = \frac{F_n}{2S_k} \tag{40}$$

In the 8th pipeline stage pure registers for synchronization only are implemented for even indices of $\bar{X}_{0,2,4,6,8,10,12,14}$ and

$$H_{\substack{9,11,13,15\\8,10,12,14}} = G_{9,11,13,15} \pm G_{8,10,12,14} \tag{41}$$

The last stage contains all scaling multipliers:

$$\bar{X}_k = \frac{H_m}{4\sqrt{2}cos\left(\frac{k\pi}{32}\right)} \tag{42}$$

for the following (k,m) pairs: (1,15), (15,14), (7,13), (9,12), (3,11), (13,10), (5,9), (11,8), (14,7), (2,6), (6,5), (10,4), (4,3), (12,2).

7. Implementation of the code into a FPGA

The spectral trigger should be generated if DCT coefficients normalized to the 1st harmonics are in an arbitrary narrow range:

$$Thr_k^L \leq \xi_k = \frac{\bar{X}_k}{\bar{X}_1} = \frac{\eta_k * H_{f(k)}}{\eta_1 * H_{15}} \leq Thr_k^H \tag{43}$$

where Thr_k^L and Thr_k^H are lower and upper thresholds for each spectral index k, respectively. Altera® Library of Parameterized Modules (LPM) contains the lpm_divide routine supporting a division of fixed-point variables. However, this routine needs huge amount of logic elements and it is slow (calculation requires 14 clock cycles in order to keep sufficiently high registered performance). DSP blocks also do not support this routine. A simple conversion to

$$H_{15} \times \theta_k^L = H_{15}\left(\frac{\eta_1}{\eta_k}Thr_k^L\right) \leq H_{f(k)} \leq H_{15}\left(\frac{\eta_1}{\eta_k}Thr_k^H\right) = H_{15} \times \theta_k^H \tag{44}$$

Fig. 8. The pipeline internal structure of 16-point DCT FPGA routine. Signal from the ADC propagates though the (horizontal) shift register $x_{15},...,x_0$. Simultaneously, the DCT coefficients are calculated in vertical chains in 9 clock cycles. Each rectangle corresponds to a single clock procedure (a logic block). 16-point DCT "engine" utilizes 35 multipliers, 45 adders, 32 sub-tractors and (16 + 38) shift registers. $H_{0,1} = G_{0,1}/4$, a division is not implemented, two low significant bits are ignored. The width of data is extended in consecutive pipeline stages from N at the shift register $x_{15},...,x_0$, even to N+8 in the H routine.

allows implementation of fast multipliers from the DSP blocks and calculation of products in a single clock cycle. θ_k^L and θ_k^H are lower and upper scaled thresholds respectively, which are set as external parameters.

According to (44) the calculation of a sub-trigger needs two multipliers, two comparators and an AND gate. The multiplier stage of an embedded multiplier block supports 9×9 or 18×18 bit multipliers. Depending on the data width or operational mode of the multiplier, a single embedded multiplier can perform one or two multiplications in parallel. Due to wide data busses embedded multiplier blocks do not use the 9×9 mode in any multiplication. Each multiplier utilizes two embedded multiplier 9-bit elements. The full DCT procedure needs the calculation of all coefficients 70 DSP blocks. However, the scaling of $\bar{X}_k$ in the last pipeline chain is no longer needed. It is moved to the thresholds according to (44). Removing last pipeline chain reduces amount of DSP blocks to 40. Sub-triggers routines (Fig. 9) need 2 DSP blocks each. The chip EP3C40F324I7 selected for the 4th generation of the 1st level SD trigger contains 252 DSP 9-bit multipliers. So, for 3-fold coincidences and an implementation of 3 "engines" the single DCT "engine" can support only 11 independent DCT coefficients (Szadkowski, 2011). Sub-triggers $A_k^{0,1,2,3}$, $B_k^{0,1,2}$, $C_k^{0,1}$ and D_k^0 are generated for the patterns A_k, B_k, C_k and D_k (k = 2,4,6) from Fig. 3, respectively. Sub-triggers are synchronized to each other in shift registers in order to put simultaneously on an AND gate (Fig. 11). In order to keep a trigger rate below the boundary deriving from the limited radio bandwidth, additionally the amplitude of the jump is verified. If the jump is too weak, a veto comparator disables the AND gate. Thus, if spectral coefficients ξ_k match pattern ranges for each time bins selected by multiplexer totally in 4 consecutive time bins and if veto circuit is enabled the final trigger is generated. A delay time for the veto signal depends on the type of shape, which is an interest of an investigation. For the single time bin of the rising edge the veto is delayed on 3 clock cycles, for the investigated pattern corresponding to the three time bins of the rising edge the maximal ADC value appears 2 clock cycles later in comparison to the previous case, so the veto should be delayed on a single clock cycle only.

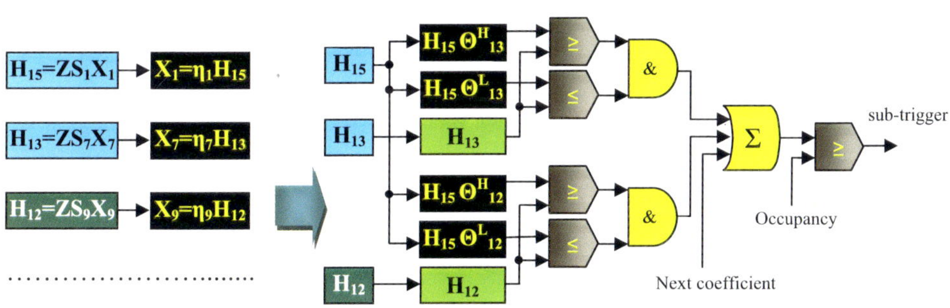

Fig. 9. The structure of sub-triggers. The DCT coefficients $\bar{X}_k$ are not directly calculated. They have been replaced by a boundary of the acceptance lane: upper and lower thresholds $H_{15} \times \theta_k^H$ and $H_{15} \times \theta_k^L$, respectively. Signals between that thresholds (two comparators + AND gate) generate preliminary sub-triggers, which are next summed and compared with the arbitrary Occupancy level. If an amount of "fired" preliminary sub-triggers is above the selected Occupancy, the final sub-trigger is generated for the next processes. It is enabled/disabled depending on the veto variable, verifying the minimal amplitude of the input signals to keep the trigger rate on the reasonable level and to prevent the saturation of the transmission channel.

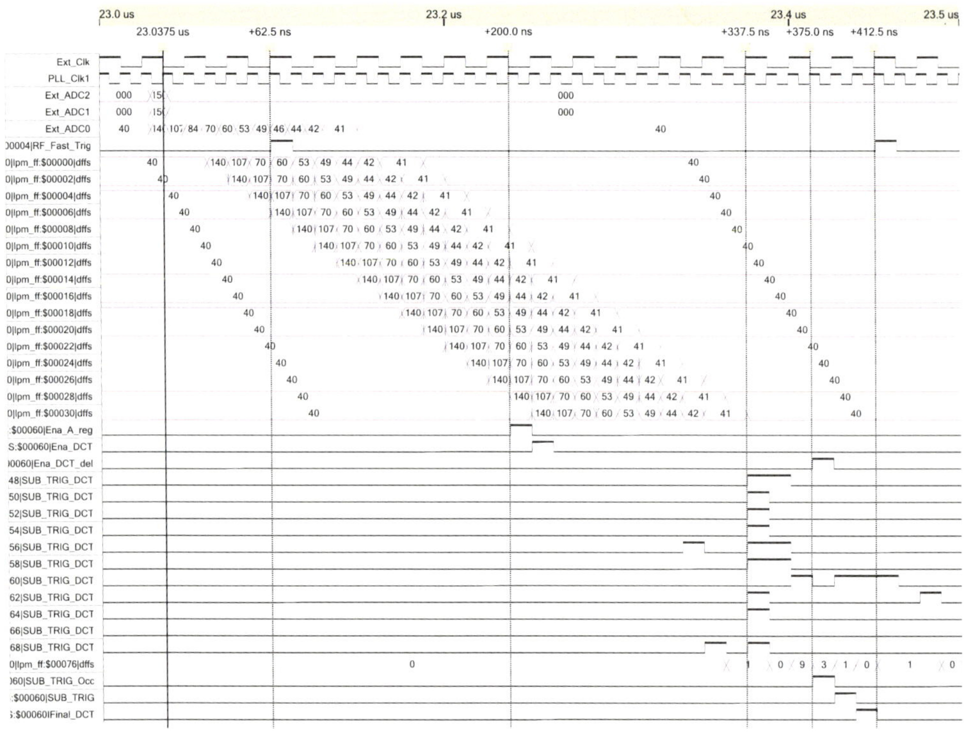

Fig. 10. Simulation of the 1-fold spectral trigger simultaneously with the 3-fold threshold trigger. The length of the shift registers = 16. Data in the Ext_ADC0 channel corresponds to a muon signal with a 1-time-bin rising edge, 11-time-bins attenuation tail and with a constant pedestal = 40 ADC-counts. Together with the begin of the muon peak (at 23.075 μs), two neighboring channels Ext_ADC1,2 are driven artificially to 150 ADC-counts to generate the standard threshold trigger based on the 3-fold coincidence. The internal PLL clock = 80 MHz. The internal standard threshold trigger appears 5 clock cycles later (+62.5 ns). The nodes lpm_ff:$00000 | dffs - lpm_ff:$00030 | dffs correspond to the shift register $x_{15},...,x_0$. The system is tuned for the Shape_A recognition (two 1st time bins on the pedestal level). Ena_A_reg is generated (+200 ns = 16 clock cycles) due to the amplitude of the signal (140 ADC-counts) is above the veto threshold. It is delayed next 15 cycles to be synchronized with SUB_TRIG_Occ. Sub-triggers are generated 27 clock cycles (+337.5 ns) after the rising edge. A calculation of the Occupancy takes next two clock cycles. 29 clock cycles after the rising edge due to a coincidence of the Occupancy and Ena_DCT_del (inversion of the veto) the SUB_TRIG is generated. Finally it appears in the same position as 3-fold coincidence threshold trigger 31 clock cycles later. Final_DCT trigger corresponds to the possible coincidence with a neighboring DCT "engines". If the standard threshold trigger(based on 3-fold coincidence) appears next any triggers are ignored though 768 clock cycles.

The 16-point DCT with 16-stage shift register for 100 MHz sampling can cover 150 ns time window. For the horizontal or very inclined showers this interval is sufficient for the analysis. However, for the higher sampling frequency, when the time window may turn out too short, the shift register may be extended from 16 to 24 stages and the eight samples for the higher indices may be taken from the last 16 shift register nodes according to the Fig. 11. The samples with higher indices correspond to the exponentially attenuated tail and the analysis of the tail is lest critical than the rising edge, where samples are analyzed with a full speed.

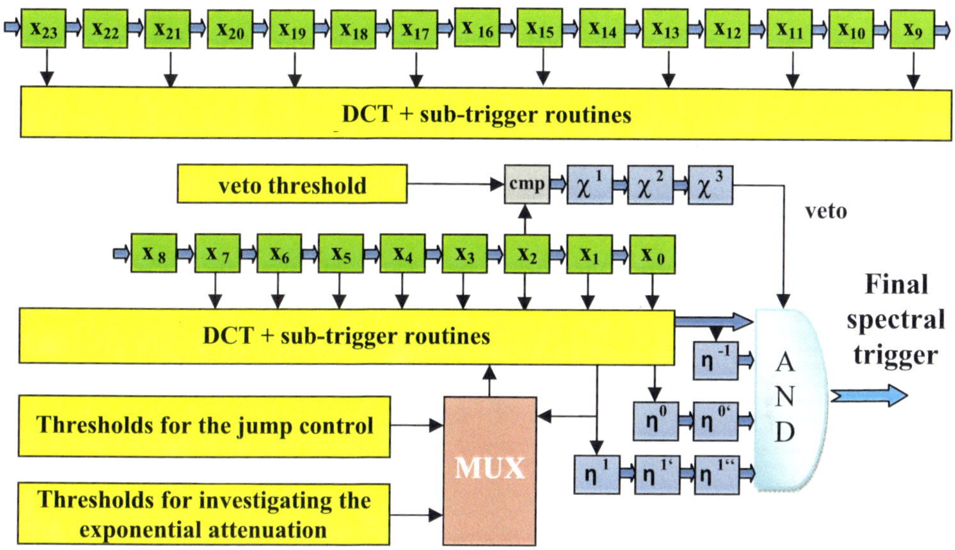

Fig. 11. A scheme of the final spectral trigger. The shift register presented here has an extended length = 24 stages to cover longer time window. However, for a sampling frequencies $f_s \leq 100$ MHz 16 stages and T ≥ 150 ns the window is wide enough for an analysis of horizontal showers. If signal shifted in the register chain matches the expected patterns for 4 consecutive time bins i.e. corresponding to ADC shapes in Fig. 3 (1st row, 3 first graphs. The 4th pattern is exactly the same as the 3rd one. The amplitude of the signal decreases, but the DCT coefficients remain the same (still an exponential attenuation).

3 DCT trigger "engines" have been successfully merged with the Auger code working with 100 MHz sampling. The final code utilizes only 38gives an opportunity to add new, sophisticated algorithms. The slack reported by the compiler corresponds to a maximal sampling frequency 112 MHz, which gives a sufficient safety margin for a stable operation of the system. For sufficiently high amplitudes of the ADC samples the Threshold trigger will be generated 32 clock cycles earlier than the spectral trigger (24 clock cycles of propagation in the shift registers + 8 clock cycles of performance in the DCT chain). If the Threshold trigger has been already generated, the next triggers are inhibited for 768 time bins necessary to fulfill memory buffers (see Fig. 7 in (Szadkowski, 2005a)). Because the Threshold trigger (sensitive to bigger signals) has a higher priority than the spectral trigger, ADC samples will not be delayed for the Threshold trigger in order to synchronize it with the spectral one. The system uses 10-bit resolution (standard Auger one). A compilation for the 12-bit resolution for the current chip EP3C40F324I7 failed, due to a lack of the DSP blocks. 12-bit system requires bigger chip EP3C55. The slack times are on the same level as for EP3C40. All pipeline routines shown in Fig. 8 are implemented in a direct mode (no pipeline mode - like i.e. in the 2nd generation of the FEB based on the ACEX family (see Fig. 2 in (Szadkowski, 2005a)) or for the FFT implementation in the Cyclone family (Fig. 2 in (Szadkowski, 2005b)). So, a performance

of a signal requires a single clock cycle only. All routines are fast enough to work with 100 MHz sampling without an additional pipeline stages and they do not introduce an additional latency.

8. Accuracy

10-bit resolution of FADC in the high-gain channels (responsible for a trigger generation) implies the ranges of $\bar{X}_k$ coefficients given in the 2nd column of Table 1. Multiplications of integer values N by real scaling factors sf give floating-point results. In order to keep possible high speed of calculation and not to utilize resources spendthrift the fixed-point algorithm of processing has been chosen. N×sf were approximated on each pipeline stage again to the integer value. For almost all scaling factors: sf ≤ 1, N×sf has a representation of the same or less amount of bits. For sf ≥ 1, N×sf extends the representation on 1 or 2 bits. This approximation introduces errors. However, the width of the data in the internal pipeline stages is extended from the N at the shift register $x_{15},...,x_0$, to N+1, N+2, N+3, N+4, N+5, N+7, N+8 in routines A, B, C, D, E, F, G, respectively (Fig. 8). This reduces approximation errors mostly to the LSB, apart the $\bar{X}_{15}$. This coefficient will not be used for a trigger.

k	range of $\bar{X}_k$	LSB	2nd bit	3rd and more	k	range of $\bar{X}_k$	LSB	2nd bit	3rd and more
0	0...4092	0.0%	0.00%	0.00%	8	± 2041	0.0%	0.00%	0.00%
1	± 2521	13.1%	0.00%	0.00%	9	± 12224	23.8%	1.55%	0.00%
2	± 2581	8.7%	0.00%	0.00%	10	± 4557	12.8%	0.00%	0.00%
3	± 2914	13.1%	0.00%	0.00%	11	± 7519	17.7%	0.00%	0.00%
4	± 2348	4.8%	0.00%	0.00%	12	± 5671	11.5%	0.00%	0.00%
5	± 4019	15.1%	0.00%	0.00%	13	± 9605	24.3%	2.00%	0.00%
6	± 3045	8.6%	0.00%	0.00%	14	± 12978	26.9%	2.86%	0.00%
7	± 10032	23.1%	1.10%	0.00%	15	± 25597	30.9%	25.08%	6.83%

Table 1. Ranges of $\bar{X}_k$ coefficients and relative errors for least significant bits of $\bar{X}_k$. For k ≤ 14 the errors appear practically only in the LSB.

According to above estimations, the configuration with 3 "engines" does not support all ξ_k sub-triggers due to limited amount of DSP blocks. However, for the next generation of the water Cherenkov detectors array, where probably only a single PMT will be used, 3 "engines" will be implemented to investigate and to detect 3 different shapes of FADC traces corresponding to i.e. different rise times of the rising edge.

9. Preliminary tests

Analysis of Auger ADC traces of very inclined showers shows that the maximum of the signal is mostly reach in a single time bin. The attenuation factor for a tail is in the range of β = (0.2 - 0.5). Fig. 12 shows shapes of signals with various attenuation factors with two first time bins on a pedestal level. For simplicity it has been set on zero. It does not reduce the generality of analysis, because the pedestal is irrelevant for DCT (k ≥ 1). The corresponding DCT coefficients are shown in upper Fig. 3 (Shape_A). After a single clock cycle, when data is shifted in the registers chain, shifted signal with only one time bin on the pedestal level determines a new set of the DCT coefficients shown in lower Fig. 3 (Shape_B). Pattern, which is going to be recognized, can be selected by a setting of DCT coefficient in the DCT engines.

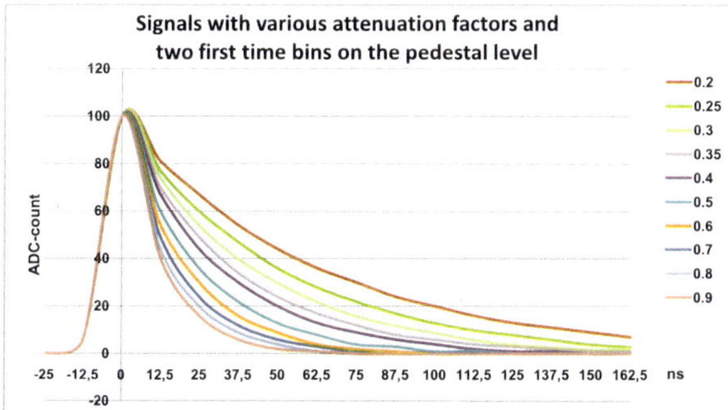

Fig. 12. Shapes of signals with various attenuation factors and two first time bins on the pedestal level

All signals with first two time bins on the pedestal level for sure will be with only one time bin on the pedestal level in the next clock cycle. But, not vice versa. A signal with only a single time bin on the pedestal level before sharp rising edge can have significant contribution in the 2nd time bin before rising edge and it will not be recognized by a pattern recognition procedure tuned on the Shape_A. A procedure recognizing Shape_A is more restrictive and gives lower trigger rate than for the Shape_B. Due to limited amount of the DSP blocks only 11 DCT coefficients can be analyzed simultaneously. For the Shape_A the $\bar{X}_4$ and $\bar{X}_{10}$ are ignored and for the Shape_B : $\bar{X}_6$ and $\bar{X}_{14}$, respectively, as weakly sensitive on changes of signal shapes. The trigger based only on the DCT pattern recognition gives too high rate, due to a contribution of very week signals with also appropriate shape, but usually treated as noise. In order to reduce and control the trigger rate, the veto threshold has been introduced. The calculation of the DCT coefficients in the pipeline chain and next the calculation of sub-triggers in multipliers and comparators block takes 12 clock cycles. The signal is synchronized with the DCT sub-triggers delayed the same time to be compared with the veto threshold, simultaneously with a generated DCT sub-triggers. If the signal is above the sum of the veto threshold and the pedestal, the sub-triggers are enabled to generated a final spectral trigger. The condition that all 11 DCT coefficients were inside the acceptance lane is too strong. The shapes are not ideal, noise introduces additional shape distortions. Similarly as in the ToT trigger only a part of "fired" sub-triggers (Occupancy ≤ 11 = max. number of sub-triggers) is enough to generate the final spectral trigger.

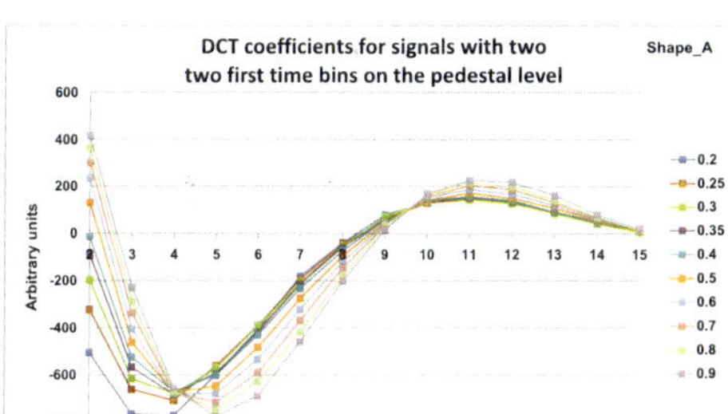

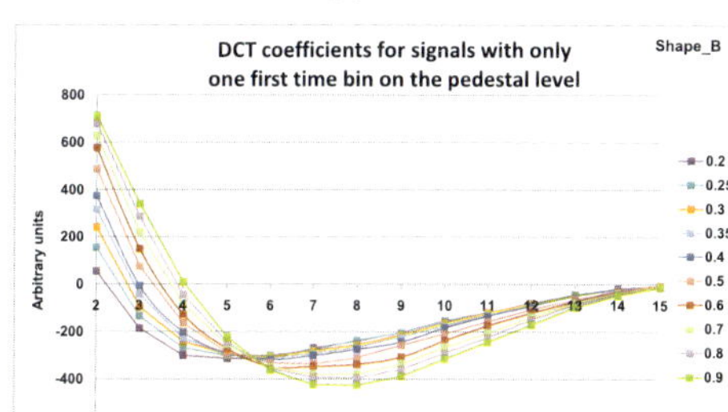

Fig. 13. Coefficients for signals with various attenuation factors and two first time bins (left) and only one time bin (right) on the pedestal level

Although the spectral trigger is being developed for the future and for a single detection channel (a single PMT), the DCT trigger in the Auger surface detector has been tested in a 2-fold coincidences of any 3 PMTs, to be close as possible for a comparison of the results with the standard Auger data.

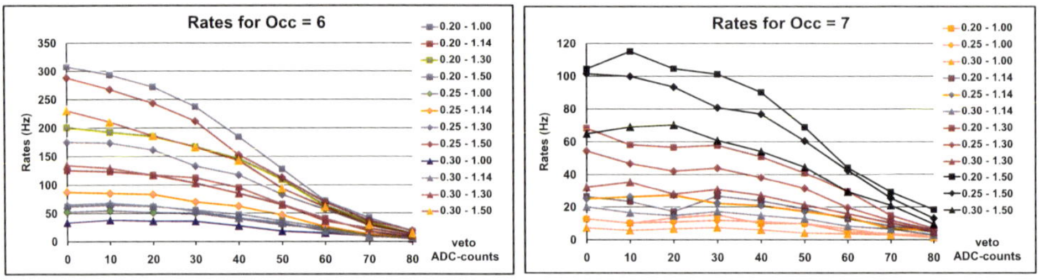

Fig. 14. Rates for various acceptance lanes (from 0.20 - 1.00 till 0.30 - 1.50) and the veto threshold (0 - 80 ADC-counts) for the Occupancy = 6 (left) and 7 (right), respectively. Due to a possible saturation of a transmission channel, the rate of the spectral trigger should not exceed ca. 40 Hz.

Fig. 14 show the trigger rate for the Occupancy = 6 and 7, respectively. The T1 trigger rate is calibrated to ca. 100 Hz. Generally, the trigger rate for Occ = 6 is too high. In order not to saturate the microcontroller and the transmission chain the total (standard Auger + spectral) trigger rate should not exceed 150 Hz. This gives max. 40 - 50 Hz for the spectral trigger only. The Occupancy = 7 with a range of attenuation factors limited to β = (0.20 - 1.14) gives a trigger range on the reasonable level. The Occupancy = 8 reduces the trigger rate below 1 Hz and seems to be too restrictive. The FPGA contains internal counters counting the trigger rate and a contribution of DCT sub-triggers to the final trigger. The required trigger rate range can be set remotely from the Central Data Acquisition System (CDAS). The FPGA automatically tunes the veto threshold to get the required trigger rate. If the veto threshold is above 60 ADC-counts (ca. 1.2 VEM) the acceptance lane is modified. The attenuation factor β from the left side of the range is increased/decreased in the range of (0.20 - 0.40) by the fixed right boundary β = 1.3.

Fig. 15 show three calibration processes, when either initial parameters has been set ideally (B) or they have to be tuned to get required trigger rate (A and C). The tuning process typically does not exceed 3 minutes. In contrary to the standard Auger tuning procedure, when the thresholds for the Threshold trigger are calculated by the external microcontroller located on the Unified Board (UB), the thresholds for the acceptance DCT lane are initially calculated and next stored in the ROM inside the FPGA and they are only multiplexed. This allows a full autonomous FPGA calibration process without a support by any external microcontroller. The new Front-End Board samples analog signals with 80 MHz. Data is written via a left port in the dual-port RAM. Stored data are next read via the right port with 40 MHz. The new board is seen by the rest of electronics as the standard one. Only a additional flag informs the system on the type of the trigger. Internal FPGA counters allow counting a contribution of the DCT coefficients to the final spectral trigger. Fig. 15d shows a relative contribution of the DCT coefficients for the Shape_A. The contribution of the $\bar{X}_5$ and $\bar{X}_9$ is a little bit lower than the rest ones. For the $\bar{X}_9$ the acceptance lane (compare Fig. 13A) is relatively narrow, so the lower contribution is not strange. $\bar{X}_5$ is probably more sensitive on signal noise and possible signal distortions. The graph A (Fig. 15) shows a process when the trigger rate is initially too

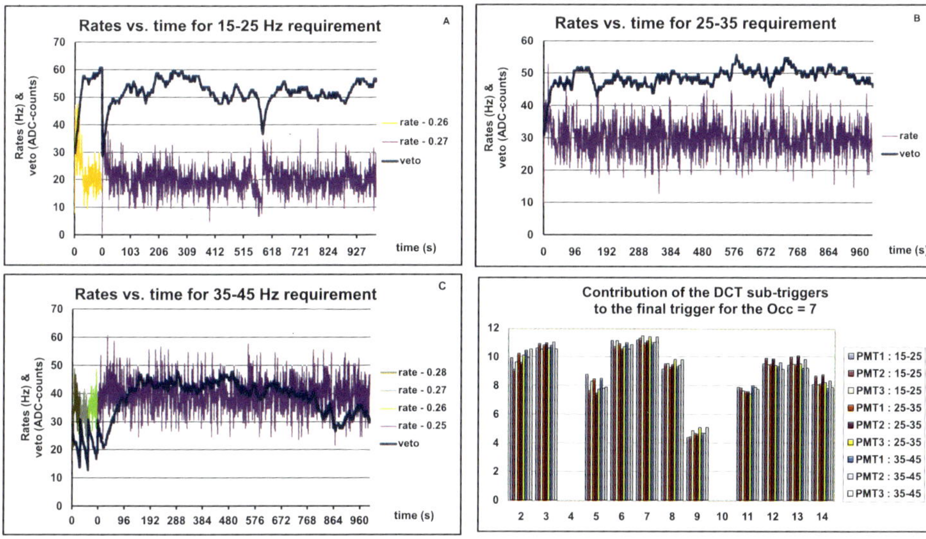

Fig. 15. Rates obtained in the test detector during an auto-calibration process (graphs A, B and C). A comparison of a contribution of "fired" DCT coefficients generating the sub-triggers for all three PMT channels and for various trigger rates requirements (right-down). There are no significant differences in a contribution of a fixed coefficients for different PMT and various configuration of trigger rate requirements

high and the attenuation factor has to be increased (the acceptance lane is narrowed down). The graph B shows the process, when the initial parameters are optimal and the acceptance lane is not modified (only the veto threshold is tuned). The graph C shows the process when the initial parameters give too low trigger rate and the acceptance lane is changed three times.

10. Conclusion

The pattern recognition technique implemented parallel with the standard threshold detection may improve an efficency of a registration of rare events, especially for a single PMT in the surface detector, when the coincidence technique cannot be longer used.
The optimized algorithm of the spectral trigger based on the Discrete Cosine Transform with veto and auto-calibration procedure has been successfully implemented into the FPGA and showed the perfect stability in the real detector. Measurements in the test detector confirmed assumption for a selection of limited amount of DCT coefficients and a stability of algorithm for arbitrarily selected acceptance lane of the spectral trigger rate.
Although 6 surface detectors from the Pierre Auger Observatory have been used for the tests, the spectral trigger is being developed more generally for future ground EAS arrays using - other than the present Pierre Auger Observatory - only one PMT per station.

11. Acknowledgement

The author would like to thank the Pierre Auger Collaboration for being allowed to use a PAO infrastructure and a test-hexagon and for getting the data made available. The successful installation and preliminary tests of the new Front End Boards with the DCT trigger would

not have been possible without the strong commitment and effort from the technical staff in Malargüe.
This work was funded by the Polish Committee of Science under KBN Grant No. N N202 200239.

12. References

Abraham J., et al., [Pierre Auger Collaboration] (2004). Properties and Performance of the Prototype Instrument for the Pierre Auger Observatory. *Nucl. Instr. Meth.*, Vol. A523, (May 2004) pp. (50-95) ISSN: 0168-9002

Abraham J. et al., [Pierre Auger Collaboration] (2010). Trigger and aperture of the surface detector array of the Pierre Auger Observatory *Nucl. Instr. Meth.*, Vol. A613, (Jan. 2010) pp. (29-39) ISSN: 0168-9002

Aglietta M., et. al, (2005). Response of the Pierre Auger Observatory Water Cherenkov Detectors to Muons, *Proceedings of 29th Int. Cosmic Rays Conerence*, pp. 100-104, ISBN, Pune, India, Aug. 2005

Arai Y.; Agui T. & Nakajima M. (1988). A fast DCT-SQ scheme for images. *Trans. IEICE*, Vol. E-71, (Nov. 1988) pp. (1095-1097)

Szadkowski Z. (2005). The concept of an ACEX® cost-effective first level surface detector trigger in the Pierre Auger Observatory *Nucl. Instr. Meth*, Vol. A551, (Oct. 2005) pp. (477-486) ISSN: 0168-9002

Szadkowski Z.; Becker K-H. & Kampert K-H. (2005). Development of a New First Level Trigger for the Surface Array in the Pierre Auger Observatory based on the Cyclone™ Altera® FPGA. *Nucl. Instr. Meth*, Vol. A545, (June 2005) pp. (793-802) ISSN: 0168-9002

Szadkowski Z. (2006). 16-point Discrete Fourier Transform based on the Radix-2 FFT algorithm implemented into Cyclone™ FPGA as the UHECR trigger for horizontal air showers. *Nucl. Instr. Meth*, Vol. A560, (May 2006) pp. (309-316) ISSN: 0168-9002

Szadkowski Z. (2009). A spectral 1st level FPGA trigger for detection of very inclined showers based on a 16-point Discrete Cosine Transform for the Pierre Auger Experiments. *Nucl. Instr. Meth*, Vol. A606, (July 2009) pp. (330-343) ISSN: 0168-9002

Szadkowski Z. (2011). Trigger Board for the Auger Surface Detector with 100 MHz Sampling and Discrete Cosine Transform. *IEEE Trans. Nucl. Science*, Vol. PP 99, (March 2011) pp. (1-9) ISSN: 0018-9499

Permissions

The contributors of this book come from diverse backgrounds, making this book a truly international effort. This book will bring forth new frontiers with its revolutionizing research information and detailed analysis of the nascent developments around the world.

We would like to thank Dr. Christian Cuadrado-Laborde, for lending his expertise to make the book truly unique. He has played a crucial role in the development of this book. Without his invaluable contribution this book wouldn't have been possible. He has made vital efforts to compile up to date information on the varied aspects of this subject to make this book a valuable addition to the collection of many professionals and students.

This book was conceptualized with the vision of imparting up-to-date information and advanced data in this field. To ensure the same, a matchless editorial board was set up. Every individual on the board went through rigorous rounds of assessment to prove their worth. After which they invested a large part of their time researching and compiling the most relevant data for our readers. Conferences and sessions were held from time to time between the editorial board and the contributing authors to present the data in the most comprehensible form. The editorial team has worked tirelessly to provide valuable and valid information to help people across the globe.

Every chapter published in this book has been scrutinized by our experts. Their significance has been extensively debated. The topics covered herein carry significant findings which will fuel the growth of the discipline. They may even be implemented as practical applications or may be referred to as a beginning point for another development. Chapters in this book were first published by InTech; hereby published with permission under the Creative Commons Attribution License or equivalent.

The editorial board has been involved in producing this book since its inception. They have spent rigorous hours researching and exploring the diverse topics which have resulted in the successful publishing of this book. They have passed on their knowledge of decades through this book. To expedite this challenging task, the publisher supported the team at every step. A small team of assistant editors was also appointed to further simplify the editing procedure and attain best results for the readers.

Our editorial team has been hand-picked from every corner of the world. Their multi-ethnicity adds dynamic inputs to the discussions which result in innovative outcomes. These outcomes are then further discussed with the researchers and contributors who give their valuable feedback and opinion regarding the same. The feedback is then collaborated with the researches and they are edited in a comprehensive manner to aid the understanding of the subject.

Apart from the editorial board, the designing team has also invested a significant amount of their time in understanding the subject and creating the most relevant covers. They scrutinized every image to scout for the most suitable representation of the subject and create an appropriate cover for the book.

The publishing team has been involved in this book since its early stages. They were actively engaged in every process, be it collecting the data, connecting with the contributors or procuring relevant information. The team has been an ardent support to the editorial, designing and production team. Their endless efforts to recruit the best for this project, has resulted in the accomplishment of this book. They are a veteran in the field of academics and their pool of knowledge is as vast as their experience in printing. Their expertise and guidance has proved useful at every step. Their uncompromising quality standards have made this book an exceptional effort. Their encouragement from time to time has been an inspiration for everyone.

The publisher and the editorial board hope that this book will prove to be a valuable piece of knowledge for researchers, students, practitioners and scholars across the globe.

List of Contributors

Zlatka Nikolova, Georgi Iliev, Miglen Ovtcharov and Vladimir Poulkov
Technical University of Sofia, Bulgaria

Nikolai Nefedov
Nokia Research Center, ISI Lab, Swiss Federal Institute of Technology Zurich (ETHZ), Switzerland

Rameez Asif, Chien-Yu Lin and Bernhard Schmauss
Chair of Microwave Engineering and High Frequency Technology (LHFT), Erlangen Graduate School in Advanced Optical Technologies (SAOT), Friedrich-Alexander University of Erlangen-Nuremberg (FAU), Cauerstr. 9, (91058) Erlangen, Germany

Ya Liu
National Time Service Center, Chinese Academy Sciences, Xi'an, Shaanxi, China Key Laboratory of Time and Frequency Primary Standard, Institute of National Time, Service Center Chinese Academy of Sciences, Xi'an, Shaanxi, China

Xiao Hui Li and Wen Li Wang
National Time Service Center, Chinese Academy Sciences, Xi'an, Shaanxi, China

Yusuke Kanno
Hitachi LTD., Japan

Miguel A. Rodríguez-Hernández
ITACA, Universitat Politècnica de Valencia, Spain

Antonio Ramos and J. L. San Emeterio
Lab. Ultrasonic Signal, Systems and Technologies, CSIC, Madrid, Spain

A. Castillo Atoche and P. Perez Muñoz
Mechatronic Department, Engineering School, Autonomous University of Yucatan, Mexico

J. Estrada Lopez and S. Soto Aguilar
Computer Engineering Dept., Mathematics School, Autonomous University of Yucatan, Mexico

Cristian Vidal Silva
University of Talca, Business Informatics Administration, Chile

Carolina del Río
University of Talca, Business Administration, Chile

Andrew Philominraj
University of Talca, Language Program, Chile

George C. Valley and T. Justin Shaw
The Aerospace Corporation, United States

Baba Tatsuro
Toshiba Medical Systems Corporation, Japan

Heinz G. Göckler
Digital Signal Processing Group, Ruhr-Universität Bochum, Germany

Masaaki Nagahara
Kyoto University, Japan

Agnese Augello, Orazio Gambino, Vincenzo Cannella, Roberto Pirrone and Salvatore Gaglio
DICGIM - University of Palermo, Palermo, Italy

Giovanni Pilato
ICAR - Italian National Research Council, Palermo, Italy

Dmitry Rublev, Vladimir Fedorov and Oleg Makarevich
Technological Institute of Southern Federal University, Russia

Amir Sabbagh Molahosseini
Kerman Branch, Islamic Azad University, Iran

Keivan Navi
Shahid Beheshti University, Iran

Yoshifumi Ukita
Yokohama College of Commerce, Japan

Toshiyasu Matsushima
Waseda University, Japan

Paul Pukite and Steven Bankes
BAE Systems, USA

Zbigniew Szadkowski
University of Łód´z, Department of Physics and Applied Informatics, Faculty of High Energy Astrophysics, Łód´z, Poland